T0291068

Origins of Life

Origins of Life

Musings from Nuclear Physics, Astrophysics and Astrobiology

Vlado Valkovic

CRC Press is an imprint of the
Taylor & Francis Group, an **informa** business

First edition published 2022
by CRC Press
6000 Broken Sound Parkway NW, Suite 300, Boca Raton, FL 33487-2742

and by CRC Press
2 Park Square, Milton Park, Abingdon, Oxon, OX14 4RN

CRC Press is an imprint of Taylor & Francis Group, LLC

Library of Congress Cataloging-in-Publication Data
Names: Valković, Vlado, author.
Title: Origins of life: musings from nuclear physics, astrophysics
and astrobiology / Vladivoj Valkovic.
Description: First edition. | Boca Raton: CRC Press, 2022. |
Includes bibliographical references and index.
Identifiers: LCCN 2021024074 (print) | LCCN 2021024075 (ebook) |
ISBN 9781032010571 (hardback) | ISBN 9781032019741 (paperback) |
ISBN 9781003181330 (ebook)
Subjects: LCSH: Life—Origin.
Classification: LCC QH325 .V35 2022 (print) | LCC QH325 (ebook) |
DDC 576.8—dc23
LC record available at https://lccn.loc.gov/2021024074
LC ebook record available at https://lccn.loc.gov/2021024075

ISBN: 9781032010571 (hbk)
ISBN: 9781032019741 (pbk)
ISBN: 9781003181330 (ebk)

DOI: 10.1201/9781003181330

Typeset in Minion Pro
by codeMantra

To my family, my wife Georgia (Đurđa),

my sons Ozren and Marin,

my grandchildren Kristian, Ida and Luka.

Contents

Preface

There are two points that are fundamental in my thinking about the problem of the origin of life: The first point is that the chemical evolution of galaxies changes the chemical composition (relative abundances of chemical elements) of stars, interstellar gas and dust. The growth of chemical element concentrations in a galaxy provides a clock of its aging. The second point is that the living matter on the planet Earth requires only some elements for its existence. Life, as we know, is (H-C-N-O)-based and is relying on the number of bulk (Na, Mg, P, S, Cl, K, Ca) and trace elements (Li, B, F, Si, V, Cr, Mn, Fe, Co, Ni, Cu, Zn, As, Se, Mo, W). All organisms use trace elements, which are parts of proteins with unique coordination, catalytic and electron transfer properties.

Therefore, to shed some light on the problems of where and when life originated, I am putting forward a hypothesis:

It originated when two-element abundance curves, that of the living matter and that of the Universe, coincided. This coincidence occurring at a particular location (redshift) could indicate the phase of the Universe when the life originated, T_{origin}.

Interstellar material also contains small micron-sized particles in addition to gas. Gas-grain interaction leads to the formation of complex molecules. Interstellar molecular clouds and circumstellar envelopes form factories of complex molecular synthesis. Surface catalysis on interstellar particles enables molecule formation and chemical pathways that cannot proceed in the gas phase owing to reaction barriers.

A high number of molecules that are used in contemporary biochemistry on the Earth are found in the interstellar medium and interplanetary dust particles. Therefore, my second hypothesis is as follows: *Life originated in an interstellar molecular cloud with the critical role of dust particles and properties of their environment.* This scenario would imply that life originated long before the formation of the planet Earth.

If life originated long before the Earth's formation, we have to hypothesize that life was brought to the Earth in the process of its creation! The role of cosmic dust grains in this process must be considered. Boron played a vital role in this process since its primary purpose has been to provide thermal and chemical stability in hostile environments.

Our hypotheses define this time, T_{origin}, as a time when conditions were right for life to originate in the primitive form and that happened only once in the history of the Universe. Our preliminary considerations, by comparing the elemental concentrations in the solar neighborhood and that of the living matter, indicate $T_{origin} = (4 \pm 1) \times 10^9$ years after the Big Bang for the origin of life on the carbonaceous dust particles in the interstellar molecular clouds, IMCs. The life could have survived the planet formation processes and evolved in the habitable zones of stars into the LUCA. During this process, the primitive organisms had to adjust their essential element concentrations to adapt to new environments. This should be seen in the concentration factor dependence on environment properties, in particular due to adaptation to different magnetic and gravitational fields, and the availability of essential elements. For any other time, $T \neq T_{origin}$, no additional such events could occur.

Therefore, an additional, third, hypothesis can be defined: *Because of Universe aging, life originated only once.* Since the first eukaryotic common ancestor (FECA) already possessed complex, eukaryote-typical features which it inherited from the LUCA, we could assume these to be transferred to the last eukaryotic common ancestor (LECA) by the FECA in subsequent evolution.

The dust that formed our solar system already contained L-type amino acids and D-type sugars; therefore, the life on the planet Earth, irrespective of its origin, had to be chiral. Exposure to cosmic rays and magnetic fields as well as nuclear physics laws made life chiral. This can be generalized to planets in any other solar system. At this point, I can put forward the fourth hypothesis: *Chirality is a sine qua non condition for the emergence of life.*

I have presented numerous examples of observations, experiments and theoretical considerations which have been used to synthetize the four hypotheses concerning the origin of life. The Universe was born only with plenty of hydrogen isotopes and some helium and lithium, while all other elements were formed as Universe aged through star formation processes, their lives and deaths, which resulted in dust clouds, a birthplace of new generation of stars. During this process of Universe aging, the chemical element abundance curve has been changing in such a manner that at some time T_{origin}, it coincided with the abundance curve of the living matter.

Our two hypotheses, first and third, define this time, T_{origin}, as a time when conditions were right for life to originate in the primitive form and that happened only once in the history of the Universe. The other two hypotheses, second and fourth, define the dust particles in the molecular clouds and their environment as a place where this has happened. Exposure to cosmic rays and magnetic fields as well as nuclear physics laws made the life chiral.

I hope that this book, in its seven chapters, provides sufficient background showing the arguments which led me to these hypotheses. Also, each chapter is accompanied by an extensive list of references, as well as the list for additional readings, which will allow the potential reader to study any of the mentioned phenomena in depth.

Lastly, I want to express my gratitude to Dr. Jasmina Obhodas, who in spite of her numerous duties, both at home and work, was patient enough to make all the drawings presented in this book. Additionally, talking to her, my thoughts became more transparent, which resulted in better text. One could say if there are no clear words, the ideas are also confused.

Introduction

The present-day thinking about life's phenomenon could be summarized in four points: (i) Life can acquire essential elements against the concentration gradient. (ii) For that, life requires a flow of energy. (iii) Life is self-replicating. (iv) Life occurs in water. All of this describes how life on the Earth sustained rather than how it originated.

The primary purpose of this book is to prepare the ground for coordinated efforts aiming to answer the question: where and when life originated. This question represents a challenging problem that has been approached by a scientific, philosophical and religious point of view. We hope that this text can be of some help to all of them.

The material in the book is organized into seven chapters. In the first one, the chemical element abundances are discussed. The element abundance curve is the result of element synthesis, which started with the p–p chain at the beginning of the Universe, some 13.75×10^9 years ago. The elemental composition of the Universe remained unchanged until the formation of first stars, approximately $(100–250) \times 10^6$ years after the Big Bang, when nucleosynthesis in stars commenced. The first stars formed from the collapse of a gas cloud made up of hydrogen, helium and the elements created in the Big Bang's aftermath. These stars used hydrogen as fuel, creating heavier carbon and oxygen elements and up to the element iron through the nuclear fusion processes. The C/O ratio is fixed early, with the chain of coincidences playing an important role, namely the stability of beryllium isotope, the existence of a favorable energy level in ^{12}C and the non-existence of a favorable level in ^{16}O. These fine-tuned conditions are responsible for our existence and the possible presence of any carbon-based life in the Universe.

A red giant star explodes at the end of its life and, as a supernova, spreads its material into interstellar space, including most of the carbon produced in the core of the red giant star. The high energies of such an explosion create additional elements heavier than iron. Some metals, such as silicon and iron, combine with oxygen to form minerals, i.e., dust. The final abundances of elements are determined in these supernova explosions.

A large quantity of light elements found in the present Universe is thought to have been restored by billions of years of cosmic ray (mostly high-energy protons)-induced breakup of heavier elements in interstellar gas and dust particles. For example, it is accepted that the main production sites for ^{9}Be and ^{10}B are spallation processes in the interstellar medium between α-particles and protons and heavier nuclei such as carbon, nitrogen and oxygen.

All the solar system objects, including the Sun and all planets, are formed from a gaseous nebula with well-defined chemical and isotopic composition. A small part of the giant molecular cloud collapsed because of gravitation, which resulted in the solar system's formation. Consequently, the solar system abundances of chemical elements are somewhat similar to those found in most stars and interstellar material in our neighborhood. In corresponding parts of other galaxies, minor variations (approx. a factor of 3) may occur in the relative amounts of hydrogen and helium, and carbon and heavier elements.

Two physical processes influenced the solar abundances over time. The first is settling of elements from the solar photosphere into the Sun's interior; the second is the decay of radioactive isotopes that contribute to an element's overall atomic abundance. Therefore, the curve representing element abundances is changing with time. The changes can be observed on the timescale of approximately 10^9 years; this is not true only for the solar system, but also for the rest of our galaxy (stars and interstellar matter) and probably the whole of the Universe.

The second chapter discusses galaxies, galactic magnetic fields and cosmic rays. After the Big Bang, the Universe was composed of subatomic particles and radiation only for some short period. What happened next is still unknown – the first possibility is those small particles slowly got together and gradually formed stars, star clusters and eventually galaxies. The second possibility is that the Universe first organized as large clumps of matter that later subdivided into galaxies.

The Milky Way, our galaxy, is typical: It has hundreds of billions of stars, and enough gas and dust to make billions of stars. It has at least ten times as much dark matter as all the stars and gas put together. All of this is held together by only one force: gravity. The Milky Way has a spiral shape similar to more than two-thirds of all known galaxies. At the center of the spiral, a tremendous amount of energy and, occasionally, flares are generated. An immense gravity is required to explain the movement of stars and the energy expelled. The astronomers concluded that the center of the Milky Way is a supermassive black hole.

Other galaxies have elliptical shapes, and a few have unusual shapes such as toothpicks or rings. The Hubble Ultra-Deep Field view shows this diversity. Hubble observed a tiny patch of sky (one-tenth the moon's diameter) for one million seconds (11.6 days) and found approximately 10,000 galaxies of all sizes, shapes and colors. This historical view is two separate images taken by Hubble's Advanced Camera for Surveys and the Near-Infrared Camera and Multi-object Spectrometer. These two images reveal galaxies that are too faint to be seen by ground-based telescopes, or even in Hubble's previous faraway looks taken in 1995 and 1998.

Magnetic fields are the most crucial player in the interstellar medium (ISM) of many galaxies, including spiral, barred, irregular and dwarf galaxies. Magnetic fields contribute significantly to the total pressure, which balances the ISM against gravity. They may affect the gas flows in spiral arms, round bars and galaxy halos. Magnetic fields enable removing angular momentum from the protostellar cloud during its collapse, making it essential for the onset of star formation. The energy from supernova explosions is distributed within the ISM by magnetohydrodynamic turbulence. This magnetic reconnection is one possible heating source for the ISM and halo gas. In addition, the density and distribution of cosmic rays in the ISM are controlled by magnetic fields. It appears that magnetic fields are present in all galaxies and galaxy clusters.

Although the initial discovery of cosmic rays dates back to more than a century, their origin is still one of the most enduring mysteries in physics. Since then, much work has been done, trying to prove that both the acceleration mechanism and site are well understood. Still, no definite proof has been obtained: Despite the impressive progress of both theory and observations, the evidence supporting the commonly accepted interpretation is only circumstantial. Regardless of the acceleration mechanism's details, cosmic rays and magnetic fields in the galaxy are crucial factors in the equilibrium of interstellar gas.

The third chapter discusses the critical subject of galactic chemical evolution. In principle, it is possible to determine the chemical composition of a galaxy as a function of position and time by measuring chemical element abundances of stars with different birthplaces and ages. This procedure can be done because their atmospheres represent the composition of the gas from which they were formed. Such studies may give valuable information about the chemical evolution of galaxies and even about the structure of the matter in the very early phases of the Universe. Besides, the growth of chemical element abundances in a galaxy provides a clock for galactic aging. For example, older stars contain less iron, on average, than younger stars.

Of particular interest are damped Lyman-alpha (DLA) systems or damped Lyman-alpha absorption systems, concentrations of neutral hydrogen gas, as detected in the spectra of quasars. DLA systems are interesting probes for objects at large redshifts. These objects are not luminous enough to be observed otherwise. DLA systems are predominantly neutral, making it easier to measure heavy elements abundance. These systems can be observed at relatively high redshifts of 2–4; they contained most of the Universe's neutral hydrogen at that time. DLA systems are believed to be associated with the early stages of galaxy formation. The high neutral hydrogen column densities of DLA systems are also typical of sightlines in the Milky Way and other nearby galaxies. DLA systems are observed in absorption rather than by emission of their stars. Therefore, they offer the opportunity to study the properties of the gas in the early galaxies directly.

The fourth chapter is concerned with the phenomenon of living matter. Living species are capable of taking the required chemical elements against the concentration gradient. The vast majority of essential trace elements serve as crucial components of the enzyme system or proteins with vital functions. Only L-amino acids are found in proteins in living organisms and D-sugars in RNA and DNA.

The position of essential elements within the periodic table of elements and their relative concentrations contains some answers about the origin of element requirements and possibly the development of life. The assumption is that vital body chemistry should bear similarities to the primordial chemical environment. No organism will depend on a rare element for its existence, provided a more abundant element can play the same role. In explaining why a particular element has been selected for an essential biochemical function, one of the possible approaches is as follows: If an element has not been selected, this may be because its abundance in the available environment is too low, either on the absolute scale or in comparison with some other elements that can play the same role. The life adoption illustrates this to the more abundant member in the following examples: K *vs* Rb, Mg *vs* Be, S *vs* Se, Cl *vs* Br, H *vs* F, Ca *vs* Sr, Na *vs* Li and Si *vs* Ge.

The environment that should be compared to the composition of living organisms is that with which the organism has an intimate contact. Contrary to all expectations, the Universal (cosmic) abundance curve of elements is in best agreement with the distribution of essential elements within the periodic table. Essential elements are the most abundant elements, with trace elements almost all being grouped in the secondary abundance peak, around Fe. This region is the region of the maximum nucleon binding energy in the nucleus, the fact that is responsible for the rise in the Universal abundance curve.

To estimate the closeness of different abundance curves, one should evaluate relations between elements and groups for life-essential elements and various media. The chemical elements abundance ratios for selected groups of elements for a particular medium and living matter, corrected by the values of concentration factors, should be close to one in spacetime region when and where life originated.

The relation of the uptake of essential elements to yield or growth may be considered as a definition of essentiality. There is a rather narrow range of adequacy of element concentration in the organisms. Smaller concentrations result in different abnormalities induced by deficiencies, which are accompanied by pertinent specific biochemical changes. Higher levels result in toxicity. In plants, it is possible to have, under severe deficiency conditions, a decrease in the concentration of an element, which results in a small increase in growth. The concentration factor (C_f= *Element concentration in organism or cell / Element concentration in cell environment*) might depend on an external magnetic field and other parameters characteristic of the environment in which life originated. In any case, the distribution and abundance of essential bulk and trace elements in the living matter should reflect the abundance curve of these elements in the environment at the time life originated.

The fifth chapter tries to summarize the efforts done to determine the origin of life's time and place. As we know, life is (H-C-N-O)-based and relies on the number of bulk and trace elements. Of interest is a hypothesis that speculates that it originated when the essential trace element abundance curve of the last universal common ancestor (LUCA) and the galactic abundance curve in the Fe region coincided.

Interstellar molecular clouds are factories for complex molecular synthesis. In addition to gas, interstellar material also contains small micron-sized particles. Gas-phase and gas-grain interactions lead to the formation of complex molecules. Surface catalysis on solid interstellar particles enables molecule formation and chemical pathways that do not proceed in the gas phase due to reaction barriers. A large number of molecules, needed for the contemporary biochemistry on the Earth, are found to be present in the interstellar medium, planetary atmospheres and surfaces, comets, asteroids, meteorites and interplanetary dust particles. Alternatively, large quantities of extraterrestrial material could be delivered via comets and asteroids to young planetary surfaces during the heavy bombardment phase or its formation. Observing organic matter's formation and evolution in space is crucial to determine the prebiotic reservoirs available to the early Earth.

An extrapolation of the function, showing genetic complexity of organisms' dependence on time, also indicates that life is older than the Earth; it began before the Earth was formed. Life may have started from systems with single heritable elements that are functionally equivalent to a nucleotide. Extrapolation back to just one DNA base pair suggests the time of life's origin to be $(9.7 \pm 2.5) \times 10^9$ years ago. This statement could be tested by comparing elemental abundances of distant galaxies observed at z (redshift) values corresponding to this age (or some other) with the LUCA's essential trace element requirements.

There are other scenarios; in one scenario described in the literature, the origin of life in the galaxy, specifically the start of galactic habitability, is taken to have occurred $(8.5 \pm 5) \times 10^9$ years ago. The range of uncertainty allows for the most extreme conceivable values from the first stars in the Universe to the origin of life on Earth.

The Universe was born only with plenty of hydrogen isotopes and some helium and lithium, while all other elements were formed as the Universe aged through star formation processes, their lives and deaths, which resulted in dust clouds as a birthplace of a new generation of stars. During this process of the Universe aging, the chemical element abundance curve has been changing in such a manner that at the time T_{origin}, it coincided with the abundance curve of the living matter.

Our hypotheses define this time, T_{origin}, as a time when conditions were right for life to originate in the primitive form and that happened only once in the history of the Universe. Exposure to cosmic rays and magnetic fields as well as nuclear physics laws made life chiral. Our preliminary considerations indicate $T_{origin} = 4 \pm 1 \times 10^9$ years after the Big Bang for the origin of life on the carbonaceous dust particles in the IMCs. The life could have survived the planet formation processes and evolved in the habitable zones of stars into the LUCA. During this process, the primitive organisms had to adjust their essential element concentrations to adapt to new environments. This should be seen in the concentration factor dependence on environment properties, in particular due to adaptation to different magnetic and gravitational fields, and the availability of essential elements. For any other time, $T \neq T_{origin}$, no additional such events could occur. Therefore, one can conclude that, because of Universe aging, life originated only once.

In this scenario, life originated long before the origin of the Earth. Therefore, one has to assume that life was brought to the Earth in the process of its formation! The role of cosmic dust mineral grains in this process must be considered. Boron played a vital role in this process since its primary role has been to provide thermal and chemical stability in hostile environments. A comparison of galactic cosmic-ray abundances with solar system nuclei abundances shows 10^5 discrepancies in boron abundances, indicating that most of the boron is of secondary origin. The low boron abundance in the solar system is because boron cannot be created from carbon spallation. Boron is primarily synthesized in the ISM. The boron atoms produced by spallation reactions are stably locked within interstellar graphite grains. Such a large difference in boron abundance indicates that the decision on boron essentiality has probably been made outside the solar system.

The primordial process that turns enormous clouds of cosmic dust into newborn planets has recently been observed directly over millions of years. A protoplanet in the making around a young star (2×10^6 years old), LkCa15, in the neighborhood of Taurus,

450 light-years from the Earth, has been spotted. So far, nearly 2,000 exoplanets have been discovered and confirmed. These could be locations where life might sustain and evolve, providing the appropriate duration of habitability conditions.

Multiverse cosmological models and the anthropic principle are described in some detail in Chapter 6. Part of anthropic reasoning that has attracted plenty of attention is its use in cosmology to explain the so-called fine-tuning of the Universe. This "tuning" refers to the fact that there is a set of values for cosmological or fundamental physical parameters. Had they been very slightly different, the Universe would not be suitable for intelligent life. Here is an example of such "fine-tuning": the early expansion speed in the classical Big Bang model. Had the expansion speed been very slightly higher, the Universe would have expanded too rapidly and no galaxies would have formed. The very low-density hydrogen gas would get more and more dispersed with time passing. It is reasonable to assume that in such a Universe, presumably, life could not evolve. Had the early expansion speed been very slightly less, then the Universe would have collapsed soon after the Big Bang and again there would have been no life. Having just the right conditions for life, our Universe appears to be balancing on a knife's edge.

Invoking the multiverse and the anthropic principles in an attempt to explain fine-tuning is still regarded with high suspicion, or even hostility, among physicists, although it has some notable apologists. There is a consensus that such explanations should not impede searches for more satisfying answers to the observed physical laws and parameters' nature.

Multiverse theories raise severe philosophical problems about the nature of reality and the nature of consciousness and observation. Attempts to sharpen the discussion and provide a more rigorous treatment of concepts such as the number of universes, the probability measures in parameter space and objective definitions of infinite sets of universes have not progressed far. Nevertheless, the multiverse idea has probably earned a permanent place in physical science, and the new physical theories will be considered in the future.

The existence of parallel universes looks like something from science fiction writers, which is not relevant to modern theoretical physics. However, the assumption that we live in a "multiverse" made up of an infinite number of parallel universes has long been considered possible. This assumption is still a matter of vigorous debate among physicists. The problem now is to find a way to test the theory. This approach should include searching the sky for signs of collisions with other universes.

The last chapter of this book lists some of the experiments (the list is by no means complete!), which could be done in the laboratory. These experiments should be done in addition to astrophysical and astrobiological observations:

A. We propose to test a hypothesis that life originated when the essential trace element abundance curve of the LUCA and the galactic abundance curve in the Fe region coincided. One possibility is that this coincidence occurred in the solar neighborhood in the Milky Way galaxy and corresponding regions of other galaxies. One should use observational data to study the evolution of chemical element abundances in galaxies to determine the time when the Universal element abundance curve coincided with the element abundance curve of the LUCA.

The genetic code has transmitted the latter's characteristic properties, while the Universe element abundance curve changed as the galaxies aged.

B. An experimental study of nuclear reactions that might influence the abundance of light elements (Li, Be, B), particularly B, on cosmic dust particles, prime candidates for the origin of life site.
   i. The (n, 2n) reactions on light elements, in particular ^{10}B(n, 2n)^{9}B reaction. This study might contribute to a better understanding of element synthesis. It is well known that BBN gives high ^{7}Li cosmic abundance compared to observation. There are many possible improvements to the model, including a chain of reactions with ^{9}B participation. Using a 14 MeV neutron beam, one should plan to determine the first excited state's location in ^{9}B. Differential cross sections should be measured for the reaction ^{10}B(n, 2n)^{9}B, Q=−8.4371 MeV. Using the proposed experimental setup, one could measure n–n coincidences with two neutron detectors placed outside the cone of a tagged neutron beam. From time-of-flight measurements, deduce the neutron energies and calculate the ^{9}B missing mass spectrum. This approach could lead to much more precise information about low-lying states of ^{9}B.
   ii. The (n,^{2}n) reaction on deuterium followed by ^{2}n (two neutrons in final state interaction geometry)-induced reactions on Li isotopes: reaction n+d→p+n+n for E_n=14 MeV, shows a strong n–n final state interaction. The two neutrons can, in principle, interact with the target nucleus as a single projectile. The two neutrons (^{2}n) induced in a reaction on Li isotopes (^{6}Li and ^{7}Li) could result in the following chain of reactions:
      i. ^{6}Li+^{2}n→^{8}Li→^{8}Be+e+ν followed by^{8}Be →2α, and
      ii. ^{7}Li+^{2}n→^{9}Li→^{9}Be+e+ν followed by^{9}Be → n+2α.

C. Establish possible dependence of concentration factors for essential trace elements on the magnetic field intensity. The magnetic field's effects on the essential trace element's concentration factor values should be measured by varying field intensity within the low field intensity interval. For example, simple single-cell organisms such as *Mucor bacilliform* could be used in the experiment. Also, cyanobacteria (blue-green algae) *Chroococcidiopsis* sp., or *Chromatium* sp. could be considered for use in the investigation. These experiments might result in the concentration factor dependence on magnetic field intensity, whose functional form might give some clue on the properties of the site of origin of life. Living cells possess electric charges exerted by ions or free radicals, which act as endogenous magnets; these magnets can be affected by an exogenous magnetic field, which can orient unpaired electrons.

D. Preferential destruction of enantiomers. Assuming that amino acids could be synthesized on dust particles in interstellar space, the observed optical activity may result from cosmic ray bombardment. High energy polarized protons in cosmic rays may be able to preferentially destroy one isomer because of significant asymmetry in proton (in cosmic radiation)–proton (in amino acid) scattering.

E. Since interstellar dust appears to play a critical role in forming interstellar molecules, molecules may be formed on or in grain surfaces. One could study the growth of single-cell microorganisms in such a medium (with varying concentration factors of essential trace elements).

Each chapter of this book has an extensive list of references and the list for additional reading, which allows the potential reader to study any of the mentioned phenomena in depth.

Author

Dr. Vladivoj (Vlado) Valkovic, a retired professor of physics, is a fellow of the American Physical Society and Institute of Physics (London). He has authored 22 books (from *Trace Elements* (Taylor & Francis, Ltd., 1975) to *Radioactivity in the Environment* (Elsevier, 1st Edition 2001, 2nd Edition, 2019)) and more than 400 scientific and technical papers in the research areas of nuclear physics and applications of nuclear techniques to trace element analysis in biology, medicine and environmental research. He has lifelong experience in the study of nuclear reactions induced by 14 MeV neutrons. This research has been done through coordination and works on many national and international projects, including US–Croatia bilateral, NATO, IAEA, EU-FP5, EU-FP6 and EU-FP7 projects.

He had worked as a professor of physics at Rice University, Houston, Texas; as a Head of Physics-Chemistry-Instrumentation Laboratory at IAEA, Vienna, Austria; and as Laboratory head and scientific advisor at the Ruđer Bošković Institute, Zagreb, Croatia.

1

Chemical Elements Abundances

1.1 Element Synthesis

1.1.1 Big Bang Nucleosynthesis

Big Bang nucleosynthesis (BBN) is very different from stellar nucleosynthesis that produces heavier elements. It is a non-equilibrium process that took place over a few minutes in an expanding radiation-dominated plasma with high entropy (10^9 photons per baryon) and many free neutrons (Burles et al., 1999).

In the first 100 seconds after the Big Bang, when the Universe was very hot ($T \geq 10^{10}K$), any nucleus heavier than proton and neutron formed at this stage was immediately dissociated by high-energy photons. For example, deuterons produced from the reaction $p+n \rightarrow d+\gamma$ were destroyed in the reaction $d+\gamma \rightarrow p+n$. When the temperature dropped to $T \approx 10^9K$, deuterons started to accumulate, and further reactions proceeded to produce more massive nuclei. This process lasted until $\sim 10^3$ seconds after the Big Bang, when almost all remaining free neutrons (with a half-life of about 14.7 minutes) have decayed via reaction $n \rightarrow p+e^-+\nu_e$. The process of nuclei production during this period is called Big Bang nucleosynthesis, and the produced nuclei are called primordial nuclei.

By now, it is accepted that the element synthesis in the Universe starts with the p–p chain illustrated by a sequence of nuclear reactions as follows:

$$\begin{aligned} &^{1}H + ^{1}H \rightarrow ^{2}H + e^{+} + \nu_e \\ &^{2}H + ^{1}H \rightarrow ^{3}He + \gamma \\ &^{3}He + ^{3}He \rightarrow ^{4}He + ^{1}H + ^{1}H \end{aligned} \quad \text{(pp-1)}$$

or

$$\begin{aligned} &^{3}He + ^{4}He \rightarrow ^{7}Be + \gamma \\ &^{7}Be + e^{-} \rightarrow ^{7}Li + \nu_e \\ &^{7}Li + ^{1}H \rightarrow ^{4}He + ^{4}He \end{aligned} \quad \text{(pp-2)}$$

DOI: 10.1201/9781003181330-1

or

$$^{7}Be + ^{1}H \rightarrow ^{8}B + \gamma$$

$$^{8}B \rightarrow ^{8}Be + e^{+} + \nu_e \qquad \text{(pp-3)}$$

$$^{8}Be \rightarrow ^{4}He + ^{4}He$$

The relative importance of pp-1 and pp-2 chains (branching ratios) depends on conditions of H-burning (T, ρ and abundances). The transition from pp-1 to pp-2 occurs at temperatures above 1.3×10^{7} K. Above 3×10^{7} K, the pp-3 chain dominates over the other two, but another process takes over in this case.

The helium produced is called the "ash" of this thermonuclear "burning". It cannot participate in the fusion reactions at these temperatures or even substantially higher temperatures. A nuclear physics prevents this process in our Universe; namely, stable isotopes (of any element) having atomic masses of 5 or 8 do not exist.

A step, which might be of interest, is

$$^{1}H + ^{2}H \rightarrow ^{3}H + e^{+} + \nu_e$$

or

$$^{2}H + ^{2}H \rightarrow ^{3}H + ^{1}H \qquad (Q=4.03\ \text{MeV})$$

which can lead to one or more neutron-producing reactions

$$^{2}H + ^{2}H \rightarrow ^{3}He + n \qquad (Q=2.45\ \text{MeV})$$

$$^{2}H + ^{3}H \rightarrow ^{4}He\ (3.5\ \text{MeV}) + n\ (14.1\ \text{MeV}) \qquad (Q=17.590\ \text{MeV})$$

$$^{3}H + ^{3}H \rightarrow ^{4}He + n + n \qquad (Q=11.3\ \text{MeV})$$

The reaction $^{2}H+^{3}H\rightarrow^{4}He+n$ is peaked in the total cross section ($\sigma_T=5$ barns) at a deuteron energy of approximately 100 keV ($=1.16\times10^{9}$ K), while the reaction $^{2}H+^{2}H\rightarrow^{3}He+n$ is peaked in the total cross section at a deuteron energy of around 2.0 MeV.

The existence of the 400 keV resonance in ^{7}Li in the reaction: $^{3}H+^{4}He\rightarrow^{7}Li+\gamma$ (Alpher and Herman, 1950) could help overcome the non-existence of nuclei with atomic weights 5 and 8. However, no such resonance was found to exist, and attempts to bridge the gap at mass 5 were abandoned. Only at extremely high temperatures (order 10^{8} K) can this bottleneck be overcome by an unlikely reaction. The fusion of two ^{4}He nuclei takes place at those temperatures, and the result is highly unstable ^{8}Be. However, it is formed at a fast enough rate resulting in a minimal equilibrium concentration of ^{8}Be at any moment (Salpeter, 1952). According to the BBN theory, the universal abundances of ^{2}H,^{3}He,^{4}He and ^{7}Li are fixed in the first few minutes, as schematically shown in Figure 1.1, indicating the relative mass fractions as a function of time and temperature (see, for example, Burles et al., 1999).

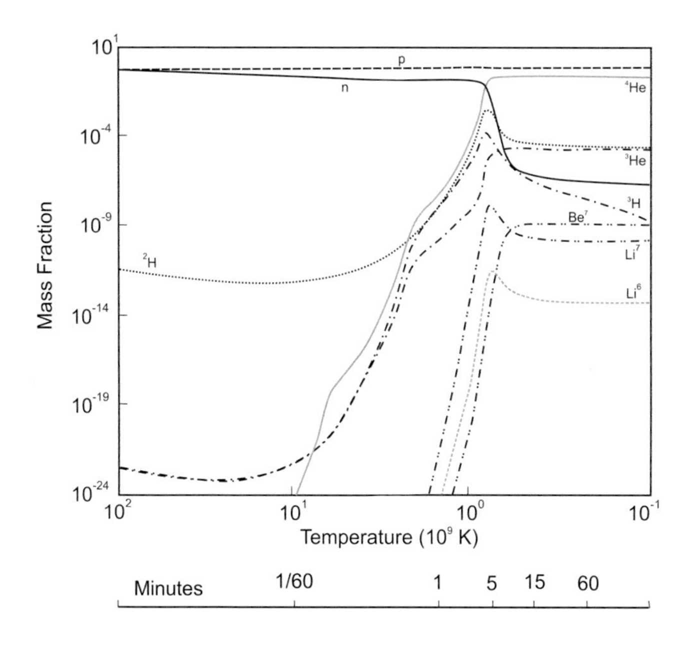

FIGURE 1.1 Schematic representation of the relative abundances of light elements as a function of temperature and time after the Big Bang.

The elemental composition of the Universe remained unchanged until the first stars were formed 100–250 million years later and until nucleosynthesis has commenced in stars. Note that stars are classified into seven main types in the order of decreasing temperatures: O, B, A, F, G, K and M. Therefore, for all of the light elements, systematic errors are a dominant limitation to the precision with which primordial abundances can be inferred.

In their report, Kneller and McLaughlin (2003) considered the possible contribution of the effect of bound di-neutron upon BBN. Although the bound state of two neutrons does not exist, the known effects of the final state interaction in reactions with three particles in the final state suggest that it may result in reactions of the type (^{2}n, x) on some light nuclei.

1.1.2 Li–Be–B Abundance Depletion

During BBN, nearly all neutrons end bound in ^{4}He, which is the most stable light element. In addition, heavier nuclei are not formed in any significant quantity. There are two reasons: the absence of stable nuclei with mass number 5 or 8 (which impedes nucleosynthesis via n+^{4}He, p+^{4}He or ^{4}He+^{4}He reactions), and the large Coulomb barriers for other reactions. The solar system shows the very low relative abundance of Li, Be and B compared to other elements (see Figure 1.2).

The so-called cosmological lithium problem is one of the most important unresolved problems in nuclear astrophysics (Cyburt et al., 2004). It refers to the large discrepancy

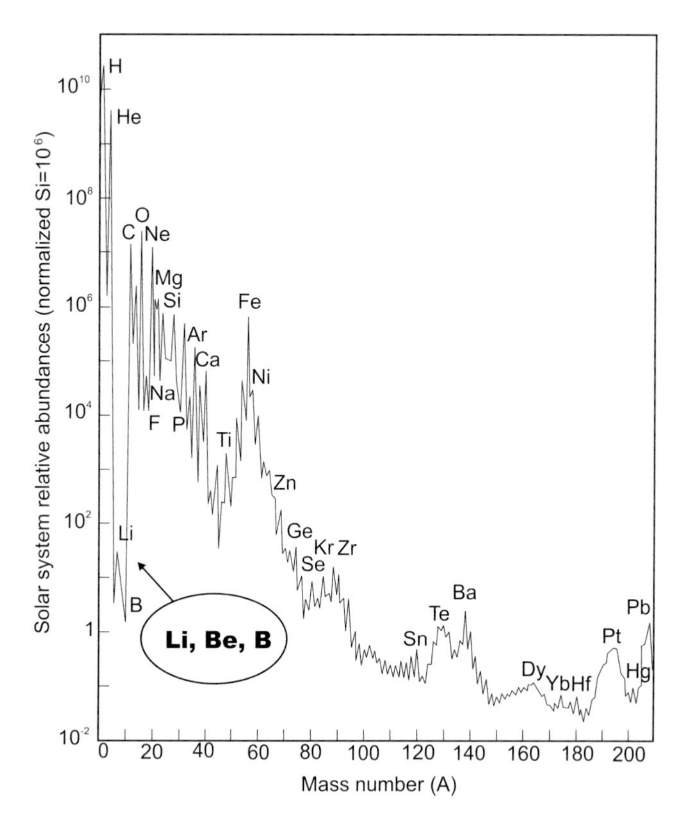

FIGURE 1.2 Very low abundances of elements Li, Be and B in the solar system. (After Lamia 2013.)

between the abundance prediction by the standard BBN theory for primordial ^{7}Li and the value deduced from measurements, as seen in the so-called Spite plateau for halo stars. The predictions of the BBN theory successfully reproduce the observations of all primordial abundances except for ^{7}Li. The abundance of Li isotope is overestimated by more than a factor of 3. Among the possible resolutions to this discrepancy are (i) ^{7}Li depletion in the atmosphere of stars, (ii) systematic errors originating from the choice of stellar parameters – most notably the surface temperature – and (iii) systematic errors in the nuclear cross sections used in the nucleosynthesis calculations.

The primordial ^{7}Li is mainly produced by β-decay of ^{7}Be ($t_{1/2}$ = 53.2 days). Therefore, the abundance of ^{7}Li is essentially determined by the production and destruction of ^{7}Be. The results of the measurements of reactions on ^{7}Be induced by charged particles rule out the removal of ^{7}Be during BBN. The neutron-induced reactions can also play a role in the destruction of ^{7}Be, in particular the ^{7}Be(n, α)^{4}He reaction in the energy range of interest for BBN, in particular between 20 and 100 keV (Barbagallo et al., 2014). Their calculations have shown that the ^{7}Be(n, α)^{4}He reaction may account for a substantial reduction in the primordial ^{7}Li abundance (a factor of 2), thus partially solving the ^{7}Li problem. On the other hand, Broggini et al. (2012) claim that it is unlikely that the ^{7}Be+n destruction rate is underestimated by factor 2.5 required to solve the problem. According to them, new unknown resonances in ^{7}Be+d and ^{7}Be+α could potentially produce significant effects, assuming suitable resonant levels in ^{9}B and ^{11}C.

During BBN, any injection of extra neutrons around the time of the ^{7}Be formation (T$\approx$50 keV) can reduce the amount of ^{7}Li (produced by ^{7}Be decay). If the extra neutron supply is one of the non-standard processes involved during the BBN, the suppression of lithium abundance below Li/H$\leq 1.9\times10^{-10}$ leads to overproduction of deuterium, D/H$\geq 3.6\times10^{-5}$, a value well outside the error bars of recent observations (Coc et al., 2014).

According to Pospelov and Pradler (2010), the most economical solution to the lithium discrepancy would be an astrophysical mechanism that depletes lithium from the photic zone in the atmosphere of Population II stars. Nuclear physics is unlikely to be responsible for the solution to the lithium problem, as all main reactions participating in creation and destruction of lithium are well known. However, particle physics models with unstable or annihilating particles may have an important effect on the Li abundance. For example, an injection of neutrons (whatever the particle physics mechanism) can reduce the amount of ^{7}Be and therefore ^{7}Li abundance.

Nowadays, it is accepted that the main production sites for ^{9}Be and ^{10}B are spallation processes on the interstellar medium (ISM). They are produced by nuclear reaction between α-particles and protons and heavier nuclei such as carbon, nitrogen and oxygen (C, N and O). At the same time, ^{11}B may have an extra production channel via neutrino-induced reactions (Primas, 2009). Irradiation processes in the early solar system and their importance in Li, Be and B formation have been discussed in detail by Chaussidon and Gounelle (2006). Main destruction sites are stellar interiors in the temperature range $(2–5)\times10^6$ K. High abundances of boron are observed in type I carbonaceous chondrites. It may be due to the presence of graphite grains in the primitive solar nebula, which have been irradiated by high-energy nucleons at some stage in their history. The boron atoms, thus produced by spallation reactions, are stably locked within interstellar graphite grains (Ramadurai and Wickramasinghe, 1975).

This point is further illustrated by a comparison between galactic cosmic-ray abundances and nuclei abundances in the solar system (George et al., 2009), as shown in Figure 1.3. The relative elemental abundances in the solar system are representative of galactic cosmic-ray sources, so the 10^5 discrepancy in boron abundances shows that most of the boron is of secondary origin. It was the Cosmic Ray Isotope Spectrometer (CRIS) (see Stone et al., 1998) on board NASA's Advanced Composition Explorer (ACE) spacecraft that provided information on the elemental composition of galactic cosmic ray (GCR) shown in Figure 1.3 and Table 1.1. Table 1.1 presents the relative GCR abundances at 160 MeV/nucleon for the CRIS solar minimum and maximum periods, normalized to Si$\equiv$1,000. The intensity of silicon at 160 MeV/nucleon during solar maximum is lower by a factor of 0.27 than that measured at solar minimum, allowing for a comparison of the absolute intensity levels in the two periods. The composition is energy dependent, and the authors chose energy because the CRIS sensitivity for all species between boron and nickel overlaps at this point.

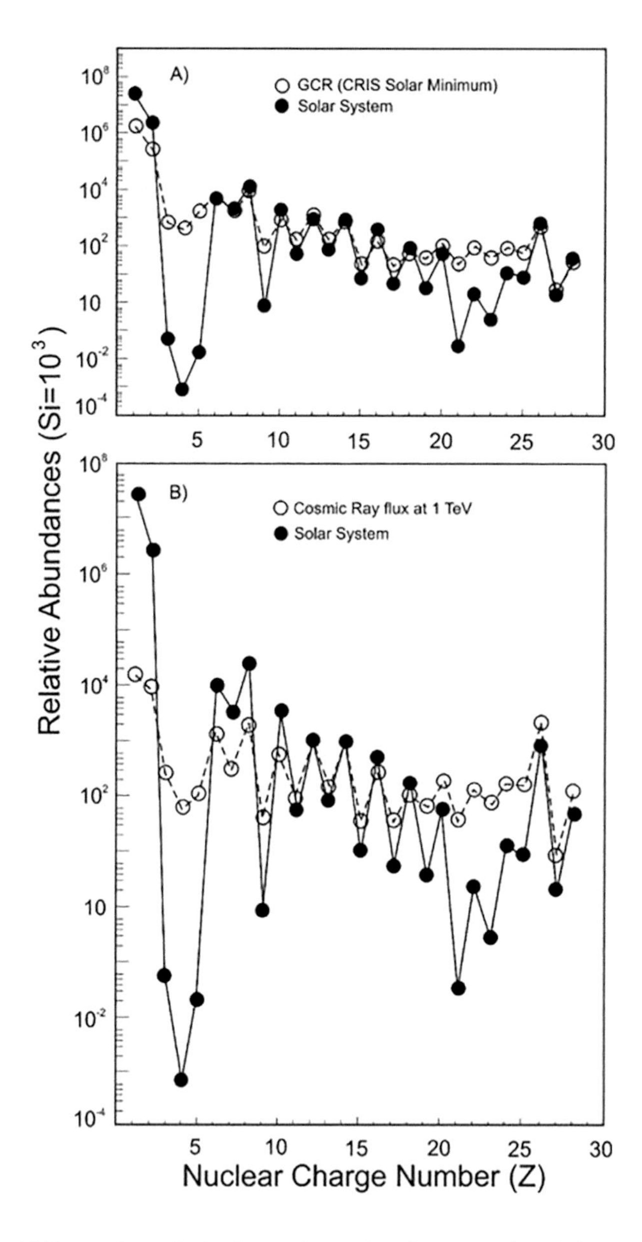

FIGURE 1.3 (a) Comparison of galactic cosmic-ray abundances at solar minimum (open circles) with solar system nuclei abundances (black circles), after George et al. (2009). In the solar system, boron cannot be created from the spallation of carbon. (b) The chemical composition of cosmic rays relative to silicon and iron at 1 TeV, and in the solar system, as a function of nuclear charge Z, following Wiebel-Sooth (1999).

TABLE 1.1 CRIS Relative Elemental Abundances at 160 MeV/Nucleon (After George et al., 2009)

Element	Solar Minimum/Solar Maximum
B	1,803.8/1,986.4
C	7,337.0/6,780.2
N	1,713.7/1,836.1
O	7,082.6/6,520.6
F	101.8/123.6
Ne	998.7/1,050.4
Na	189.6/211.5
Mg	1,368.2/1,367.3
Al	202.7/226.5
Si	1,000.0
P	26.2/34.2
S	157.0/181.2
Cl	24.9/38.4
Ar	58.8/78.5
K	41.6/62.5
Ca	124.8/155.8
Sc	26.0/35.2
Ti	100.4/125.6
V	45.7/54.7
Cr	98.8/109.7
Mn	61.4/71.4
Fe	653.7/742.1
Co	3.7/4.6
Ni	27.8/33.8

Note: Values are normalized to Si. Statistical uncertainties are less than 1% for elements lighter than Si and (1%–8%) for heavier elements. The absolute intensity for silicon at 160 MeV/nucleon is $(107.4 \pm 3.3) \times 10^{-9}$ (cm^2 s sr MeV/nucleon)$^{-1}$ for solar minimum and $(29.1 \pm 0.9) \times 10^{-9}$ (cm^2 s sr MeV/nucleon)$^{-1}$ for solar maximum.

Boron is primarily synthesized in the ISM. This fact is best illustrated in Figure 1.4 that shows the range of Li, Be and B abundances vs. metallicity (Fe/H) as a scatter plot (after Lamia, 2013). Although the number of observations for lithium is large compared with those of beryllium and, more evident, with boron, by inspection of Figure 1.4, the following can be concluded:

- At lower metallicity, lithium abundances exhibit the so-called Li-plateau (Spite and Spite, 1982). This plateau indicates Li formation during the primordial nucleosynthesis period.

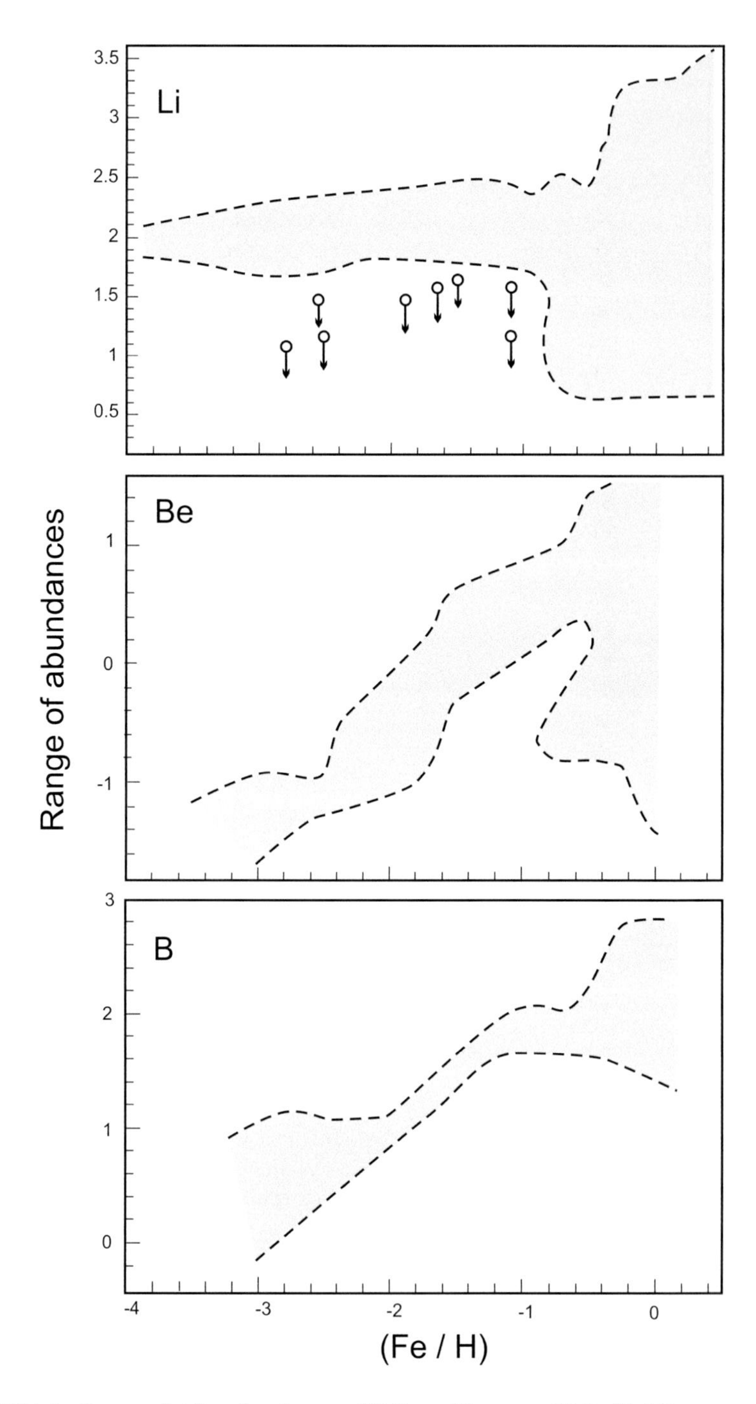

FIGURE 1.4 Ranges of stellar abundances of Li, Be and B vs. metallicity (Fe/H).

- Beryllium and boron abundances are strongly related to metallicity, thus suggesting their production mainly via a synthesis occurring in a continuously evolved ISM, i.e., GCR's nucleosynthesis.
- Beryllium and boron abundances do not exhibit any plateau, indicating a very scarce contribution from primordial nucleosynthesis.

It should be noted that Be and B can be produced via the spallation process (>100 MeV) from energetic protons and α-particles in galactic cosmic rays colliding with C, N and O in ISM (Fields et al., 2013).

Thus, the present-day interstellar boron abundance provides a direct constraint for models that employ various schemes to account for the chemical evolution of the light elements. Unfortunately, measurements of gas-phase interstellar boron abundances can yield only a lower limit to the total interstellar abundance since some depletion is expected even in the lowest density warm gas. Ritchey et al. (2011) find a gas-phase boron abundance for the warm diffuse ISM (from the mean abundance in six lowest density sightlines) of $B/H = (2.4 \pm 0.6) \times 10^{-10}$. This result is in agreement with the value of $(2.5 \pm 0.9) \times 10^{-10}$ obtained by Howk et al. (2000), which was based on the analysis for only one line of sight. Both results indicate a depletion of 60% relative to the solar system (meteoritic) abundance of $(6.0 \pm 0.6) \times 10^{-10}$ (Lodders, 2003).

The evidence for depletion is twofold. First, a density-dependent effect is seen in the gas-phase abundance data for each element examined. Second, the difference between the solar system abundance and the mean abundance in the least reduced sightlines for a particular element increases with the condensation temperature of the element. This situation does not necessarily mean that the set of solar system abundances is the most appropriate cosmic abundance standard against which to measure interstellar depletion. Indeed, the abundances in F and G disk dwarfs of solar metallicity or hot B stars may be more suitable for such comparisons.

Although the lithium abundance decreases as a function of stellar age, as shown in Figure 1.5 presenting results of many investigations, there are deviations from this general behavior. For example, in the recent study of planets around the solar twin HIP 68468, Meléndez et al. (2016) found that the Li abundance in HIP 68468 (1.52 ± 0.03 dex) is much higher (four times; 0.6 dex) than expected for its age. The decay of lithium with age is supported by stellar evolution models, including transport processes beyond merely the mixing length convection. Lithium is depleted and not produced in stars like the Sun; therefore, the enhanced lithium abundance of HIP 68468 is probably due to external pollution. Carlos et al. (2016) have also recently identified two stars with increased lithium abundances and suggested that they may have been polluted with lithium by planet ingestion, which might also be the case of 16 Cyg A (Meléndez et al., 2016).

Recently, Jofré et al. (2015) have announced the finding of a new exceptional young lithium-rich giant star, designated KIC 9821622. They used the data obtained with the high-resolution GRACES at the Gemini North telescope in Hawaii. They were able to determine the chemical abundances of 23 elements of this star and to reveal its mass, radius and age. According to their findings, KIC 9821622 is a lithium-abundant, intermediate-mass giant star (about 1.64 solar masses). It is located at the red giant branch near the luminosity bump, approximately 5,300 ly from the Earth. Lithium-rich

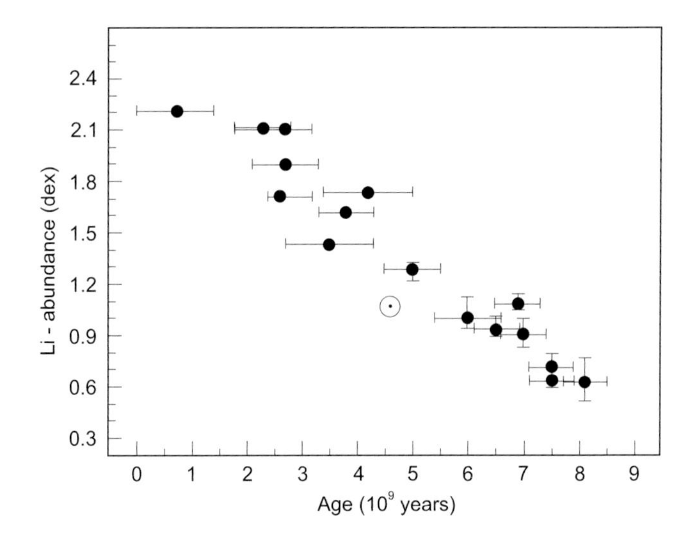

FIGURE 1.5 Lithium abundances versus age for a variety of stars, as summarized by Meléndez et al. (2016). An open circle indicates the value of our Sun.

stars are very scarce. It is assumed that only 1%–2% of all observed giants have the same amount of this element as KIC 9821622.

It has been established that KIC 9821622 also shows a high abundance of carbon, nitrogen and oxygen, aside from being lithium-rich. It is established that stars destroy most of their lithium soon after formation, consuming it at nuclear fusion temperatures. This element is not normally remade, so the question is how giant stars become rich in lithium remains. The explanation offered by Jofré et al. (2015) is that the lithium-rich nature could be the result of fresh internal lithium that is formed near the luminosity bump. Another possibility is that it could be caused by the merger with planets or brown dwarfs. This process could also provide a sudden fresh supply of lithium. In any case, KIC 9821622 is a unique and fascinating object. It deserves further scrutiny to reveal the real mechanism behind the observed anomalous abundances.

In the case of boron, two studies, in particular, provide useful non-solar abundance standards:

i. Cunha et al. (2000) derived baron abundances for 14 near-solar metallicity F and G dwarfs from Goddard High-Resolution Spectrograph (GHRS) spectra of the B_I $\lambda 2497$ line.
ii. Venn et al. (2002) compiled and updated several B-star abundances, notably the boron abundances obtained from the B_{III} $\lambda 2066$ line by Proffitt and Quigley (2001).

The average boron abundance in both of these samples, excluding any stars showing evidence of light element depletion, is $B/H=3\times10^{-10}$. Boesgaard et al. (2004) obtained a similar value (i.e., $B/H=2.6\times10^{-10}$ at $Fe/H=0.0$) from their fit to the boron abundances in 20 solar-type dwarf stars of the galactic disk. All dwarf stars showed undepleted beryllium, implying that these stars have also retained their full initial abundances of boron. If such a value is representative of the present-day ISM, then boron would seem to be only lightly depleted along the lowest density interstellar sightlines.

However, many of the B-type stars described in the Venn et al. (2002) sample may have had some of their initial boron destroyed through rotationally induced mixing (though the most severe cases were not included in deriving the above mean value). The highest boron abundances in Venn et al. (2002) are consistent with the solar system value.

Furthermore, the boron abundances in the Cunha et al. (2000) sample of F and G stars exhibit a clear positive correlation with the abundance of oxygen (Smith et al., 2001), indicating that the mean value of the sample may not reflect the present-day abundance in the ISM. Additionally, the most metal-rich star in the Cunha et al. (2000) sample (HD 19994) has a B/H ratio that is virtually identical to the meteoritic value. Since this star likely formed quite recently, the solar system abundance of boron may well represent the present interstellar value.

The establishment of the relationship between the abundances of boron and oxygen over both disk and halo metallicities is necessary if the origin and evolution of galactic boron are to be understood. The oxygen abundance is a more appropriate metallicity indicator than [Fe/H] in this context because the evolutionary histories of boron and oxygen are closely linked. The spallation production of ^{10}B and ^{11}B results, at least in part, from interactions with ^{16}O, either as an interstellar target for energetic protons and α-particles or as an accelerated particle spalled from ambient interstellar H and He (Ramaty et al., 1997). Galactic oxygen is a product of helium burning in massive stars and is released into the ISM by core-collapse supernovae (SNe), which may also produce boron during the ν-process (Woosley et al., 1990).

Critical insights into the mechanisms responsible for boron production may thus be gained by tracking the dependence of B/H on O/H over the lifetime of the galaxy. The available data result in the trend displayed in Figure 1.6, in which Ritchey et al. (2011) have adopted the halo boron and oxygen abundances of Tan et al. (2010) – open circles in Figure 1.6. These authors derived new non-local thermodynamic equilibrium corrections for B_I and applied their results to existing observations of boron in metal-poor stars. Also included in the figure are the mean disk abundances of Cunha et al. (2000) and Venn et al. (2002) as well as the solar system and interstellar values. A simple linear fit to the data yields a slope of 1.5 ± 0.1, implying that a mixture of primary and secondary nucleosynthetic processes contributes to the galactic evolution of boron.

Data acquired by Prochaska et al. (2003) place constraints on nucleosynthetic processes in the early Universe. Their observations help distinguish between processes where B production relates to the metallicity of the galaxy and processes that are independent of it. Some other theories [such as neutrino spallation in the carbon shells of supernovae (Woosley et al., 1990) and the spallation of C and O nuclei accelerated by supernovae onto local interstellar gas (Cassé et al., 1995)] predict that the B/O ratio will remain constant with O/H metallicity. In contrast, some other theories (such as p, n accelerated onto interstellar C, N and O seed nuclei) predict that the B/O ratio will increase with increasing metallicity. The current observation of a solar B/O ratio being approximately 1/3 argues for B synthesis independent of metallicity. Future considerations could test this

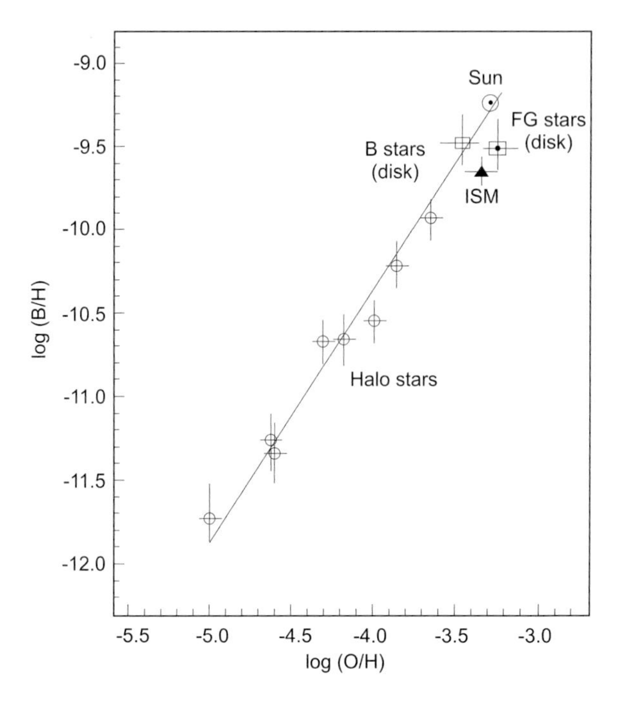

FIGURE 1.6 Galactic evolutionary trend of boron vs. oxygen, after Ritchey et al. (2011). For more explanations, see the text.

hypothesis in other galaxies. They may allow a measurement of the $^{10}B/^{11}B$ isotopic ratio, which will be critical to determining the relative importance of neutrino and cosmic-ray spallation (Burbidge et al., 1957).

1.1.3 Nucleosynthesis in Stars

A minimal concentration of ^{8}Be can start undergoing reactions with other ^{4}He nuclei to produce an unstable excited state of the mass-12 carbon. Some of these excited ^{12}C nuclei emit a gamma-ray fast enough and become stable before they disintegrate. This sequence of nuclear processes is called the triple-alpha process. Its result is a combination of three α-particles (that is, $3\times{}^{4}He$ nuclei) into a ^{12}C nucleus (see Figure 1.7).

From the inspection of ^{12}C and ^{16}O energy levels, it could be concluded that the nuclei "remember" the formation process. This "memory" explains why the cluster model describes so well the low-lying excited states!

Carbon is mainly produced in the core of a red giant star. It is stored there until the end of the star's life. It will still be present when the star explodes as a supernova and spreads its material into interstellar space. The sum of $^{12}C+\alpha$ masses is 7.1695 MeV. Therefore, the reaction $^{12}C(\alpha, \gamma)^{16}O$ should occur through a known nonresonant level at 7.1169 MeV. However, the 7.1169 level is just below $^{12}C+\alpha=7.1695$ MeV, and therefore, the resonance cannot happen. With the introduction of 7.65 MeV resonance, the carbon yield has increased by a factor of about 10^{7} compared to $3\alpha \rightarrow {}^{12}C$.

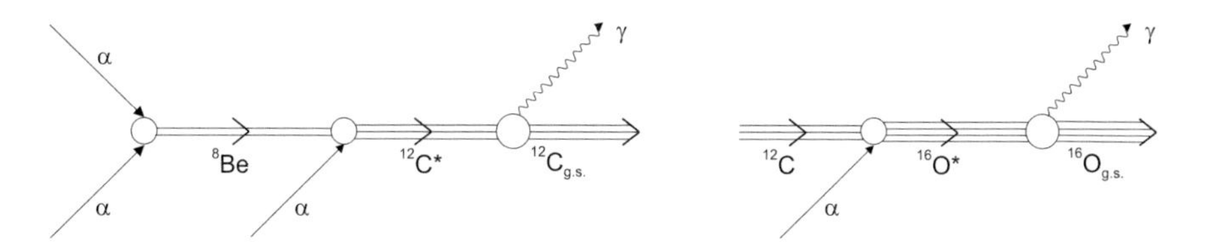

FIGURE 1.7 The triple-α and $^{12}C(\alpha, \gamma)^{16}O$ reactions. These reactions co-occur during the helium-burning stage of stellar evolution.

The rate of the triple-α reaction affects the synthesis of heavy elements in the Ga–Cd range in proton-rich neutrino-driven outflows of core-collapse supernovae (Wanajo, 2006; Fröhlich et al., 2006; Wanajo et al., 2011). At the beginning, outflows contain only protons and neutrons which subsequently combine into α-particles, then ^{12}C and some heavier elements. Although the most dominant is $3\alpha \rightarrow {}^{12}C$ reaction involving 7.5 MeV state, the very high nucleon densities may cause the importance of proton and neutron scattering processes altering the effective width of 7.5 MeV state (Turan and Kozlowsky, 1969; Beard et al., 2017). The results of these calculations indicate that, for very high temperatures and particle densities $>10^8$ g/cm^3, the rates for inelastic scattering to the ground state of ^{12}C can exceed the rates of spontaneous radiative transitions by a significant factor.

In a recent paper, Jin et al. (2020) speculated that in proton-rich neutrino-driven outflows of a core-collapse supernova, proton and neutron scattering processes enhance the triple-α reaction rate by an order of magnitude. In such a way, heavy proton-rich isotopes are produced by ν–p process in the innermost material ejected by supernovae. Because the in-medium contribution to the triple-α reaction rate must be present at high densities, this effect needs to be included in supernova nucleosynthesis models (Jin et al., 2020).

The following sequence of reactions can envision the making of ^{10}B nucleus:

$$^{8}Be + p \rightarrow {}^{9}B + \gamma$$

$$^{9}B + p \rightarrow {}^{10}C + \gamma$$

$$^{10}C \rightarrow {}^{10}B + e^{+} + \nu$$

Of interest is the ^{9}B level structure – the first excited state can help in resolving some of the dilemmas? Some information about it may be obtained by $^{10}B(n, 2n)^{9}B$ reaction (Q=−8.438 MeV).

At lower temperatures, the p–p chain dominates, but with an increase in temperature, there is a sudden transition to the dominance of the CNO cycle (Weizsäcker, 1938; Bethe, 1939). CNO cycle has an energy production rate that varies strongly with temperature.

$$^{12}C + p \rightarrow {}^{13}N + \gamma$$

$$^{13}N \rightarrow {}^{13}C + e^{+} + \nu_e$$

$$^{13}C + p \rightarrow {}^{14}N + \gamma$$

$$^{14}N + p \rightarrow ^{15}O + \gamma$$

$$^{15}O \rightarrow ^{15}N + e^{+} + \nu_e$$

$$^{15}N + p \rightarrow ^{12}C + ^{4}He$$

The fusion of hydrogen to helium in the p–p chain and the CNO cycle requires temperatures around 10^7 K or higher. Only at those temperatures, plasma will contain enough hydrogen ions with high enough velocities to tunnel through the Coulomb barrier at sufficient rates.

The conclusion that our Universe is expanding and cooling is entirely consistent with the hot Big Bang model. Many authors have studied the star formation rate (SFR) in high-redshift galaxies as a function of redshift (see Lilly et al., 1996; Giavalisco et al., 2004; Stanway et al., 2003, and others). There is little doubt that SFR rises from the present to about $z \approx 1$. What happens in the redshift range $z \approx 1$–6 is still somewhat controversial (Livio, 2008).

If a star had sufficient mass potentially enough, C would accumulate enabling temperature and density to allow C nuclei fusion into Ne nuclei. Two outer shells would surround the carbon-burning core: the innermost burning He, and the outermost burning H. The sequence of the central core collapsing and increasing temperature continues for some time. The result is a further round of fusion and formation of more shells. As the fusion process continues, the result is the increase in Fe concentration in the core of the star, followed by the core contraction, and the temperature increases. Fe nucleus has the highest binding energy per nucleon, and it is the most stable of all atomic nuclei. Because of this, when it undergoes nuclear reactions, it does not release energy, but instead, it absorbs. This property results in no release of energy to balance the force of gravity. However, this decrease in the internal pressure, together with gravity, increases the intensity of core collapse. In this process, the Fe nuclei in the central portion of the core are broken into alpha particles, protons and neutrons and compressed even further. However, they cannot be infinitely compressed. Eventually, the outer layers of material are repealed off the compressed core and are thrown outward. The consequence of this series of events is an enormous collision between the collapsing outermost layers and recoiling core layers.

Under the extreme conditions of this collision, the formation of the most massive elements is made possible because of two factors. First, the temperature increases to very high levels, not achievable even in the most massive stars. The consequence is high kinetic energies of all nuclei present, making them very reactive. Second, there is a high concentration of neutrons ejected from the core during the supernova because of the breaking of iron nuclei in the central core. Surrounding nuclei capture these neutrons and afterward decay into a proton by the emission of an electron and an antineutrino. Because of the large neutron flux created during a supernova explosion, the neutron capture/decay sequence can be repeated many times. This process of adding protons forms increasingly more massive nuclei. Although the described conditions exist for only a short time, the highest mass nuclei can be formed. During the explosion, the outer layers of the star, enriched with the higher mass nuclei, are blown off into space. This material will later be part of nebulas and subsequently incorporated into new stars. The same cycle will be repeated in this medium. Each star uses up more of the H and

TABLE 1.2 Basic Nuclear Reaction Links

	N − 2	N − 1	N	N + 1	N + 2
Z+2				(α,n)	(α,γ)
Z+1		(p,n)	(p,γ)		(α,p)
Z		(γ,n)	[Z, N]	(n,γ)	
Z−1	(p,α)		(γ,p)	(n,p)	
Z−2	(γ,α)	(n,α)			

He from the early Universe and creates more considerable amounts of the higher-mass elements. The essential nuclear reaction links are shown in Table 1.2.

It is believed that the rapid neutron-capture process makes half of all elements in the Universe heavier than iron (see Burbidge et al., 1957). Where this happens, since the enormous neutron fluxes are required, it has not been determined by certainty. Most of the evidence pointed to neutron star mergers as a probable r-process site. Element strontium, made by neutron capture, was identified by the re-analysis of the spectra recorded from kilonova AT2017gfo (which was found following the discovery of the neutron star merger GW170817) by Watson et al. (2019). This detection of a neutron-capture element establishes the origin of rapid neutron-capture process (r-process) elements in neutron star mergers and indicates that neutron stars are made of neutron-rich matter.

Although neutron-capture cross sections have been measured for most stable nuclei, fewer results exist for radioactive isotopes. Also, the statistical model predictions have significant uncertainties. As an example, we mention the work by Shusterman et al. (2019), who exposed radioactive nucleus ^{88}Zr to the intense neutron flux of a nuclear reactor and determined ^{88}Zr thermal neutron-capture cross section of $861{,}000 \pm 69{,}000$ barns. This value is the second-largest thermal neutron-capture cross section ever measured. The only other nuclei known to have values greater than 10^5 barns are ^{135}Xe (2.6×10^6 barns) and ^{157}Gd (2.5×10^5 barns).

1.1.4 Heavy Nuclides Production

At the beginning of the development of the theory of nucleosynthesis (Burbidge et al., 1957), the foundation of the solar system composition, in the heavy element domain, was explained by the introduction of three sorts of heavy nuclides called s-nuclides, p-nuclides and r-nuclides. These processes correspond to the "topology" of the chart of the nuclides, which exhibits three categories of stable heavy nuclides:

i. those located at the bottom of the valley of nuclear stability, called the s-nuclides,
ii. those situated on the neutron-deficient side, named p-nuclides and
iii. those located on the neutron-rich side of the valley, named the r-nuclides.

These three different mechanisms are needed to account for the production of three types of stable nuclides. The processes involved are referred to as the s-, r- and p-processes. In the thermonuclear framework, the neutron deficiency of the p-nuclides forbids their production in neutron-capture chains of the s- or r-types. Figure 1.8 displays in a very schematic way some possible nuclear routes through which seed s-nuclides (r-nuclides) can be transformed into p-nuclides.

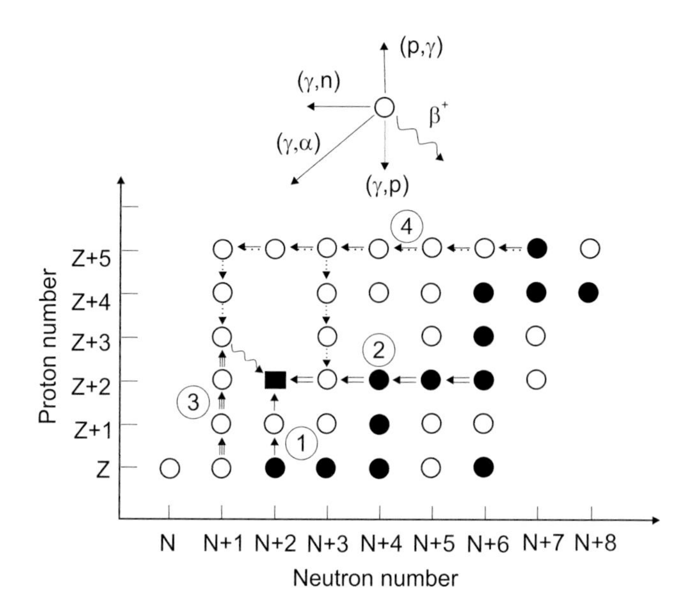

FIGURE 1.8 Representation of simple nuclear routes through which seed s-nuclides or r-nuclides (black dots) can be transformed into p-nuclides (black square). Open dots represent unstable nuclei. Routes (1) and (2) are made of a succession of (p, γ) and (γ, n) reactions leading directly to the p-nuclide. Slightly more complicated chains involve (p, γ) reactions followed by β-decays (Route (3)), or a combination of (γ, n) and (γ, p) or (γ, α) and β-decays (Route (4)). More complicated patterns, involving combinations of the proposed routes, can also be envisioned. The p-nuclide destruction channels are not indicated. (After Arnould and Goriely, 2003.)

However, the final abundances of elements are determined in supernova explosions! See Figures 1.9 and 1.10 for supernovae remnants in the Milky Way. Supernovae are classified based on the presence or absence of certain features in their optical spectra taken near maximum light. The present classification of supernovae is presented in many places; we recommend the article by Turatto (2003). The first two main classes of SNe were identified based on the presence or absence of hydrogen lines in their spectra: SNe of type I did not show H lines, while those with the obvious presence of H lines were called type II. In the mid-1980s, it was realized that there is a class of SNeI physically distinct from the others. The objects of the new class were characterized by the presence of He I lines; they were called SNeIb, while the original type was renamed to SNeIa. The new class further branched into another variety SNeIc, based on the absence of He I lines.

It is assumed that more substantial quantities of lighter elements in the present Universe have been restored by cosmic-ray bombardment. This process lasted billions of years and resulted in cosmic-ray (mainly high-energy protons) bombardment of heavier elements in interstellar gas and dust, resulting in their breakup.

At a distance of about 20,000 ly, G292.0+1.8 (shown in Figure 1.10) is one of the only three supernova remnants in the Milky Way assumed to contain large amounts of oxygen. These oxygen-rich supernovas are one of the primary sources of the heavy elements (heavy element term is used for all elements other than hydrogen and helium) necessary

FIGURE 1.9 Historical supernova remnant Cassiopeia A, located 11,000 ly away. (The picture was taken (January 7, 2013) by NASA's Nuclear Spectroscopic Telescope Array, or NuSTAR. Blue indicates the highest energy X-ray light, where NuSTAR has made the first resolved image ever of this source. Red and green colors show the lower end of NuSTAR's energy range, which overlaps with NASA's high-resolution Chandra X-ray Observatory. (Image credit: NASA/JPL-Caltech/DSS.)

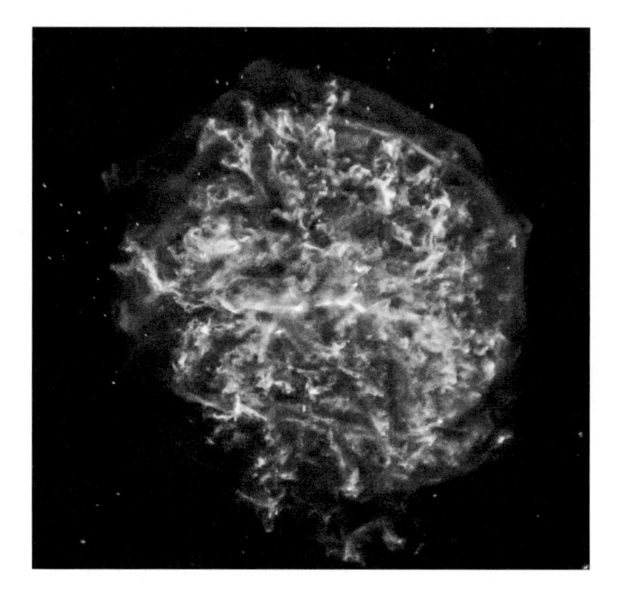

FIGURE 1.10 G292.0+1.8 is one of the only three supernova remnants in the Milky Way. (Image credit: NASA/CXC/SAO.)

to form planets and eventually sustain life. Chandra X-ray image shows a rapidly expanding, intricately structured, debris field. In addition to oxygen (yellow and orange), other elements such as magnesium (green) and silicon and sulfur (blue) are also present that were made in the star before it exploded.

In recent publications (Driver et al., 2016), the authors presented the Galaxy And Mass Assembly (GAMA) Panchromatic Data Release (PDR), representing over 230 deg^2 of imaging with photometry in 21 bands extending from the far-UV to the far-IR. These data complement their spectroscopic campaign of over 300,000 galaxies. They are compiled from observation data from several facilities including (i) GALaxy Evolution eXplorer, (ii) Sloan Digital Sky Survey, (iii) Visible and Infrared Survey Telescope for Astronomy (VISTA), (iv) Wide-field Infrared Survey Explorer, (v) Herschel, with the GAMA regions surveyed by VLT Survey Telescope (VST) and (vi) Australian Square Kilometre Array Pathfinder (ASKAP). The compiled data are processed to a common astrometric solution. In such a way, photometry is obtained for 221,373 galaxies with $r<19.8$ mag. Sophisticated online tools are provided to access and download data cutouts, or the full mosaics of the GAMA regions in each band. The focus was on the reduction and analysis of the data obtained by VISTA Kilo-degree Infrared Galaxy and comparison to earlier data sets (i.e., 2MASS and UKIDSS) before combining the data and examining their integrity. Having derived the 21-band photometric catalogue, they fit the data by the energy balance code called MAGPHYS. All of these data are then used to obtain the first entirely empirical measurement of the 0.1–500 μm energy output of the Universe. Exploring the cosmic spectral energy distribution across three time intervals (0.3–1.1, 1.1–1.8, and $1.8–2.4\times10^9$ years), it was found that the Universe is now generating $(1.5\pm0.3)\times10^{35}$ h_{70} W/Mpc^{-3}, down from $(2.5\pm0.2)\times10^{35}$ h_{70} W/Mpc^{-3} 2.3×10^9 years ago (Driver et al. 2016).

More importantly, the authors report significant and smooth evolution in the integrated photon escape fraction at all wavelengths. The UV escape fraction was found to be increasing from 27% (18%) at $z = 0.18$ in NUV (FUV) to 34% (23%) at $z = 0.06$.

This measurement of the energy generated within a large portion of space is done more precisely than ever before. The analysis done by Driver et al. (2016) represents the most comprehensive assessment of the energy output of the nearby Universe. The results obtained confirm that the energy produced in a section of the Universe today is only about half what it was 2×10^9 years ago; also, this fading is occurring across all wavelengths from the ultraviolet to the far-infrared. In conclusion, this implies that the Universe is slowly dying!

Neutron clusters can be produced in both young and mature Universes. In the young Universe, during BBN, in the period of transition from quark–gluon plasma to baryonic matter, the probability of neutron cluster formation (in a bound state or unbound state) is probably increased. This possibility would have led, for example, to Li destruction via reactions involving two neutron clusters (even in an unbound state, two neutrons interacting in the final state, fsi) by the following series of nuclear reactions (Valkovic et al., 2018):

i. $^6Li+^2n\rightarrow^8Li\rightarrow^8Be+e+\bar{\upsilon}$ followed by $^8Be\rightarrow2\alpha$.
ii. $^7Li+^2n\rightarrow^9Li\rightarrow^9Be+e+\bar{\upsilon}$ followed by $^9Be\rightarrow n+2\alpha$.

The ^2n symbol represent neutron–neutron final state interaction (fsi) as observed in the $n+d \rightarrow p+n+n$ reaction for $E_n=14$ MeV bombarding energy.

In mature Universe, when all periodic table nuclei are already made, two scenarios are possible. In the first one, heavy nuclei can decay by emission of neutron clusters. For example, Gulkanyan and Margaryan (2014) searched for the octa-neutron using alpha spectroscopy of ^{252}Cf decays. In another experiment of interest, Bystritsky et al. (2017) described an effort to register neutron nuclei ($N \geq 6$) using a multi-neutron detector especially constructed for this purpose. The authors claim that their results indicate the presence of hexa-neutron and octa-neutron emission in the decay of ^{238}U nuclei.

In the second, neutron clusters might be produced by heavy nuclei collisions in the processes involving objects with significant abundances of heavy nuclei. This process could be measured in the laboratory by heavy-ion accelerators accelerating heavy ions to relativistic energies; what is usually measured is neutron multiplicity in heavy-ion collisions. The curves showing cross section vs. neutron numbers depend on types of colliding ions, their energies and other experimental parameters. Figure 1.11 (after Jahnke, 1986) shows the partial cross sections for the emission of 1–30 neutrons in ^{20}Ne (390 MeV)-induced reactions on several heavier target nuclei from ^{100}Mo to ^{238}U. The measured distributions peaked at eight neutrons for ^{100}Mo and at 20 neutrons for ^{238}U.

1.2 Cosmic (Universal) Element Abundances

Today, scientists assume that galaxies have formed out of a primordial gas consisting of ~77% hydrogen, ~23% helium, and traces of deuterium and lithium without heavier elements. However, presently, the chemical composition of the interstellar medium in the solar neighborhood is ~70% hydrogen, ~28% helium and ~2% heavier elements. This progressive enhancement of helium and heavier chemical elements at the expense of hydrogen in the interstellar gas refers to galactic chemical evolution (Matteucci, 2001).

The factors that govern the chemical evolution of galaxies are, among others, the rate at which stars form, their mass spectrum, development through successive thermonuclear cycles and the dynamics of the gas-star system. Each generation of stars contributes to the chemical enrichment of a galaxy. This enrichment is done by processing new material in the stellar interiors and restoring to the interstellar medium a fraction of its total mass. It is done in the form of both processed and unprocessed matter, during various events such as stellar winds, planetary nebula ejection and supernova explosion. The gravitational collapse forms the next stellar generation out of this enriched gas. It "evolves, giving rise to an ongoing process which terminates when all the available gas has been consumed" (Matteucci, 2001).

To describe in detail this process of enrichment, one should know several factors:

i. how much gas per unit time is turned into stars,
ii. the initial mass function (IMF) and
iii. how much and when nuclearly processed material is restored to the ISM by each star (stellar yields).

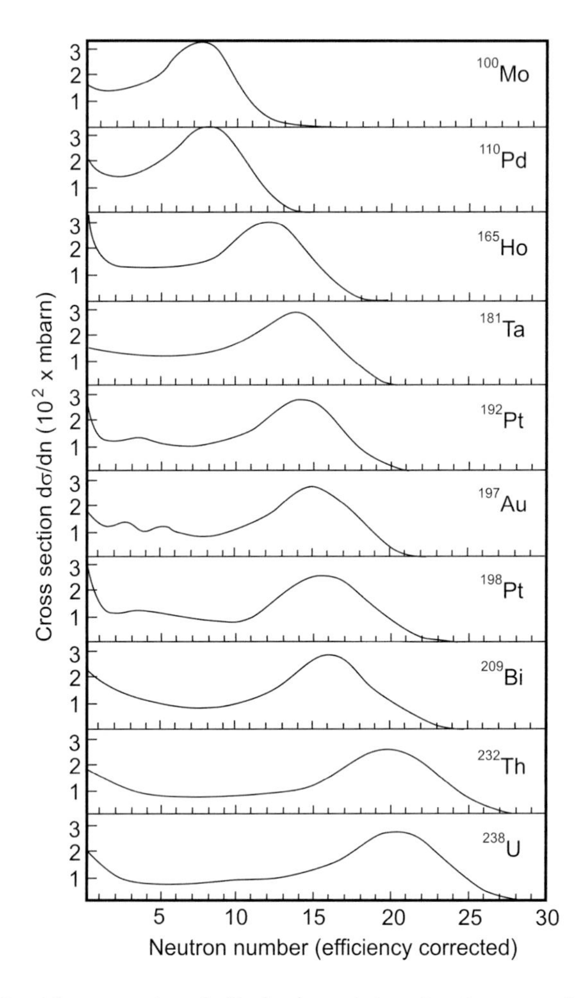

FIGURE 1.11 Partial cross sections dσ/dn for the emission of 1–30 neutrons in ^{20}Ne (390 MeV) bombardments of various targets, as indicated on the figure.

It is established that our solar system is the result of the gravitational collapse of a small part of a giant molecular cloud. We can assume that the Sun, the planets and all other objects in the solar system were formed from a hot gaseous nebula having well-established chemical and isotopic composition. The measurements showing comparatively large and widespread variations in oxygen isotopic composition have resulted in some doubts about this assumption. The establishment of huge (up to the percent level) isotope anomalies of some heavy elements in meteoritic silicon carbide (SiC) grains provides additional evidence of incomplete mixing. It also indicates a small-scale homogenization in the primordial solar nebula.

Abundances of chemical elements in the solar system are somewhat similar to those of most stars and interstellar material in our neighborhood and corresponding parts of other galaxies. Minor variations (within a factor of 3) may occur in the relative amounts of hydrogen and helium, and the relative amounts of carbon and heavier elements. These relations are the consequence of the fact that hydrogen and the bulk of helium are relics from the "Big Bang". In contrast, heavier elements (and some helium) are results of nuclear reactions taking place in stars or the interstellar medium.

Carbon and heavier elements are usually more abundant in the central regions of massive galaxies than in their outer parts or in small galaxies. This situation is the case of our galaxy as well. They are also more abundant in stars belonging to the outer spheroidal halo of our galaxy. When compared to solar system values, carbon and heavier elements, in some cases, may be deficient by a factor of up to 1,000 or more (relative to hydrogen and helium). Among all chemical elements, carbon, nitrogen, iron and elements such as barium (resulting from the "slow" neutron-capture process or s-process in the progenitor stars) can be deficient by more significant factors than oxygen, magnesium and other "α-particle" elements formed in massive stars which undergo supernova explosions after approximately 10 million years. Over- and under-abundances of various elements can also be found in some highly evolved stars as a result of internal nuclear reactions, and in the surface layers of individual stars where the diffusive separation of elements seems to have occurred.

Faisst et al. (2015) measured the relationship between the depth of four prominent rest-UV absorption complexes and metallicity for local galaxies and verified it up to z~3. They applied this relation to a sample of 224 galaxies at $3.5<z<6.0$ ($<z>=4.8$) in COSMOS, having unique UV spectra in DEIMOS, and accurate stellar masses from SPLASH are available. The galaxy population at z~5 and $\log(M/M_{sun})>9$ is characterized by a value of 0.3–0.4 dex, in units of $12+\log(O/H)$. These are lower metallicities than at z~2, but comparable to z~3.5. They found galaxies with weak/no Ly-alpha emission to have metallicities comparable to z~2 galaxies, which, therefore, may represent an evolved sub-population of z~5 galaxies. A correlation between metallicity and dust is in good agreement with local galaxies. It shows an inverse relation between metallicity and star formation rate in agreement with observations at z~2. The relation between stellar mass and metallicity (M–Z relation) is similar up to z~3.5. However, there are indications that this relation is slightly shallower, in particular for the young, Ly-alpha-emitting galaxies. They showed that a shallower M–Z relation is expected for the case of a fast increase in the stellar mass with an "e-folding time" of $1–2\times10^{8}$ years. As a consequence, the process of dust production and metal enrichment as a function of mass could be very stochastic in the first billion years of galaxy formation because of the fast evolution.

Recently, very accurate abundances of the Fe-group elements Sc through Zn ($Z=21–30$) in the bright main-sequence turnoff star HD 84937 have been derived by Sneden et al. (2016). The base of their approach was the usage of high-resolution spectra covering the visible and ultraviolet spectral regions. New laboratory transition data for 14 species of seven elements have been used. To obtain abundance ratios of high precision, the authors combined the observed abundances of lines from non-Fe atoms with Fe lines observed in HD 84937. The abundances have been determined by measurements of both neutral and ionized transitions, which were generally in agreement. They found no substantial departures from the standard LTE Saha ionization balance for this star having metallicity $[Fe/H]=-2.32$. Measured abundances of $[Co/Fe]=+0.14$ and $[Cu/Fe]=-0.83$

are in agreement with past studies of abundance trends in this and other low-metallicity stars. A detailed examination of scandium, titanium and vanadium abundances in large-sample spectroscopic surveys reveals that they are positively correlated in stars with [Fe/H] < -2; HD 84937 lies at the high end of this correlation. These trends constrain the synthesis mechanisms of Fe-group elements. They also examine the galactic chemical evolution abundance trends of the Fe-group elements, including a new nucleosynthesis model with jet-like explosion effects.

It is assumed that the first stars have formed within 200 million years after the Big Bang. The first star has not yet been discovered. However, stars with minimal amounts of elements heavier than helium (called metals by astronomers) have been found in the outer regions of the Milky Way galaxy. The first stars should be found today in the central areas of galaxies, called "bulges", because they were formed in the highest density regions. The area of Milky Way bulge underwent a rapid chemical enrichment during the first $(1\text{–}2) \times 10^9$ years, leading to a reduction in the number of early, metal-poor stars. Howes et al. (2015) report observations of extremely metal-poor stars in the Milky Way bulge, including a star with iron abundance about 10,000 times lower than the solar value without noticeable carbon enhancement. They reported that most of the metal-poor bulge stars are on tight orbits around the galactic center. This finding is in contrast to being halo stars only passing through the bulge, as expected for stars formed at redshifts higher than 15. The chemical compositions of studied stars are similar to typical halo stars of the same metallicity, although some differences are present, for example, the lower abundances of carbon.

Martig et al. (2016) show that the masses of red giant stars can be well predicted from their photospheric carbon and nitrogen abundances. This analysis should be done in conjunction with their spectroscopic stellar parameters log g, T_{eff} and [Fe/H]. This approach is qualitatively expected from mass-dependent post-main-sequence evolution. The authors established an empirical relation between these quantities in 1,475 red giants with asteroseismic mass estimates from Kepler having spectroscopic labels from Apache Point Observatory Galactic Evolution Experiment (APOGEE), Data Release 12 (DR12). They assessed the accuracy of the model and found that it predicts stellar masses with fractional rms errors of about 14% (typically 0.2 $M_\odot$). From these masses, they derived ages with rms errors of 40%. This empirical model allowed for the first time to make age determinations (in the range of $(1\text{–}13) \times 10^9$ years) for vast numbers of giant stars across the galaxy. They applied their model to ~52,000 stars in APOGEE DR12, for which no direct mass and age information was previously available. It was found that these estimates highlight the vertical age structure of the Milky Way disk. It also shows that the relation of age with [α/M] and metallicity is in agreement with the predictions of detailed studies of the solar neighborhood.

Thielemann (2013) presented an improved Fe-group composition with a discussion of the effects which enhance the composition in general and extended it to Cu and Zn; see Figure 1.12. Supernovae conditions greatly influence this abundance (see, for example, Pruet et al., 2005). Namely, as an explosion develops in the collapsed core of a massive star, neutrino emission drives convection in a hot bubble. This bubble is made of radiation, nucleons and pairs just outside a proto-neutron star. Shortly after that, neutrinos

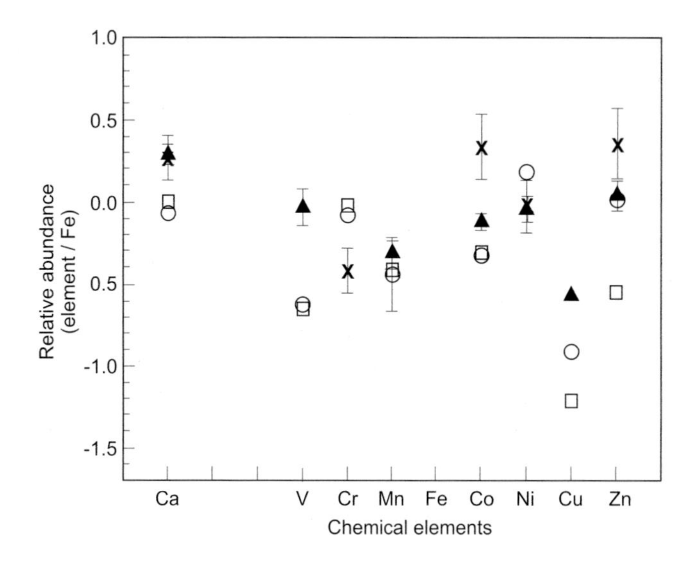

FIGURE 1.12 Abundance values for some life-essential elements resulting from a supernova explosion (based on a compilation by Thielemann, 2013). (Data presented are from Cayrel et al. (2004) – marked **X**; Gratton and Sneden (1994) – black triangles; Thielemann et al. (2012) – open squares; Fröhlich et al. (2005) – open circles.)

drive a wind like an outflow from the neutron star. In both cases, the convective bubble and the early wind, weak interactions temporarily (for at least the first second) cause a proton excess to develop in the ejected matter. The approximately 0.05–0.1 M_{solar} of material that is ejected has a composition that is of importance for later nucleosynthesis. This approach is elaborated in detail in a two-dimensional model of a supernova explosion developed by Janka et al. (2003).

Pruet et al. (2005) determined the composition of this material. They found that it is made mainly of helium and ^{56}Ni. Also, relatively rare nuclei produced by the decay of proton-rich isotopes unstable to positron emission are present. As a result, the nuclear flow will be held up by long-lived nuclei in the vicinity of ^{64}Ge. The resulting abundance pattern can be rich in a few unusual, rare isotopes such as ^{45}Sc, ^{49}Ti and ^{64}Zn. The calculations shown account for the solar abundance of ^{45}Sc and ^{49}Ti. It is expected that the primary production of these species may have noticeable signatures in the abundances of low-metallicity stars. The authors also discuss uncertainties in the nuclear physics and early supernova evolution to which abundances of exotic nuclei are sensitive (Pruet et al., 2005).

1.2.1 Cosmic Dust

First stars formed from the collapse of a gas cloud made up of hydrogen, helium and elements created in the aftermath of the Big Bang. These stars used hydrogen as a fuel to create heavier elements such as carbon and oxygen and up to the iron through nuclear fusion processes. These newly formed elements are released at the end of the star's lifetime when it collapses under its gravity and explodes as a supernova. During such an explosion, additional elements heavier than iron are produced. Combination of heavier elements, metals such as silicon and iron, with oxygen forms minerals – which is what dust is.

The origin of dust particles in galaxies is still an unresolved matter; see, for example, Gall et al. (2011), Matsuura et al. (2009), Draine (2009), and Dunne et al. (2011). Supernova explosions produce the majority of the refractory elements. Still, it is unknown how and where dust grains grow and how they avoid destruction in the harsh environments of star-forming galaxies. The recent detection of a large mass of dust clouds (0.5 $M_{\odot}$) in nearby supernova remnants (Matsuura et al., 2011; Indebetouw et al., 2014; Gomez et al., 2012) suggests in situ dust formation. However, others (Gall et al., 2014) reported rapid (40–240 days) formation of dust grains in the luminous supernova 2010jl. The detailed investigation of this dust reveals the presence of large (exceeding 1 μm) grains, which resist destruction.

Dust particles' interaction with the cosmic plasma environment leads to the accumulation of surface charge on the grains. The charging of particles depends on their size and material composition as well as on the surrounding plasma and radiation. The timescales for dust particles to attain their equilibrium surface charge are small compared to typical lifetimes of dust in the solar system. Also, tiny particles are sufficiently highly charged so that their motion is influenced by Lorentz force (Mann, 2001).

Several processes are responsible for the charging of cosmic dust particles. Among them, the most important are: (i) sticking of low-energy electrons and ions of the surrounding plasma, (ii) induced secondary electron emission, (iii) thermionic emission and (iv) photoelectric emission due to solar and interstellar radiation. Draine and Salpeter (1979) carried out a detailed study of the charging of dust grains in hot plasma with temperatures in the range of 10^4–10^9 K. This procedure is applied to dust in the cosmic, interstellar environments. Plasma density and temperature, the relative velocity of the dust and the plasma, the solar radiation and, finally, the physical properties of the dust grains determine their charging rates. Mukai (1981) calculated the surface charge of dust particles in the solar system. He showed that in the interplanetary medium, this is predominantly determined by photoelectric emission due to solar radiation. In small dust particles that reach high positive values of the surface potential, the field emission of positive ions can cause a significant mass loss and lead to the destruction of particles (Draine and Salpeter, 1979). In their report, Kimura and Mann (1998) have shown this not to be the case for dust particles in the interplanetary medium.

Here are some examples of the observed cosmic dust clouds: Cosmic dust clouds spread across a field of stars in sweeping telescopic vista near the northern boundary of Corona Australis, the Southern Crown (see Figure 1.13). Probably less than 500 ly away, it is effectively blocking light from more distant stars in the Milky Way. We should point out that the densest part of the dust cloud is about 8 ly long. A series of nebulae catalogued as NGC 6726, 6727, 6729, and IC 4812 is positioned at its tip (lower left). The origin of their characteristic blue color is the production of light from hot stars being reflected by the cosmic dust. The tiny but intriguing yellowish arc visible near the blue

FIGURE 1.13 Stars and dust across Corona Australis. Astronomy Picture of the Day 2006 April 3. (Credit: Adam Block, NOAO, AURA, NSF.)

nebulae marks the young variable star R Coronae Australis. The globular star cluster NGC 6723 is seen below and left of the nebulae. Although NGC 6723 appears to be just outside Corona Australis in the constellation Sagittarius, it lies nearly 30,000 ly away. This distance is far beyond the Corona Australis dust cloud.

The Magellanic Clouds is a common name for two galaxies (Large and Small Magellanic Clouds) that share a gaseous envelope and lie about 22° apart in the sky near the south celestial pole. The Large Magellanic Cloud is a luminous patch about 5° in diameter. The Small Magellanic Cloud measures less than 2° across. The Large Magellanic Cloud is about 160,000 ly from the Earth. The Small Magellanic Cloud is further away; it is about 190,000 ly from the Earth. Both objects are smaller than the Milky Way. The Large Magellanic Cloud is 14,000 ly in diameter, and the Small Magellanic Cloud is 7,000 ly in diameter, while the Milky Way galaxy is about 140,000 ly across. They are the nearest objects to the Milky Way galaxy, and for a long time, they were regarded as portions of the Milky Way galaxy system. Both Small and Large Magellanic Clouds are too far south to be seen from northern latitudes. Also, the irregular shapes of the objects and their numerous hot blue stars, star clusters and gas clouds make them resemble the southern Milky Way galaxy.

Feedback from massive stars played a critical part in the evolution of the Universe. This role has been done by driving powerful outflows from galaxies that enrich the intergalactic medium and regulate star formation. An essential source of outflows may be dwarf galaxies, the most numerous galaxies in the Universe. These galaxies quickly lose their star-forming material in the presence of intense stellar feedback with only small gravitational potential wells. McClure-Griffiths et al. (2018) showed that a nearby dwarf galaxy, the Small Magellanic Cloud, has outflows of atomic hydrogen extending at least 2 kpc from the star-forming bar of the galaxy. These outflows are cold (<400 K) and may have formed during a period of active star formation $(25–60)\times10^6$ years ago. The total mass of atomic gas in the outflow is equal to about 10^7 solar masses or about 3% of the entire atomic gas of the galaxy. They conducted mass flux in atomic gas alone, MHI $\approx$ 0.2–1.0 solar masses/year, which is close to one order of magnitude higher than the star formation rate. The authors suggested that most of the observed outflow would be due to the interaction of the Small Magellanic Cloud with its companion, the Large Magellanic Cloud, and the Milky Way, feeding the Magellanic Stream of hydrogen encircling the Milky Way (McClure-Griffiths et al., 2018).

Figure 1.14 shows the Large Magellanic Cloud galaxy in infrared light as seen by the Herschel Space Observatory. This mission is a European Space Agency-led mission with significant contributions from NASA and its Spitzer Space Telescope. In the combined data of all instruments, this nearby dwarf galaxy looks like a circular explosion. Rather than fire, however, those ribbons are giant waves of dust ranging tens or hundreds of light-years. Significant areas of star formation are noticeable in the center, just left of center and right. The bright center-left region is called 30 Doradus, or the Tarantula Nebula because of its appearance in visible light. The colors in this image present temperatures in the dust filtered by the Cloud. New star formation is at its earliest stages (or it is shut off) in the colder regions, while warm areas point to new stars heating surrounding dust. The coldest fields and objects are marked by red color; they correspond to infrared light measured at 250 μm by Herschel's Spectral and Photometric Imaging Receiver. Herschel's Photodetector Array Camera and Spectrometer fill out the mid-temperature bands, shown in green, taken at 100 and 160 microns. The warmest spots appear in blue, thanks to 24- and 70-micron data from Spitzer.

Seen through a cosmic dust cloud a mere 400 ly away, the Pleiades or Seven Dusty Sisters star cluster is well-known astronomical images for its striking blue reflection nebulae (see Figure 1.15). The visible starlight is scattered or reflected by the dust; however, in this figure taken in the infrared light region by the Spitzer Space Telescope, the dust itself glows. The false-color image spans about 1 degree or 7 ly at the distance of the Pleiades. The densest regions of the dust cloud are shown as yellow and red shades. Exploring this young, nearby cluster resulted in the Spitzer data revealing cool, low-mass stars, brown dwarfs or failed stars, and planetary debris disks.

Several bright blue nebulas are particularly apparent in the vast Orion Molecular Cloud Complex; see Figures 1.16 and 1.17 showing two of the most prominent reflection nebulas:

FIGURE 1.14 Large Magellanic Cloud galaxy in infrared light as seen by the Herschel Space Observatory, a European Space Agency-led mission with significant NASA contributions, and NASA's Spitzer Space Telescope. (Image credit: ESA/NASA/JPL-Caltech/STScI.)

FIGURE 1.15 Seven Dusty Sisters. NASA Astronomy Picture of the Day 2007 April 13. (Credit: NASA, JPL-Caltech, J. Stauffer (SSC, Caltech).)

FIGURE 1.16 M78 and Orion Dust Reflections. Astronomy Picture of the Day 2017 January 24. (With permission. Image Credit & Copyright: Marco Burali, Tiziano Capecchi, Marco Mancini (MTM observatory, Italy).)

FIGURE 1.17 Orion Nebula: The Cosmic Hearth. NASA Astronomy Picture of the Day 2013 February 13. (Image Credit: NASA/JPL-Caltech/UCLA.)

dust clouds lit by the reflecting light from bright embedded stars. The image center shows nebula M78, which was catalogued over 200 years ago. On the lower left is the lesser-known NGC 2071. The study of these reflection nebulas continues to understand better how the interior of stars is formed. The Orion complex is about 1,500 ly distant. It contains the Orion and Horsehead Nebulas and covers much of the constellation of Orion (NASA, 2013). The Great Nebula in Orion is an exciting place. It is visible to the naked eye, and it appears as a small fuzzy patch in the constellation of Orion. However, the image in Figure 1.17, an illusory color composite of four colors of infrared light taken with the Earth-orbiting WISE (Wide-field Infrared Survey Explorer) observatory, shows the Orion Nebula in a neighborhood of recently formed stars, hot gas and dark dust. The light behind much of the Orion Nebula (M42) is due to the stars of the Trapezium star cluster. The green glow surrounding the bright stars pictured here is their starlight reflected by dust filaments that cover much of the region. It is assumed that the present Orion Molecular Cloud Complex, which includes the Horsehead Nebula, will slowly scatter over the next 100,000 years.

Interstellar dust particles, which predate our solar system, can provide insight into the processes at the end of the life of ancient stars. The present-day interplanetary dust, the inner solar system, contains some interstellar dust particles. The majority of interplanetary dust particles in our solar system originate from comets approaching the Sun, or from the collision between asteroids in the asteroid belt. These dust particles contain clues about the makeup and formation of dust condensation nuclei which are assumed to be the first steps of planet formation from the massive dust and gas cloud surrounding a new star.

An extraordinary tape of hot gas trailing behind a galaxy like a tail has been discovered using data from NASA's Chandra X-ray Observatory; see Figure 1.18. X-ray

FIGURE 1.18 Ribbon of hot gas trailing behind a galaxy like a tail. (Image credit: X-ray: NASA/CXC/University of Bonn/G. Schellenberger et al.; Optical: INT.)

tail is a consequence of gas being stripped from the galaxy as it moves through a large, hot cloud of intergalactic gas. With a length of some 250,000 ly, it is probably the most extensive such tail ever detected. In this new composite image, shown in Figure 1.18, X-rays picture from Chandra (blue) has been combined with visible-light data (yellow) obtained by the Isaac Newton Group of Telescopes, Canary Islands, Spain. The observed tail is located in the galaxy cluster Zwicky 8338, at a distance of 7×10^8 ly from the Earth. The length of the tail is bigger than twice the diameter of the Milky Way galaxy. The tail contains gas with a temperature of about 10×10^6 degrees, about 20×10^6 degrees cooler than the intergalactic gas. At this temperature, the gas is still hot enough to glow brightly in X-rays that Chandra can detect. The general opinion is that the tail was produced as a galaxy CGCG254-021, or possibly a group of galaxies which included this massive galaxy, moved through the hot gas in Zwicky 8338. The pressure conducted by this rapid motion caused gas to be stripped away from the galaxy.

The primordial process, lasting over millions of years, that turns enormous clouds of cosmic dust into newborn planets has been observed directly for the first time. A protoplanet in the process of condensation around a young star (2×10^6 years old), LkCa15, in the neighborhood of Taurus, 450 ly from the Earth, has been spotted by Sallum et al. (2015). Protoplanetary disks with inner clearings are natural places for the study of planet formation. Some observed disks show evidence for the presence of young planets in the form of disk asymmetries or infrared sources detected within their clearings, as in the case of LkCa15 (Kraus and Ireland, 2012; Ireland and Kraus, 2014).

Let us turn our attention to molecular clouds in the star-forming regions (for a full treatment of this topic, see the textbook by Stahler and Palla, 2004). For stars to form, molecular clouds have to contract, which they can do under their gravity if there is sufficient critical mass within a particular volume. Once they collapse, they mast rid significant angular momentum to contract to a condensed prestellar object. These aspects raise two interesting questions:

i. Which process impedes the collapse of all molecular clouds on the scale of a Hubble time?
ii. Why can stars form at all, given the conservation of angular momentum?

It can be shown that the internal pressure and the rotation of molecular clouds cannot impede their collapse. The fact that magnetic fields can support a cloud against gravitational collapse is the result of "flux freezing". Typical field strengths derived from Zeeman measurements support this picture: Denser clouds or their cores possess stronger magnetic fields, reaching values of mG there. Flux freezing does not imply that the magnetic field is rigidly tied to the molecular clouds; otherwise, they would not be able to contract at all. Through a process called "ambipolar diffusion", the magnetic field manages to slip through the gas. The scattered component of the interstellar magnetic field also plays a role; if the magnetic field were entirely uniform, then the gas would simply slip along field lines, leading to a flat (pancake-like) morphology upon contraction. According to Klein (2013), this has never been observed.

A large number of cosmic dust particles enter the Earth's atmosphere. Estimates of the global input rate of cosmic dust particles into the Earth's atmosphere vary from ~3 to 300 metric tons per day; see Plane (2012). The size and velocity distributions of cosmic dust particles entering the Earth's atmosphere are uncertain. This issue has been discussed in detail by Carrillo-Sánchez et al. (2015), who showed that the relative concentrations of metal atoms in the upper mesosphere, and the surface accretion rate of cosmic spherules, provide sensitive probes of this distribution. Three cosmic dust models were selected as case studies: Of them, two are astronomical models (M1 and M2).

(M1) The first is constrained by infrared observations of the Zodiacal Dust Cloud.
(M2) The second is by radar observations of meteor head echoes.
(M3) And the third model is based on measurements made with a space-borne dust detector.

For each model, a Monte Carlo sampling method combined with a chemical ablation model is used to predict the ablation rates of Na, K, Fe, Mg and Ca above 60 km and cosmic spherule production rate. It follows that a significant fraction of the cosmic dust consists of small (<5 μg) and slow (<15 km^{-1}) particles. One such particle of cosmic dust sample from the list of Cosmic Dust Samples at Astromaterials Curation – NASA is shown in Figure 1.19.

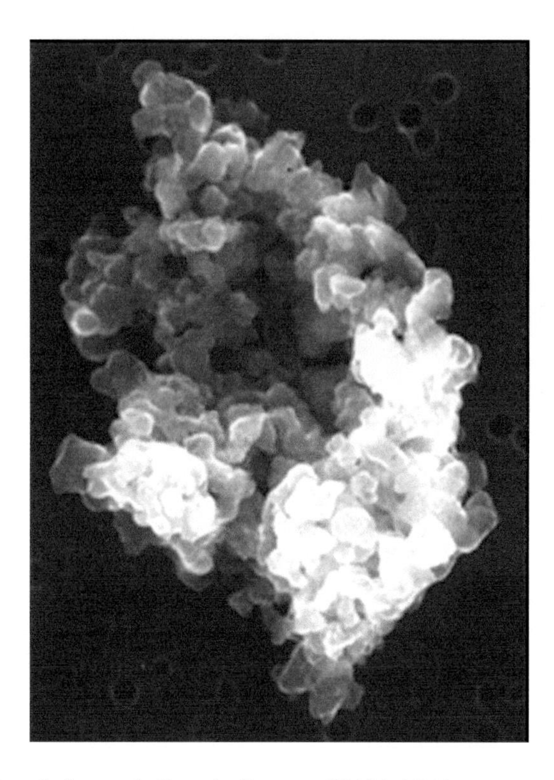

FIGURE 1.19 Cosmic dust grain 11 μm in diameter. (NASA, 2012.)

The results obtained by Carrillo-Sánchez et al. (2015), presented in Figure 1.20, show a histogram of the particle mass distributions for each of the three models. The mass distribution is expressed in terms of mass flux per decade versus the mass range from 10^{-9} to 0.1 g, which covers the bulk of the incoming daily material (Ceplecha et al., 1998). In the M2 distribution, the median input mass of the incoming dust particles is ~10 μg, with a total input rate of 110 ± 55 t/d (Love and Brownlee, 1993). For the M3 and M1 models, the mass distribution is shifted to smaller mass ranges with a median input mass of ~1 μg. The total input rate in the M1 model is 34 ± 17 t/d, although there may be 30%–50% of additional mass input from asteroids and long-period comets (Nesvorný et al., 2011). The M3 predicts an input rate of 14 ± 3 t/d (Fentzke and Janches, 2008; Janches et al., 2014).

In conclusion, the study by Carrillo-Sánchez et al. (2015) shows that a significant fraction of the cosmic dust entering the Earth's atmosphere needs to consist of small (<5 μg) and slow (<15 km/s) particles to explain the measured accretion rate of cosmic spherules at the surface. The same argument is needed to explain the significant differential ablation of the refractory meteoric metals. This effect has been observed in meteors travelling through the mesosphere and lower thermosphere. Of the three models selected for the study by Carrillo-Sánchez et al. (2015), the Zodiacal Dust Cloud model (M1) seems to do best when judged against these criteria.

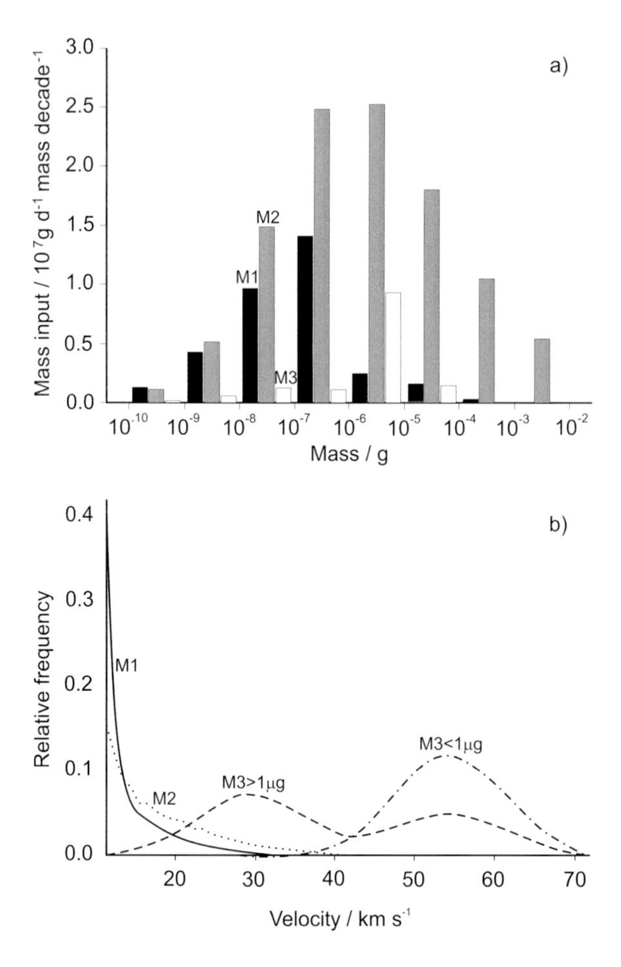

FIGURE 1.20 (a) Histogram of the particle mass distributions and (b) entry velocity distributions for the M1 (solid line), M2 (dotted line) and M3 (dashed and dashed-dotted lines) models. (After Carrillo-Sánchez et al., 2015.)

1.3 Solar System Abundances

The fact that the Earth contains heavy elements such as gold, lead or uranium (all heavier than iron) shows that the Sun is a third- or higher-generation star, preceded by at least one supernova explosion of another nearby star. Also, the carbon-to-oxygen ratio in a planet provides critical information about its primordial origins and subsequent evolution. A primordial C/O more significant than 0.8 results in an interior dominated by carbide, contrary to the silicate-dominated composition of the Earth. The atmosphere can also be different from those in the solar system. The solar C/O is 0.54.

Our solar system originated from the gravitational collapse of a small part of a massive molecular cloud. The assumption is that all objects in the solar system, the Sun and the planets were formed from a hot gaseous nebula having well-defined chemical and isotopic composition. Some doubt upon this assumption exists because of the findings of comparatively significant and widespread variations in oxygen isotopic composition. The incomplete mixing and small-scale homogenization of elements and nuclides in the primordial solar nebula deduced from the detection of huge (up to the percent level) isotope anomalies of some heavy elements in silicon carbide (SiC) grains provide additional evidence.

The proton–proton, p–p, chain and CNO cycle are the two processes for the fusion of hydrogen into helium. They are rather well understood theoretically (see Bahcall, 1989; Vinyoles et al., 2017). The only direct probe of these processes are neutrinos emitted during the involved nuclear reactions. It turns out that the p–p chain produces about 99% of the solar energy as concluded from the spectroscopic study of solar neutrinos (The Borexino Collaboration, 2018). The CNO cycle is the primary mechanism for the stellar conversion of hydrogen into helium in the Universe. The experimental findings by The Borexino Collaboration (2020) quantified the relative contribution of CNO cycle to the energy production by Sun to be approximately 1%. Additionally, their experimental data also help better estimate the metallicity of the Sun and, as such, offer a unique probe of the Sun initial condition.

Ca–Al-rich inclusions (CAIs) are the oldest dated samples of the solar system, consequently providing the direct record of its formation on an absolute timescale. This subject is discussed by Brennecka et al. (2020) in their recent research paper, in which they investigate the problem by using the isotopic composition of CAIs. CAIs are isotopically distinct from subsequently formed solids in the solar protoplanetary disk. Compared with bulk meteoritic materials, CAIs are rich in ^{16}O and more similar to the Sun than the objects formed later (McKeegan et al., 2011). The authors showed that the distinct molybdenum isotopic compositions of CAIs cover almost the entire compositional range of material that formed in the protoplanetary disk while the Sun was in transition between protostellar and pre-main-sequence phase of star formation.

The present-day solar system composition and the solar system element abundances as they were 4.56×10^9years ago are discussed in the works of Lodders et al. (2009) and Lodders (2010). Peaks of abundances are at mass numbers of closed proton and neutron "shells". These nuclear "shells" are analogous to the closed electron "shells" that characterize atomic properties. They are called the "magic numbers" for nuclear stability and are 2, 8, 20, 28, 50, 82 and 126. Nuclides with Z and N equal to these magic numbers are the ones that show the highest abundances in the functional dependence of abundance on mass numbers ($A=Z+N$). This fact is particularly visible in the light doubly-magic nuclei (equal magic Z and N): ^{4}He, ^{16}O and ^{40}Ca.

The present-day photospheric abundances are different from those in the beginning of the solar system. The processes that affected the solar abundances over time are elements settling from the solar photosphere into the Sun's interior and the decay of radioactive isotopes that contribute to the overall atomic abundance of an element.

Beyond the region of nuclides having mass number 56, element abundances decline almost smoothly with spikes at certain mass number regions. The nuclides beyond the Fe are products from neutron-capture processes. The peaks in the distribution correspond to areas where nuclide is made either by the slow neutron-capture process (s-process) taking place in red giant stars (Y and Ba regions) or by the rapid neutron-capture process (r-process) probably running in supernovae, e.g., Pt region. Here, the "slow" and "rapid" are about beta-decay timescales of the intermediate, unstable nuclei produced during the neutron-capture processes. The nuclide yields from these processes depend on the neutron energies, neutron flux, element abundance and the probability of the target-nuclei neutron-capture process (which depends on Z and N numbers). In such a way, the abundance distribution is controlled by the more stable "magic" nuclides. These nuclides serve as bottlenecks for the overall yields in the neutron-capture processes. See reports by Wallerstein et al. (1997), Woosley et al. (2002) and Sneden et al. (2008) for the reviews on stellar nucleosynthesis.

Figure 1.21 (after Lodders et al., 2009) shows the solar system abundances of the nuclides 4.56×10^9 years ago. The figure shows that nuclides with even A (shown as closed symbols) have usually higher abundances than the odd-numbered nuclides (open symbols). On the other hand, Figure 1.22 shows the present-day solar system abundances of the elements as a function of the atomic number. The large abundances of H and He are not shown to avoid scale compression in the diagram (after Lodders, 2010).

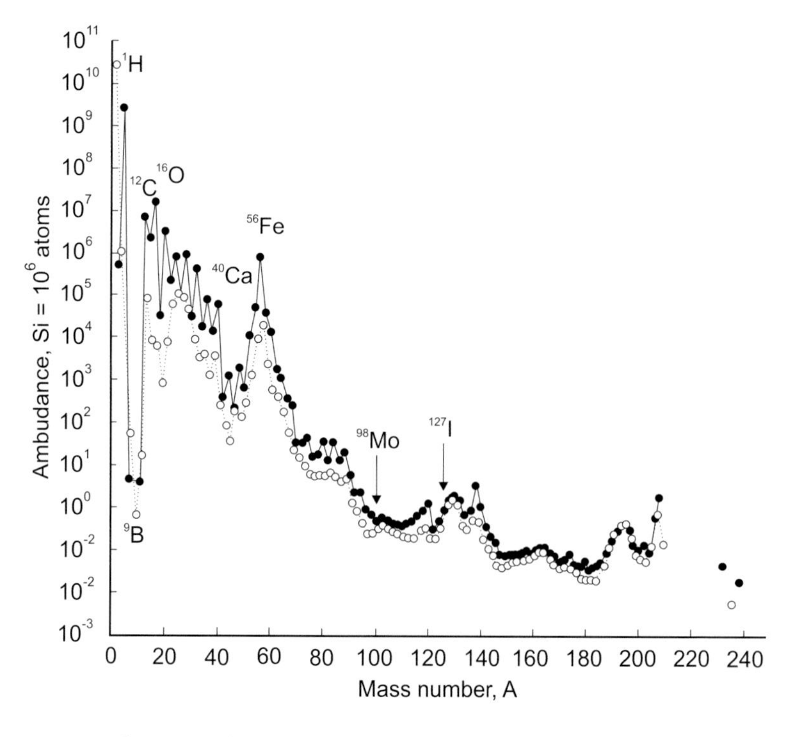

FIGURE 1.21 Solar system abundances of the nuclides 4.56×10^9 years ago. The figure shows that the nuclides with even A (shown as closed symbols) usually have higher abundances than the odd-numbered nuclides (open symbols). (After Lodders et al., 2009.)

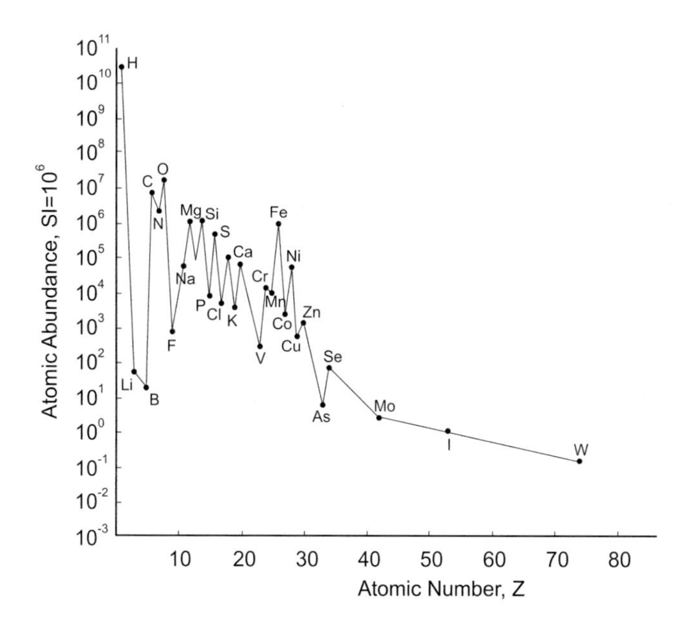

FIGURE 1.22 Present-day solar system abundances of the chemical elements essential for life as a function of the atomic number. (Modified from data in Lodders, 2010.)

The fact is that the curve describing element abundances is changing with the time. These changes can be observed on a timescale of 10^9years. This phenomenon is real not only for the solar system, but also for the rest of our galaxy (stars and interstellar matter).

High-temperature condensates found in meteorites display uranium isotopic variations ($^{235}U/^{238}U$) that complicate dating the solar system's formation whose origin remains mysterious. These variations may be due to the decay of the short-lived radionuclide ^{247}Cm ($t^{½}=1.56\times10^7$years) into ^{235}U, but they could also be due to uranium kinetic isotopic fractionation during condensation. Tissot et al. (2016) reported uranium isotope measurements of meteoritic inclusions revealing ^{235}U excesses of ~+6% compared to average solar system composition, which can only be due to the decay of ^{247}Cm. This fact allows constraining the $^{247}Cm/^{235}U$ ratio at solar system formation to $(1.1\pm0.3)\times10^{-4}$, thus providing new clues on the universality of the nucleosynthetic r-process (Tissot et al., 2016).

There are many difficulties present in determining the element abundances of distant objects caused by dust particles in the interstellar medium. Solar system abundances are somewhat similar to those found in most stars and interstellar material in our neighborhood and corresponding parts of other galaxies. Minor variations (factor of 3) may occur in the relative amounts of hydrogen and helium, and the relative amounts of carbon and heavier elements. This situation is a reflection of the fact that hydrogen and the bulk of helium are relics from the "Big Bang", whereas heavier elements (and a minority of the helium) result from nuclear reactions in stars or the interstellar medium.

1.4 Earth Abundances

The Earth is made of a variety of minerals, glasses, melts, fluids and volatiles, all left behind during the birth of the solar system. The Earth could be divided into several layers by their distinct chemical and seismic properties: mass=5.972×10^{24} kg; diameter: 12,756.3 km, which is divided into the following (see, for example, Wiesel, 2010): crust: 0–14 km, upper mantle: 40–400 km, transition region: 400–650 km, lower mantle: 650–2,700 km, “D” layer: 2,700–2,890 km, outer core: 2,890–5,150 km and inner core: 5,150–6,378 km. The exterior of the Earth is made of the solid lithosphere, liquid hydrosphere and gaseous atmosphere. The atmosphere is 320 km above the lithosphere/hydrosphere. About 70% of the Earth's surface is covered by water.

The Earth's crust varies considerably in thickness; it is thinner under the oceans and thicker under continents. The Earth's inner core and crust are solid. The outer core and mantle layers are supposed to be plastic or semi-fluid. Discontinuities separate the various layers, which are evident in seismic data. The best known of these is the Mohorovičić discontinuity between the crust and the upper mantle.

Most of the mass of the planet Earth is in the mantle, and the rest is in the core. The part Earth inhabited by life is a tiny fraction of the whole. The masses are as follows: atmosphere=5.1×10^{18} kg, oceans=1.4×10^{21} kg, crust=2.6×10^{22} kg, mantle=4.043×10^{24} kg, outer core=1.835×10^{24} kg, and inner core=9.675×10^{22} kg.

The Earth's core is probably composed mostly of iron (or nickel/iron). The temperature at the center might be hotter than the surface of the Sun. The lower mantle is made of mostly silicon, magnesium and oxygen with some iron, calcium and aluminum. The composition of the upper mantle is mainly iron/magnesium silicates, calcium and aluminum. The crust is primarily made of silicon dioxide and other silicates. The Earth's crust is made of several separate solid plates. They float around independently on top of the hot mantle below. The Earth's chemical composition is shown in Table 1.3 (after McDonough, 2001).

The outermost layer of the Earth, known as the crust, is responsible for the majority of life on the Earth. It is this layer that supports the growth of plants, the survival of animals, the structure of our land and the development of human civilization. The Earth's crust has four main components, which are referred to as the Earth's materials. These materials include minerals, rocks, soil, sediments and water. It is the combination of these materials that makes life on the Earth possible. The composition of the Earth established by integrating the diverse chemical components that exist from core to atmosphere is presented in the work by McDonough (2001).

Energetic collisions of differentiated bodies that happened early in the solar system history affected the final composition of the terrestrial planets through their partial destruction. Enstatite chondrites (EC) are the best candidates to represent the primordial terrestrial precursors as they present the most similar isotopic compositions to the Earth. Boujibar et al. (2015) showed that at low pressures, the first silicate melts are highly enriched in incompatible elements Si, Al and Na, and depleted in Mg. Loss of proto-crusts through impacts raises the Earth's Mg/Si ratio to its present value. To

TABLE 1.3 The Composition of the Earth – Concentrations Are in μg/g (ppm), Unless Stated as Indicated in % (After McDonough, 2001)

Element	Concentration/ppm	Element	Concentration/ppm	Element	Concentration/ppm
H	260	Zn	40	Pr	0.17
Li	1.1	Ga	3	Nd	0.84
Be	0.05	Ge	7	Sm	0.27
B	0.2	As	1.7	Eu	0.1
C	730	Se	2.7	Gd	0.37
N	25	Br	0.3	Tb	0.067
O (%)	29.7	Rb	0.4	Dy	0.46
F	10	Sr	13	Ho	0.1
Na (%)	0.18	Y	2.9	Er	0.3
Mg (%)	15.4	Zr	7.1	Tm	0.046
Al (%)	1.59	Nb	0.44	Yb	0.3
Si (%)	16.1	Mo	1.7	Lu	0.046
P	1210	Ru	1.3	Hf	0.19
S	6,350	Rb	0.24	Ta	0.025
Cl	76	Pd	1	W	0.17
K	160	Ag	0.05	Re	0.075
Ca (%)	1.71	Cd	0.08	Os	0.9
Sc	10.9	In	0.007	Ir	0.9
Ti	810	Sn	0.25	Pt	1.9
V	10.5	Sb	0.05	Au	0.16
Cr	4,700	Te	0.3	Hg	0.02
Mn	1,700	I	0.05	Tl	0.012
Fe (%)	31.9	Cs	0.035	Pb	0.23
Co	880	Ba	4.5	Bi	0.01
Ni	18,220	La	0.44	Th	0.055
Cu	60	Ce	1.13	U	0.015

match all significant element compositions, their model implies a preferential loss of volatile lithophile elements and recondensation of refractory elements after the impacts.

The progress in understanding mineral evolution, the Earth's changing near-surface mineralogy through time, depends on the availability of detailed information on mineral locations at a given age and geologic settings. Hazen et al. (2011) implemented a comprehensive database including the above details employing the mindat.org web site as a platform. (Note: Mindat.org is the world's largest open database of minerals, rocks, meteorites and their localities.) Their resource incorporates the software able to correlate a range of mineral occurrences and properties vs. time. The Mineral Evolution Database holds a prospect of revealing mineralogical records of important geophysical, geochemical and biological events in the Earth's history (Hazen et al., 2007).

1.4.1 Crust and Atmosphere Abundances

The amount of energy impinging on the outer part of the atmosphere, the solar flux, is 1,367 W/m^2. About 30% of this is scattered back into space by clouds, dust, the atmospheric gas molecules and the Earth's surface. Clouds and the atmosphere absorb about 19% of the solar radiation, leaving 51% of the incident energy to be absorbed by the Earth's surface. Because of thermal equilibrium, the Earth emits about 240 W/m^2 corresponding to the power issued by a black body at 255 K (–18°C), the average temperature of the atmosphere at an altitude of 5 km. The mean global surface temperature of the Earth is 13°C and is presumably the temperature required to maintain thermal equilibrium between the Earth and the atmosphere.

The present-day thinking is based on ideas first discussed by Safronov (1969, 1972) and Wetherill (1980) who advocated that in the early stages, interplanetary dust accumulated into larger bodies (planetesimals) due to the gravitational instability in a thin dust layer and subsequent coagulation of the planetesimals, which lead to the so-called planetary embryo, Moon- to Mars-sized bodies. Collision of embryos yielded finally the formation of terrestrial planets. The so-called terrestrial planets with substantial atmosphere are the Earth, Venus and Mars. Origin and evolution of their atmospheres is usually discussed as a single issue (see, for example, Hunten, 1993; Lammer et al., 2018; and Sossi et al., 2020). Compared to Venus and Mars, the modern Earth is unique in the sense that it has surface liquid water, active plate tectonics, a magnetic dynamo and an N_2-dominated atmosphere with a greenhouse effect that keeps its surface from freezing. Both Venus and Mars ended with a CO_2 atmosphere. The terrestrial planets obtained the majority of their volatiles (H_2O, CO_2, CH_4, NH_3, HCN, N_2, etc.) by impactors of volatile-rich bodies from outer solar system during the main stages of terrestrial planet formation. According to Sossi et al. (2020), an exchange between a magma ocean and vapor produced the Earth's earliest atmosphere. The Earth's present-day atmosphere (mainly N_2–O_2) differs from those of Venus and Mars (CO_2-rich). The composition of the present-day Earth's atmosphere is shown in Table 1.4. Three main groups of components are shown: major, minor and trace, separated by black lines in this table.

Figure 1.23 shows the crustal element abundance curve. The usual natural chemistry is one that occurs at the Earth's surface. However, one should note that the cosmic abundance of elements is strikingly different from the composition of the Earth's crust (see Table 1.5, after Klemperer, 2008). That of hydrogen and helium does not merely reflect this difference. The abundance ratio of carbon and oxygen, being 1/2.3 cosmically and 1/250 terrestrially, is of particular interest. If carbon is the main building block of life, it would appear that the Earth is not the optimal location for it (Klemperer, 2008)!

1.4.2 Seawater Abundances

The earth and its oceans and atmosphere grew probably together. There is a little sure knowledge in this field, so we shall not get involved in a lengthy discussion which scenario is more likely. For the readers interested in the subject of the origins of the oceans, there is a lengthy literature list; we took out arbitrary following books and papers of Turekian (1968, 2001), Kastner (1999) and Webb (2020).

TABLE 1.4 Composition of the Atmosphere

Component	Symbol	Concentration	Group
Nitrogen	N_2	78.08%	Major
Oxygen	O_2	20.95%	
Argon	Ar	0.93%	
Water	H_2O	0–4%	Minor
Carbon dioxide	CO_2	325 ppm	
Neon	Ne	18 ppm	
Helium	He	5 ppm	
Methane	CH_4	2 ppm	
Krypton	Kr	1 ppm	
Hydrogen	H_2	0.5 ppm	
Nitrous oxide	N_2O	0.3 ppm	
Carbon monoxide	CO	0.05–0.2 ppm	
Ozone	O_3	0.02–10 ppm	
Xenon	Xe	0.08 ppm	
Ammonia	NH_3	4 ppb	Trace
Nitrogen oxide	NO	1 ppb	
Sulfur dioxide	SO_2	1 ppb	
Hydrogen sulfide	H_2S	0.05 ppb	

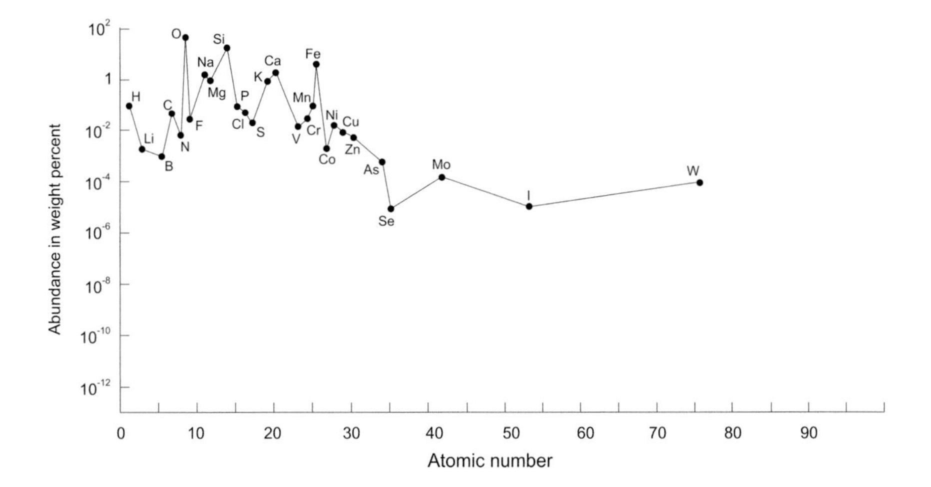

FIGURE 1.23 Earth crustal abundance curve for life-essential chemical elements.

TABLE 1.5 Values of Cosmic and the Earth's Crust Element Abundances (After Klemperer, 2008)

Element	Cosmic Abundance	Earth's Crust
H	330,000	0.22
He	46,000	–
C	100	0.19
N	31	0.002
O	235	46.6
F	0.01	0.06
Ne	91	–
Na	0.6	2.8
Mg	10.6	2.1
Al	0.8	8.1
Si	9.9	27.1
P	0.1	0.1
S	5.1	0.06
Cl	0.1	0.02
Ar	2	–
K	0.03	0.9
Ca	0.8	5.2
Ti	0.04	0.5
Cr	0.18	0.02
Fe	8	6.7
Ni	0.5	0.01

When all height contours of both sea and land are put together, the hypsographic curve results are shown in Figure 1.24 (after Anthoni, 2000) giving a quick overview of the distribution of height and depth over the planet. Horizontal axis presents the surface area of the Earth as a percentage of the total (5.2 million km^2). The seas are present on over 70%, and the land on under 30% of the Earth's surface. The amount of area above sea level is very much less than the volume of the sea. If the land was spread into the sea, the oceans would still be 3,000 m deep. Mount Everest (8,848 m) is the highest point on the land. The deepest point in the sea is in the Mariana Trench (11,034 m).

The hypothesis on the concentrations in seawater being crucial to the origin of elemental requirements has often been described in many reports (see, for example, Chappell et al., 1974; Jukes, 1974; Orgel, 1974; Banin and Navrot, 1975). The ocean is assumed to be in a steady state with sediments and the atmosphere. Chemical elements in seawater do not exist on their own, but are attracted to preferential ions of opposite charge; for example, sulfur will occur mainly as sulfate and sodium as sodium chloride.

Continental materials are unlike seawater or river water. The most common ions in seawater and river water and the elemental composition of the Earth's crust (excluding oxygen) are listed in order of abundance in Table 1.6. The detailed composition of seawater at 3.5% salinity is given in Turekian (1968); see Table 1.7. Figure 1.25 is a graphical representation of seawater (for life-essential) element concentration values.

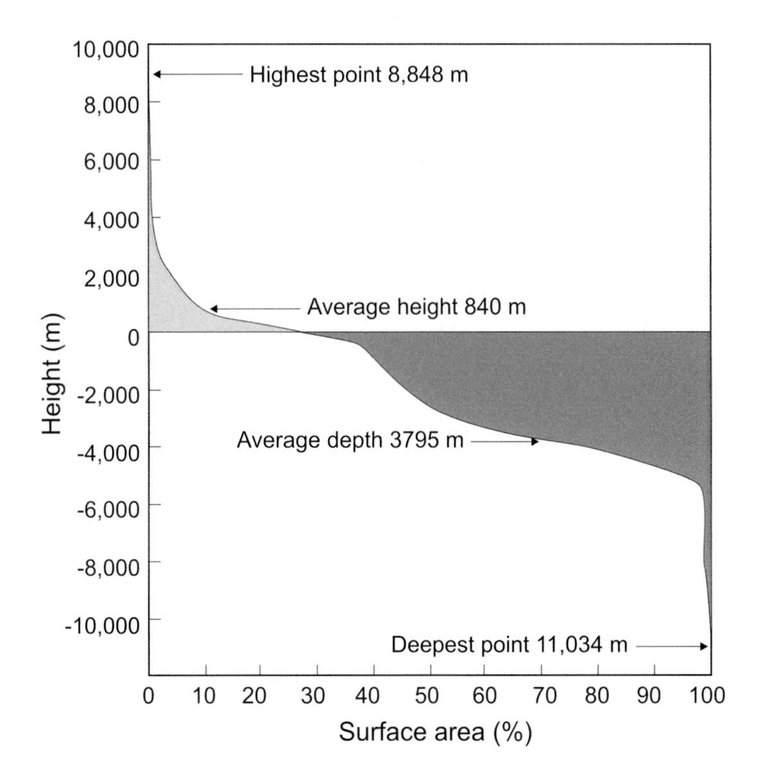

FIGURE 1.24 Distribution of the surface area, hypsographic curve, showing how the surface area of the Earth is distributed. Horizontal: % of the total Earth area ($5.2\times10^6 km^2$). Vertical: height or depth in meters. Mount Everest is the highest point on land (8,848 m), while the deepest point in the ocean is in the Mariana Trench (11,034 m).

TABLE 1.6 Most Common Ions in Waters and the Most Common Elements in the Crust

Seawater (as Ions)	Earth's Crust (as Elements)	River Water (as Ions)
CL^-	Si	HCO_3^-
Na^+	Al	Ca^{2+}
SO_4^{2-}	Fe	SiO_2
Mg^{2+}	Ca	SO_4^{2-}
Ca^{2+}	Na	Mg^{2+}
K^+	K	K^+
HCO_3^-	Mg	Na^+
Br^-	Ti	Fe^{3+}
$H_2BO_3^-$	Mn	
Sr^{2+}	P	

TABLE 1.7 Detailed Composition of the Seawater (Modified after Turekian, 1968)

Element	Concentration/ ppm	Element	Concentration/ ppb	Element	Concentration/ ppb	Element	Concentration/ ppb
H in H_2O	110,000	Sc	<0.004	Ru	0.0007	Dy	0.00091
O in H_2O	883,000	Ti	1	Rh	–	Ho	0.00022
Na in NaCl	10,800	V	1.9	Pd	–	Er	0.00087
Cl in NaCl	19,400	Cr	0.2	Ag	0.28	Tm	0.00017
Mg	1,290	Mn	0.4	Cd	0.11	Yb	0.00082
S	904	Fe	3.4	In	–	Lu	0.00015
K	392	Co	0.39	Sn	0.81	Hf	0.0008
Ca	411	Ni	6.6	Sb	0.33	Ta	0.00025
Br	67.3	Cu	0.9	Te	–	W	0.0001
He	0.0000072	Zn	5	I	64	Re	0.0084
Li	0.170	Ga	0.03	Xe	0.047	Os	–
Be	0000006	Ge	0.06	Cs	0.3	Ir	–
B	4.450	As	2.6	Ba	21	Pt	–
C	28.0	Se	0.9	La	0.0029	Au	0.011
N	15.5	Kr	0.21	Ce	0.0012	Hg	0.15
F	13	Rb	120	Pr	0.00064	Tl	–
Ne	0.00012	Sr	8 100	Nd	0.0028	Pb	0.03
Al	0.001	Y	0.013	Sm	0.00045	Bi	0.02
Si	2.9	Zr	0.026	Eu	0.0013	Th	0.0004
P	0.088	Nb	0.015	Gd	0.0007	U	3.3
Ar	0.450	Mo	10	Tb	0.00014	Pu	–

The compositions of the oceans and their biota have influenced each other through the Earth's history. Nitrogen is the only biologically essential element whose seawater concentration is clearly controlled biologically. This fact is the main reason why the stoichiometry of nitrogen (defined as its molar ratio to phosphorus), and not that of the trace nutrients manganese, iron, cobalt, nickel, copper, zinc and cadmium, is the same in seawater and the plankton. Both the primary nutrients and the trace nutrients are depleted in surface seawater as a result of quasi-complete utilization by the biota. This depletion is partly a result of the ability of marine phytoplankton to replace one trace metal by another in performing various biochemical functions. The result of this replacement is also an equalization of the availability of most essential trace elements in the surface seawater. The deep seawater is the dominant source of new nutrients to the surface. Nutrients are likely recycled with different efficiencies in the photic zone. This situation results in the difference in the stoichiometric composition of the plankton and deep seawater. The variation in the composition of the ocean and its biota helps in understanding the coupling of biochemistry and biogeochemistry in seawater (Morel, 2008).

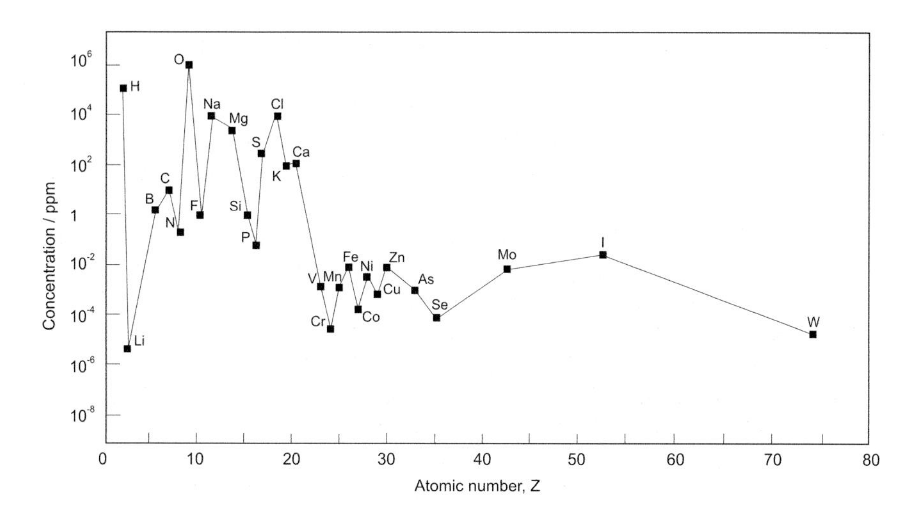

FIGURE 1.25 Elemental composition of seawater, only for life-essential elements.

Chopra and Lineweaver (2008) discussed the significant elemental abundance differences between life, the oceans and the Sun. They presented the correlation between the elemental composition of life (humans and bacteria), the Sun and seawater. Some elements such as carbon, nitrogen and phosphorus are more abundant in life than in seawater, while others such as chlorine and sodium are more abundant in seawater than in life.

References

Alpher, R. and Herman, R. 1950. Theory of the origin and relative abundance distribution of the elements. *Reviews of Modern Physics* 22:153–212.

Anthoni, J. 2000. Oceanography: Oceans - continental drift and the origins of oceans. www.seafriends.org.nz/oceano/oceans.htm.

Arnould, M. and Goriely, S. 2003. The p-process of stellar nucleosynthesis: Astrophysics and nuclear physics status. *Physics Reports* 384:1–84.

Bahcall, J. N. 1989. *Nautrino Astrophysics.* Cambridge University Press, Cambridge.

Banin, A. and Navrot, J. 1975. Origin of life: Clues from relations between chemical compositions of living organisms and natural environments. *Science* 189:550–551.

Barbagallo, M., Musumarra, A., Mengoni, A., et al. (n.TOF collaboration). 2014, Jun. 5. Measurement of $^7Be(n, \alpha)\, ^4He$ and $^7Be(n, p)\, ^7Li$ cross sections for the cosmological lithium problem. *Proposal for the ISOLDE and Neutron-Time-of-Flight Committee.* CERN-INTC-2014-049/INTC-P-417. CERN, Geneva, Switzerland.

Beard, M., Austin, S. M. and Cyburt, R. 2017. Enhancement of the triple alpha rate in a hot dense medium. *Physical Review Letters* 119:112701.

Bethe, H. A. 1939. Energy production in stars. *Physical Review* 55(5):434–456.

Boesgaard, A. M., McGrath, E. J., Lambert, D. L. and Cunha, K. 2004. Boron benchmarks for the galactic disk. *The Astrophysical Journal* 606:306–318.

Boujibar, A., Andrault, D., Bolfan-Casanova, N., Bouhifd, M. A. and Monteux, J. 2015. Cosmochemical fractionation by collisional erosion during the Earth's accretion. *Nature Communications* 6, Article number: 8295.

Brennecka, G. A., Burkhardt, C., Budde, G., Kruijer, T. S., Nimmo, F. and Kleine, T. 2020. Astronomical context of Solar System formation from molybdenum isotopes in meteorite inclusions. *Science* 370:837–840.

Broggini, C., Canton, L., Fiorentini, G. and Villante, F. L. 2012. The cosmological ^{7}Li problem from a nuclear physics perspective. *Journal of Cosmology and Astroparticle Physics* 06:1–16. Doi: 10.1088/1475-7516/2012/06/030.

Burbidge, E. M., Burbidge, G. R., Fowler, W. A. and Hoyle, F. 1957. Synthesis of the elements in stars. *Reviews of Modern Physics* 29:547–649.

Burles, S., Nollett, K. M. and Turner, M. S. 1999, Mar. 19. Big-Bang nucleosynthesis: Linking inner space and outer space. arXiv:astro-ph/9903300v1.

Bystritsky, V. M., Dudkin, G. N., Kuznetsov, S. I. and Padalko, V. N. 2017, Feb. 2017. Direct search for neutron nuclei in decay of 238-U. *International Journal of Modern Physics E* 03:1750004.

Carlos, M., Nissen, P. E. and Meléndez, J. 2016. Correlation between lithium abundances and ages of solar twin stars. *Astronomy and Astrophysics* 587:A100.

Carrillo-Sánchez, J. D., Plane, J. M. C., Feng, W., Nesvorný, D. and Janches, D. 2015. On the size and velocity distribution of cosmic dust particles entering the atmosphere. *Geophysical Research Letters* 42:6518–6525.

Cassé, M., Lehoucq, R. and Vangioni-Flam, E. 1995. Production and evolution of light elements in active star-forming regions. *Nature* 373:318–321.

Cayrel, R., Depagne, E., Spite, M. et al. 2004. First stars V - Abundance patterns from C to Zn and supernova yields in the early galaxy. *Astronomy and Astrophysics* 416:1117–1138.

Ceplecha, Z., Borovička, J., Elford, W. G., ReVelle, D., Hawkes, R., Porubčan, V. and Šimek, M. 1998. Meteor phenomena and bodies. *Space Science Reviews* 84(3–4):327–471.

Chappell, W. R., Meglen, R. R., and Runnells, D. D., 1974. Comments on "directed panspermia". *Icarus* 21:513–515.

Chaussidon, M. and Gounelle, M. 2006. Irradiation processes in early solar system. In Lauretta, D. S. and McSween, H. Y. (Eds.) *Meteorites and Early Solar System II*. University of Arizona Press, Tucson, pp. 323–340.

Chopra, A. and Lineweaver, C. H. 2008. The major elemental abundance differences between life, the oceans and the Sun. *Proceedings from 8th Australian Space Science Conference*, Canberra, 29 September–October 1, 2008, National Space Society of Australia Ltd, Sydney, Australia, pp. 49–55.

Coc, A., Pospelov, M., Uzan, J.-P., et al. 2014, May 7. Modified Big Bang nucleosynthesis with non-standard neutron sources. arXiv:1405:1718v1 [hep-ph].

Cunha, K., Smith, V. V., Boesgaard, A. M. and Lambert, D. L. 2000. A uniform analysis of boron in F and G disk dwarfs from Hubble space telescope archival spectra. *The Astrophysical Journal* 530:939–948.

Cyburt, R. H., Fields, B. D. and Olive, K. A. 2004. Solar neutrino constraints on the BBN production of Li. *Physical Review D* 69:123519.

Draine, B. T. 2009. In cosmic dust-near and far In Henning, T., Grun, E. and Steinacker, J. (Eds.) *Astronomical Society of the Pacific Conference Series*, Astronomical Society of the Pacific, San Francisco, California, USA, Vol. 414, pp. 453–472.

Draine, B. T. and Salpeter, E. E. 1979. On the physics of dust grains in hot gas. *The Astrophysical Journal* 231:77–94.

Driver, S., Wright, A. H., Andrews, S. K., et al. 2016. Galaxy and Mass Assembly (GAMA): Panchromatic data release (far-UV—far-IR) and the low-z energy budget. *Monthly Notices of the Royal Astronomical Society* 455(4):3911–3942.

Dunne, L., Gomez, H.L., da Cunha, E., et al. 2011. Herschel-ATLAS: Rapid evolution of dust in galaxies over the last 5 billion years. *Monthly Notices of the Royal Astronomical Society* 417:1510–1533.

Faisst, A. L., Capak, P. L., Davidzon, I., et al. 2015. Rest-UV absorption lines as metallicity estimator: the metal content of star-forming galaxies at z~5. 2015arXiv151200018F.

Fentzke, J. T. and Janches, D. 2008. A semi-empirical model of the contribution from sporadic meteoroid sources on the meteor input function in the MLT observed at Arecibo. *Journal of Geophysical Research* 113:A03304.

Fields, B. D., Molaro, P. and Sarkar, S. 2013. Big-Bang nucleosynthesis. In online version of Beringer, J. et al. [Particle Data Group Collaboration]. *Physical Review D* 86:010001(2012).

Fröhlich, C., Martinez-Pinedo, G., Liebendörfer, M. et al. 2006. Neutrino-induced nucleosynthesis of A>64 nuclei: The νp-process. *Physical Review Letters* 96:142502.

Gall, C., Hjorth, J. and Andersen, A. C. 2011. Production of dust by massive stars at high redshift. *The Astronomy and Astrophysics Review* 19:43.

Gall, C., Hjorth, J., Watson, D., Dwek, E., Maud, J. R., Fox, O., Leloudas, G., Malesani, D. and Day-Jones, A. C. 2014. Rapid formation of large dust grain in the luminous supernova 2010jl. *Nature* 13558:1–15.

George, J., Lave, K. A., Wiedenbeck, M. E., et al. 2009. Elemental composition and energy spectra of galactic cosmic rays during solar cycle 23. *Astrophysical Journal* 698:1666–1681.

Giavalisco, M., Dickinson, M., Ferguson, H. C., et al. 2004. The rest-frame ultraviolet luminosity density of star-forming galaxies at redshifts z>3.5. *Astrophysical Journal* 600:L103–L106.

Gratton, R. G. and Sneden, C. 1994. Abundances of neutron-capture elements in metal-poor stars. *Astronomy and Astrophysics* 287:927–946.

Gulkanyan, H. and Margaryan, A. 2014. Alpha-spectroscopy of Cf-252 decays: A new approach to searching for the octoneutron. arXiv.org:1409.1772 [nucl-ex], preprint 1628.

Hazen, R. M., Griffin, P. L., Carothers, J. M. and Szostak, J. W. 2007. Functional information and the emergence of biocomplexity. *PNAS* 104(1):8574–8581.

Hazen, R. M., Bekker, R., Bish, D. L., et al. 2011. Needs and opportunities in mineral evolution research. *American Mineralogist* 96:953–963.

Howes, L. M., Casey, A. R., Asplund, M., et al. 2015. Extremely metal-poor extremely metal poor stars from the cosmic dawn in the bulge of the Milky Way. *Nature* 527(7579):484–487.

Howk, J. C., Sembach, K. R. and Savage, B. D. 2000. The abundance of interstellar boron. *The Astrophysical Journal* 543:278–283.
Hunten, D. M. 1993. Atmospheric evolution of the terrestrial planets. *Science* 259(5097): 915–920.
Indebetouw, R., Matsuura, M., Dwek, E., et al. 2014. Dust production and particle acceleration in supernova 1987 A revealed with ALMA. *The Astrophysical Journal* 782:L2.
Ireland, A. L. and Kraus, A. L. 2014. Orbital motion and multi-wavelength monitoring of LkCa15 b. In Booth, M., Matthews, B. C. and Graham, J. R. (Eds.) *Exploring the Formation and Evolution of Planetary Systems. IAU Symposium*. Cambridge University Press, Cambridge, UK, Vol. 299, pp. 199–203.
Jahnke, U. 1986. Neutron multiplicity and high excitation energies, experiments with a 4π neutron multiplicity meter. *Journal de Physique Colloques* 47(C4):317–328.
Janches, D., Plane, J. M. C., Nesvorný, D., Feng, W., Vokrouhlický, D. and Nicolls, M. J. 2014. Radar detectability studies of slow and small zodiacal dust cloud particles: I. The case of Arecibo 430 MHz meteor head echo observations, *The Astrophysical Journal* 796:41.
Janka, H.-Th., Buras, R. and Rampp, M. 2003. The mechanism of core-collapse supernovae and the ejection of heavy elements. *Nuclear Physics A* 718:269–276.
Jin, S., Roberts, L. F., Austin, S. M., Schatz, H. 2020. Enhanced triple-α reaction reduces proton-rich nucleosynthesis in supernovae. *Nature* 588(7836):57–60.
Jofré, E. Petrucci, R., Garcia, L. and Gómez, M. 2015. KIC 9821622: An interesting lithium-rich giant in the Kepler field. *Astronomy and Astrophysics* 584:L3.
Jukes, T. H. 1974. Sea-water and the origins of life. *Icarus* 21:516–517.
Kastner, M. 1999. Oceanic minerals: Their origin, nature of their environment, and significance. *Proceedings of the National Academy of Sciences* 96(7):3380–3387.
Kimura, H. and Mann, I. 1998. Electric charge of interstellar dust in the solar system. *The Astrophysical Journal* 499:454–462.
Klein, U. 2013. *Galactic and Intergalactic Magnetic Fields*. Course astro 848, Argelander-Institut fűr Astronomie, Bonn.
Klemperer, W. 2008. Fine-tuning and interstellar chemistry. In Barrow, J. D., Morris, S. C., Freeland, S. J. and Harper, Jr., C. L. (Eds.) *Chapter 17 in Fitness of the Cosmos for Life* Cambridge University Press, Cambridge, UK, pp. 366–383.
Kneller, J. P. and McLaughlin, G. C. 2003, Dec. 15. The effect of bound dineutrons upon BBN. arXiv:astro-ph/0312388v1.
Kraus, A. L. and Ireland, M. J. 2012. LkCa15: A young exoplanet caught at formation? *The Astrophysical Journal* 745:5–16.
Lamia, L. 2013. Light element LiBeB depletion in astrophysics and related (p, α) cross section measurements via the THM. *10th Russbach School on Nuclear Astrophysics*, 10–16 March 2013 Russbach, Austria.
Lammer, H., Zerkle, A. L., Gebauer, S., et al. 2018. Origin and evolution of the atmospheres of early Venus, Earth and Mars. *The Astronomy and Astrophysics Review* 26:2 (72 p).
Lilly, S. J., Le Fèvre, O., Hammer, F. and Crampton, D. 1996. The Canada-France redshift survey: The luminosity density and star formation history of the universe to z ~ 1. *Astrophysical Journal* 460:L1–L4.

Livio, M. 2008. The interconnections between cosmology and life. In Barrow, J. D., Morris, S. C., Freeland, S. J. and Harper, Jr., C. L. (Eds.) *Chapter 7 in Fitness of the Cosmos for Life*. Cambridge University Press, Cambridge, UK, pp. 114–131.

Lodders, K. 2003. Solar system abundances and condensation temperatures of the elements. *The Astrophysical Journal* 591:1220–1247.

Lodders, K. 2010. Solar system abundances of the elements. In Goswami, A. and Reddy, B.E. (Eds.) Lecture Notes of the Kodai School on Synthesis of Elements in Stars held at Kodaikanal Observatory, India, April 29 – May 13. 2008. *Astrophysics and Space Science Proceedings*, Springer-Verlag, Berlin, Heidelberg, pp. 379–417.

Lodders, K., Palme H. and Gail, H.P. 2009. Abundances of the elements in the solar system. In Trümper, J. E. (Ed.) *Landolt Börnstein, New Series*, Vol. VI/4B, Chap. 4.4, Springer Verlag, Berlin, Heidelberg, New York, pp. 560–630.

Love, S. G. and Brownlee, D. E. 1993. A direct measurement of the terrestrial mass accretion rate of cosmic dust, *Science* 262(5133):550–553.

Mann, I. 2001. Charging effects on cosmic dust. Spacecraft charging technology. In Harris, R. A. (Ed.) *Proceedings of the Seventh International Conference held 23–27 April 2001 at ESTEC*, European Space Agency, Noordwijk, The Netherlands, ESA SP-476: 629.

Martig, M., Fouesneau, M. Rix, H.-W., et al. 2016. Red giant masses and ages derived from carbon and nitrogen abundances. *Monthly Notices of the Royal Astronomical Society* 456(4):3655–3670.

Matsuura, M., Dwek, E., Meixner, M., et al. 2011. Herschel detects a massive dust reservoir in supernova 1987A. *Science* 333:1258–1261.

Matteucci, F. 2001. *The Chemical Enrichment of Galaxies*. Kluwer Academic Publ, Dordrecht, The Netherlands, p. 293.

McClure-Griffiths, N. M., Dénes, H., J. M. Dickey, J. M., et al. 2018. Cold gas outflows from the Small Magellanic Cloud traced with ASKAP. *Nature Astronomy* 2:901–906.

McDonough, W. F. 2001. The composition of the Earth. *International Geophysics* 76:3–23.

McKeegan, K. D., Kallio, A., Heber V., et al. 2011. The oxygen isotopic composition of the sun inferred from captured solar wind. *Science* 332:1528–1532.

Meléndez, J., Bedell, M., Bean, J. L., et al. 2016, Oct. 31. The solar twin planet Search-V. Close-in, low-mass planet candidates and evidence of planet accretion in the solar twin HIP 68468. *Astronomy & Astrophysics* manuscript no. hip68468_2016_09_29.

Morel, F. M. M. 2008. The co-evolution of phytoplankton and trace element cycles in the oceans. *Geobiology* 6(3):318–324.

Mukai, T. 1981. On the charge distribution of interplanetary grains. *Astronomy & Astrophysics* 99:1–6.

NASA. 2012. *Cosmic Dust Sample Investigator's Guidebook. JSC-66466*. NASA Johnson Space Center, Houston, TX.

NASA. 2013. NASA Astronomy Picture of the Day 2013 February 13.

NASA. 2015. Image for the day. nasa_subscriptions@service.govdelivery.com.

Nesvorný, D., Janches, D., Vokrouhlický, D., Pokorný, P., Bottke, W. F. and Jenniskens, P. 2011. Dynamical model for the zodiacal cloud and sporadic meteors. *The Astrophysical Journal* 743(2):129–144.

Orgel, L. E., 1974. Reply: "Comments on 'directed panspermia'" and "seawater and the origin of life". *Icarus* 21:518.
Plane, J. M. C. 2012. Cosmic dust in the Earth's atmosphere, *Chemical Society Reviews* 41(19):6507–6518.
Pospelov, M. and Pradler, J. 2010. Big bang nucleosynthesis as a probe of new physics. *Annual Review of Nuclear and Particle Physics* 60:539–568.
Primas, F. 2009. Beryllium and boron in metal-poor stars. In Charbonnel, C., Tosi, M., Primas, F. and Chiappini, C. (Eds.) *Light Elements in the Universe, Proc, IAU Symposium No. 268*. Cambridge University Press, ISBN 9780-521-76506-0.
Prochaska, J. X., Howk, J. C. and Wolfe, A. M. 2003. The elemental abundance pattern in a galaxy at z=2.626. *Nature* 423:57–59.
Proffitt, C. R. and Quigley, M. F. 2001. Boron abundances in early B stars: Results from the B III resonance line in IUE data. *The Astrophysical Journal* 548:429–438.
Pruet, J., Woosley, S. E., Buras, R., et al. 2005. Nucleosynthesis in the hot convective bubble in core-collapse supernovae. *The Astrophysical Journal* 623(1):325–336.
Ramadurai, S. and Wickramasinghe, N. C. 1975. The mystery of the cosmic boron abundance. *Astrophysics and Space Science* 33:L41–L44.
Ramaty, R., Kozlovsky, B., Lingenfelter, R. E. and Reeves, H. 1997. Light elements and cosmic rays in the early galaxy. *The Astrophysical Journal* 488:730–748.
Ritchey, A. M. Federman, S. R., Sheffer, Y. and Lambert, D. L. 2011. The abundance of boron in diffuse interstellar clouds. *The Astrophysical Journal* 728:70 (37 p).
Safronov, V. S. 1969. Evolution of protoplanetary cloud and formation of Earth and the planets. Akad. Nauk SSSR Moscow. English translation, NASA ttf–667.
Safronov, V. S. 1972. Evolution of protoplanetary cloud and formation of Earth and the planets. English translation, Israel Program for Scientific Translation, Keter Publishing House, Jerusalem.
Sallum, S., Follette, K. B., Eisner, J. A., et al. 2015. Accreting protoplanets in the LkCa15 transition disk. *Nature* 527:342–344.
Salpeter, E. E. 1952. Nuclear reactions in stars without hydrogen. *Astrophysical Journal* 115:326–328.
Shusterman, J. A., Scielzo, N. D., Thomas, K. J., et al. 2019. The surprisingly large neutron capture cross-section of ^{88}Zr. *Nature* 565:328–330.
Smith, V. V., Cuncha, K. and King, J. R. 2001. The observed trend of boron and oxygen in field stars of the disk. arXiv: astro-ph/0104136v1.
Sneden, C., Cowan, J.J. and Gallino, R. 2008. Neutron-capture elements in the early galaxy. *Annual Review of Astronomy and Astrophysics* 46:241–288.
Sneden, C., Cowan, J. J., Kobayashi C., et al. 2016. Iron-group abundances in the metal-poor main-sequence turnoff star HD~84937. *The Astrophysical Journal* 817(1) article id. 53: 16 p.
Sossi, P. A., Burnham, A. D., Badro, J., Lanzirotti, A., Newville, M. and O'Neill. 2020. Redox state of Earth's magmaocean and its Venus-like early atmosphere. *Science Advances* 6:eabd1387.
Spite, F. and Spite, M. 1982. Abundance of lithium in unevolved halo stars and old disk stars - Interpretation and consequences. *Astronomy & Astrophysics* 115:357–366.

Stahler, W. and Palla, F. 2004. *The Formation of Stars*. Whiley-VCH, Weinheim, 865 p.
Stanway, E. R., Bunker, A. J. and McMahon, R. G. 2003. Lyman break galaxies and the star formation rate of the Universe at $z \approx 6$. *Monthly Notices of the Royal Astronomical Society*. 342:439–445.
Stone, E. C., Cohen, M. S., Cook, W. R., et al. 1998. The cosmic-ray isotope spectrometer for the advanced composition explorer. *Space Science Reviews* 86 (1–4):285–365.
Tan, K., Shi, J. and Zhao, G. 2010. A non-local thermodynamic equilibrium analysis of boron abundances in metal-poor stars. *The Astrophysical Journal* 713:458–468.
The Borexino Collaboration. 2018. Comprehensive measurement of pp-chain solar neutrinos. *Nature* 562:505–510.
The Borexino Collaboration. 2020. Experimental evidence of neutrinos produced in the CNO fusion cycle in the Sun. *Nature* 587:577–582.
Thielemann, F.-K. 2013. Nucleosynthesis of massive stars and their supernovae: An attempt to put the finger on open questions. *Presented at 10th Russbach School on Nuclear Astrophysics*, March 10–16, 2013. Russbach, Austria. http://russbach-wks.sciencesconf.org/conference/russbach-wks/asconafkt.pdf.
Thielemann, F., Käppeli, R. Winteler, C., et al. 2012. r-process in jet ejecta of magnetorotational core collapse supernovae. *Proceedings of the XII International Symposium on Nuclei in the Cosmos (NIC XII)*. August 5–12, 2012. Cairns, Australia. Published online at http://pos.sissa.it/cgi-bin/reader/conf.cgi?confid=146, id.61.
Tissot, F. L. H., Dauphas, N. and Grossman, L. 2016. Origin of uranium isotope variations in early solar nebula condensates. *Science Advances* 2:e1501400.
Turan, J. W. and Kozlowsky, B. Z. 1969. The enhancement of the $3^4He \rightarrow ^{12}C$ reaction rate in dense matter by inelastic scattering processes. *The Astrophysical Journal* 158:1021–1032.
Turatto, M. 2003. Classification of Supernovae. arXiv:astro-ph/0301107v1.
Turekian, K. K. 1968. *Oceans*. Prentice-Hall, New York.
Turekian, K. K. 2001. *Origin of the Oceans*. Academic Press. Doi:10.1006/rwos.2001.0417.
Valkovic, V., Sudac, D. and Obhodas, J. 2018. The role of 14 MeV neutrons in light element nucleosynthesis. *IEEE Transactions on Nuclear Science* 65(9):2366–2371.
Venn, K. A., Brooks, A. M., Lambert, D. L., Lemke, M., Langer, N., Lennon, D. J. and Keenan, F. P. 2002. Boron abundances in B-type stars: A test of rotational depletion during main-sequence evolution. *The Astrophysical Journal* 565:571–586.
Vinyoles, N., Serenelli, A. M., Villante, F. L., et al. 2017. A new generation of standard solar models. *The Astrophysical Journal* 835:202 (16 p).
von Weizsäcker, C. F. 1938. Über Elementumwandlungen in Innern der Sterne II. *Physikalische Zeitschrift* 39:633–645.
Wallerstein, G., Iben, I., Parker, P., et al. 1997. Synthesis of the elements in stars: Forty years of progress. *Reviews of Modern Physics* 69:995–1084.
Wanajo, S. 2006. The rp-process in neutrino-driven winds. *The Astrophysical Journal* 647:1323–1340.
Wanajo, S., Janka, H.-T. and Kubono, S. 2011. Uncertainties in the νp-process: Supernova dynamics versus nuclear physics. *The Astrophysical Journal* 729:46 (18 p).
Watson, D., Hansen, C.J., Selsing, J., et al. 2019. Identification of strontium in the merger of two neutron stars. *Nature* 574:497–500.

Webb, P. 2020, Jul. Introduction to oceanography. Pressbooks, licenced under a Creative Commons Attribution 4.0 International Licence.
Wetherill, G. W. 1980. Formation of terrestrial planets. *Annual Review of Astronomy and Astrophysics* 18:77–113.
Wiebel-Sooth, B., and Biermann, P. L. 1999. *Cosmic Rays. Landolt-Börnstein*, vol. VI/3c, Springer Verlag, Berlin Heidelberg, Germany, pp. 37–90.
Wiesel, W. E. 2010. *Modern Astrodynamics*, second edition. Aphelion Press, Beavercreek, OH.
Woosley, S. A., Hartmann, D. H., Hoffman, R. D. and Haxton, W. C. 1990. The nu-process. *The Astrophysical Journal* 356:272–301.
Woosley, S.E., Heger, A. and Weaver, T.A. 2002. The evolution and explosion of massive stars Rev. *Reviews of Modern Physics* 74:1015–1072.

Additional reading

Abbott, R., Abbott, T. D., Abraham, S., et al. (LIGO Scientific Collaboration and Virgo Collaboration). 2020. GW190521: A binary black hole merger with a total mass of 150 M. *Physical Review Letters* 125(10):101102 (1–17).
Akeson, R. L., Chen, X. and Ciardi, D. 2013. The NASA exoplanet archive: Data and tools for exoplanet research. *Publications of the Astronomical Society of the Pacific* 125:989–999.
Boesgaard, A. M. 1976. Stellar abundances of lithium, beryllium, and boron. *Publications of the Astronomical Society of the Pacific* 88:353–366.
Boyd, R. N., Brune, C. R., Fuller, G. M. and Smith, C. J. 2010. New nuclear physics for big bang nucleosynthesis. arXiv:1108.0848v1 [astro-ph.CO] 04 Aug 2010.
Bystritsky, V. M., Dudkin, G. N., Emets, E. G., et al. 2017. Astrophysical S-factor of T(^{4}He, γ)^{7}Li reaction at Ecm= 15.7 keV. *Physics of Particles and Nuclei Letters* 14(4):560–570.
Bystritsky, V. M., Dudkin, G. N., Nechaev, B. A., et al. 2018. Investigation of the D(^{3}He, p)^{4}He reaction in the astrophysical energy region of 18–30 keV. *JETP Letters* 107(11):665–670.
Charbonnel, C. and Talon, S. 2005. Influence of gravity waves on the internal rotation and Li abundance of solar-type stars. *Science* 309:2189–2191.
Coc, A. 2013. Primordial nucleosynthesis. 11th international conference on nucleus-nucleus-nucleus collisions. *Journal of Physics: Conference Series* 420:012136.
Coc, A., Pospelov, M., Uzan, J.-P. and Vangioni, E. 2014. Modified Big Bang nucleosynthesis with non-standard neutron sources. arXiv:1405.1718v1 [hep-ph 7 May 2014.]
Côté, B., Eichler, M., López, A. Y., et al. 2021. ^{129}I and ^{247}Cm in meteorites constrain the last astrophysical source of solar r-process elements. *Science* 371:945–948.
Denissenkov, P. A. 2010. A model of magnetic braking of solar rotation that satisfies observational constraints. *Astrophysical Journal* 719(1):28–44.
Do Nascimento, J. D., Jr., Castro, M., Meléndez, J., Bazot, M., Théado, S., Portode Mello, G. F. and de Medeiros, J. R. 2009. Age and mass of solar twins constrained by lithium abundance. *Astronomy and Astrophysics* 501:687–694.
Fumagalli, M., Cantalupo, S., Dekel, A., Morris, S. L., O?Meara, J. M., Prochaska, J. X. and Theuns, T. 2016. MUSE searches for galaxies near very metal-poor gas clouds

at z~3: new constraints for cold accretion models. *Monthly Notices of the Royal Astronomical Society* 462 (2):1978–1988.

Fumagalli, M., O'Meara, J. M. and Prohaska, J. X. 2011. Detection of pristine gas two billion years after the big bang. *Science* 334(6060):1245–1249.

Gomez, H. L., Krause, O., Barlow, M.J., et al. 2012. A cool dust factory in the crab nebula: A Herschel study of the filaments. *The Astrophysical Journal* 760:96.

Haenecour, P., Howe, J. Y., Zega, T. J., Amari, S., Lodders, K., José, kaji, K., Sunaoshi, T. and Muto, A. 2019, Apr. 29. Laboratory evidence for co-condensed oxygen- and carbon-rich meteoritic stardust from nova outbursts. *Nature Astronomy Letters.* Doi: 10.1038/ s41550-019-0757-4.

Kobayashi, C., Karakas, A. I. and Lugaro, M. 2020. The origin of elements from carbon to uranium. *The Astrophysical Journal* 900:179 (33p).

Knoche, K., Sprute, L., Behrmann, W., et al. 1992. Total neutron multiplicities as a measure of the violence in heavy ion reactions. *Zeitschrift für Physik A. - Hadrons and Nuclei* 342:319–327.

Kovács, O. E., Bogdán, A., Smith, R. N., Kraft, R. P. and Forman, W. R. 2018. Detection of the missing baryons toward the sightline of H1821+643, arXiv:1812.04625v1 [astro-ph.CO] 11 Dec. 2018.

Leung, S.-C. and Nomoto, K. 2018. Explosive nucleosynthesis in near-chandrasekhar-mass white dwarf models for type Ia supernovae: Dependence on model parameters. *The Astronomical Journal* 861: 143 (37 p).

Norman, R., Bhalekar, A. A., Bartoli, S. B., Buckley, B., Dunning-Davies, J., Rak, J. and Santilli, M. 2017. Experimental confirmation of the synthesis of neutrons and nautroids from hydrogen gas. *American Journal of Modern Physics* 6(4-1):85–104.

Rodrigues, M. R. D., Wada, R., Hagel, K., et al 2011. Neutron multiplicity from primary hot fragments produced in heavy ion reactions near Fermi energy. *Journal of Physics: Conference Series* 312 (2011):082009.

Schlager, H., Grewe, V. and Roiger A. 2012. Chemical composition of the atmosphere. In: Schumann, U. (Ed.) *Atmospheric Physics, Research Topics in Aerospace*, Springer-Verlag, Berlin Heidelberg, pp. 17–35.

Stave, S., Ahmed, M. W., France III, R. H., Henshaw, S. S., Müller, B., Perdue, B. A., Prior, R. M., Spraker, M. C. and Weller, H. R. 2011. Understanding the $^{11}B(p, \alpha)\alpha\alpha$ reaction at the 0.675 MeV resonance. *Physics Letters B* 696:26–29.

Vangioni-Flam, E. and Cassé, M. 1999. Cosmic lithium-beryllium-boron story. *Astrophysics and Space Science* 265:77–86.

Vangioni-Flam, E., Cassé, M. and Audouze, J. 1999. Lithium-beryllium-boron: Origin and evolution. arXiv:astro-ph/9907171.

Wanajo, S. and Janka, H-T. 2011, Dec. 07. The r-process in the neutrino-driven wind from a black-hole torus. arXiv:1106.6142v2 [astro-ph.SR].

Xiong, D. R. and Deng, L. 2009. Lithium depletion in late-type dwarfs. *Monthly Notices of the Royal Astronomical Society* 395:2013–2028.

2

Galaxies, Galactic Magnetic Fields and Cosmic Rays

2.1 Introduction

The radiation and subatomic particles were the only components of the Universe after the Big Bang. What happened immediately after the Big Bang is not known. One possibility is those small particles slowly teamed up and gradually formed stars, star clusters and eventually galaxies. The other option is that the Universe first organized itself as large clumps of matter that later subdivided into galaxies.

According to the current thinking about the cosmic structure formation, it is assumed that the precursors of the massive structures in the Universe began to form shortly after the Big Bang. This process happened in regions corresponding to the most considerable fluctuations in the cosmic density field (Springel et al., 2005; Cole et al., 2008; Behroozi et al., 2013). The observation of these structures during their period of active growth and assembly, the first few hundred million years of the Universe, is challenging. This approach requires surveys that are sensitive enough to detect distant galaxies that act as signposts of these structures and wide enough to capture rarest objects. Consequently, very few such objects have been detected so far (Riechers et al., 2013; Vieira et al., 2013). Marrone et al. (2018) have recently reported observations of a far-infrared-luminous object at redshift 6.9 (less than 0.8×10^9years after the Big Bang) that was discovered in a wide-field survey (Strandet et al., 2017). High-resolution imaging shows the object to be a pair of extremely massive star-forming galaxies. The larger of them is forming stars at a rate of 2,900 solar masses per year, and it contains 270×10^9 solar masses of gas and 2.5×10^9 solar masses of dust. It is more massive than any other known object at a redshift of more than 6. The companion galaxy probably triggers the rapid star formation at a projected separation of 8 kpc. This merging companion hosts 35×10^9 solar masses of stars and has a star formation rate of 540 solar masses per year. However, it has an order of magnitude of less gas and dust than its neighbor and physical conditions similar to those observed in lower-metallicity galaxies in the nearby Universe (Cormier et al., 2015). These objects indicate the presence of a dark matter halo with a mass of more than 100×10^9 solar masses. Presumably, this appearance is among the rarest dark matter haloes that should exist in the Universe at this epoch (Marrone et al., 2018).

DOI: 10.1201/9781003181330-2

Our galaxy, called the Milky Way, is a typical one. It contains hundreds of billions of stars and enough gas and dust to make billions of more stars. It also has at least ten times as much dark matter as all the stars and gas in it put together. All of this is held together by gravity. More than two-thirds of the known galaxies are similar to the Milky Way, and they also have a spiral shape. At the center of the spiral, an enormous amount of energy is generated, and occasionally, flares are also generated. Since immense gravity would be required to explain the movement of stars and the energy expelled, the astronomers concluded that the center of the Milky Way is a supermassive black hole.

Other galaxies have elliptical shapes, and a few have unusual shapes such as toothpicks or rings. The Hubble Ultra-Deep Field (HUDF) view shows this diversity. Hubble observed a tiny patch of sky (one-tenth the diameter of the moon) for 1 million seconds (11.6 days) and found approximately 10,000 galaxies of all sizes, shapes and colors. This historical view is two separate images taken by Hubble's Advanced Camera for Surveys (ACS) and the Near-Infrared Camera and Multi-object Spectrometer (NICMOS). Both images show galaxies that are very faint and cannot be seen by ground-based telescopes, or even in Hubble's previous faraway looks, called the Hubble Deep Fields (HDFs), taken in 1995 and 1998.

Galaxy Zoo (http://www.galaxyzoo.org) is now arguably the world's best-known online citizen science project and is undoubtedly the one with the most significant number of publications based on citizen scientists' input. It started in July 2007, with a data set of a million galaxies imaged by the Sloan Digital Sky Survey, which still provides some of the images on the site today. The project received more than 50×10^6 inputs during its first year, contributed by more than 150,000 people. The Galaxy Zoo's web site relaunched an updated design in September 2014. In this version, new imaging from Sloan, giving the best view of the local Universe, was combined with the most distant images yet from Hubble's CANDELS survey. In the CANDELS survey, the new Wide Field Camera 3 (WFC3) is used, which was installed during the final shuttle mission to Hubble to enable taking ultra-deep images of the Universe.

The Galaxy Zoo science program has contributed to a diverse set of topics, primarily focused on the nearby and intermediate-redshift Universe. Some recent highlights include the most extensive studies of galaxy mergers (Darg et al., 2010), tidal dwarf galaxies, dust lanes in early-type galaxies (Kaviraj et al., 2012) and bars in disk galaxies (Masters et al., 2010, 2012) in the nearby Universe to date. One of the unique aspects of Galaxy Zoo over automated morphological measurements is the possibility of accidental discoveries. These have included the "Green Pea" galaxies (a class of compact extremely star-forming galaxies in the local Universe (Cardamone et al., 2009)) and perhaps the most famous "Hanny's Voorwerp" (Lintott et al., 2009) along with a survey of similar active galactic nuclei (AGN) – ionized gas clouds (Keel et al., 2012). The existence of a large sample of galaxies with both color and morphological information has led to the critical realization that color, not morphology, is most strongly correlated with the environment (Bamford et al., 2009). This leads to intriguing subclasses of galaxies such as red spiral galaxies and blue ellipticals (Schawinski et al., 2009).

The discovery of the bimodality in the color–magnitude and color–mass diagrams resulted in the color space between the two central populations – the green valley – being viewed as the crossroads of galaxy evolution. The galaxies in the green valley represent the transition from the blue cloud of star-forming galaxies to the red sequence of quenched, passively evolving galaxies. In the first approximation, all galaxies were presumed to follow similar evolutionary tracks across the green valley, with a rapid transition suggested by the relatively small number of galaxies in the green valley in comparison with the blue cloud or red sequence.

Schawinski et al. (2013) used SDSS+GALEX+Galaxy Zoo data to study the quenching of star formation in low-redshift galaxies. They show that the green valley between the blue cloud of star-forming galaxies and the red sequence of quiescent galaxies in the color–mass diagram is not a single transitional state through which most blue galaxies evolve into red galaxies. Instead, an analysis that takes morphology into account makes clear that only a small population of blue early-type galaxies scurry across the green valley after the morphologies are transformed from disk to spheroid, and star formation is quenched rapidly. In contrast, direction from the blue region (these are the main-sequence star-formers) toward the red sequence, at higher masses, shows no sign of a green valley (in the sense of a color bimodality).

Schawinski et al. (2013) illustrated two crucial findings: (i) Both early- and late-type galaxies span almost the entire u-r color range. In the morphology-sorted plots, we can see small numbers of blue early-type and red late-type galaxies (e.g., Schawinski et al., 2009; Masters et al., 2010). (ii) The green valley is a distinct region only in the all-galaxies plot. Most early-type galaxies occupy the red sequence, with a long tail (10% by number) to the blue cloud at relatively low masses; this could represent blue galaxies transiting rapidly through the green valley to the red sequence. More strikingly, the late types form a single, unimodal distribution peaking in the majority of blue star-forming galaxies that have significant disks, and they retain their late-type morphologies as their star formation rates decline very slowly.

Schawinski et al. (2013) summarize a range of observations that lead to these conclusions, including UV–optical colors and halo masses, which both show a striking dependence on the morphological type. The authors interpreted these results in terms of the evolution of cosmic gas supply and gas reservoirs. They conclude that late-type galaxies are consistent with a scenario where the infinite supply of gas is shut off, perhaps at a critical halo mass, followed by slow exhaustion of the remaining gas over several 10^9 years, driven by secular or environmental processes. In contrast, early-type galaxies require a scenario where the gas supply and gas reservoir are destroyed virtually instantaneously, with rapid quenching accompanied by a morphological transformation from disk to a spheroid. This gas reservoir destruction could be the consequence of a significant merger, which in most cases transforms galaxies from disk to elliptical morphology, and mergers could play a role in inducing black hole accretion and possibly AGN feedback (Schawinski et al., 2013).

Here we should mention the SAMI Galaxy Survey by the Sydney-Australian Astronomical Observatory Multi-object Integral-Field Spectrograph. Its third and final data release is published (Croom et al., 2021), and the data are available from Astronomical Optics' Data Central Service at https://datacentral.org.au/. In Data Release 3, data for the full sample of 3068 unique galaxies observed are presented. This includes the SAMI cluster sample of 888 unique galaxies for the first time. For each galaxy, there are two primary spectral cubes covering the blue (370–570 nm) and red (630–740 nm) optical wavelength ranges at spectral powers of R=1,808 and R=4,304, respectively. For each galaxy, complete 2D maps from parameterized fitting to the emission-line and absorption-line spectral data are included. These maps provide information on gas ionization and kinematics, stellar kinematics and populations, and more.

Several observational studies have been performed to detect deviations from isotropy in the orientation of galaxies following approaches first devised by Hawley and Peebles (1975). If caught, the presence of alignments and specific preferred directions can give information on the origin and evolution of the galaxies and their relation to large-scale structures. The arrangement might arise from the large-scale environmental influences during galaxy formation or development. Cosmic magnetic fields are present on scales of galaxy clusters and larger (Ratra, 1992). Effects of seed magnetic fields from initiation, axionic fields post-inflation, and cosmic strings are possible candidates that could affect alignment in galaxies even on scales larger than galaxy clusters (Taylor and Jagannathan, 2016).

In their paper, Taylor and Jagannathan (2016) presented the measurement of the distribution of angles for radio jet position of galaxies over an area of 1 square degree in the ELAIS N1 field. Giant Metrewave Radio Telescope observed ELAIS N1 at 612 MHz with the rms noise level of 10 μJy and angular resolution of 6×5 arcsec. The image taken contains a total of 65 resolved radio galaxy jets. The established spatial distribution shows a prominent alignment of jet position angles along a "filament" of about 1°. They examined the possibility that the apparent alignment could arise from an underlying random distribution. They found that the probability of chance alignment was less than 0.1%. Data analysis was done by the angular covariance method, and it indicated the presence of spatial coherence for position angles on scales $>0.5°$. The deduced angular scales correspond to a comoving scale of >20 Mpc at a redshift of 1. The proposed alignment of the spin axes of massive black holes gives rise to the radio jets suggesting the presence of large-scale spatial coherence in angular momentum. Their results reinforce prior evidence for large-scale spatial alignments of quasar optical polarization position angles.

Recently, Hirschauer et al. (2016) have reported a discovery of the most metal-poor gas-rich galaxy known: AGC 198691, also known as Leoncino (little lion). It is the so-called dwarf galaxy, and it is only 1,000 ly in diameter and composed of only several million stars. The galaxy is blue due to the presence of recently formed hot stars but rather dim, having the lowest luminosity level observed in a system of this type. Leoncino is a member of the "local Universe", a region of space about 1×10^9 ly from the Earth. According to estimations, it contains several million galaxies, of which only a small portion has been catalogued. Leoncino, identified in 2005, is a galaxy recognized to possess the lowest metal abundance. It has been estimated to have a 29% lower metal abundance. The elemental content of metal-poor galaxies is very close to that of the early Universe. Hirschauer et al. (2016) find the dwarf galaxy AGC 198691 to be an extremely metal-deficient system with an oxygen abundance of $12+\log(O/H)=7.02\pm0.03$. This value makes AGC 198691 the lowest-abundance star-forming galaxy known in the local Universe.

FIGURE 2.1 Hubble views cosmic collision between spiral galaxies (Jäger, 2018). (Credit: ESA/Hubble, NASA.)

Hubble enabled the views of the cosmic collision between spiral galaxies (Jäger, 2018). Figure 2.1 shows an image of the peculiar galaxy NGC 3256. The picture is taken with the WFC3, and the Advanced Camera for Surveys (ACS) installed on the NASA/ESA Hubble Space Telescope. The galaxy shown is about 100 million ly from the Earth and is the result of a past galactic merger. NGC 3256 is an ideal target for the investigation of starbursts that have been triggered by galaxy mergers. This distorted galaxy shown is the relic of a collision between two spiral galaxies. It is estimated that this event occurred 500 million years ago.

Recently, Hashimoto et al. (2019) have observed signals of oxygen, carbon and dust from a galaxy in the early Universe 13 billion years ago by using the radio telescope ALMA (Atacama Large Millimeter/submillimeter Array). From the analysis of measured signals, the authors concluded that the observed galaxy is two galaxies merging, making it the earliest example of merging galaxies yet discovered.

The Sloan Digital Sky Survey III, in short SDSS-III, has released new data, presented in the form of the most massive three-dimensional map of our Universe so far. This map displays 1.2 million galaxies over a quarter of the sky and a volume of space of 650 cubic billion ly. Hundreds of scientists worked together for 10 years to make this map and to use its data for the most precise estimate of dark energy, the force causing an accelerating expansion of our Universe. Several papers describing these results were published in the July 14, 2016, issue of Monthly Notices of the Royal Astronomical Society.

FIGURE 2.2 One slice through a new map of the large-scale structure of the Universe. The image shown contains 48,741 galaxies, about 3% of the full data set used to create the entire map. The gray patches are small regions without survey data. Image via Daniel Eisenstein and the Sloan Digital Sky Survey III (SDSS-III).

The image in Figure 2.2 covers about one-twentieth of the sky. This slice of the Universe is 6×10^9 ly wide (thus, we see some of these galaxies 6 billion years into the past), 4.5×10^9 ly high and 5×10^8 ly thick. Each dot in the picture at the top indicates the position of a galaxy. The color indicates the distance from the Earth. It is ranging from yellow on the near side of the slice to purple on the far side. This new map helps to reveal the vast scale of our Universe, which is thought to contain more than one 100×10^9 galaxies.

Data about the galaxies shown in the new map are obtained by measurements performed within the program of SDSS-III called the Baryon Oscillation Spectroscopic Survey (BOSS). The distribution of galaxies is mapped by detecting the imprint of baryon acoustic oscillations in the early Universe. The BOSS web site explains how the sound waves that propagate in the early Universe mark a characteristic scale on cosmic microwave background fluctuations.

A parsec (symbol: pc) is a unit of length used to measure large distances to objects outside our solar system; it corresponds to the distance at which one astronomical unit covers an angle of one arcsecond.

Note: 1 pc ≈ 206,264.81 au (astronomical unit, the distance from the Sun to the Earth),

$$\approx 3.0856776\times10^{16}\,\text{m},$$

$$\approx 3.2615638\ \text{ly}$$

A distance of 1,000 pc is equal to 3,262 ly and is denoted by the kiloparsec (kpc). A distance of one million parsecs is denoted by the megaparsec (Mpc).

These fluctuations have evolved into today's walls and voids of galaxies, meaning this baryon acoustic oscillation (BAO) scale is visible among galaxies today.

Shaped by a continuous competing of dark matter and dark energy, the map revealed by BOSS allows measurements of the expansion rate of the Universe and thus the determination of the amount of dark matter and dark energy.

2.2 Black Holes

This discussion should start by mentioning the work by Penrose, who won half the 2020 Nobel Prize in Physics for proving that under very general conditions, the collapsing matter would trigger the formation of a black hole (Penrose, 1965). This rigorous result opened up the possibility that the astrophysical process of gravitational collapse, which occurs when a star runs out of its nuclear fuel, would lead to the formation of black holes in nature. He was also able to show that at the heart of a black hole must lie a physical singularity – an object with infinite density, where the laws of physics simply break down. At the singularity, our concepts of space, time and matter fall apart. Penrose invented new mathematical concepts and techniques while developing this proof. Equations that Penrose derived in 1965 have been used by physicists studying black holes ever since. In fact, just a few years later, Stephen Hawking, together with Penrose, used the same mathematical tools to prove that the Big Bang cosmological model had a singularity at the initial moment. These are results from the celebrated Penrose–Hawking singularity theorem. The fact that mathematics demonstrated that astrophysical black holes might exist in nature energized the quest to search for them using astronomical techniques. Indeed, since Penrose's work in the 1960s, numerous black holes have been identified.

Supermassive black holes (SMBHs) are huge, and they contain millions or even billions of times the mass of the Sun. In today's Universe, they are found in the center of nearly all large galaxies (see, for example, Vayner et al., 2017), including the Milky Way. The central SMBH of the Milky Way galaxy has a mass of four million solar masses. The first black hole seeds were formed when the Universe was younger than $\sim 0.5 \times 10^9$ years. These seeds are thought to play an important role in the growth of early ($z \sim 7$) SMBHs. Lots of progress has been made in the understanding of the formation and growth of black hole seeds; however, their observational characteristics remain unexplored. Consequently, no detection of such sources has been confirmed so far. There are three types of black holes: (i) stellar-mass black hole, $M_{bh} = (1\text{–}100) \times M_{solar}$; intermediate-mass black hole, $M_{bh} = (10^2\text{–}10^5) \times M_{solar}$; and supermassive black hole, $M_{bh} = (10^5\text{–}10^9) \times M_{solar}$.

The stellar-mass black hole in the Cygnus X-1 (X-ray binary system) is the first black hole ever detected; it is one of the closest black holes to the Earth. Using the updated measurements for the black hole's mass and its distance away from the Earth, Miller-Jones et al. (2021) were able to conclude that it has a mass 21 times the mass of the Sun. Simultaneously, they confirmed that Cygnus X-1 is spinning with a speed close to the speed of light, faster than any other black hole found to date.

Stars lose mass to their surrounding environment through stellar winds that blow away from their surface. However, to make a black hole this heavy and rotating so quickly, we need to dial down the amount of mass that bright stars lose during their lifetimes. Therefore, it could be concluded that the black hole in the Cygnus X-1 system began life as a star approximately 60 times the mass of the Sun and collapsed tens of thousands of years ago. Incredibly, it's orbiting its companion star every five-and-a-half days at just one-fifth the distance between the Earth and the Sun. The formation of such a high-mass black hole in a high-metallicity system (within the Milky Way) constrains wind mass loss from massive stars (Miller-Jones et al., 2021).

The first direct visual evidence – a photographic image in the "light" of radio waves – of a SMBH was announced on April 10, 2019, in a special issue of The Astrophysical Journal Letters (The Event Horizon Telescope Collaboration, 2019a–f). It was an observation of the enormous black hole in the center of the galaxy M87, 55×10^6 ly from the Earth. The black hole itself cannot be seen in the image. The black holes are "black" because no light can escape them, and thus the holes themselves are invisible. Instead, the picture shows the black hole's "shadow", a bright ring formed as light bends in the intense gravity around the hole. This black hole positioned in the center of the M87 galaxy is thought to be some 6.5×10^9 times more massive than our Sun.

Two main theories are explaining the formation of SMBHs in the early Universe. In the first scenario, one assumes that the seeds grow out of black holes, having a mass of about 10–100 times greater than our Sun, a possible result of the collapse of a massive star. Next, the black hole seeds grew by mergers with other small black holes and by pulling in gas from their surroundings. In such a way, they would have to grow at an unusually high rate to reach the mass of SMBHs already discovered in the 10^9-year-old Universe.

The recent findings, as reported by Pacucci et al. (2016), support another scenario where at least some very massive black hole seeds with 10^5 times the mass of the Sun formed directly when an enormous cloud of gas collapses. In this scenario, the growth of the black holes would be jump-started and would proceed more quickly. Pacucci et al. (2016) presented a novel method for the identification of black hole seed candidates in deep multi-wavelength surveys. They assumed that these sources are characterized by a steep spectrum in the infrared region (1.6–4.5 μm). The method used identified only two objects with robust X-rays found in the CANDELS/GOODS-S survey with a redshift $z \gtrsim 6$. To fit their infrared spectra with only a stellar component would require enormous star formation rates ($\gtrsim$2,000 $M_{\odot}$/y). The selected objects represent the most promising black hole seed candidates, possibly formed via the direct collapse black hole scenario, with predicted mass $>10^5$ $M_{\odot}$. This result is one of the best photometric observations of high-z sources available to date; further progress will come from spectroscopic and more in-depth X-ray data. Upcoming observatories, such as NASA JWST (James Webb Space Telescope), will significantly expand the scope of this type of work. There is a lot of discussions over which path these black holes take. The work by Pacucci et al. (2016) suggests convergence on one answer, where black holes start big and grow at the standard rate, rather than starting small and growing at a fast pace. It is now assumed that SMBHs in the early Universe were produced by the direct collapse of a massive gas cloud.

In their paper, Most et al. (2019) discuss signatures of quark–hadron phase transition, which occur during neutron stars merging into the black hole. The authors show that the phase transition of neutrons into quarks appearing throughout the merger will result in deviation of the gravitational signal until the newly formed massive neutron star collapses under its weight to form a black hole (see also Bauswein et al., 2019).

When black holes collide and merge, they release gravitational waves, which travel across the Universe, eventually reaching our detectors on the Earth. For all the dozens of black hole mergers that we have witnessed so far, the gravitational wave signature is exactly what general relativity predicts. In spite of this, there are efforts trying to describe black holes in the framework of string theory where black holes are replaced by string-theoretic horizon-scale microstructure, fuzzballs (Mayerson, 2020).

Recently, a team of astronomers led by the University of Arizona have observed a luminous quasar 13.03×10^9 ly from the Earth – the most distant quasar discovered to date. Dating back to 670 million years after the Big Bang, when the Universe was only 5% its current age, the quasar is powered by a SMBH equivalent to the combined mass of 1.6×10^9 M_{Sun} and more than 1,000 times brighter than our entire Milky Way galaxy. The quasar, called J0313-1806, is seen as it was when the Universe was only 670 million years old and is providing astronomers with valuable insight into how massive galaxies – and the SMBHs at their cores – formed in the early Universe (NRAO, 2021).

2.3 Galaxies in the Universe

The large-scale structure of the Universe is a complex web of clusters, filaments and voids. Galaxies congregate in clusters and along filaments and are missing from large regions referred to as voids. These structures are seen in maps derived from spectroscopic surveys that reveal networks of structure that are interconnected with no clear boundaries. Extended regions with a high concentration of galaxies are called "superclusters", although this term is not precise.

Only some galaxies occur alone or in pairs; however, they are more often parts of more significant associations known as groups, clusters and superclusters. Galaxies in such groups regularly interact and even merge in a dynamic cosmic process of interacting gravity. Mergers cause gases to flow toward the galactic center, which can trigger phenomena such as rapid star formation.

Recently, hundreds of hidden nearby galaxies have been studied for the first time (see Staveley-Smith et al., 2016), shedding light on a mysterious gravitational anomaly called the Great Attractor. Although it is just 250 million ly from the Earth (very close in astronomical terms), the new galaxies had been hidden from our view until now by our galaxy. The authors used CSIRO's Parkes radio telescope equipped with an innovative receiver. With this instrument, they were able to see behind the stars and dust of the Milky Way and explore a previously unexplored region of space. This discovery may help explain the Great Attractor region. This object is drawing the Milky Way and hundreds of thousands of other galaxies toward it with a gravitational force equivalent to 10^{15} Suns.

FIGURE 2.3 Hubble views merging galaxies in Eridanus. (Image credit: ESA/Hubble & NASA; Acknowledgment: Judy Schmidt. From: http://www.nasa.gov/image- feature/goddard/2016/hubble-views-merging-galaxies-in-Eridanus.)

The research by Staveley-Smith et al. (2016) identified several new structures that could help explain the movement of the Milky Way. This discovery includes three galaxy concentrations (named NW1, NW2 and NW3) and two new clusters (called CW1 and CW2).

The NASA/ESA Hubble Space Telescope image shows a galaxy known as NGC 1487, positioned at a distance of about 30 million ly away in the southern constellation of Eridanus (see Figure 2.3).

In this picture, we are viewing an event rather than seeing it as a celestial object: We are observing two or more galaxies in the act of merger forming a single new galaxy. Each galaxy has lost almost all signatures of its original appearance, as stars and gas were thrown by gravity in an intricate cosmic vortex. Unless one galaxy is very much bigger than the other, galaxies are always disrupted by the violence of the merging process. Therefore, it is complicated to determine what the original galaxies looked like and how many of them there were. In this case, we may be seeing the merger of several dwarf galaxies that were previously members of a small group.

The subject of the NASA/ESA Hubble Space Telescope image shown in Figure 2.4 is called NGC 3597. This object is the product of a collision between two large galaxies and is slowly developing into a vast elliptical galaxy. This type of galaxy has become more and more common as the Universe has changed. The initially small galaxies merge and progressively build up into larger galactic structures over time.

FIGURE 2.4 Hubble views a galactic mega-merger. (Image credit: ESA/Hubble & NASA; Acknowledgement: Judy Schmidt. From http://www.nasa.gov/image-feature/goddard/2016/hubble-views-a-galactic-mega-merger.)

The NGC 3597 is located approximately 150 million ly away in the constellation of Crater (the cup). Studies of NGC 3597 help in learning more about how elliptical galaxies form. Many elliptical galaxies began their lives far in the history of the Universe. Older elliptical galaxies are called "red and dead" because these galaxies are not anymore producing new, bluer stars and are thus full of ancient and redder stellar populations.

2.3.1 Redshift, Distance and Age of Galaxies

When determining the distance of far galaxies, astronomers typically give the value purely in terms of its redshift, often marked as z. To be able to calculate the redshift z of an object, one needs to look for an emission or absorption line, such as those of hydrogen. Afterward, the observed wavelength of the line from the object is compared with the standard (not redshifted) line. The number known as z is the difference between the observed and standard wavelengths divided by the standard wavelength.

If there is no redshift, then there is no difference between the observed and standard lines; hence, the z is zero. A positive number thus gives redshift; if the number is bigger, the bigger is the z. Technically, there is no limit to the value z can have, but the highest we have observed is about $z=12$. A bigger z also means a greater distance. The light of a distant galaxy is redshifted more than the light of a closer galaxy because of the expansion of the Universe. So the galaxy with the greatest redshift is the most distant.

$$z = \Delta\lambda/\lambda = \sqrt{(1+v/c)/(1-v/c)} - 1$$

The reason astronomers usually talk about redshift instead of distance is that the measured z is purely an observational result. The fact is that a bigger z means a greater distance; however, the exact distance depends on the model you use for the Universe. The model used can be relatively accurate for determining distances, but by using z, one does not have to assume any model.

There is a relation between redshift, distance, physical and angular sizes, and parameters of the cosmological models which can be computed. Cosmological models are all characterized by a set of settings, including the present Hubble constant, H_0, the dark matter, m, and dark energy densities, Ω_Λ. For more information about the age–redshift relationship, see the paper by Wei et al. (2015). Several online calculators are available at NASA NED web site: http://ned.ipac.caltech.edu/. Although these calculators are ready to use, one can also use a calculator based on the nomogram method.

The paper-and-pencil calculator proposed by Pilipenko (2013) is designed for the ΛCDM cosmological model using the recent values of cosmological parameters from the Planck mission: $H_0=67.15$ km/s/Mpc, dark energy density $\Omega_\Lambda=0.683$ and matter density parameter $\Omega_m=0.317$ (Planck collaboration, 2013). The calculator contains the following quantities:

- z – redshift;
- H_0 – the present Hubble constant, km/s/Mpc;
- age – age of the Universe, in 10^9 years;
- time – lookback time, in 10^9 years;
- size 1" – the physical size of an object which is seen as a 1" arc on the sky, kpc;
- angle 1 kpc – the angular size of a rod with physical size 1 kpc, arcsec.

To use this calculator, one should find a known value on a respective vertical scale. All the other values are located on the same horizontal level. The calculators are available for three redshift intervals: $z<20$, $z<1$ and $z<0.1$. The space between major (labeled) marks on each vertical scale is always divided into ten equally spaced intervals of the denoted value. The computer code used to produce these calculators is publicly available and can be found at http://code.google.com/p/cosmonom/. Figure 2.5 shows a modified redshift lookup table with only z, H, age and time columns.

The mass–metallicity relation has been studied by Erb et al. (2006) at $z \sim 2.2$ and by Maiolino et al. (2008) and Mannucci et al. (2009) at $z=3$–4. Their findings show monotonic evolution resulting in metallicity decrease with redshift at a given mass. The same authors (Erb et al., 2006; Erb, 2008; Mannucci et al., 2009) have also studied the relation between metallicity and gas fraction, i.e., the effective yields, obtaining clear evidence of the presence of infall in high-redshift galaxies. More recently, Cresci et al. (2010) have studied a small sample of three-disk galaxies at $z=3$. They found evidence of "positive" metallicity gradients. Lower metallicities were in more active regions of star formation. The presence of these gradients in galaxies with very regular dynamics can be understood as a consequence of accretion of metal-poor gas, producing a new episode of star formation in older, more metal-rich galaxies. Their observations of low-metallicity

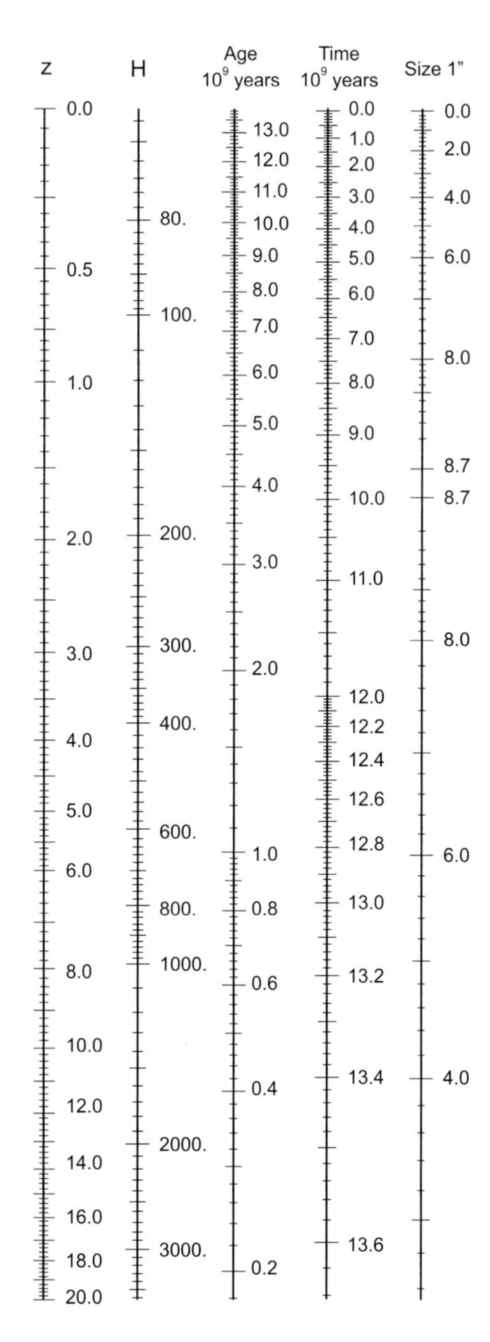

FIGURE 2.5 A redshift lookup table for the Universe. (Modified from Pilipenko 2013.)

regions in three galaxies at $z \approx 3$, therefore, provide evidence for the actual presence of accretion of metal-poor gas in massive high-z galaxies. They are capable of sustaining high star formation rates avoiding frequent mergers of already evolved and enriched subunits. This scenario was already indirectly suggested by other observational studies of gas-rich disks at $z \approx 1$–2 (Förster Schreiber et al., 2009; Tacconi et al., 2010). The reviews of the variation of metallicity as a function of a gas fraction (Mannucci et al., 2009) and star formation rate (Mannucci et al., 2010) also suggest this scenario.

Tacconi et al. (2010) surveyed molecular gas in samples of massive star-forming galaxies at mean redshifts $<z>$ of 1.2 and 2.3 when the Universe was 40% and 24%, respectively, of its current age. Their measurements reveal that distant star-forming galaxies were indeed gas-rich and that the star formation efficiency is not very much dependent on the cosmic epoch. The fraction of cold gas relative to total galaxy baryonic mass at $z=2.3$ and $z=1.2$ is about 44% and 34%, respectively, three to ten times higher than in today's massive spiral galaxies (Leroy et al., 2005). Instead, a slight decrease between $z \approx 2$ and $z \approx 1$ could probably be explained by a mechanism of quasi-continuous replenishment of fresh gas to the young galaxies.

2.3.2 Some Examples of the Observed Galaxies

Here we present some cases of the observed galaxies; most, if not all, galaxies have a SMBH at their center. It is not yet known whether the galaxies formed around pre-existing black holes or the holes formed after those galaxies had come into being. Astronomers suspect that their central black holes help regulate galaxies' rates of star formation. For example, Schlegel et al. (2016) have recently described a double-arc-like X-ray structure lying approx. 15–30″ (≈ 0.8–1.7 kpc) south of the NGC 5195 nucleus visible in the merged exposures of long Chandra pointings of M5l; see Figure 2.6. The curvature and orientation of the arcs argue for a nuclear origin. They interpret the arcs as episodic outbursts from the central SMBH. The arcs are radially spaced by ≈15″, but are rotated relative to each other by ≈ 30°, requiring episodic activity. They also found a slender H_α-emitting region just outside the outer edge of the outer X-ray arc, suggesting that the X-ray-emitting gas plowed up and displaced the H_α-emitting material from the galaxy core. Star formation may have commenced in that arc. H_α emission is present at the inner arc, but appears more complex in structure. In contrast to an explosion expected to be azimuthally symmetric, the X-ray arcs suggest a focused outflow. Schlegel et al. (2016) concluded that NGC 5195 represents the nearest galaxy exhibiting ongoing, large-scale outflows of gas, in particular, two episodes of a focused outburst of the SMBH. The observation has implications for SMBH feedback and the subsequent galaxy evolution.

Previous studies suggest that the growth of SMBHs may be fundamentally related to host galaxy stellar mass (M*). To investigate this SMBH growth–M_x relation in detail, Yang et al. (2017) calculated the long-term accretion rate of SMBH as a function of M* and redshift [BHAR(M*, z)]. Calculation is done over ranges of log(M*/M) = 9.5–12 and $z=0.4$–4. Their high-quality survey data constrain BHAR(M*, z) by the stellar mass function and the X-ray luminosity function. At a given M*, BHAR is higher at high redshift. This redshift dependence is stronger in more extensive systems (for log(M*/M) ≈ 11.5, BHAR is three decades higher at $z=4$ than at $z=0.5$), possibly due to AGN feedback. The

FIGURE 2.6 The Whirlpool Galaxy M51, also known as NGC 5194, is one of the brightest and most picturesque galaxies in the sky. The smaller galaxy, NGC 5195, appearing here to the right, is well behind M51. (Credit: NASA, ESA, S. Beckwith (STScI), and the Hubble Heritage Team (STScI/AURA).)

results indicate that the ratio between BHAR and average star formation rate (SFR) rises toward high M* at a given redshift. The BHAR/SFR dependence on M* does not support the scenario that SMBH and galaxy growth are in lockstep. The authors calculated SMBH mass history [$M_{BH}(z)$] based on their BHAR(M*, z) and the M*(z) from the literature and found that the M_{BH}–M* relation has weak redshift evolution since $z \approx 2$. The M_{BH}/M^* ratio is higher toward massive galaxies. It rises from ≈1/5,000 at log $M^* \leq 10.5$ to ≈1/500 at log $M^* \geq 11.2$. The authors predicted that M_{BH}/M^* ratio at high M* is similar to that observed in local giant ellipticals, suggesting that SMBH growth from mergers is unlikely to dominate overgrowth from accretion.

Figure 2.7 shows the ring-like swirls of dust filling the Andromeda Galaxy. This picture is the recent image taken by the Herschel Space Observatory, a European Space Agency mission with NASA participation (see http://www.esa.int/SPECIALS/Herschel/index.html, and http://www.herschel.caltech.edu, http://www.nasa.gov/herschel). To make them easier to see, the colors in this image have been enhanced. The colors represent real variations in the data. The coldest clouds are bright in the longest wavelengths and are colored red here, while the warmer ones are bluish. These observations reported were made by Herschel's Spectral and Photometric Imaging Receiver (SPIRE) instrument. The Andromeda Galaxy, also called Messier 31 (M31), lies 2×10^6 ly away. It is the closest large galaxy to our own Milky Way. It is estimated to have up to 10^{12} stars, whereas the Milky Way contains several 100×10^9 stars. Recent evidence suggests Andromeda's overall mass may be less than the mass of the Milky Way when dark matter is included.

FIGURE 2.7 Andromeda Galaxy. (Image credit: ESA/NASA/JPL-Caltech/NHSC.)

In the presentation at the 227th meeting of the American Astronomical Society, Comerford (2016) described the event in which the galaxies hit into one another and merge. The result of such a merger is a galaxy, at least for a time, with two central black holes. Generally, galaxies with two black holes in their cores put out more than ten times as much light as those with one. This extra light is created by additional amounts of material sucked into their twinned center. One consequence of all this light is the ionization of much more of the surrounding gas and then pushing of this ionized gas out of the way. That provides a second way that the behavior of a galaxy's black-hole-inhabited core can turn off star formation in the rest of the system.

Dual AGN and offset AGN are <10 kpc separated SMBH pairs created during galaxy mergers. Although few dual and offset AGN have been discovered to date, they are ideal probes of the link between galaxy mergers and AGN activity. Using Chandra and HST observations, Comerford (2016) has discovered six galaxies that host either dual AGN or offset AGN. These galaxies have two stellar bulges visible in the HST images, and the Chandra observations reveal which bulges host AGN. Most of the dual and offset AGN are created in significant mergers of galaxies, and these AGN are ten times more luminous than single AGN that is not in mergers. This fact is observational support of the theoretical prediction that the most luminous AGN are triggered in significant mergers. It was also found that the AGN in the less massive stellar bulge accrete at a higher ratio than the AGN in the more massive galactic bulge, which is an effect reproduced by simulations of AGN triggering during galaxy mergers.

Magnetic fields in spiral galaxy IC 342 (shown in Figure 2.8) have been studied in Beck (2015); it was found that IC 342 hosts a diffuse radio disk with an intensity that decreases exponentially with increasing radius. The frequency dependence of synchrotron emission indicates energy-dependent propagation of the cosmic-ray electrons, probably by the streaming instability. The equipartition strength of the total magnetic field in the primary spiral arms is around 15 μG, while that of the ordered field is about 5 μG.

FIGURE 2.8 Spiral galaxy IC 342, also known as Caldwell 5; data obtained from NASA's Nuclear Spectroscopic Telescope Array, or NuSTAR. Date taken: January 7, 2013; ID nustar130107a. (Credit NASA/JPL-Caltech.)

Multi-telescope study of this galaxy (see Beck, 2015; Moss et al., 2015) revealed the magnetic structure of this galaxy. The result showed a vast, helically twisted loop around the main spiral arm of the galaxy. Such an arrangement, never seen before in a galaxy, is adequately strong to be able to influence the flow of gas around the spiral arm. Spiral arms cannot be formed by gravitational forces alone. This IC 342 image indicates that magnetic fields also play an essential role in shaping spiral arms. The new measurements explain yet another aspect of this galaxy: a bright central region that is hosting a black hole that is producing new stars at a high rate. A steady inflow of gas from the galaxy's outer regions into its center is required to maintain a high speed of star production. Since the magnetic field lines at the inner part of the galaxy point toward the galaxy's center, they support an inward flow of gas.

Another compelling case is a galaxy about 23 million ly away showing ongoing fireworks. This enormous light show involves a giant black hole, shock waves and vast reservoirs of gas in NGC 4258, also known as M106, a spiral galaxy like the Milky Way. It is characterized by two extra spiral arms that glow in X-ray, optical and radio light. These features, or anomalous arms, are not aligned with the plane of the galaxy, but instead intersect with it; see Figure 2.9. The unusual arms are seen in this new composite image of NGC 4258. In the figure, X-rays from NASA's Chandra X-ray Observatory are marked blue. Radio data from the NSF's Karl G. Jansky Very Large Array are marked purple. Optical data from NASA's Hubble Space Telescope are marked yellow, and infrared data from NASA's Spitzer Space Telescope are marked red.

Díaz-Santos et al. (2015) reported the observation of the galaxy named WISE J224607.55−052634.9 (or W2246−0526 for short) at $z=4.601$ (12.4×10^9 ly from the Earth)

FIGURE 2.9 A spiral galaxy NGC 4258, also known as M106. Release date: July 2, 2014. Color code: X-ray (blue); optical (gold, blue); infrared (red); radio (purple). (Image credit: X-ray: NASA/CXC/Caltech/P.Ogle et al; optical: NASA/STScI; IR: NASA/JPL-Caltech; radio: NSF/NRAO/VLA.)

with $L=3.5\times10^{14}\ L_{\odot}$, which is the most luminous galaxy known in the Universe and hosts a deeply buried active galactic nucleus – a SMBH. This galaxy was discovered using the Wide-field Infrared Survey Explorer (WISE) in 2015. It is classified as a hot dust-obscured galaxy based on its luminosity and dust temperature. The authors presented spatially resolved ALMA [C II] 157.7 μm observations of W2246-0526, which provides a unique insight into the kinematics of its ISM. The ratio of measured [C II] to far-infrared is $\approx 2\times10^{-4}$, suggesting the ISM conditions which could be compared with only the most obscured, compact starbursts and AGN in the local Universe today. The spatially resolved [C II] line is very uniform and very broad, 500–600 km/s width. It extends through the entire galaxy over about 2.5 kpc, with only a modest shear. Such a large, homogeneous velocity dispersion is a result of a highly turbulent medium; i.e., W2246-0526 is unstable because of the injection of the energy and momentum into the ISM. The gas is blown away from the system isotropically on its road to become an unobscured quasar. W2246-0526 provides an extraordinary laboratory to study and model the properties and kinematics of gas in an extreme environment under strong feedback, at a time when the Universe was one-tenth of its current age: a system pushing the limits that can be reached during galaxy formation (see Figure 2.10).

The largest known gravitationally bound entities in the Universe are clusters of galaxies. These clusters comprise up to several thousand galaxies, the majority of them being dwarf galaxies. As an example, we mention here the Virgo Cluster, which contains some 250 massive galaxies and more than 2,000 smaller galaxies. Tiny assemblies of galaxies are referred to as groups. For instance, the local group comprises two massive galaxies, the Milky Way and M31, and a third one, M33, being somewhat less massive. Additional members are Large and Small Magellanic Clouds and some 30 other dwarf galaxies with

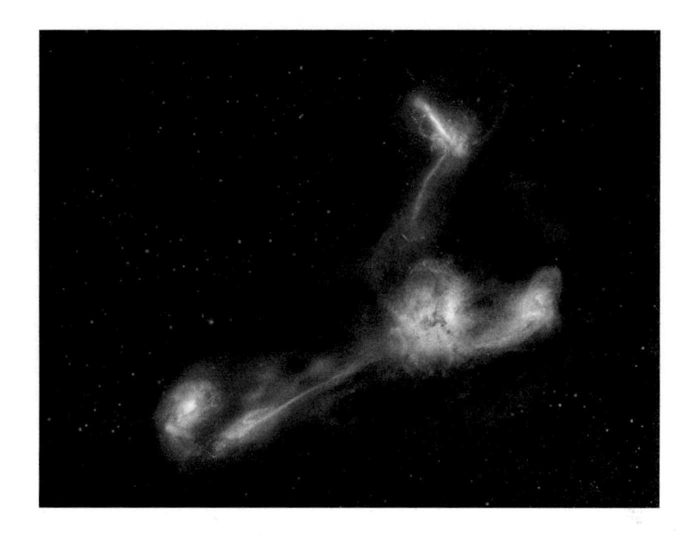

FIGURE 2.10 Artist's concept of galaxy WISE J224607.55–052634.9 (or W2246–0526 for short), the most luminous galaxy known. Discovered by NASA's space-based Wide-field Infrared Survey Explorer (WISE) in 2015. New research suggests that there's turbulent gas across the whole galaxy, the first example of its kind. (Image credit: NRAO/AUI/NSF, S. Dagnello.)

masses down to what is comparable to the most massive globular clusters. The cluster mass is, as is well known, dominated by dark matter, as inferred from the galaxies' velocity dispersion, the temperature of the hot gas, and gravitational lensing.

Recently, Chu et al. (2016) have reported the discovery of the most distant massive galaxy cluster. This was accomplished by a group of astronomers from MIT, the University of Missouri, the University of Florida, and elsewhere, who have detected a massive, sprawling, churning galaxy cluster that formed only 3.8×10^9years after the Big Bang. Located 10×10^9 ly from the Earth and potentially comprising thousands of individual galaxies, the megastructure is about 250×10^{12} times more massive than the Sun, or 10^3 times more massive than the Milky Way galaxy (see Figure 2.11).

The cluster, called IDCS J1426.5+3508 (or IDCS 1426), is the most massive cluster of galaxies yet discovered in the first 4×10^9years after the Big Bang. Cluster IDCS 1426 is undergoing a significant amount of reversal. The astronomers have observed a bright knot of X-rays positioned off-center in the cluster, resulting from the cluster's core shift from its center some hundred thousand light-years. It is assumed that the core may have been moved from its position by a collision with another massive galaxy cluster, resulting in the chaotic gas movement within the group. Such an accident may explain how IDCS 1426 formed so quickly in the early Universe, at a time when individual galaxies were only beginning to take shape (Brodwin et al., 2016). The authors reported the relatively stable distribution of the gas-to-total mass ratio; they said that metals were not detected in the intracluster medium (ICM) of this system. The estimated upper limit on the metallicity argues that this system is probably still in the process of enrichment of its ICM. The cluster has a dense, low-entropy core, offset by $\approx$30 kpc from the X-ray centroid. Therefore, it is one of the few "cool-core" clusters discovered at $z>1$, and the first

FIGURE 2.11 A massive galaxy cluster that was formed only 3.8×10^9 years after the Big Bang. The group shown is the most massive cluster of galaxies yet discovered in the first 4×10^9 years after the Big Bang. (Image credit: NASA, European Space Agency, University of Florida, University of Missouri and the University of California.)

known cool-core cluster at $z>1.2$. The observed offset suggests that this cluster has had a relatively recent (500×10^6 years) merger/interaction with another large system (Brodwin et al., 2016).

Oesch et al. (2016) have recently presented Hubble WFC3/IR slitless grism spectra of a remarkably bright $z\geq10$ galaxy candidate, GN-z11. This galaxy was initially identified from CANDELS/GOODS-N imaging data. A significant spectroscopic continuum break was detected at $\lambda=1.47\pm0.01$ μm. A combination of the grism and photometric data rules out all reasonable lower-redshift values for this source. The only logical solution is that this continuum break is the Ly-α break redshifted to $z_{grism} = 11.09^{+0.08}_{-0.12}$, just $\approx400\times10^6$ years after the Big Bang. This observation extends the current spectroscopic frontier by 150×10^6 years. This extension is well before the Planck (instantaneous) cosmic reionization peak at $z\sim8.8$, demonstrating that galaxy buildup was well underway early in the reionization epoch at $z>10$. GN-z11 is remarkably and unexpectedly luminous for a galaxy at such an early time: Its UV luminosity is 3× larger than $L_\odot$ measured at $z\sim6$–8. The Spitzer IRAC detections up to 4.5 μm of this galaxy are consistent with a stellar mass of $\sim10^9\ M_\odot$. This spectroscopic redshift measurement suggests that JWST will be able to similarly and easily confirm such sources at $z>10$ and characterize their physical properties through detailed spectroscopy.

Furthermore, WFIRST, with its wide-field near-IR imaging, would find large numbers of similar galaxies and contribute significantly to JWST's spectroscopy if it is launched early enough to overlap with JWST (Oesch et al., 2016). This far-away galaxy, named GN-z11, existed a mere 400 million years after the Big Bang, or about 13.3×10^9 years ago. Since the light from such a distant galaxy must travel a considerable distance to reach the Earth, we are seeing this galaxy as it looked over 13×10^9 years ago.

Hagen et al. (2016) have recently provided evidence that UGC 1382, long believed to be a passive elliptical galaxy, is a giant low-surface-brightness (GLSB) galaxy that rivals the

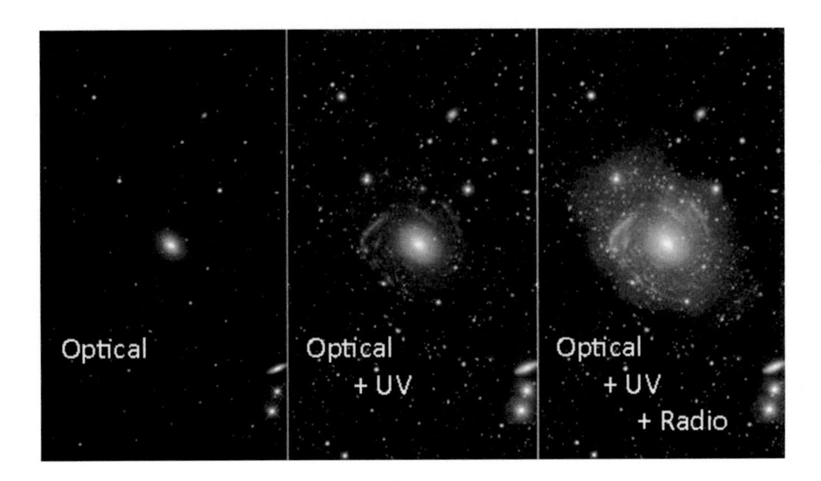

FIGURE 2.12 UGC 1382, one of the biggest known galaxies in the Universe, as seen in optical, optical+UV and optical+UV+radio spectra. (Credit: NASA/JPL/Caltech/SDSS/NRAO.)

archetypical GLSB Malin 1 in size (see Figure 2.12). Like other GLSB galaxies, UGC 1382 is made of two disks: an inner high-surface-brightness disk surrounded by an extended low-surface-brightness (LSB) disk. The central part is a lenticular system with an effective radius of 6 kpc. The LSB disk also has an effective range of ~38 kpc and an extrapolated central surface brightness of ~26 mag/arcsec^2. Both components of UGC 1382 have a combined stellar mass of $\sim 8\times10^{10}$ M_{Sun} and are embedded in a massive (10^{10} M_{Sun}) low-density (<3 M_{Sun}/pc^2) HI disk with a radius of 110 kpc, making this one of the most massive isolated disk galaxies known. The system resides within a massive dark matter halo of at least 2×10^{12} M_{Sun}. The system is possibly part of a small group. It could be that its low-density environment plays a role in the formation and retention of the giant LSB and HI disks. The authors constructed a model of the spectral energy distributions with the result that the LSB disk is probably older than its lenticular component. The galaxy UGC 1382 has UV–optical colors typical of galaxies transitioning through the green valley. It is assumed that within the LSB disk are spiral arms forming stars at very low efficiencies. The gas depletion time rate of $\sim 10^{11}$years suggests that UGC 1382 may be a very long-term resident of the green valley. Hagen et al. (2016) proposed that the formation and evolution of the LSB disk in UGC 1382 are best explained by the accretion of gas-rich LSB dwarf galaxies.

Comerford et al. (2017) presented the discovery of an active AGN that is turning off and on again in the z=0.06 galaxy SDSS J1354+1327. This episodic nuclear activity is, according to the authors, a result of discrete accretion events. These events could have been triggered by a past interaction with the companion galaxy that is currently located 12.5 kpc away. The authors initially targeted SDSS J1354+1327 because the Sloan Digital Sky Survey spectrum has narrow AGN emission lines that exhibit a velocity offset of 69 km/s relative to systemic. To determine the properties of the galaxy and measure its velocity-offset emission lines, they observed SDSS J1354+1327 with several instruments. The list contained Chandra/ACIS, Hubble Space Telescope/WFC3, Apache Point Observatory optical long-slit spectroscopy and Keck/OSIRIS integral-field spectroscopy. They found a ~10 kpc cone of photoionized gas south of the galaxy center and a ~1 kpc semi-spherical front of shocked gas. This gas is supposed to be responsible for the velocity offset in the emission lines north of the galaxy center. The authors interpreted these two outflows as the result of two separate AGN accretion events. The first outburst created the southern discharge, and then $<10^5$ years later, the second outburst formed the northern shock front. The galaxy SDSS J1354+1327 has the most persuasive evidence for an AGN that has turned off and then on again, and it fits into the broader context of AGN flickering that includes observations of AGN light echoes.

There are many similar-looking objects in the sky. For example, Tremblay et al. (2018) described a galaxy-scale fountain of cold molecular gas pumped by a black hole. They presented observations made by ALMA and Multi-Unit Spectroscopic Explorer. These observations show the brightest cluster galaxy in Abell 2597, which is a nearby ($z=0.0821$) cool-core cluster of galaxies. The obtained data map the kinematics of approximately three billion solar mass filamentary nebulae that span the innermost 30 kpc of the galaxy's core. Its warm ionized and cold molecular components are both cospatial and comoving. This fact is consistent with the assumption that the optical nebula follows the warm envelopes of cold molecular clouds drifting in the velocity field of the hot X-ray atmosphere. The clouds are not in dynamical equilibrium. Instead, they show evidence for inflow toward the central SMBH, outflow together with the jets it launches, and hot bubbles those jets inflate. The proposed scenario is in agreement with a galaxy-spanning "fountain", in which cold gas clouds drain into the accretion reservoir of a black hole, thus powering jets and bubbles that uplift a cooling plume of low-entropy multiphase gas. This gas may stimulate additional cooling and accretion as part of an automatic feedback loop. All proposed velocities are below the escape speed from the galaxy. Therefore, these clouds should fall back to the galaxy center from which they came, keeping the fountain long-lived. Their speculations are consistent with the predictions of chaotic cold accretion, precipitation and stimulated feedback models and may, therefore, trace processes fundamental to galaxy evolution at effectively all mass scales (Tremblay et al., 2018).

The key ingredient of our understanding of the early formation of massive galaxies is the studies of dusty star-forming galaxies (DSFGs) at high redshift selected at submillimeter wavelengths (submillimeter galaxies, or SMGs) over the past two decades (Riechers et al., 2017). The hyper-luminous DSFGs (hyper-luminous infrared galaxies, or HyLIRGs) represent some of the most luminous, massive galaxies in the early Universe, emerging from compact regions only a few kiloparsecs in diameter. While most of the DSFG population is thought to be somewhat heterogeneous, the HyLIRGs are likely significant mergers of gas-rich galaxies, and they may also be associated with protoclusters of galaxies, which represent some of the most over-dense environments in the early Universe. Because of the high dust content, most of the stellar light in DSFGs is subject to dust extinction, making their identification for the high redshifts very difficult. Many DSFGs were found and reported at $z=2$–3.5 relatively early. The first examples at $z>4$ were identified more than a decade after the initial discovery of this galaxy population.

After the Herschel Space Observatory launching, it became possible to develop color selection techniques for a systematical search of the most distant DSFGs in large-area surveys like the Herschel Multi-tiered Extragalactic Survey (HerMES). Dominik et al. (2017) extended the search to z~6. They report the detection of ADFS-27, a dusty, starbursting merger at a redshift of z=5.655 observed by the ALMA. ADFS-27 event was selected from Herschel/SPIRE and APEX/LABOCA data as an extremely red "870 μm riser", demonstrating the utility of this technique to identify some of the highest redshift dusty galaxies. The 870 μm dust continuum emission was resolved into two components, 1.8 and 2.1 kpc in diameter, separated by 9.0 kpc, with comparable dust luminosities, suggesting an ongoing major merger. Based on the infrared luminosity of $L_{IR \approx} 2.4 \times 10^{13}$ $L_{\odot}$ it was concluded that this system is a binary HyLIRG, the most distant of its kind presently known. The discovery of this system is consistent with a significantly higher space density than previously thought for the most luminous dusty starbursts within the first 10^9 years of cosmic time, easing tensions regarding the space densities of z~6 quasars and massive quiescent galaxies at $z \geq 3$ (Riechers et al., 2017).

Galaxies in the early Universe, bright at submillimeter wavelengths, are forming stars at a rate that is 1,000 times higher than the Milky Way. A large fraction of the new stars forms in the central kiloparsec of the galaxy, a region that is similar in size to the massive, quiescent galaxies found at the peak of cosmic star formation history and the cores of the present-day giant elliptical galaxies. The physical properties inside these compact starburst cores are poorly understood because probing them at relevant spatial scales requires exceptionally high angular resolution. Tadaki et al. (2018) reported observations with a linear resolution of 550 pc of gas and dust in the submillimeter-bright galaxy at a redshift of z=4.3 when the Universe was less than 2×10^9 years old. They resolved the spatial and kinematic structure of the molecular gas inside the heavily dust-obscured core and showed that the underlying gas disk is clumpy and rotationally supported (that is, its rotation velocity is larger than the velocity dispersion). Analysis of the molecular gas mass suggests that the starburst disk is gravitationally unstable. This fact implies that the self-gravity force of the gas is more significant than the effects produced by disk differential rotation and the internal pressure due to stellar radiation feedback. Because of the gravitational instability in the disk, the molecular gas would be consumed by star formation on a timescale of 100×10^6 years, which is comparable to gas depletion times in merging starburst galaxies (Tadaki et al., 2018).

The embryonic galaxy named SPT0615-JD, which existed when the Universe was just 500 million years old, was spotted in an intensive survey deep into the Universe by NASA's Hubble and Spitzer space telescopes. Although only a few other primitive galaxies have been seen at this early epoch, they have mostly all looked like red dots because of their small size and tremendous distances. However, in this case, the gravitational field of a massive foreground galaxy cluster not only amplified the light from the background galaxy, but is also smearing the image of it into an arc about 2 arcseconds long (Webster, 2018). In this case, NASA's Hubble Space Telescope managed to capture a rare up-close view of the farthest and oldest galaxy known to human. This particular galaxy is so old that it's nearly as early as the Universe itself. The SPT0615-JD galaxy could be among the first galaxies created after the Big Bang.

Le Fèvre et al. (2020) described observations and sample properties of 118 star-forming galaxies at $4<z<6$, from a survey called the ALPINE-ALMA [C II]. The ALMA is a single telescope of revolutionary design, composed of 66 high-precision antennas located on the Chajnantor plateau, at 5,000 m altitude in northern Chile (ALMA, 2020). It is the largest astronomical project in existence. The ALMA Large Program to Investigate (ALPINE) features 118 galaxies observed in the [C II] 158 μm line and far-infrared (FIR) continuum emission during the period of rapid mass assembly. In the investigated sample, authors found a surprisingly wide range of galaxy types, including 40% mergers, 20% extended and dispersion-dominated, 13.3% rotating disks and 10.7% compact, with the remaining 16% too faint to be classified.

This diversity of types indicates that several physical processes are at work for the assembly of mass in these galaxies, mainly for galaxy merging. While galaxy merging is commonly associated with starbursts above the main sequence, at least up to $z \sim 3$, merging systems in ALPINE at $z \sim 4.7$ lie mainly on the main sequence, making merging also a dominant process for normal star-forming galaxies at the studied epoch (Le Fèvre et al., 2020).

Due to the recent advances in instrumentation on the Hubble Space Telescope and large ground-based telescopes, a number of high-redshift galaxies have been discovered. While the observation in the previous example was done on a ground-based telescope, the work by Jiang et al. (2020) was done based on Hubble Space Telescope imaging data. They reported a probable detection of three ultraviolet emission lines from GN-z11, photometrically selected as a luminous star-forming galaxy candidate at redshift $z>10$. The lines selected were $[C_{III}]$ $\lambda1907$, $C_{III}]$ $\lambda1909$ doublet and $O_{III}]$ $\lambda1666$ at $z=10.957 \pm 0.001$. At that time, the Universe was only $\sim 420 \times 10^6$ years old, or ~3% of its current age. This makes GN-z11 the most distant galaxy known to date. Observations of even higher redshifts are needed to probe the earliest phases of the formation and evolution of the first galaxies (Jiang et al., 2020).

2.3.3 Milky Way Galaxy

Studies of the nearby Universe encompass a region of approximately 1×10^9 ly in radius, over which the effects of cosmic evolution are small. Within this volume, galaxies and associated objects are essentially frozen in their present-day configurations.

The local group of galaxies to which the Milky Way galaxy belongs has over 30 galaxies that are considered to be in the local group, and they are spread over a diameter of nearly 10 million ly, with the center of them being somewhere between the Milky Way and M31. M31 and the Milky Way are the most massive members of the local group, with M33 being the third largest. Both M31 and the Milky Way have dwarf galaxies associated with them. M31, the Andromeda Galaxy, has two small satellite galaxies, M32 and M110. The dynamics of the local group are changing, and some astronomers speculated that one day the two large spirals in it (M31 and the Milky Way) might collide and merge to form a giant elliptical galaxy. It is also possible that the local group may one day merge with the next nearest big galaxy cluster, the Virgo Cluster.

According to Tully et al. (2014), the supercluster of galaxies that includes the Milky Way is 100 times bigger in volume and mass than previously thought. They have mapped the enormous region and given it the name Laniakea – Hawaiian for "immeasurable heaven".

The Milky Way galaxy is a spiral galaxy with a diameter of ~30 kpc. Most of the bright stars reside in a disk (~1 kpc thick) surrounded by a bright central bulge. The Sun is located on the lower arm of the disk and is about 8.5 kpc away from the center. A vast dimmer halo surrounds the entire disk. The space between the stellar systems in the Milky Way contains ISM, including ionized gas, molecular gas, dust and cosmic rays, with an average density of ~1 nucleon/cm^3 (Strong and Moskalenko, 1998; Gilmore et al., 1989).

The solar neighborhood is the part of the Milky Way galaxy associated with a cylinder centered at the Sun and perpendicular to its disk. There is no definition of the exact radius of this region; for example, European Southern Observatory (ESO, 2020) has published a 3D map of stellar systems in the solar neighborhood assuming a radius of only 12.5 ly. Some investigators consider solar neighborhood to be stars within about 15 ly of the Sun (containing 56 stars in 38 systems), and some within 25 pc (1.0 pc=3.26 ly, ly). Some even use terms like the immediate solar neighborhood (within about 5 pc) and the extended solar neighborhood (within a few hundred pc). In all cases, the solar neighborhood contains mostly baryonic matter (stars and gas); the amount of dark matter is negligible inside it. It is also estimated that some 80% of matter is in stars and 20% in gaseous material.

Considering a sample of 12,000 main-sequence and subgiant stars, Binney et al. (2000) concluded that the age of the solar neighborhood is $(11.2 \pm 0.75) \times 10^9$ years with very small sensitivity to variations in the assumed metallicity distribution of old disk stars. This age is only 10^9 years younger than the age of the oldest Milky Way globular clusters.

Our galaxy is probably almost as old as the Universe itself since the oldest known halo star is estimated at 13.2×10^9 years old (Frebel, 2007). This ancient star is metal-poor and lacking in the heavy elements needed for planetary formation. In contrast, the oldest brown dwarfs, intermediates between stars and planets, are estimated to be 10×10^9 years old (Schilbach et al., 2009). The thin galactic disk, where conditions are more favorable for life, is expected to be between 6.5 and 11.1×10^9 years old (del Peloso, 2005).

This age has been determined using Th/Eu nucleocosmochronology (del Peloso et al., 2005). In their work, [Th/Eu] abundance ratios for a sample of 20 disk dwarfs/subgiants of F5 to G8 spectral type with $-0.8 \leq [Fe/H] \leq +0.3$ were adopted for this analysis. They developed a galactic chemical evolution model that includes the effect of refuse, which is composed of stellar remnants (white dwarfs, neutron stars and black holes) and low-mass stellar formation residues (terrestrial planets, comets, etc.). Both model inputs are contributing to a better fit to observational constraints. Two galactic disk ages were estimated, by comparing the literature data on Th/Eu production and solar abundance ratios to the model $((8.7+5.8/-4.1) \times 10^9$ years) and by comparing [Th/Eu] vs. [Fe/H] curves from the model to our stellar abundance ratio data $((8.2 \pm 1.9) \times 10^9$ years), yielding the final average value of $(8.3 \pm 1.8) \times 10^9$ years. The first galactic disk age was determined via Th/Eu nucleocosmochronology and corroborates the most recent white dwarf ages determined via cooling sequence calculations, which indicate a low age ($\leq 10 \times 10^9$ years) for the disk.

Milky Way galaxy evolution has been studied from different aspects. For example, Toyouchi and Chiba (2014) studied the detailed properties of the radial metallicity gradient in the stellar disk of our galaxy, intending to obtain information about its chemical and structural evolution. For this purpose, they selected and analyzed approximately 18,500 disk stars from two data sets called SDSS and HARPS (High-Accuracy Radial Velocity Planetary Searcher). As a part of these surveys, they examined the metallicity gradient, Δ[Fe/H]/ΔR_g, along with the guiding-center radii, R_g, of stars, and its dependence on the [α/Fe] ratios. These studies are done to determine time evolution and the original metallicity distribution of the gas disk from which those stars formed. In both sample sets, the thick-disk candidate stars characterized by high [α/Fe] ratios ([α/Fe] > 0.3 in SDSS; [α/Fe] > 0.2 in HARPS) are found to show a positive Δ[Fe/H]/ΔR_g. In contrast, the thin-disk candidate stars characterized by lower [α/Fe] ratios show a negative one. Note that the "α-element" is simply a convenient phrase used for some even-Z elements (O, Mg, Si, S, Ca and Ti). It reflects the observation of increased abundance of these elements at low metallicity. It does not signify that these elements are all products of a single nuclear reaction chain that occurs in the same astrophysical environment. Furthermore, they found that the relatively young thin-disk population characterized by much lower [α/Fe] ratios ([α/Fe] < 0.2 in SDSS; [α/Fe] < 0.1 in HARPS) shows notably a flattening Δ[Fe/H]/ΔRg with decreasing [α/Fe], in contrast to the old one with higher [α/Fe] ratios ([α/Fe] ~ 0.2 in SDSS; [α/Fe] ~ 0.1 in HARPS). Toyouchi and Chiba (2014) discuss the possible implication for the early disk evolution of these findings, in particular galaxy formation accompanying the rapid infall of primordial gas in the inner disk region, generating a positive metallicity gradient. The subsequent chemical evolution of the disk resulted in a flattening effect of the metallicity gradient at later epochs.

It is established that galaxies possess hot gaseous halos containing a large amount of mass being an integral part of galaxy formation and evolution. The Milky Way has a 2×10^6 K halo detectable in emission and by absorption in the O_{VII} resonance line against bright background AGN, and for which the best current model is an extended spherical distribution. Using XMM-Newton Reflection Grating Spectrometer data, Hodges-Kluck et al. (2016) measured the Doppler shifts of the O_{VII} absorption lines toward an ensemble of AGN. These Doppler shifts represent boundary conditions on the dynamics of the hot halo. Their results rule out a stationary halo at about 3σ and a co-rotating halo at 2σ and present a best-fit rotational velocity of $v_\phi = 183 \pm 41$ km/s or an extended halo model. The results suggest that the hot gas rotates with an angular momentum comparable to that in the stellar disk. The authors examined the possibility of a model with a kinematically distinct disk and spherical halo. The disk must contribute less than 10% of the column density, implying that the Doppler shifts probe motion in the extended hot halo is consistent with the emission line X-ray data.

The Milky Way offers a uniquely detailed view of galactic structure, and it represents a prototypical spiral galaxy. However, recent observations suggest that the Milky Way is atypical in some respect. It has an undersized SMBH at its center. Additionally, it is surrounded by a very low-mass, excessively metal-poor stellar halo, and it has an unusually sizeable nearby satellite galaxy, the Large Magellanic Cloud (LMC). Cautun et al. (2019) show that the LMC is on a collision course with the Milky Way with which it will merge in $\left(2.4^{+1.2}_{-0.8}\right)\times10^9$ years (68% confidence level). This long-overdue event will be catastrophic and will probably restore the Milky Way to normality. Using the EAGLE galaxy formation simulation, Cautun et al. (2019) predicted that, as a result of the merger, the central SMBH would increase in mass by up to a factor of 8. The galactic stellar halo would undergo an equally impressive transformation, becoming five times more massive. The new stars would come predominantly from the disrupted LMC, but a sizeable number of stars would be ejected from the stellar disk on to the hallo. The post-merger stellar halo would have a median metallicity of [Fe/H] = −0.5 dex, a typical value of other galaxies of similar mass to the Milky Way. At the end of this assumed event, the Milky Way would become a real benchmark for spiral galaxies, at least temporarily (Cautun et al., 2019).

More knowledge about the Milky Way galaxy has recently been collected by Gaia satellite (see http://sci.esa.int/gaia/). The main goal of the Gaia satellite mission is to make the most extensive, most precise three-dimensional map of our galaxy from the survey of one percent of the galaxy's population of 100×10^9 stars (Gaia Collaboration, 2016a). During this process, Gaia satellite detected and accurately measured the motion of individual stars in orbit around the center of the galaxy. Much of this motion was assigned upon each star during its birth. Studying this motion allows astronomers to look back in time to when the galaxy was first forming. The construction of a detailed map of the stars by Gaia represents a crucial tool for studying the formation of the Milky Way (Gaia Collaboration, 2016b, 2018, 2020).

The results provided by Gaia satellite tremendously helped the recently developed modelling frameworks to connect the observed information to the properties of progenitor satellites. Satellite galaxy accretion histories are often expressed in terms of merger trees. In particular, the use of globular clusters to trace the assembly history of the Milky Way has been popular. In their recent paper, Kruijssen et al. (2020) used the age–metallicity of galactic globular clusters to infer the formation and assembly history of the Milky Way by the use of their hydrodynamical cosmological simulation model. Their efforts resulted in the partial reconstruction of the Milky Way merger tree, identifying five specific accretion events. This has been done by combining the resulting predictions with the constraints on the assembly history of the Milky Way (Kruijssen et al., 2019b).

Kruijssen et al. (2020) used the E-MOSAICS simulations (see Kruijssen et al., 2019a) to quantify the stellar masses and accretion redshifts of five satellite galaxies that have been accreted by the Milky Way. These five satellites are (i) Kraken (progenitor of the low-energy globular cluster), (ii) Gaia Enceladus, (iii) the progenitor of the Helmi streams, (iv) Sequoia and (v) Sagittarius. They are the most massive objects that the Milky Way accreted since $z=4$. These processes of accretion were quick, considering the mass involved.

The formation and evolutionary processes of galaxy bulges are still unclear, and the presence of young stars in the bulge of the Milky Way is intensively debated (Ferraro et al., 2020). The discovery reported in their paper define a new class of stellar systems orbiting the Milky Way bulge. The characteristics of these systems are as follows:

- They are indistinguishable from the genuine globular cluster by appearance.
- They have metallicity and abundance patterns compatible with those observed in the bulge field stars.

- They host an old stellar population formed in an early epoch of the Milky Way galaxy assembling.
- They also host a young stellar population, several to many 10^9 years younger than their old population. These structures are likely remnants of massive primordial systems that formed in situ and contributed to generate the bulge ~12×10^9 years ago. The authors named them "bulge fossil fragments, BFFs" (Ferraro et al., 2020).

The authors suggested that other surviving remnants could still be hidden in the heavily obscured regions of the Milky Way bulge.

The possibility of collision of Milky Way galaxy with the nearby Andromeda Galaxy, M31, the nearest large spiral galaxy to the Milky Way, has been discussed for some time. Recently (on February 7, 2019), the European Space Agency (ESA) provided the newest predictions of this collision, using data from its Gaia satellite. Gaia has collected information about the motions of stars within both the Andromeda Galaxy and the Triangulum Galaxy (M33), the third large galaxy in our local group. The data contain some information about the Andromeda Galaxy's collision course with the Milky Way (van der Marel et al., 2019). The reported data imply that the M31 orbit toward the Milky Way is somewhat less radial than previously inferred and strengthen arguments that M33 may be on its first infall into M31. In any case, the Andromeda Galaxy will probably deliver more of a glancing blow to the Milky Way than a head-on collision.

2.3.3.1 Chemical Evolution of the Milky Way Galaxy

Spitoni and Matteucci (2015) presented detailed chemical evolution models for the Milky Way and M31 galaxies. They discuss the effect of radial gas flow on the chemical evolution of these two galaxies. Their models follow in detail the growth of several chemical elements (H, He, C, N, O, α-elements, Fe-peak elements) in space and time. The contribution of the supernovae of different types of chemical enrichment is taken into account also.

Figure 2.13 shows the results for the abundance gradient for oxygen in the presence of radial gas flows of the Milky Way. The best model of Mott at al. (2013) is shown for the Milky Way compared with the data from Cepheids, curve (a). Spitoni and Matteucci (2015) found that an inside-out formation of the disk, with no threshold in the surface gas density for the star formation rate coupled with radial gas inflows of variable speed, can reproduce the observed abundance gradients. Besides, Figure 2.13 also shows the effects of different parameters on models without radial gas flows, compared with the data from Cepheids. Spitoni and Matteucci (2015) examined the following settings: inside-out formation, the threshold in gas density for the SFR and the star formation efficiency (SFE). The model with inside-out formation, threshold and constant SFE well reproduced the abundance gradient up to 14 kpc. If, instead, an inside-out structure for the thin disk is not assumed, and the timescale of infall is kept constant, the present-day abundance gradients provided by the model are too flat in the inner part of the disk even if a threshold in the gas density is assumed. The threshold influences mostly the outer gradient. The model with inside-out formation but no limit shows an abundance gradient too flat between 6 and 12 kpc, and in the outer parts of the disk, it even increases, clearly at variance with the observational data. These data agree with what was found by

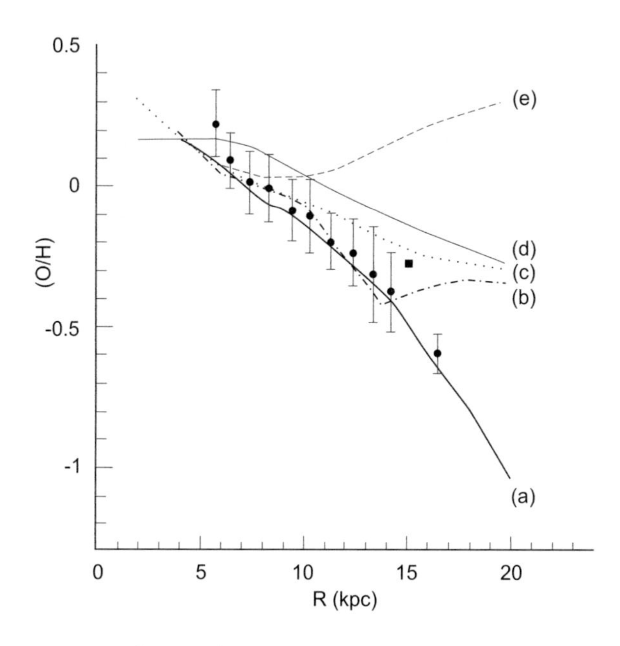

FIGURE 2.13 Oxygen gradient in the Milky Way. The effects on the oxygen abundance gradients of several parameters characterize the chemical evolution in the absence of radial flows. Lines are predictions of different models discussed in Spitoni and Matteucci (2015). The line "e" is the model without threshold, with inside-out formation. The model characterized by the absence of inside-out formation is indicated by the line "d". The model with variable SFE, inside-out formation and threshold is represented by the line "b". The line "c" represents the model with inside-out formation, threshold and constant SFE. The data are from Luck and Lambert (2011).

Chiappini et al. (2001). Thus, it can be concluded that a threshold in the gas density seems to be necessary to have the right trend of the gradients in the outer parts of the disk in a model without radial gas flows. The model, which assumes a variable SFE, a threshold and inside-out formation, provides an excellent fit to the observed present-day abundance gradients from 4 to 14 kpc in the Milky Way. However, beyond this distance, the slope predicted by the models is too flat and inconsistent with the observations (Spitoni and Matteucci, 2015).

The evolution of chemical element abundances in a galaxy provides a clock for galactic aging. It is partially the case for the solar neighborhood. There is a relation between the age and metal abundances of stars: On average, older stars contain less iron than younger stars.

The galactic center has been subject to a great many studies. Because the galactic center is the nearest galactic nucleus of a spiral galaxy, we can observe most details of its structure and activity. Several observers have mapped the structures of the molecular clouds in the region, mainly in the ^{12}CO and ^{13}CO lines. These surveys had sufficient sensitivity and identified several impressive structures in the high-velocity range of $v \geq 30$ km/s. However, they strongly suffer from the foreground and background contaminations in the low-velocity range of $v \leq 30$ km/s (Tsuboi et al., 1999).

Oka et al. (1998) presented a high-resolution CO survey of the galactic center region with the 2×2 focal-plane array receiver on the 45 m telescope in Nobeyama Radio Observatory. They have collected about 44,000 $^{12}C^{16}O$ (J=1-0) spectra and over 13,000 $^{13}C^{16}O$ (J=1-0) spectra with a 34″ (1.4 pc) grid spacing. The ^{12}CO mapping area is covering almost the full extent of the molecular gas concentration in the galactic center. The CO images show a very complex distribution and kinematics of molecular gas in the galactic center. Its large-scale behavior can be assigned to the well-known coherent features, while bright CO emission arises from several compact clouds with large velocity widths ($\Delta v \geq 30$ km/s). Filaments, arcs and shells characterize the small-scale structure of molecular gas. The molecular gas kinematics there may be a result of the violent release of kinetic energy by several supernova explosions and/or Wolf–Rayet stellar winds.

Large-scale infrared surveys done recently have been revealing stellar populations in the inner galaxy seen through strong interstellar extinction in the disk. For example, classical Cepheids, with their period–luminosity and period–age relations, are useful markers of galactic structure and evolution. Matsunaga et al. (2016) reported the discovery of nearly 30 classical Cepheids toward the bulge region and discussed the significant impact of the reddening correction on distance estimates for these objects. They are assuming that the four Cepheids in the nuclear stellar disk (NSD) are located at the distance of the galactic center. Assuming further that the near-infrared extinction law (wavelength dependency of the interstellar extinction) is not systematically different between the NSD and other bulge lines of sight, most of the other Cepheids presented are located further than the galactic center. It suggests a lack of Cepheids in the inner 2.5 kpc region of the galactic disk except for the NSD. Reports of recent radio observations show a similar distribution of star-forming areas.

GALAH, which stands for Galactic Archaeology with HERMES (High Efficiency and Resolution Multi-Element Spectrograph), is a large-scale stellar spectroscopic survey of the Milky Way, designed to deliver complementary chemical information to a large number of stars covered by the European Space Agency Gaia mission (see Buder et al., 2018). In this publication, the authors presented the GALAH second public data release (GALAH DR2) containing 342,682 stars. For all of these stars, the GALAH collaboration provides stellar parameters and abundances for up to 23 elements. A combination of the GALAH DR2 catalogue information with the kinematic information from Gaia space observatory will enable additional galactic archaeology studies.

Buder et al. (2018) concentrated on the open cluster M67, for which we have observed turnoff, subgiant, red giant branch and red clump stars with a mean iron abundance of [Fe/H] $= -0.01 \pm 0.08$ and a radial velocity of 34.08 ± 0.85 km/s. M67 is a very old open cluster located in the constellation of Cancer. Estimations of its age vary between 3.2 and 5 billion years, with recent valuations putting it at 4 billion. This makes it one of the oldest known open clusters. Consequently, M67 contains a variety of stellar types, including many Sun-like stars, red giants and white dwarfs. Element abundances of all of the stars considered in the open cluster M67 are shown in Table 2.1.

When assessing the scatter within dwarfs and giants of the cluster separately, there is a lower scatter for several elements, which are a more appropriate measure of their internal precision. The precision ranges from the highest (0.04–0.08 dex) for elements Fe, Al,

TABLE 2.1 Element Abundances of the Open Cluster M67 (After Buder et al., 2018)

Element Abundances	Number of Stars	Mean Value
[Fe/H]	156	−0.01±0.08
[Li/Fe]	4	0.92±1.58
[C/Fe]	49	0.05±0.09
[O/Fe]	131	−0.03±0.15
[Na/Fe]	134	0.09±0.09
[Mg/Fe]	143	−0.00±0.12
[Al/Fe]	122	−0.05±0.09
[Si/Fe]	155	−0.04±0.12
[K/Fe]	113	0.25±0.17
[Ca/Fe]	137	0.03±0.12
[Sc/Fe]	153	0.09±0.09
[Ti/Fe]	142	0.00±0.06
[V/Fe]	129	0.13±0.14
[Cr/Fe]	144	0.02±0.10
[Mn/Fe]	141	0.06±0.09
[Co/Fe]	9	−0.04±0.15
[Ni/Fe]	147	0.17±0.17
[Cu/Fe]	121	−0.03±0.08
[Zn/Fe]	154	0.03±0.14
[Y/Fe]	156	0.17±0.15
[Ba/Fe]	122	0.18±0.21
[La/Fe]	31	0.06±0.23
[Eu/Fe]	5	−0.02±0.19

Sc, Ti, V and Cu; high (0.08–0.12 dex) for elements C, Na, Si, Cr and Mn; intermediate (0.12–0.16 dex) for elements O, Mg, K, Ca, Co, Ni, Zn and Y; to low (above 0.16 dex) for elements Li, Ba, La and Eu for the stars observed in M67 (Buder et al., 2018).

2.4 Galactic Magnetic Fields

Magnetic fields are a significant agent in the interstellar medium of the spiral, barred, irregular and dwarf galaxies. Magnetic fields contribute to the total force, which balances the effect of gravity on ISM. They may force the gas to flow in spiral arms, round bars and galaxy halos. Magnetic fields are essential in the onset of star formation since they play a vital role in the removal of angular momentum from the protostellar cloud during its collapse. Also, MHD turbulence distributes energy generated during supernova explosions within the ISM. Magnetic fields are responsible for heating the ISM and halo gas. They also control the density and distribution of cosmic rays in the ISM. Magnetic fields exist in all galaxies and galaxy clusters; recent measurements show that a weak magnetic field is present even in the smooth low-density intergalactic medium.

Magnetic fields play an essential role in the formation and stabilization of spiral structures of galaxies. Still, the interaction of interstellar gas with magnetic fields has not yet been fully understood. In particular, the phenomenon of "magnetic arms" in the space between material arms is a mystery (Beck, 2015). The origin of magnetic fields in AGN is still a matter of speculation. The most natural path of arguing is probably connected with the rotational energy of the central black hole.

Observations of the galactic magnetic fields can be done in the optical range via starlight being polarized by interstellar dust grains. As a result of the magnetic field presence, these elongated grains can be aligned in such a way that the major axis is perpendicular to the field lines. Data obtained by measurements of many stars show a general picture of the magnetic field in the Milky Way near the Sun. Aligned dust grains also emit polarized infrared light, which is a handy indicator of magnetic fields in dust clouds in the Milky Way. Zeeman splitting of radio spectral lines allows the measurement of relatively strong fields in dense gas clouds in the Milky Way. Observations in external galaxies by those three techniques are still tough to perform. The method of measuring synchrotron emission is the most powerful one and can be applied over the whole Milky Way.

The total strength of the magnetic field can be determined from the intensity of total synchrotron emission. This procedure can be done by assuming energy balance (equipartition) between magnetic fields and cosmic rays. This assumption seems valid on broad spatial and timescales; however, deviations were observed on local levels within galaxies. The typical average equipartition strength of the magnetic field for spiral galaxies is about 10 μG or 1 nT. This field is much weaker than the Earth's magnetic field, which has an average strength of 0.3 G (30 μT). Radio-faint galaxies M31 and M33 (our Milky Way's neighbors) have weaker fields (about 5 μG), contrary to gas-rich galaxies with high star formation rates (such as M51, M83 and NGC 6946), which have 15 μG on average. In prominent spiral arms, where cold gas and dust are concentrated, the field strength can be up to 25 μG. The strongest total fields (50–100 μG) were found in starburst galaxies, for example in M 82, and nuclear starburst regions, for instance in the centers of NGC 1097 and of other barred galaxies.

Galactic fields are sufficiently strong to be essential for galaxy dynamics. They push mass inflow into the centers of galaxies, and they play an important role in the formation of spiral arms and the rotation of gas in the outer regions of galaxies. Most importantly, magnetic fields provide the angular momentum required for the collapse of gas clouds and, hence, the formation of new stars.

Beck and Wielebinski (2013) reported the average equipartition strength of the total fields for a sample of 74 spiral galaxies to be $B = (9 \pm 2)$ μG. The average strength of 21 bright galaxies observed since 2000 is $B = (17 \pm 3)$ μG. Dwarf galaxies host fields of similar strength as spirals if their star formation rate per volume is similarly high. Blue compact dwarf galaxies are radio bright with equipartition field strengths of 10–20 μG. Spirals with moderate star-forming activity and moderate radio surface brightness such as M31 and M33, our Milky Way's neighbors, have $B \approx 6$ μG. In "grand-design" galaxies with massive star formation, such as M51, M83 and NGC6946, a typical average strength of the whole field is $B \approx 15$ μG (Beck and Wielebinski, 2013, and references therein).

2.4.1 Origin of the Magnetic Fields

How the first magnetic fields in the Universe originated is still not known (Widrow, 2002). It is assumed that protogalaxies were probably already magnetic due to field ejection from the first stars or jets produced by the first black holes. It should be recognized that a primordial field of a young galaxy is hard to maintain because a galaxy rotates in such a way that its angular velocity decreases with a radius so that the magnetic field lines get sharply wound up. The result of such motion is that field lines with opposite polarity may cancel via magnetic reconnection. This effect suggests the existence of a mechanism that sustains and organizes the magnetic field.

The most promising mechanism capable of this action is the dynamo, which transfers mechanical energy into magnetic energy (e.g., Beck and Hoernes, 1996; Brandenburg and Subramanian, 2005). According to these authors, a suitable configuration of the fluid or gas flow can generate a strong magnetic field (stationary or oscillating) from a weak seed field. Seed fields could have been produced in the early Universe at phase transitions, or in shocks in protogalactic halos (Biermann battery), or due to fluctuations in the protogalactic plasma.

To construct a model of efficient dynamo for astronomical objects such as stars, planets or galaxies, they must possess turbulent motions and non-uniform rotation; the model is called, in this case, an alpha-Omega dynamo. Such a dynamo would generate large-scale regular fields, even if the seed field were rough. The field structure obtained by such dynamo models is characterized by azimuthal symmetry and vertical symmetry perpendicular to the disk plane. Information about this type of mode can be obtained from the pattern of polarization angles and Faraday rotation in multi-wavelength radio observations. Various kinds of modes can be induced in the same object.

For spherical bodies such as stars or planets or galactic halos, the most reliable mode is a double torus on the equator with a reversal across the equatorial plane, surrounded by a dipolar field. In flat objects such as galactic disks, the most appropriate mode is a single torus in the plane axisymmetric spiral shape, without reversals. This mode is surrounded by a weaker quadrupolar field (even vertical symmetry). This mode is frequently observed. The next pattern of bisymmetric spiral shape with two field reversals in the disk, possibly excited by gravitational interaction, seems to be rare. The next higher quadri-symmetric mode, weaker than the axisymmetric mode, can be excited by spiral arms and was found in many spiral galaxies. A large-scale field reversal, as seen in the Milky Way, is either a distortion of dynamo action by gravitational interaction with a companion galaxy, or a relic from early times when the magnetic field was still chaotic (Beck, 2007).

The magnetic field forms delicate spiral patterns in almost every galaxy, even in flocculent and bright irregular types that lack any spiral optical structure (Beck and Wielebinski, 2013). This situation is regarded as a strong argument for the action of galactic dynamos. Helical fields are also observed in the central regions of galaxies. In galaxies having massive spiral arms, the magnetic field lines run parallel to the optical arms concentrated at the inner edge or between the spiral arms. In several galaxies, for example, NGC 6946, the field forms independent magnetic arms between the arms. In some galaxies, the field pattern seems to follow the gas flow. Since the gas rotates faster than the spiral, a shock wave could occur in the cold gas, which has a small sound speed; at the same time, the warm, diffuse gas is only slightly compressed. Since the observed compression of the field in spiral arms and bars is also tiny, the ordered field is coupled to the hot gas and is strong enough to affect the flow of the heated gas.

Squire and Bhattacharjee (2015) proposed a new mechanism for a turbulent field dynamo in which the magnetic fluctuations from a small-scale dynamo drive the generation of large-scale magnetic fields. It is in total contrast to the common idea that small magnetic fields should be harmful to massive dynamo action. These dynamos occurring in the presence of large-scale velocity shear do not require net helicity. Off-diagonal components of the turbulent resistivity tensor can produce it as the magnetic analog of the "shear-current" effect. Given the inevitable existence of nonhelical small-scale magnetic fields in turbulent plasmas, as well as the generic nature of velocity shear, the suggested mechanism may help explain the generation of large-scale magnetic fields across a wide range of objects.

The standard model of the origin of galactic magnetic fields assumes the amplification of seed fields via dynamo or turbulent processes to the level consistent with the present observations (Parker, 1955; Zweibel and Heiles, 1997; Ryu et al., 2008). Although other mechanisms may also operate (Schlickeiser and Shukla, 2003; Miniati and Bell, 2011), currents from misaligned pressure and temperature gradients (the so-called Biermann battery process) inevitably accompany the formation of galaxies in the absence of a primordial field. Driven by geometrical asymmetries in shocks (Miniati et al., 2000) associated with the collapse of protogalactic structures, the Biermann battery is believed to generate tiny seed fields to a level of about 10^{-21} G (Kulsrud et al., 1997; Xu et al., 2008). With the advent of high-power laser systems in the past two decades, a new area of research has opened, in which, using simple scaling relations (Ryutov et al., 1999; Ryutov et al., 2000), astrophysical environments can effectively be reproduced in the laboratory (Remington et al., 1999; Remington et al., 2006). Gregori et al. (2012) described the generation of scaled protogalactic seed magnetic fields in laser-produced shock waves by the Biermann battery effect. They show that these results can be scaled to the intergalactic medium, where turbulence, acting on timescales of around 700×10^6 years, can amplify the seed fields (Cho and Vishniac, 2000; Sur et al., 2010) sufficiently to affect galaxy evolution.

According to Naoz and Narayan (2013), a possible explanation for the observations of magnetic fields in the Universe is that a seed magnetic field was generated by some unknown mechanism early in the life of the Universe. Many dynamos later amplified this field in nonlinear objects such as galaxies and clusters. They propose that a primordial magnetic field is generated in the early Universe on purely linear scales. Vorticity was induced by scale-dependent temperature fluctuations or by the varying speed of sound in the gas. Extra free electrons left over after recombination are responsible for the generation of a magnetic field via the so-called Biermann battery process. The battery is functional even in the absence of any relative velocity between matter and gas at the time of recombination. The presence of a relative velocity modifies the predicted magnetic field. At redshifts of the order of a few tens, they estimate a root-mean-square field strength of order 10^{-25} to 10^{-24} G on comoving scales –10 kpc. (Comoving coordinate is a coordinate system in which an observer is comoving with the Hubble flow. Only for these observers in the comoving coordinates, the Universe is isotropic.) This field, which is generated from linear perturbations, is amplified significantly after reionization and further boosted by dynamo processes during nonlinear structure formation.

The process of the formation of molecular clouds, which serve as stellar nurseries in galaxies, is not well understood. Some cloud formation models suggest that a large-scale galactic magnetic field has no consequence on the individual clouds because its turbulence and rotation may change the orientation of its individual magnetic field (Dobbs, 2008; Hartmann et al., 2001). On the other hand, galactic fields could be strong enough to impose their direction upon individual clouds (Passot et al., 1995; Shetty and Ostriker, 2006), thereby regulating cloud accumulation and fragmentation (Li et al., 2011) and affecting the rate and efficiency of star formation (Price and Bate, 2008). In their paper, Li and Henning (2011) reported observations of the magnetic field orientation of six giant molecular cloud complexes in the near, almost face-on, galaxy M33.

The hadronic constituent of galaxies consists of gas and stars, with only a minor contribution of dust. These components rotate in a gravitational potential made up of dark matter. In spiral galaxies, the gas has a neutral and an ionized part. The neutral part is made up of (in order of increasing density) the molecular, the cold and the warm phase. The ionized component consists of the warm and the hot phase. In elliptical galaxies, the bulk of the gas is ionized. An exception to this is neutral gas that has been captured from the surroundings or neighboring galaxies. The ionized gas in galaxies immediately implies a high conductivity, which in the absence of turbulence would result in the long-lived sustainment of magnetic fields.

Spiral galaxies possess stellar and gaseous disks in which density waves can be excited, depending on the depth of their gravitational potentials. Dwarf irregulars have shallow gravitational potentials, and their kinematics is, therefore, possibly influenced by local disturbances caused by star formation and subsequent supernovae. Large elliptical galaxies have their baryonic mass dominated by the stellar constituent, the hot gas forming a hot corona.

2.4.2 Milky Way Galaxy and Its Magnetic Field

Today, it is known that the Milky Way galaxy contains an ordered, large-scale magnetic field of the value of 10^{-10} Tesla (0.1 nT). This field configuration has been explored even in the past through various methods of analysis of starlight polarization, modelling pulsar and Faraday rotation, and Zeeman splitting of hydroxyl masers into areas of star formations. Zeeman splitting of atomic or molecular lines indicates the direction and magnitude of the magnetic field component in a cloud of gas.

Any spiral galaxy similar to the Milky Way has three essential parts of its visible matter: (i) the disk (containing the spiral arms), (ii) the halo and (iii) the nucleus or central bulge. Each part contains different components of the interstellar medium, which include hydrogen and helium gases and dust grains. Because of the varying density of the galaxy's components, the magnetic field ranges from 0.5 to 2 nT, which could be represented in a schematic diagram showing the field's direction along the spiral arms.

The magnetic field of a galaxy is not strong in comparison with that of the Earth, which is approximately 30 μT in average. Even so weak, it can act on the dust particles and cause rotation of their axes to line them so that their short axes are parallel to the direction of the field, which is aligned with the Milky Way band and the galactic plane. Polarization measurements confirm this pattern.

The interstellar magnetic field intensity of the Milky Way has been determined using several methods that allowed obtaining valuable information about the amplitude and spatial structure of it. Grasso and Rubinstein (2001) reported the average field strength to be 3–4 μG. Such an intensity corresponds to the energy equipartition between the magnetic field, the cosmic rays confined to the galaxy and the small-scale turbulent motion (Kronberg, 1994).

$$\rho_m = \frac{B^2}{8\pi} \approx \rho_t \approx \rho_{CR}$$

The magnetic energy density is almost equal to the energy density of the cosmic microwave background radiation (CMBR). The magnetic field keeps its orientation on scales of the order of few kpc, comparable with the galactic size. Two field reversals have been observed between the galactic arms, indicating that the galaxy field morphology may be symmetrical. Similar magnetic fields have been observed in several other spiral galaxies. Equipartition fields were observed in some galaxies such as M33. In some other galaxies, such as the Magellanic Clouds and M82, the magnetic field is stronger than the equipartition threshold. The observational situation of the spatial structure of the galactic fields is quite confused since some galaxies have an axially symmetrical geometry, some other asymmetrical ones, and others no recognizable field structure (Zweibel and Heiles, 1997).

Voyager 1, during its escape from the solar system and entrance to the interstellar medium, directly measured the galactic magnetic field to be about 4 μG (Burlaga et al., 2013). The galactic magnetic field has a high level of inhomogeneity. The turbulence motion of the plasma in the ISM leads to the turbulence motion of the magnetic field. The field lines are frozen in the ISM plasma and move together with the plasma. Recent observations (Armstrong et al., 1995) on the local ISM show the turbulence wave number spectrum can be described by the Kolmogorov model, $E(k) \sim k^{-5/3}$ (Kolmogorov, 1991).

Recently, an international team of astrophysicists has released an original map of the entire sky, showing the magnetic field shaping of the Milky Way galaxy. This effort could help our understanding of the birth of the Universe. The team – which includes researchers from the University of British Columbia and the Canadian Institute for Theoretical Astrophysics (CITA) at the University of Toronto – created the map using the data from the Planck space telescope. The Planck telescope has made a map of the cosmic microwave background (CMB) radiation, the record of the Universe at only 380,000 years after the Big Bang. This adventure started in 2009. With its high-frequency instrument, Planck space telescope detects the light scattered by microscopic dust particles within our galaxy. It identifies the non-random direction in which the light waves vibrate – i.e., polarization. This polarized light indicates the orientation of the magnetic field lines.

Our galaxy has a large-scale magnetic field, which is 100,000 times weaker than the magnetic field at the Earth's surface. All of this, together with other results, is described in more detail in Planck Collaboration Reports (2015a and 2015b).

Magnetic field amplitude near the galactic center has an uncertainty of two orders of magnitude. On the scale of approximately 100 pc, fields of roughly 1,000 μG have been reported. These findings imply a magnetic energy density more than 10,000 times stronger than it is typical for the galaxy. The assumed pressure equilibrium between the various phases of the galactic center: the interstellar medium with its turbulent molecular gas and the contested "very hot" plasma, on the one hand, and the magnetic field, on the other hand, suggests fields of approximately 100 μG over around 400 pc size scales. Finally, assuming equipartition, fields of only about 6 μG have been inferred from radio observations for 400 pc scales. Crocker et al. (2010) reported a compilation of data that reveal a breakdown in the region's nonthermal radio spectrum, which is attributable to a transition from bremsstrahlung to synchrotron cooling of the in situ cosmic-ray electron population. They showed that the spectral break requires that the galactic center field be at least 50 μG on 400 pc scales to prevent the synchrotron-emitting electrons from producing too much gamma-ray emission under the existing constraints. According to other considerations, a magnetic field of 100 μG is proposed. The meaning of this value is that $>10\%$ of the galaxy's magnetic energy is contained in only $<0.05\%$ of its volume.

A map of the entire sky offers a remarkably detailed picture (see Figures 2.14 and 2.15) of the magnetic fields that shape the Milky Way. The map includes field lines that run parallel to the plane of the galaxy, including high loops associated with nearby clouds of gas and dust.

Planck telescope detects light from microscopic dust particles within our galaxy. The density of the dust is very low. A volume of space 100 m× 50 m×50 m (equal to a large sports stadium or arena) would contain one grain. The magnetic field is responsible for the coupling of the motions of gas and dust in space between stars and so plays a role in star formation and the dynamics of cosmic rays.

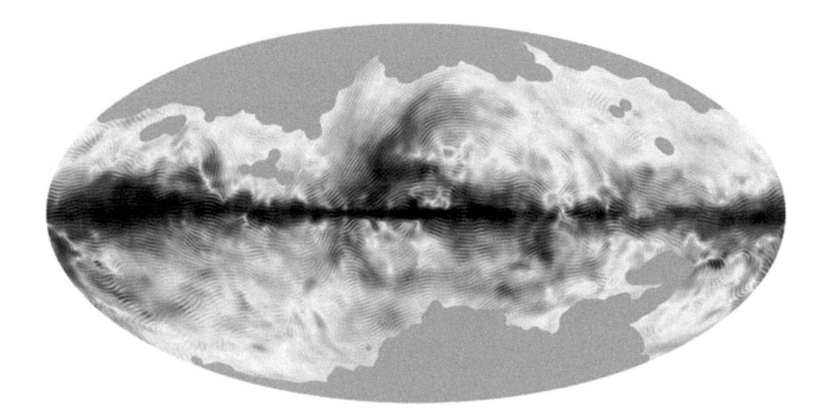

FIGURE 2.14 The magnetic field of our Milky Way galaxy as seen by ESA's Planck satellite on May 2014. (Photo credit: ESA and the Planck Collaboration.)

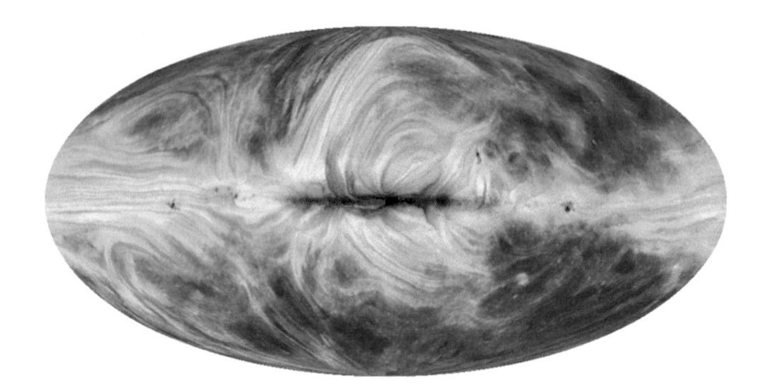

FIGURE 2.15 Magnetic field lines traced by synchrotron radiation at 30 GHz in February 2015. (Acknowledgment: ESA and the Planck Collaboration.)

Spectroscopic determination of masses (and implied ages) for red giants has recently been reported by Ness et al. (2016). The most fundamental parameter of a star is its mass. For red giant stars, it also implies a stellar evolution age. Stellar masses and ages have never been derived directly from the spectra of red giants. Using the APOGEE Kepler sample of stars (the APOKASC sample), with high-quality spectra and asteroseismic masses, they were able to build a data-driven spectral model using The Cannon (Ness et al., 2015) to infer the stellar mass and age from stellar spectra. They determined stellar masses to 0.07 dex from APOGEE DR12 spectra of red giants; these imply age estimates accurate to 0.2 dex (40%). The Cannon constrains the ages foremost from spectral regions with particular absorption lines, elements whose surface abundances reflect mass-dependent dredge-up. The authors delivered an unprecedented catalogue of 85,000 giants (including 20,000 red-clump stars) with mass and age estimates, spanning the entire disk (from the galactic center to $R \approx 20$ kpc). Such stellar age constraints across the Milky Way open up new avenues in galactic archaeology, including a possible assumption that older Milky Way stars lie near the center of the galaxy, while subsequent generations formed further out.

Data to be released by scientists from the Planck collaboration should allow astronomers to separate with high confidence any possible foreground signal from our galaxy from the tenuous, primordial, polarized signal from the CMB to study the origin and evolution of the Universe. In March of 2014, scientists from the BICEP2 collaboration reported the detection of such a signal. The data from Planck will enable a much more detailed investigation of the early history of the Universe. These data will cover the period from the accelerated expansion when the Universe was much less than 1 second old to the time when the first stars were born, several hundred million years later.

Planck (launched in 2009) is an ESA's mission to observe the first light at the beginning of the Universe. Planck was selected in the year 1995 as the third Medium-Sized Mission of ESA's Horizon 2000 Scientific Programme. Later, it became part of its Cosmic Vision Programme. It was designed to measure the temperature and polarization anisotropies of the CMB radiation field over the whole sky. This measurement was supposed to be done with unprecedented sensitivity and angular resolution. Planck is continuously

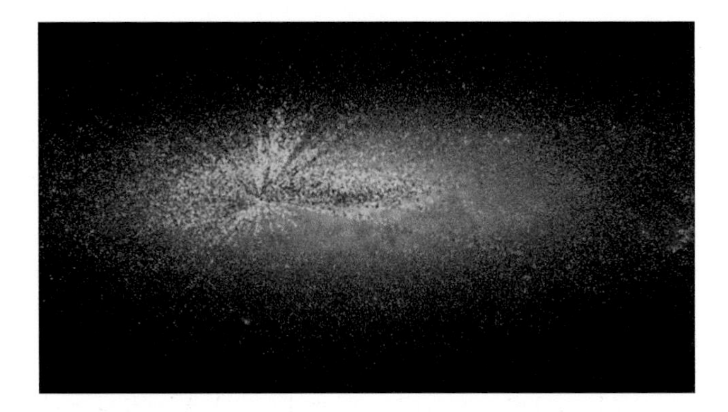

FIGURE 2.16 Artist's conception of the Milky Way. Colored dots show stars. The red dots show stars that formed early when the Milky Way was young, while the blue dots show stars that formed later when the Milky Way was mature. The color scale indicates how many billion years have passed since those stars formed. (Image: G. Stinson (Max Planck Institute for Astronomy, MPIA). Sloan Digital Sky Survey (SDSS), January 8, 2016.)

testing theories of the early Universe and the origin of cosmic structure. It is also a significant source of data relevant to many cosmological and astrophysical problems. The Planck Science Team directs its scientific mission development.

Figure 2.16 shows an artist's conception of the Milky Way, with colored dots indicating the galaxy's age. Red dots show stars that formed when the Milky Way was young and small. Blue dots show stars that formed more recently when the Milky Way was big and mature (Ness et al., 2016). This Milky Way age-map uses the data of more than 70,000 stars. It extends halfway across our galaxy to a distance of 50,000 ly away. It seems that our galaxy grew up by growing out.

Knowing the mass of a red giant star, we determine its age. This calculation can be done by using the fusion clock inside every star. In other words, there's a known relationship between age and mass in these stars. Finding the masses of giant stars have historically been very difficult, however, and that's why the astronomers combined ground-based and space-based observations to pin down the enormous red masses and, subsequently, their ages.

A class of slowly rotating neutron stars is called magnetars. They are highly magnetized young neutron stars that occasionally produce enormous bursts and flares of X- and γ-rays. They spin with periods between ~2 and 12 seconds that are thought to be powered by their decaying ultra-strong magnetic fields. More than 2,600 pulsars have been found so far, but only 31 magnetars or magnetar candidates are currently known (see the McGill Magnetar Catalog). Most of them are galactic magnetars, many of which are located in the inner region of the Milky Way. Typical surface dipolar magnetic fields of magnetars are in the range of $\sim 10^{14}$–10^{15} G, which is higher than the $\sim 10^{12}$ G fields of rotation-powered pulsars. They are characterized by transient X- and gamma-ray outbursts, which are hallmark features of their emission and have resulted in the discovery of the majority of magnetars. The PSR J1745-2900 is one of only four magnetars with detectable radio pulsations. PSR J1745-2900 is unique among the population of magnetars because of its proximity to the 4×10^6 M_* black hole, Sagittarius A* (Sgr A*), at the galactic center. This discovery of a rare magnetar near the galactic center may suggest that the environment around Sgr A* is more conducive to magnetar formation. These observations of PSR J1745-2900 also provide a valuable probe into the interstellar medium near the galactic center, which may shed light on why previous searches for radio pulsars within ~10 arcmin of Sgr A* have been unsuccessful. It is an accepted belief that these searches may have been hindered by scattering-induced pulse broadening of the pulsed radio emission as a result of large electron densities along the line of sight (Pearlman et al., 2018).

Fast radio bursts (FRBs) are repeating millisecond-duration bursts of radio waves arriving from cosmological distances (Petroff et al., 2019). A leading model for repeating FRBs is that they are extragalactic magnetars powered by their intense magnetic fields. This would imply that these FRBs must have radio luminosities many orders of magnitude larger than those seen from known galactic magnetars. The CHIME/FRB Collaboration (2020) reported the detection of an extremely intense radio burst from the galactic magnetar SGR 1935+2154 using the FRB project. The fluence of this two-component bright radio burst and the estimated distance to SGR 1935+2154 together imply a burst at 400–800 MHz of approximately 3×10^{34} erg, which is three orders of magnitude higher than the burst energy of any radio-emitting magnetar detected so far. Such a burst coming from a nearby galaxy would be indistinguishable from a typical FRB.

2.4.3 Magnetic Fields in Galaxy Clusters

Observations of several Abell clusters (Kim et al., 1991), some of which have a measurable X-ray emission, resulted in valuable information on fields in clusters of galaxies. The phenomenological equation well describes the magnetic field strength in the intercluster medium (ICM):

$$B_{ICM} \approx 2\mu G(L/10\,\text{kpc})^{-1/2}(h_{50})^{-1}$$

where L is the length of the reversal field and h_{50} is the reduced Hubble constant. The typical values for L are 10–100 kpc, which corresponds to field amplitudes of 1–10 μG. The concrete case of the Coma Cluster (Feretti et al., 1995) can be fitted with a core magnetic field $B \approx 8.3h_{100}^{1/2}$ G tangled at scales of about 1 kpc. An example of clusters with a strong magnetic field is the Hydra A cluster, for which the rotation measurements (RMs) imply a 6 μG field coherent over 100 kpc superimposed with a tangled field of strength ~30 μG (Taylor and Perley, 1993). A big set of high-resolution images of radio sources embedded in galaxy clusters shows evidence of strong magnetic fields in the central cluster regions (Eilek, 1999). The typical central field strength is ~10–30 μG, with peak values being as large as ~70 μG. It is known that for such large fields, the magnetic pressure exceeds the gas pressure derived from X-ray data. This fact suggests that magnetic fields may play a significant role in cluster dynamics.

Loeb and Mao (1994) showed the existence of a discrepancy between the mass estimate for the Abell cluster 2218 derived by gravitational lensing and that estimated from X-ray observations. This discrepancy can be explained by the pressure support produced by a magnetic field having a strength of ~50 μG. It is still not evident that the apparent decrease in the magnetic field strength in the outer region of clusters is due to the intrinsic field structure, or it is a spurious effect due to the reduction in the gas density. Observations also show evidence for a filamentary spatial arrangement of the field. It seems entirely plausible that all galaxy clusters are magnetized. These observations are a severe challenge to most of the models proposed to explain the origin of galactic and cluster magnetic fields (Grasso and Rubinstein, 2001).

Gravitational lensing experiments are powerful tools with varied applications in cosmology and galaxy evolution. They are only sensitive to total mass distribution independent of the dynamical state or light emission. The so-called weak lensing studies have proved to be very useful because of the wealth of information provided by this technique and the large number of systems where it can be applied (Gurri et al., 2020). These authors developed a new experimental design for precision measurements of the effect of weak lensing. Because of the consideration of velocity information for the background source, their approach is complementary to conventional approaches based only on shape information. Therefore, it is well applicable to galaxy–galaxy weak lensing studies at low redshift.

Let us turn to the magnetic field strength of high-redshift objects. The high-resolution RMs of very distant quasars have allowed probing magnetic fields in the distant past. Kronberg and his team make the most significant measurements (see Kronberg, 1994 and references therein). The RMs of the radio emission of the quasar 3C 191, at $z=1.945$, presumably due to a magnetized gas at the same redshift, are consistent with a field strength range (0.4–4) μG. The field maintains its prevailing direction over at least ~15 kpc, which is comparable with the typical galaxy size. The magnetic field of a young spiral galaxy at $z=0.395$ was determined by performing RMs of the radio emission of the quasar PKS 1229-021 lying behind the galaxy at $z=1.038$. The magnetic field amplitude was estimated to be in the range of (1–4) μG. Even more impressive was the finding of field reversals with distance almost equal to the separation of spiral arms, in a way quite similar to that observed in the Milky Way (Grasso and Rubinstein, 2001).

The distant quasars' radio emission is also a constraint on the intensity of magnetic fields in the intergalactic magnetic fields (IGM), which we may suppose to permeate the entire Universe. Translation of RMs into an estimation of the field strength is quite tricky for media for which the density of ionized gas and field coherence length are not well known. Nevertheless, some new limits can be derived based on well-known estimates for the Universe ionization fraction and adopting some reasonable values for the magnetic coherence length. Assuming an aligned magnetic field, Kronberg (1994) estimated that $B_{IGM}<10^{-11}$ G. However, a field aligned on cosmological scales is unlikely. As seen previously, the most significant reversal scale in galaxy clusters is at most 1 Mpc. By adopting this scale as the typical cosmic magnetic field coherence length and applying the $RM(z_s)$ up to $z_s \sim 2.5$, Kronberg (1994) found the less strict limit $B_{IGM} \approx 10^{-9}$ G for the present magnetic field strength.

In their work, Grasso and Rubinstein (2001) discussed in great many detail all aspects of magnetic fields in the early Universe. They start by providing the reader with a short overview of the current state-of-the-art observations of cosmic magnetic fields. They present arguments in favor of the primordial origin of magnetic fields in the galaxies and the clusters of galaxies. The only promising way of testing this idea is to look for possible signatures of magnetic fields in the temperature and polarization anisotropies of the CMB radiation. With this purpose in mind, they provide a review of the most relevant effects of magnetic fields on CMB radiation.

A period of the Universe history when primordial magnetic fields may have produced observable consequences is the Big Bang nucleosynthesis. Grasso and Rubinstein (2001) discussed three main effects: (i) the impact of the magnetic field energy density on the Universe expansion; (ii) the modification produced by a strong magnetic field of the electron–positron gas thermodynamics; and (iii) the change of the weak processes keeping neutrons and protons in chemical equilibrium. All these effects provide a variation in the final neutron-to-proton ratio and hence in the relative abundances of light relic elements. The impact of the field on the Universe expansion rate was proven to be dominant, although the others cannot be neglected. Also, the non-gravitational effects of the magnetic field can exceed those on the expansion rate in the regions where the magnetic field intensity may be larger than the mean value for the Universe. This situation could have produced fluctuations in baryon-to-photon ratio as well as fluctuations in the relic neutrino temperature. The BBN upper limit on primordial magnetic fields of $B_0 \leq 7 \times 10^{-7}$ G looks less reliable compared to other restrictions which come from the Faraday rotation measurements (RMs) of distant quasars or the global isotropy of the CMB radiation. However, the authors have shown that this conclusion is not correct if magnetic fields are complicated. The reason is that BBN probes lengths that are of the Hubble horizon size at BBN time (corresponding to approximately 100 pc today), while CMB radiation and the RMs probe much larger scales. Also, constraints derived from the analysis of effects taking place at different times may not be directly comparable if the magnetic field evolution is not adiabatic.

2.4.4 Magnetic Fields in Galactic Clouds

Magnetic fields of molecular clouds have been discussed in detail in the textbook by Stahler and Palla (2004). Observations of magnetic fields in molecular clouds are essential for understanding their role in the evolution of dense clouds and in the star formation process. For stars to form, molecular clouds have to contract, which they can do under their gravity if there is sufficient critical mass within a specific volume. Once they collapse, they must get rid of significant angular momentum to contract to a condensed prestellar object. It is the magnetic fields that support a cloud against gravitational collapse by a process called "flux freezing". Typical field strengths derived from Zeeman measurements support this picture: Denser clouds or their cores possess stronger magnetic fields, reaching values of mG. Flux freezing does not imply that the magnetic field is rigidly tied to the molecular clouds. Otherwise, they would not be able to contract at all. Through a process called "ambipolar diffusion", the magnetic field manages to slip through the gas.

An extensive discussion of the observations of magnetic fields in molecular clouds is presented by Crutcher (2003). He described the techniques for measuring magnetic fields in molecular clouds by presenting three examples of observational results. He concluded that both turbulence and strong magnetic fields are important in the physics of molecular clouds and in the star formation process.

We should note the importance of the random component of the interstellar magnetic field. If the magnetic field were entirely uniform, then the gas would slip along field lines, leading to a flat (pancake-like) morphology upon contraction, which has never been observed (see Table 2.2 adapted from Klein, 2013).

Theory and observations have recently converged on an acceptable description of interstellar grain alignment based on radiative processes. The radiative alignment torque (RAT) theory allows specific, testable predictions for realistic interstellar conditions (Andersson et al., 2015). The continued action of the radiative torques over the precession period aligns the angular momentum of the grain with the magnetic field, yielding the observed alignment of the grain and the magnetic field. However, this fact alone does not determine whether the angular momentum shall align parallel or perpendicular to the magnetic field.

Magnetic fields can influence the dynamics of the interstellar medium from hot diffuse gas to the star-forming cores. Interstellar medium polarization is caused by aligned dust grains. The mechanism responsible for the alignment of the dust grains requires an understanding of the fraction of the total grain population that contributes to the polarization in different environments. The total grain population is made up of large (~0.01–1 μm) and very small (~0.001–0.01 μm) grains and polycyclic aromatic hydrocarbon (PAH) molecules. For polarization measurements, the large grains are most relevant because grains smaller than ~0.05 μm are typically not aligned in the interstellar medium (Andersson et al., 2015).

The dust is usually assumed to consist of silicates, amorphous carbon and small graphite particles of simple shapes such as spheres, spheroids or cylinders. In the dark and cold parts of molecular clouds, grain aggregates as well as mantles of volatile ice materials may form.

TABLE 2.2 Magnetic Field Strengths Inferred from Zeeman Absorption Measurements in Galactic Clouds

Object	Type	$B_{\parallel}$ (μG)
Ursa Major	Diffuse cloud	+10
L204	Dark cloud	+4
NGC 2024	GMC clump	+87
Bernard 1	Dense core	−27
S106	HII region	+200
Sgr A/West	Molecular disk	−3,000
W75N	Maser	+3,000
DR21	HII region	−360

Source: Adapted from Klein (2013).

We should mention here a related subject of observed dust particle alignment in the solar magnetic field (Kolokolova et al., 2015). They observed circular polarization (CP) produced by scattering of sunlight on cometary dust in 11 comets and observed values from 0.01% to 0.8%. The most plausible reason why the light scattered by cometary dust becomes circularly polarized is the alignment of the dust particles by the solar magnetic field.

2.4.5 Enhanced Magnetic Activity of Superflare Stars

A buildup of charged particles inside the star's plasma is a trigger for solar flares seen on stars. That energy is sometimes released in vast explosions of radiation, spanning the electromagnetic spectrum. Some stars are known as flare stars, as they have predictable activity levels. The most massive known solar flare was the Carrington event in AD 1859 (Carrington, 1859; Hodgson, 1859). This flare and associated coronal mass ejection were so intensive that they caused worldwide auroras. They allowed telegraphs to operate on the currents induced by the accompanying geomagnetic storm (Muller, 2014).

As there were no X-ray measurements of the Carrington event, it is not clear how large it was compared with the most massive flares observed during the space age, which are classified according to their peak X-ray flux. Estimates based on magnetometer data predict that the Carrington event, with a total energy of up to 5×10^{32} erg, was likely larger than any solar flare observed in the space age (Cliver and Dietrich, 2013). On the other hand, the hazard caused by the Carrington event was minimal compared with that potentially posed by, for example, a 2×10^{34} erg superflare (Karoff et al., 2016).

A flare star is a variable star that undergoes unpredictable dramatic increases in brightness for a few minutes. It is believed that the flares on flare stars are analogous to solar flares in that they are caused by the magnetic energy stored in the stars' atmospheres. The brightness increase is across the full spectrum, from X-rays to radio waves. Some time ago, Schaefer et al. (2000) identified nine so-called superflares, defined as flares with energies ranging from 10^{33} to 10^{38} erg. These stars are called superflare stars. Superflares are large explosive events on the surfaces of stars, one to six orders of magnitude larger than the largest flares observed on the Sun. Due to the large amount of energy released during these events, it has been speculated whether the underlying mechanism is equivalent to solar flares, which are caused by magnetic reconnection in the solar corona.

According to Karoff et al. (2016), superflares on solar-like stars may arise from at least three different mechanisms, apart from coronal magnetic reconnection:

i. Star–star interactions, where a close binary companion tidally spins up an F or G main-sequence star.
ii. Star–disk interactions, with the dipole magnetic field of the central star connected to the co-rotating disk.
iii. Star–planet interactions, taking place in two different ways, either similar to the star–star mechanism through disruption of interconnecting field lines or tidal interaction between the star and the planet.

Karoff et al. (2016) analyzed the observations of 5,648 solar-like stars, including 48 superflare stars, made with the LAMOST telescope. (LAMOST is a short name for Large Sky Area Multi-Object Fiber Spectroscopic Telescope.) The telescope, also known as the Guo Shoujing Telescope after a 13–14th-century Chinese scientist, is located at Xinglong Observing Station of the National Astronomical Observatories, about 100 miles northeast of Beijing, China. Their observations show that superflare stars are characterized by more massive emissions compared to other stars, including the Sun. However, superflare stars with activity lower than, or comparable to, the Sun do exist. This fact suggests that solar flares and superflares probably share the same origin. The substantial ensemble of solar-like stars included in their study enables detailed and robust estimates of the relationship between activity and the occurrence of superflares.

These observations also result in a question: Can our Sun create a superflare? Superflare stars with activity lower than, or comparable to, the Sun do exist. The results indicate that superflares on the solar-like stars with Sun-like rotation are an order of magnitude less probable than superflares on solar-like stars in general. Of interest is the frequency of past solar flares recorded in geological archives. Geological archives represented by cosmogenic nuclides ^{10}Be and ^{14}C in ice cores and tree rings can be used for the evaluation of the flare frequency through the so-called solar particle events. During massive solar flares, protons are accelerated to energies sufficiently high to produce cosmogenic nuclides when they reach the Earth's atmosphere (see, for example, Schrijver et al., 2012 for a recent review).

Several geological archives indicate a break, or roll-over, in the distribution of solar particle events at energies around 10^{33} erg (Kovaltsov and Usoskin, 2014; Usoskin and Kovaltsov, 2012). The discussion of this effect has received renewed attention with discoveries based on the studies of ^{14}C in Japanese tree rings, resulting in evidence that the Sun had a superflare with an energy larger than 10^{33} erg in AD 775 (Miyake et al., 2012; Melott and Thomas, 2012; Usoskin et al., 2013; Cliver et al., 2014; Jull et al., 2014; Miyake et al., 2015). The indications of the roll-over effect in the flare frequency weighted for magnetic activity is also in agreement with recent estimates, which predicts that sunspot groups larger than those historically reported would be needed for the Sun to produce a superflare with an energy larger than $\sim 6 \times 10^{33}$ erg (Aulanier et al., 2013).

2.4.6 Magnetic Field of Our Sun

The knowledge of the origin of the magnetic system of the Sun is critical for understanding the nature of the solar system. The Sun's magnetic field governs the solar events that cause space weather on the Earth, such as auroras. It generates the interplanetary magnetic field and radiation through which our spacecraft travelling through the solar system must go. Presently, it is not known exactly where in the Sun the magnetic field is created. It could be close to the surface or deep inside the Sun – or over a wide range of depths. However, for sure, the answers lie in the fact that the Sun is a giant magnetic star made of a material that moves in agreement with the laws of electromagnetism.

A complete understanding of the solar magnetic field, including knowing exactly its origin and its structure deep inside the Sun, is not yet mapped out. However, it is well established that the solar magnetic system is subject to an approximately 11-year activity cycle. With every eruption, the magnetic field of the Sun smooths out slightly until it reaches its purest state. At that point, the Sun experiences solar minimum, when solar explosions are least frequent. From that point, the magnetic field of the Sun grows more complicated over time until it peaks at solar maximum, which is some 11 years after the previous solar maximum. At solar maximum, its magnetic field has a very complicated shape with lots of small structures throughout. These are the observed active regions. At a solar minimum, the magnetic field is weaker and concentrated at the poles. It has a very smooth structure that does not form sunspots.

The magnetic field of our Sun and stars can be probed in a rather precise and direct manner. Namely, in the presence of a magnetic field, the energy levels of atoms, ions and molecules are split into more levels. It causes spectral transition lines also to be split into more than one line, with the amount of splitting proportional to the magnetic field intensity. This effect is the so-called Zeeman effect, and the corresponding increase in the number of spectral lines is called Zeeman splitting. Thus, one can infer the presence of magnetic fields from the observation of Zeeman splitting in the spectrum and measure the strength of the field by measuring the amount of Zeeman splitting quantitatively.

The changes in the Sun's magnetic field are illustrated in Figure 2.17 (from http://www.nasa.gov/feature/goddard/2016/understanding-the-magnetic-sun). The figure shows a side-by-side comparison of how the magnetic fields changed, grew and subsided from January 2011 to July 2014. One can see that the magnetic field is much more concentrated near the poles in 2011, 3 years after solar minimum. In July 2014, the solar maximum structure was much more complex, with closed and open field lines knocking out all over, which is ideal condition for solar explosions. More about this event can be found on http://www.nasa.gov/mission_pages/sunearth/index.html.

Cameron and Schüssler (2015) argued that sunspots and the number of other phenomena occurring in the 11-year solar activity cycle are a consequence of the emergence of magnetic flux at the solar surface. The observed orientations of bipolar sunspot groups suggest that they originate from the toroidal (azimuthally orientated) magnetic flux in the envelope of the Sun. They show that the net toroidal magnetic flux generated by differential rotation within a hemisphere of the convection zone is given by the emerged magnetic flux at the solar surface and thus can be calculated using the observed

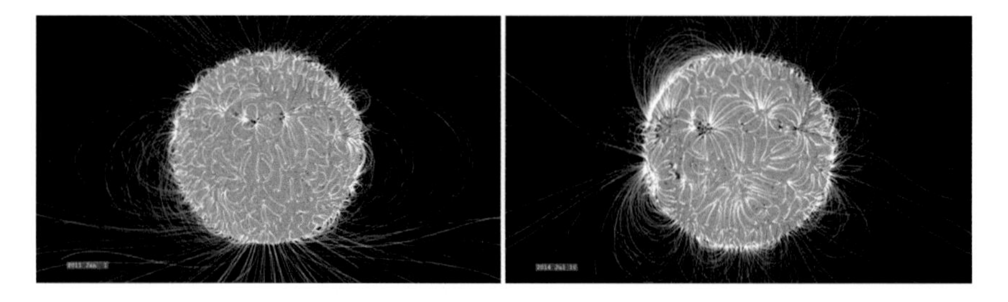

FIGURE 2.17 The comparison of the relative complexity of the solar magnetic field between January 2011 (left) and July 2014 (right). (Credit: NASA's Goddard Space Flight Center/Bridgman.)

magnetic field distribution. The primary source of the toroidal flux is the roughly dipolar surface magnetic field at the polar caps peaking around the minima of the activity cycle.

Measurement of the light output from sunspots, obtained by masking off the light from parts of the Sun not in the sunspot, shows significant Zeeman splitting of the spectral lines. Therefore, sunspots are connected with strong magnetic fields. Furthermore, when sunspots come in pairs, one sunspot tends to have a magnetic field polarity that is opposite to that of the other sunspot. That is, one of them behaves magnetically like the north pole of a bar magnet, and the other one behaves magnetically like the south pole of a bar magnet. During a given sunspot cycle, the leading groups of sunspots in the northern hemisphere of the Sun all have the same polarity. In contrast, the same is true of sunspots in the southern hemisphere, except that the standard polarity is reversed. During the next sunspot cycle, the previously noted regularities change themselves: The polarity of the leading spots in both hemispheres is the opposite of what it was in the last cycle.

The preceding considerations show that the solar magnetic field has a 22-year cycle, exactly twice that of the sunspot cycle, since the polarity of the field returns to its original value every two sunspot cycles. Thus, the initial period governing solar activity is the 22-year magnetic cycle, and the sunspot cycle (which is precisely half that) is just a particular manifestation of the magnetic cycle. The magnetic field has an essential role in most aspects of the active Sun (sunspots, prominences, flares, the solar wind, the nature of the corona), so the 22-year magnetic cycle is defining the periodicity of the Sun's activity.

The basic concept of the large-scale solar dynamo involves a cycle during which the poloidal and toroidal fields are mutually generated by one another (Charbonneau, 2010; Spruit, 2011). The winding of the poloidal field can produce a toroidal field by differential rotation. A reversed poloidal field is a result of the formation of magnetic loops in the toroidal field, which become twisted by the Coriolis force produced by solar rotation. In turn, the reversed poloidal field becomes the source of a reversed toroidal field. In this way, the 11-year cycle of solar activity is the result of 22 years of magnetic polarity.

Cameron and Schüssler (2015) calculated the magnetic flux at the solar surface of the northern hemisphere (shown in Figure 2.18), as well as its counterpart for the southern hemisphere, to evaluate the net toroidal flux in both hemispheres of the Sun as a function of time. The calculation begins with zero flux in February 1975, the start time of the synoptic observations. This effect is near solar activity minimum, during which time one expects the toroidal flux to change sign.

The result in Figure 2.18 shows that the modulus of the toroidal flux generated from the polar fields reaches peak values on the order of 1×10^{23} to 6×10^{23} Mx (1 Wb$=10^8$Mx) per hemisphere during recent activity cycles. Note that the flux generated for the new period first has to cancel the opposite-polarity flux from the previous cycle so that it reaches its peak value around the activity maximum of the new cycle.

2.5 Cosmic Rays

The initial discovery of cosmic rays dates back to more than a century ago (1912). However, their origin remains one of the most enduring mysteries in physics; see more at http://www.astrobio.net/topic/deep-space/cosmic-evolution/origin-of-cosmic-rays-

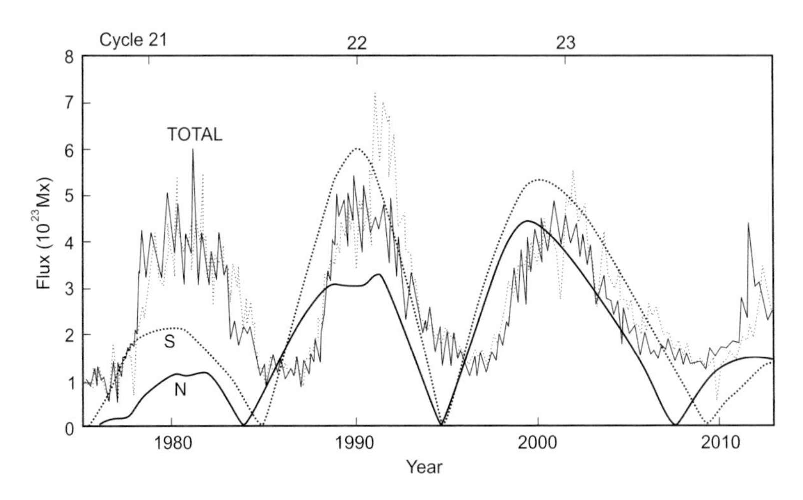

FIGURE 2.18 Observed magnetic flux at the solar surface and calculated net toroidal flux; Mx stands for maxwell. Hemispheric total unsigned surface flux (dashed lines) and the net toroidal flux modulus of the three solar cycles are shown. The solid line refers to the northern hemisphere, and the dotted line to the southern hemisphere. (After Cameron and Schüssler, 2015.)

a-mystery/#sthash.A4ZSPjSM.dpuf. Their identification as particles rather than radiation dates back to about 20 years later. After 20 more years, also the first suggestion that they were associated with supernova remnants was in place. The primary mechanism behind their acceleration was suggested some 40 years ago. Much work was done since then to prove that both the acceleration mechanism and site are well understood. Still, no definite proof has been obtained: Despite the impressive progress of both theory and observations, the evidence in support of the commonly accepted interpretation is only circumstantial (Amato, 2014). According to Fermi (1954), cosmic rays (regardless of the details of the acceleration mechanism) and magnetic fields in the galaxy are critical factors in the equilibrium of interstellar gas.

Cosmic rays (CR) also play an important role in plasma astrophysics. The enrichment with heavy elements and the magnetization of the intracluster and the intergalactic media have been done very early by the galactic winds from starburst dwarf galaxies. The origin of cosmic rays is, to some extent, still a matter of speculation. While supernova (SN) shock fronts may have well-produced particles with energies up to 10^{14} eV, those with higher energies ($>10^{14}$ eV) still pose a problem in this respect. AGN are candidates for being factories, while the highest energies are likely to be provided by γ-ray bursts (GRBs). These so-called ultra-high-energy cosmic rays (UHECRs) have energies of up to $E \approx 10^{21}$ eV, and their sites of origin must be searched for within the so-called GKZ horizon (Greisen, 1966; Zatsepin and Kuz'min, 1966).

Cosmic-ray electrons are essential because they reveal much of the relativistic plasmas by synchrotron radiation. We must keep in mind also that each relativistic electron is accompanied by a proton, with its ~2,000 times larger rest mass.

Let us briefly discuss the process of particle acceleration by multiple shocks in the ISM. This mechanism was first treated by E. Fermi (1949) and is referred to as the second-order Fermi acceleration. In Fermi's original picture, charged particles are reflected from "magnetic mirrors" associated with irregularities in the galactic magnetic field. (First-order Fermi acceleration happens when only head-on collisions occur.) The particles may get trapped in a "magnetic bottle" and are reflected forward and backward between moving clouds, which have a frozen-in magnetic field. It could be shown that this process establishes a power-law energy distribution of the accelerated particles, as is observed.

To calculate the radiation spectrum of an ensemble of relativistic electrons, one needs to know their energy spectrum. This spectrum has been measured in the Earth's vicinity by studying the CRs raining down onto the atmosphere. From such measurements, a power law of the following form has been found:

$$N(E)dE = A \cdot E^{-k} dE$$

Here, A is a "constant", referring to the local conditions near the Earth. It contains the local number density of relativistic particles per energy interval. The power-law index is generally found to be $k=2.4$.

In Figure 2.19, the measured energy spectrum of cosmic rays near the Earth is displayed. The energies of particles producing the galactic synchrotron radiation are in the range of <1 GeV to several tens of GeV. Unfortunately, the measured spectrum is strongly modulated by the solar wind below a few GeV, which explains the deviation from the power law there. Hence, nothing is known about the shape of the spectrum at the lowest CR energies. At the highest energies, there are changes in the spectrum called "knee" (at $>10^{15}$ eV) and "ankle" (at $>10^{18}$ eV). The particles with the highest recorded energies ($>10^{20}$ eV, the so-called ultra-high-energy cosmic rays, or UHECRs) are a real enigma, their origin being unknown. The complete spectrum is shown in Figure 2.20 presenting the CR all-particle spectrum observed by different experiments above 10^{11} eV, as reported by Nagano and Watson (2000). The differential flux in units of events per area, time, energy and the solid angle was multiplied with E^3 to project out the steeply falling character. The first "knee" can be seen at $E \simeq 4\times10^{15}$ eV, the "second knee" at $\simeq 3\times10^{17}$ eV, and the "ankle" at $E \simeq 5\times10^{18}$ eV; see also Biermann and Sigl (2000) for more information.

Even though magnetic fields cannot have any dynamical significance on large scales, they may play their role in smaller, local scales. For instance, they must have a strong influence on the star formation process. Another area in which magnetic fields must play a cardinal role is cosmic-ray propagation and containment within a galaxy. Cosmic rays are produced in supernovae and are accelerated by their shock fronts. Since they consist of charged particles (p, e^-, ions), they are tightly coupled to the magnetic field, which in turn governs their propagation. If the CR pressure is sufficiently high, exceeding that of other pressure terms (also magnetic), they may escape from a galaxy in a galactic wind (for details, see Klein, 2013).

The magnetic field is coupled to the gas via the ionized (thermal) component. Therefore, all of these constituents form a disk (with different scale heights), which is subject to hydrostatic equilibrium unless it is overpressured by strongly enhanced star formation and subsequent supernova activity.

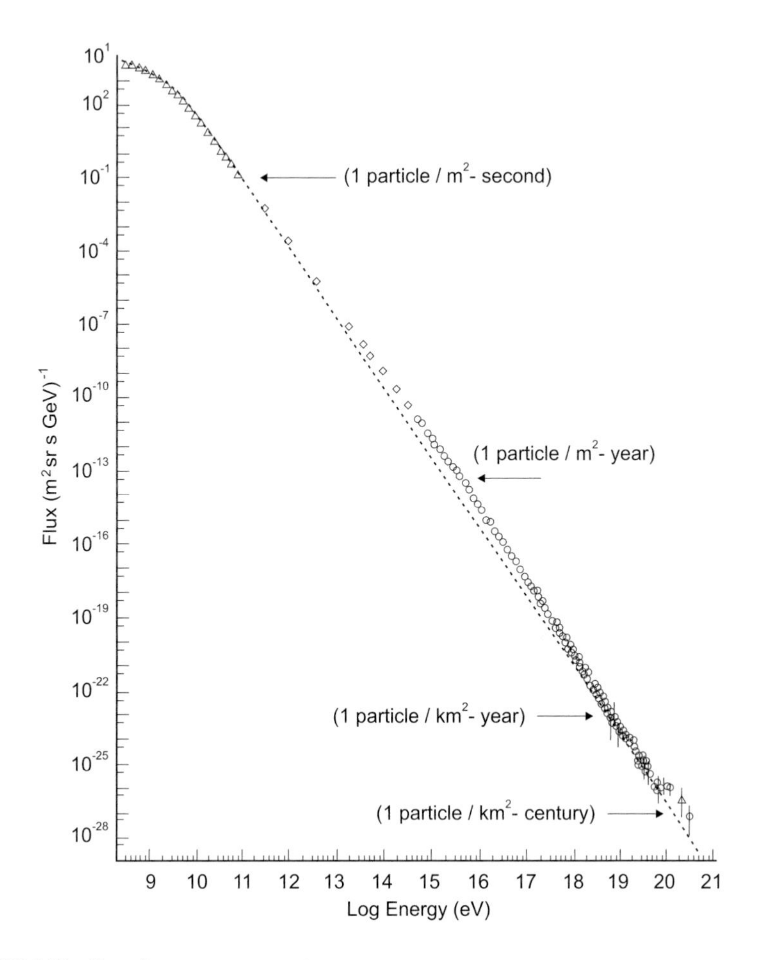

FIGURE 2.19 Cosmic-ray energy spectrum.

2.5.1 Galactic Cosmic Rays

Galactic cosmic rays (GCRs) are coming from outside the solar system, but generally from within our Milky Way galaxy. GCRs are atomic nuclei from which all electrons have been stripped during their high-speed passage through the galaxy. They have probably been accelerated inside the last few million years and have travelled many times across the galaxy being trapped by the galactic magnetic field. The GCRs have been accelerated to almost the speed of light, probably by supernova remnants. As they travel through the thin gas of interstellar space, some of the GCRs interact and emit gamma-rays, which is the information we obtain on their passage through the Milky Way and other galaxies (http://helios.gsfc.nasa.gov/gcr.html).

The elemental makeup of GCRs has been studied in a great many detail. It is very similar to the composition of the Earth and the solar system. However, studies of the composition of the isotopes in GCRs may indicate that the seed population for GCRs

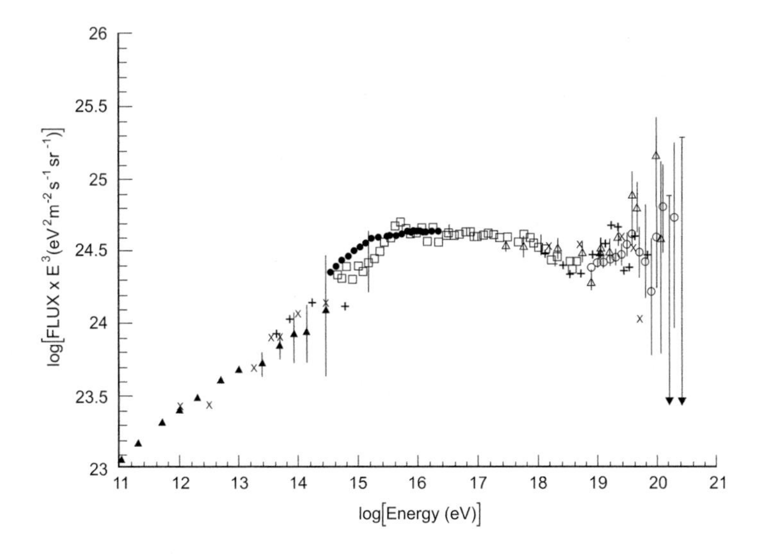

FIGURE 2.20 The CR all-particle spectrum observed by different experiments above 10^{11}eV, from Nagano and Watson (2000). The differential flux in units of events per area, time, energy and the solid angle was multiplied with E^3 to project out the steeply falling character. The first "knee" can be seen at $E \simeq 4\times10^{15}$eV, the "second knee" at approx. 3×10^{17}eV, and the "ankle" at $E \simeq 5\times10^{18}$eV.

is not from the interstellar gas nor the debris of giant stars that went supernova. This problem is an area of the current study.

For some time already, the HESS observatory in Namibia, run by an international collaboration of 42 institutions from 12 countries, has been active in mapping the center of our galaxy in very-high-energy gamma-rays. The measured gamma-rays result in cosmic-ray bombardment of nuclei in the innermost region of the galaxy. A detailed analysis of the HESS data reveals, for the first time, a source producing cosmic radiation at energies never observed before in the Milky Way. Our galaxy's central SMBH provides cosmic rays of energies 100 times higher than those achieved at the largest particle accelerator on the Earth (see HESS Collaboration, 2016).

GCRs reach energies of a few peta-electronvolts, i.e., 10^{15} electronvolts (Berezinskii et al., 1990). This fact implies that our galaxy contains peta-electronvolt accelerators ("PeVatrons"), but all proposed models of galactic cosmic-ray accelerators encounter difficulties elaborating exactly these energies (Malkov and Drury, 2001). Dozens of galactic accelerators that can accelerate particles to energies of tens of tera-electronvolts (of the order of 10^{13} electronvolts) were found in recent γ-ray observations (Hillas, 2013). However, none of the presently known accelerators, not even the handful of shell-type supernova remnants commonly believed to supply most GCRs, has shown the specific tracers of peta-electronvolt particles, namely power-law spectra of γ-rays extending continuously, without a cutoff or a spectral break, to tens of tera-electronvolts (Aharonian, 2013). HESS Collaboration (2016) reported deep γ-ray observations with arcminute angular resolution of the region surrounding the galactic center, which shows the expected tracer of the presence of peta-electronvolt protons within the central 10 pc of the galaxy. They proposed that the supermassive black hole Sagittarius A* is linked to this PeVatron. Sagittarius A* went through active phases in the past, as illustrated by X-ray outbursts (Clavel et al., 2013) and outflow from the galactic center (Su et al., 2010). Sagittarius A*'s current rate of particle acceleration is not sufficient to provide a substantial contribution to GCRs. However, Sagittarius A* could have plausibly been more active over the last 10^6–10^7 years and therefore should be considered as a possible alternative to supernova remnants as a source of peta-electronvolt GCRs (HESS Collaboration, 2016).

An important observable for an understanding of ultra-high-energy cosmic rays (UHECRs) is the energy spectrum. Aab et al. (2020), the Pierre Auger Collaboration, have recently reported a measurement above 2.5×10^{18} eV (215,030 events), with more than 16,000 events having energy beyond 10^{19}eV. The details of the measurement are reported in the companion paper (Pierre Auger Collaboration, 2020). The reported data constrain the luminosity density that is continuously emitting sources must inject into space in UHECRs to supply the observed energy density. This amounts to approximately 6×10^{44} erg/Mpc^3/y above 5×10^{18}eV at the redshift of zero. Classes of extragalactic sources that match such rates in the gamma-ray band include AGN and starburst galaxies (Dermer and Razzaque, 2010).

2.5.2 Acceleration of Cosmic Rays

As already mentioned, at present, it is not yet clear where and how cosmic rays are accelerated. In the majority of the existing models based on the energetic estimate, acceleration is usually believed to occur in the vicinity of stars. Indeed, the energy of pulsar rotation of the stellar wind of spectral types O and B stars and, in particular, the energy released in the supernovae explosion are sufficient to accelerate particles up to relativistic energies and to provide a necessary cosmic-ray flux in the galaxy. Radiation generated by these particles detects the presence of accelerated particles in the vicinity of the stars mentioned above. Even in the atmosphere of an ordinary star such as the Sun, there occur acceleration processes, and accelerated particles with energies up to several tens of GeV are directly observed on the Earth during solar flares.

According to Dogiel et al. (1987), the vicinity of stars is not the only place in the galaxy where cosmic rays can be accelerated. Gamma-ray studies have revealed an anomalously large gamma-ray flux from some giant molecular clouds. Knowing the mass of the molecular cloud, one can calculate the cosmic-ray density, which in some clouds turned out to be several times (and even by several orders of magnitude) more significant than in the surrounding space (Morfill, 1984).

It is clear that if there exists a mechanism transforming the turbulent energy into the energy of the accelerated particle, this could lead to a substantial increase in the cosmic-ray density. The paper by Dogiel et al. (1987) is devoted to the investigation of such a possibility. Equations of particle motion are derived for weakly ionized turbulent plasma containing a magnetic field. In this case, particle energies change due to fluctuating electromagnetic fields. Particle motion in coordinate and momentum space is shown to be described by the diffusion approximation. The solution of these equations is obtained for both magnetized and unmagnetized particles. Magnetized particle is shown to be effectively accelerated under certain conditions. The distribution functions have been calculated for the states of galactic giant molecular clouds. The cosmic-ray density may be higher inside clouds than in the intercloud medium. The possible influence of this process on gamma-emissivity of the galaxy, on the chemical composition of cosmic rays, on the flux of galactic antiprotons and positrons and on the galactic radio emission is discussed. According to Dogiel et al. (1987), the mean density of protons accelerated in clouds in interstellar space is equal to 1%–10% of the total interstellar density of these particles. Thus, acceleration in clouds does not substantially affect the energy balance in the interstellar space.

Next, we shall mention some more recent papers on the origin and acceleration of cosmic rays. First, let us say the GALPROP program (Strong and Moskalenko, 1998; Moskalenko and Strong, 1998) performs the calculations of cosmic-ray propagation for nuclei, electrons and positrons and computes γ-ray and synchrotron emission in the same framework. The 3D spatial approach, which includes a realistic distribution of interstellar gas, distinguishes it from leaky-box calculations. Initially, they considered that the complexity of the spatial model would preclude using cosmic-ray reaction networks with more than a few species, and their results have been obtained using a simplifying "weighted cross sections" method. In the paper by Strong and Moskalenko (1999), the cosmic-ray propagation code GALPROP has been generalized to include fragmentation networks of arbitrary complexity. The code can provide an alternative to leaky-box calculations for full isotopic abundance calculations and has the advantage of adding the spatial dimension, which is essential for radioactive nuclei. Even for stable nuclei, it has the benefit of a physically based propagation scheme with a spatial distribution of sources instead of an ad hoc path length distribution. It also facilitates tests of re-acceleration.

According to HESS Collaboration (2006), the origin of GCRs (with energies up to 10^{15} eV) remains unclear. However, it is widely believed that they originate in the shock waves of expanding supernova remnants. Currently, the best approach to the investigation of their acceleration and propagation is observing the gamma-rays produced during the interaction of cosmic rays with interstellar gas. In their paper, observations of an extended region of very-high-energy (VHE, >100 GeV) gamma-ray emission that correlated spatially with a complex structure of giant molecular clouds in the central 200 pc of the Milky Way were reported. The hardness of the gamma-ray spectrum and the conditions in those molecular clouds indicate that the cosmic rays producing the gamma-rays are likely to be protons and nuclei rather than electrons. The energy associated with cosmic rays could have come from a single supernova explosion around 10,000 years ago.

The recent investigations into the galactic plane with space- and ground-based detectors revealed several high-energy and very-high-energy γ-ray sources associated with supernova remnants, both young and middle-aged. These results imply that the actual production of relativistic particles is most likely through the process of shock acceleration. The interpretation of γ-ray data from several representatives of young supernova remnants (SNRs) within the so-called hadronic models requires proton spectra, which extend to 100 TeV, and total energy released in accelerated protons and ions $W_{CR} \geq 10^{50}$ erg (Aharonian, 2013). This effect can be treated as a support for the SNR paradigm of GCRs. However, hadronic models are challenging to handle and represent non-trivial challenges. Moreover, in many cases, data on γ-ray spectra can successfully be explained also by the inverse Compton scattering of directly accelerated electrons. Deep spectroscopic and morphological studies of SNRs using the Cherenkov Telescope Array (CTA) over four decades in energy, from 30 GeV to 300 TeV, promise a breakthrough regarding the identification of radiation mechanisms (Aharonian, 2013).

There is relatively rapid spatial and temporal variability in the X-ray radiation from some molecular clouds near the galactic center. This effect shows that this emission component is due to the reflection of X-rays generated by a source being luminous in the past, probably the central supermassive black hole, Sagittarius A* (Clavel et al., 2013). The study of the evolution of the molecular cloud reflection features is, therefore, a pivotal element to reconstruct Sagittarius A*'s past activity. In their work, Clavel et al. (2013) aimed to study this emission on small angular scales to characterize the short-time source outbursts. They used Chandra high-resolution data collected from 1999 to 2011 to study the most rapid variations detected so far, including those of clouds between 5′ and 20′ from Sagittarius A* toward positive longitudes. Their systematic spectral imaging analysis of the reflection emission, notably of the Fe K_α line at 6.4 keV and associated 4–8 keV continuum, allowed them to characterize the variations down to 15″ angular scale and 1-year timescale. They revealed for the first time abrupt changes of few years only and, in particular, a low peaked emission, with a factor of 10 increase followed by a comparable decrease that propagates along the filaments of the Bridge region cloud. This 2-year feature contrasts with the slower 10-year linear variations they revealed in all the other molecular structures of the region. Based on column density constraints, Clavel et al. (2013) argued that these two different behaviors are unlikely to be a result of the same illuminating event. The variations are probably due to a highly variable active phase of Sagittarius A* sometime within the past few hundred years, which was characterized by at least two luminous outbursts of a few-year timescale and during which the luminosity of Sagittarius A* went up to at least 10^{39} erg/s.

According to Bell et al. (2013), GCR acceleration to the knee in the spectrum at a few PeV is possible only with cosmic rays escaping the SNR amplifying the magnetic field ahead of a supernova remnant shock. A model formulated with the electric charge carried by the escaping cosmic rays predicts the maximum cosmic-ray energy and the energy spectrum of cosmic rays released into the surrounding medium. They claim that historical SNR such as Cas A, Tycho and Kepler may be expanding too slowly to accelerate cosmic ray to the knee at present.

In the paper by Istomin (2013), it is shown that the relativistic jet emitted from the center of the galaxy during its activity had the power and energy spectrum of accelerated protons sufficient to explain the current distribution of cosmic rays in the galaxy. Proton acceleration happens on the light cylinder surface formed by the rotation of a massive black hole carrying into rotation the radial magnetic field and the magnetosphere. This effect is observed in gamma-ray, X-ray and radio bands bubble above and below the galactic plane. The size of the bubble sets the time when the jet's started, ~2.4×10^7 years ago. The jet worked for more than 10^7 years, but less than 2.4×10^7 years.

The Compton Gamma Ray Observatory (CGRO) is the second of NASA's "Great Observatories" (the first being the Hubble Space Telescope). The Energetic Gamma Ray Experiment Telescope (EGRET) gives the highest energy gamma-ray window for CGRO. EGRET's energy range is from 30 MeV up to 30 GeV. EGRET is 10–20 times bigger and more sensitive than previous detectors operating at these high energies and has already made detailed observations of high-energy processes associated with diffuse gamma-ray emission, gamma-ray bursts, cosmic rays, pulsars and active galaxies known as gamma-ray blazars. The image shown in Figure 2.21 is an all-sky map in galactic coordinates obtained by EGRET of gamma-ray emission at energies above 100 MeV. The brightest emission is colored yellow in this false-color image, while the faintest emission is blue. The whole plane of the Milky Way galaxy is seen as a strong source of diffuse and resolved emission.

The belief that GCRs are accelerated at supernova remnant shocks still needs a conclusive proof. The most substantial support to this idea is probably the fact that supernova remnants are observed in gamma-rays, which are expected as the result of the hadronic interactions between the cosmic rays accelerated by the shock and the ambient gas. However, in most cases, leptonic processes can also explain the observed gamma-ray emission. This fact implies that the detections in gamma-rays do not necessarily mean that supernova remnants accelerate cosmic-ray protons.

To better understand this problem, the multi-wavelength emission (from radio to gamma-rays) from individual supernova remnants has been studied by Cristofari et al. (2013). In some cases, it was possible to attribute the gamma-ray emission to one of the

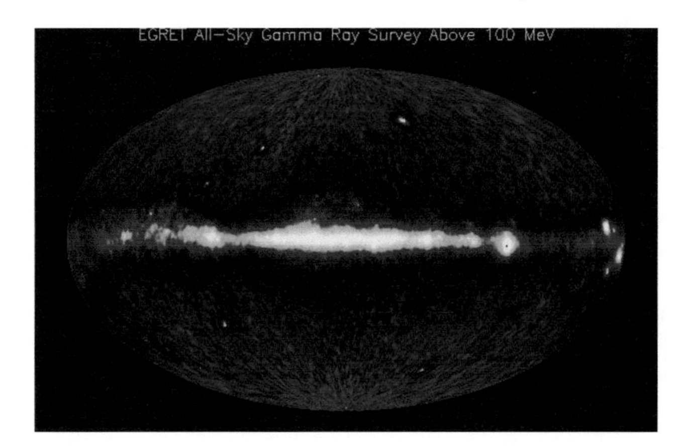

FIGURE 2.21 This map of the sky in gamma-ray light is based on the data taken with the EGRET instrument (in this case, showing light with energy>100 MeV). Aside from the apparent bright point sources, the strong emission along the center of this image is the Milky Way. Beyond our galaxy, a much fainter, extragalactic emission can be seen (blue areas in this image). (Credit: Compton Gamma Ray Observatory Team.)

two processes (hadronic or leptonic). The authors adopted a different approach, and instead of a case-by-case study, they aimed for a population study and computed the number of supernova remnants expected to be seen in TeV gamma-rays above a given flux, assuming that these objects indeed are the sources of cosmic rays. The predictions are found to match well with the current observational results, thus providing a new consistency check for the supernova remnant postulate for the origin of GCRs.

Moreover, hints are presented, suggesting that particle spectra significantly steeper than E^{-2} are produced at supernova remnants. Finally, they expect that several of the supernova remnants detected by HESS in the survey of the galactic plane could exhibit a gamma-ray emission dominated by hadronic processes (i.e., neutral pion decay). The fraction of the measured remnants for which the leptonic emission dominates over the hadronic one depends on the proposed values of the physical parameters and can be as high as roughly a half.

Star-forming factories in galaxies produce compact clusters and loose associations of young massive stars. Fast radiation-driven winds and supernovae input their tremendous kinetic power into the interstellar medium in the form of highly supersonic and superalfvenic outflows. Apart from gas heating, collisionless relaxation of fast plasma outflows results in fluctuating magnetic fields and energetic particles. The energetic particles comprise a long-lived component, which may contain a sizeable fraction of the kinetic energy released by the winds and supernova ejecta and thus modify the magnetohydrodynamic flows in the systems. Bykov (2014) presented a concise review of observational data and models of nonthermal emission from starburst galaxies, superbubbles and compact clusters of massive stars. The efficient mechanisms of particle acceleration and amplification of fluctuating magnetic fields with a wide dynamical range in starburst regions were discussed. Sources of cosmic rays, neutrinos and multi-wavelength nonthermal emission associated with starburst regions, including potential galactic "PeVatrons", were reviewed in the global galactic ecology context.

In their new paper, van Weeren et al. (2017) discussed the electron re-acceleration at galaxy cluster shocks. They reported the discovery of a direct connection between a radio relic and a radio galaxy in the merging galaxy cluster Abell 3411–3412. This discovery was accomplished by combining radio, X-ray and optical observations. Their finding indicates that fossil relativistic electrons from AGN are re-accelerated at cluster shocks. Cluster mergers are considered to be the most energetic events in the Universe after the Big Bang, releasing energies up to $\sim 10^{64}$ erg on 10^{9}-year timescales. Almost all of the gravitational energy released during cluster merger events is converted into thermal energy. However, a small fraction ($<1\%$) of the energy dissipated at shocks could be channeled into the acceleration of cosmic rays. In the presence of magnetic fields, cosmic-ray electrons would then emit synchrotron radiation, which can be observed with radio telescopes. According to van Weeren et al. (2017), the origin of the large-scale magnetic fields and the nature of particle acceleration processes that operate in these dilute cosmic plasmas are still open questions.

In the end, we should mention the review by Blasi (2013), which mainly focuses on supernova remnants as the most credible sources of GCRs. He reviewed the main aspects of the modern theory of diffusive particle acceleration at supernova remnant shocks, with particular attention to the dynamical reaction of accelerated particles on the shock and the phenomenon of magnetic field amplification at the shocks. The escape of cosmic rays from the sources is discussed as a necessary step to determine the spectrum of cosmic rays at the Earth. He also discussed the phenomenon of cosmic-ray acceleration at shocks propagating in partially ionized media and the implications of this phenomenon on the widths of the Balmer line emissions.

2.5.3 Elemental Composition of Cosmic Rays

Precise determination of the cosmic-ray fluxes and compositions is of crucial importance for the understanding of astrophysical phenomena taking place in the galaxy. Moreover, data from space missions are needed since the accurate determination of the chemical composition for balloon-borne experiments is intrinsically limited to the region below a few hundred GeV per nucleon because of uncertainties in the atmospheric corrections due to the secondaries produced by cosmic rays interacting with the atmosphere.

A review of the existing data on the chemical composition of cosmic rays from space-borne experiments, with a particular focus on the data produced by the PAMELA experiment (Picozza et al., 2007), was presented by Boezio and Mocchiutti (2012). Here, we shall limit the discussion to only one problem, namely the boron–carbon ratio.

Iron is the most abundant heavy nuclei beyond silicon in the primary cosmic rays. Iron is assumed to be mostly produced and accelerated in astrophysical sources. Iron nuclei interact much more with the interstellar medium during propagation because of a significantly larger cross section than those of lighter nuclei (He, C, O, Ne, Mg and Si). Precise knowledge of the iron spectrum in the GV–TV rigidity region provides important information on the origin, acceleration and propagation processes of cosmic rays in the galaxy. Aguilar et al. (2017, 2020) reported the precision measurements of the primary cosmic-ray He, C and O fluxes and Ne, Mg and Si fluxes with the Alpha Magnetic Spectrometer (AMS) experiment on the International Space Station. These measurements revealed an identical rigidity dependence of the He, C and O fluxes above 60 GeV and their deviation from a single power law above ~200 GeV. The AMS results also revealed unexpected differences in the rigidity dependence of the Ne, Mg and Si fluxes compared to He, C and O fluxes.

Recently, Aguilar et al. (2021) have reported the observation of new properties of primary Fe cosmic rays in the rigidity range from 0.65 GV–3.0 TV with 62×10^4 iron nuclei collected by the AMS. Above 80.5 GeV, the rigidity dependence of the cosmic-ray Fe flux is identical to the rigidity dependence of the primary cosmic-ray He, C and O fluxes, with the Fe/O flux ratio being constant at 0.155 ± 0.006. This shows that Fe and, He, C and O belong to the same class of primary cosmic rays, which is different from the primary cosmic-ray Ne, Mg and Si class.

2.5.3.1 Boron-to-Carbon Ratio

Precision measurements of the boron–carbon ratio present in cosmic rays were carried out in the range of 1–670 GeV/n by using the first 30 months of flight data of AMS-02 located in the International Space Station (Sun, 2015). His data, together with previously

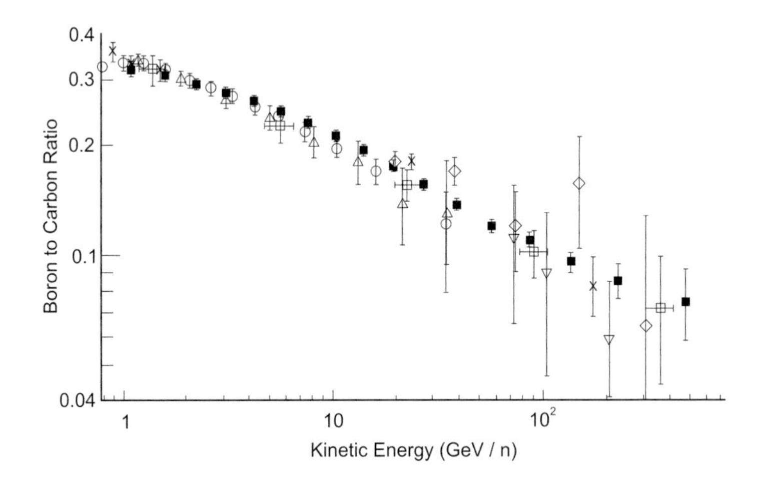

FIGURE 2.22 Comparison of the boron-to-carbon ratios as reported in different measurements. (After Sun, 2015.)

reported B/C ratio measurements, are shown in Figure 2.22. The figure shows two satellite-based experiments in the 1980s, HEAO (Engelmann et al., 1990) and CNR (Swordy et al., 1990); measurements were performed on B and C at about 0.5–50 and 50–300 GeV/n. In the 2000s, three balloon-based experiments ATIC (Panov et al., 2007), CREAM (Ahn et al., 2008) and TRACER (Obermeier et al., 2011) extended the measurements to higher energy. The test flight of AMS (AMS-01, Aguilar et al., 2010) in 1998 measured B/C ratio up to 50 GeV/n.

In conclusion, the B/C ratio measured by Sun (2015) is found to be ~0.32 at 1 GeV/n, and it decreases to be ~0.08 at 500 GeV/n. The B/C spectrum follows a falling power law. This fact agrees with the assumption that boron is a secondary particle from the spallation of heavier elements in collision with the interstellar medium.

2.5.4 Neutrinos

In 1931, Pauli made another important contribution to physics: the theoretical prediction of the neutrino. The discovery arose from the beta-decay research carried out in the late 1920s that indicated that a small amount of energy and momentum dissipates when an atomic nucleus emits a beta particle. This loss clearly violated the laws of conservation. In order to explain the anomaly while keeping the conservation laws intact, Pauli suggested the existence of an uncharged particle of little or no mass that conducted energy and momentum away from the atomic nucleus, thus explaining the apparent loss.

Pauli's theoretical particle later came to be called the neutrino (a term coined by Enrico Fermi) and was first discovered experimentally in the late 1950s. A particle fitting the expected characteristics of the neutrino was announced by Clyde Cowan and Fred Reines (a founding member of Super-Kamiokande; UCI professor emeritus and recipient of the 1995 Nobel Prize in Physics for his contribution to the discovery). This neutrino is later determined to be the partner of the electron.

In 1962, experiments at Brookhaven National Laboratory and CERN, the European Laboratory for Nuclear Physics, made a surprising discovery that neutrinos associated with muon do not behave the same as those produced with electrons. They have, in fact, discovered a second type of neutrino (the muon neutrino).

In 1968, the first experiment detecting (electron) neutrinos produced by the Sun's burning (using a liquid chlorine target deep underground) reported that less than half the expected neutrinos were observed. This is the origin of the long-standing "solar neutrino problem". The possibility that the missing electron neutrinos may have transformed into another type (undetectable to this experiment) was soon suggested, but the unreliability of the solar model on which the expected neutrino rates are based was initially considered a more likely explanation.

In 1978, the tau particle was discovered at SLAC, the Stanford Linear Accelerator Center. It was soon recognized to be a heavier version of the electron and muon, and its decay exhibits the same apparent imbalance of energy and momentum that led Pauli to predict the existence of the neutrino in 1931. The existence of a third neutrino associated with the tau is hence inferred, although this neutrino has yet to be directly observed.

In 1987, Kamiokande, another large water detector looking for proton decay and IMB, detected a simultaneous burst of neutrinos from SN 1987A. In 1989, Kamiokande become the second experiment to detect neutrinos from the Sun and confirmed the long-standing anomaly by finding only about one-third of the expected rate.

Three types of neutrinos are known; there is strong evidence that no additional neutrinos exist unless their properties are unexpectedly very different from the known types. Each type or "flavor" of the neutrino is related to a charged particle (which gives the corresponding neutrino its name). Hence, the "electron neutrino" is associated with the electron, and the two other neutrinos are associated with heavier versions of the electron called the muon and the tau. (Elementary particles are frequently labeled with Greek letters to confuse the layman.) Table 2.3 lists the known types of neutrinos (and their electrically charged partners).

Neutrinos interact only very weakly with matter, but giant detectors have succeeded in detecting small numbers of astrophysical neutrinos. Aside from a diffuse background, only two individual sources have been identified: the Sun and a nearby supernova in 1987. A multi-team collaboration detected a high-energy neutrino event whose arrival direction was consistent with a known blazar – a type of quasar with a relativistic jet oriented directly along our line of sight. The blazar, TXS 0506+056, was found to be undergoing a gamma-ray flare, prompting an extensive multi-wavelength campaign. Motivated by this discovery, the IceCube Collaboration examined lower-energy neutrinos detected

TABLE 2.3 Three Neutrino Types

Neutrino	ν_e	ν_μ	e_τ
Charged partner	Electron (e)	Muon (μ)	Tau (τ)

over the previous several years, finding an excess emission at the location of the blazar. Thus, blazars are a source of astrophysical neutrinos (IceCube Collaboration, 2018).

A generic prediction of the standard Big Bang cosmology is the existence of a relic neutrino background through the entire Universe. The experimental confirmation of neutrino flavor oscillation (Fukuda et al., 1998) suggests that at least two of the three kinds of neutrinos have nonzero masses. However, their absolute masses and the hierarchy among the three eigenstates remain unknown, and a precise measurement of the neutrino mass by terrestrial experiments is still required (Yoshikawa et al., 2020).

Researchers from the IceCube Neutrino Observatory have spotted a diffuse flux of neutrinos with energies up to peta-electronvolt (10^{15} eV); see Aartsen et al. (2013). South Pole Neutrino Observatory – IceCube is operated by the University of Wisconsin–Madison and is supported by the National Science Foundation. The IceCube, the instrument with which this discovery was made, consists of a cubic kilometer of ice, instrumented with phototubes that detect the light produced by neutrinos interacting with particles in the ice. Roughly a gigaton in mass, the IceCube sits 1.5 km beneath the Antarctic surface near the South Pole (Gaisser and Halzen, 2014; Halzen and Klein, 2008). The energy spectrum of neutrinos detected by the IceCube displays a much flatter power law and diverges significantly from the atmospheric neutrino background at energies beyond ~100 TeV. The extragalactic origin of the detected high-energy neutrinos is deduced from their approximately isotropic arrival directions. As possible sources of these high-energy neutrinos, the following objects have been considered (Mészáros, 2018; Mészáros, 2017):AGN, galaxy clusters, starburst galaxies and cosmic-ray bursts. The small number of TeV–PeV neutrino events being collected by the IceCube and other detectors makes it challenging to associate neutrinos with specific sources. Let us mention that the IceCube and some 20 telescopes combined forces (see IceCube Collaboration, 2018; IceCube Collaboration et al., 2018) to establish a spatial and time coincidence between the production of a 300 TeV neutrino observed at the IceCube and the generation of gamma-ray flares seen from the blazar TXS0506+056. The blazar is assumed to be one of the sources of high-energy neutrinos. Blazars represent a class of AGN with powerful relativistic jets pointed close to our line of sight (Urry and Padovani, 1995). They are prominent candidate sources of such a high-energy neutrino emission (Mannheim, 1995; Halzen and Zas, 1997; Mucke et al., 2003; Murase, 2017; Petropoulou and Dermer, 2016; Guepin and Kotera, 2017).

2.5.5 Cosmic Rays–Dust Particles Interaction

Collisions between cosmic ray's energetic protons and α-particles and carbon, nitrogen and oxygen in the interstellar medium have been considered (Reeves, 1994) to be the primary source of lithium, beryllium and boron. The process is the fragmentation of the larger nuclei, although this mechanism is unable to account for the observed abundances of the isotopes ^{7}Li and ^{11}B in the solar system. The detection of an excess of γ-rays, as reported by Bloemen et al. (1994), in the direction of a star-forming region in the Orion cloud has been interpreted (Bykov and Bloemen, 1994) as arising from the excitation of carbon and oxygen nuclei ejected from supernovae. The process responsible is supernovae collisions with the surrounding gas, which is primarily molecular and atomic hydrogen. Cassé et al. (1995) investigated the consequences of the two-body interactions of the ejected carbon and oxygen nuclei (and the α-particles discharged with them) with the hydrogen and helium in the surrounding gas. They showed that these interactions offer a way to make lithium, beryllium and boron independent of the heavy elements' abundances in the surrounding environment. Such supernova-driven interactions, in combination with the effect of GCRs, can explain the observed solar system abundances of these light elements.

Bloemen et al. (1994) reported evidence for cosmic-ray-induced gamma-ray lines. The authors declare the detection of γ-ray emission from the Orion complex in the 3–7 MeV range by the COMPTEL telescope aboard the Compton Gamma Ray Observatory. This emission can be associated with the 4.44 and 6.13 MeV nuclear de-excitation lines of $^{12}C^*$ and $^{16}O^*$, respectively, which are predicted to be the strongest γ-ray lines originating from the interaction of energetic particles with present matter. However, the observed flux of $(1.0 \pm 0.15) \times 10^{-4}$ photon cm^{-2}/s (3–7 MeV) is much larger than anticipated. There is good circumstantial evidence that the authors' findings indicate strongly enhanced abundances of C and O in low-energy cosmic rays (~10 MeV/nucleon), rather than high fluxes of cosmic-ray protons and α-particles. The positrons resulting from energetic particle interactions in the Orion region should produce 511 keV annihilation radiation that is detectable by OSSE.

Tatischeff and Kiener (2004) studied gamma-ray lines from cosmic-ray interactions with interstellar dust grains. Observations of nuclear interaction γ-ray lines from the interstellar medium would provide a unique tool to study galactic cosmic-ray ions at nonrelativistic energies, as well as the physical conditions of the emitting regions. If the lines produced in the gaseous phase are expected to be significantly Doppler-broadened, some lines produced in interstellar dust grains can be very narrow because some of the excited nuclei can stop in solid materials before emitting γ-rays. The latter are prime candidates for detection with γ-ray telescopes having a high spectral resolution. Using the Monte Carlo method similar to that described in Ramaty et al. (1979), they have calculated in detail the shapes of these four lines as they are produced by cosmic-ray interactions with the ISM, namely 847, 1369, 1779 and 6129 keV from $^{56}Fe^*$,$^{24}Mg^*$,$^{28}Si^*$ and $^{16}O^*$, respectively. They show that the line shapes are mainly sensitive to relatively large interstellar grains, with radii ≥ 0.25 μm. Line fluxes from the inner galaxy are then predicted.

In earlier work, Lal and Venkatavaradan (1967) discussed the activation of cosmic dust by cosmic-ray particles. In their work, the isotopic changes induced in cosmic dust by nuclear interactions of solar and galactic cosmic rays are calculated for several radioisotopes suitable for use as signatures of cosmic dust in terrestrial environments. The increase in the concentration of these radioisotopes is a function of the composition and size distribution of cosmic dust, in addition to the cosmic dust mass influx rate and the flux of low-energy particle radiation in space. The monitor radioisotopes are produced efficiently only in dust grains of certain preatmospheric size intervals.

Deuterium (D)-rich interplanetary dust particles (IDPs) are the most primitive extraterrestrial materials available for laboratory studies in terms of mineralogy, chemistry and isotopic compositions. Transmission electron microscopy analysis of one D-rich

IDP shows, as reported by Keller et al. (2000), it to be a highly porous object. It consists of crystalline grains and glass with embedded metal and sulfides (GEMS) enclosed within a matrix of amorphous carbonaceous material. Crystalline grains are Mg-rich pyroxene and olivine, and FeNi sulfides. The non-solar H isotopic anomaly found in this IDP provides direct evidence for the incorporation of partially preserved cold molecular cloud material predating the solar system. Their results indicate that organic molecules associated with the carbonaceous matrix of the IDP are the most likely D carriers in the particle. GEMS is a principal constituent of this D-rich IDP. The physical and chemical properties of GEMS show many similarities to the properties of interstellar amorphous silicates inferred from observations, including size (diameters of 0.1–0.5 μm), solar abundances for heavy elements, presence of FeNi metal and an amorphous silicate matrix. Infrared transmission spectra from GEMS-rich IDPs are similar to astronomical data for interstellar silicates. The association of GEMS and mineral grains with D-rich matter of probable interstellar origin suggests that these inorganic materials are themselves interstellar grains.

In the work by Altobelli et al. (2016), Cosmic Dust Analyzer onboard the Cassini spacecraft was used to detect 36 interstellar dust (ISD) grains as they passed by Saturn, and the grains' elemental abundances were measured. The results show that these grains lack carbon-bearing compounds and have been homogenized in the interstellar medium into silicates with iron inclusions. The authors determined the mass distribution of 36 interstellar grains, their elemental composition and a lower limit for the ISD flux at Saturn. Mass spectra and grain dynamics suggest the presence of magnesium-rich grains of silicate and oxide composition, partly with iron inclusions. Major rock-forming elements (magnesium, silicon, iron and calcium) are present in cosmic abundances, with only small grain-to-grain variations, but sulfur and carbon are depleted. The ISD grains in the solar neighborhood appear to be homogenized, likely by repeated processing in the interstellar medium.

Altobelli et al. (2016) discussed the composition of ISD grains. Owing to the high velocity of ISD impacts (typically above 20 km/s), the energy densities upon impact are high enough to vaporize the solid grain and yield cation spectra that are dominated by elemental ions rather than molecular ones. The main elemental cation peaks in all ISD spectra (going from lower to higher atomic masses) are carbon, oxygen, sodium/magnesium, potassium/calcium and iron. Owing to the relatively small mass resolution of the CDA, signatures of Na/Mg and K/Ca cannot be easily separated in most spectra. However, the positions of the maxima of the respective peaks indicate that Mg+ and Ca+ dominate the adjacent species in almost every ISD spectrum. Although not forming individual peaks, Al^+ and Si^+ are visible as an extended flank on the Mg signature. In one particular spectrum, cations of Cr are indicated. The Fe signature is very broad in every ISD spectrum; therefore, although not detectable, minor amounts of Ti, V, Mn and Ni are possible (Altobelli et al., 2016).

The results obtained by Altobelli et al. (2016) indicate that a homogeneous population of grains with roughly average cosmic abundances (concerning Mg:Si:Ca:Fe ratios), possibly including metallic nanophase iron, is the primary constituent of the LIC (local interstellar cloud)-ISD. This fact is emphasizing the importance of recondensation

TABLE 2.4 Summary of ISD Candidates (After Westphal et al., 2014)

Particle No.	Mass/Diameter	Composition	Structure	Capture Speed (km/s)
1	$(3.1 \pm 0.4)\ 10^{-12}$g	Mg_2SiO_4+$MgAl_{2+}$Fe+Cr, Mn, Ni, Ca	Low density (0.7 g/cm)	<<10
2	$(4.0 \pm 0.7)\ 10^{-12}$g	Mg_2SiO_4+Mg silicate+Al, Cr, Mn+Fe	Low density (0.4 g/cm)	<<10
3	$\sim 3 \times 10^{-12}$g	Possible Si+C		>15
4	0.28 μm crater	(Mg+Fe) / Si=3.3	Single particle with chemical zoning	>10
5	0.37 μm crater	Silicate (Mg:Fe:Si)+FeS	Single-particle or nanoscale aggregate	~5–10
6	0.39 μm crater	Silicate (Mg:Fe:Si)+Fe, Ni (metals and sulfides)	Two-particle aggregate	~5–10
7	0.46 μm crater	Silicate (Mg:Fe:Si)+Fe metals+Fe, Ni sulfide	Nanoparticle aggregate	~5–10

processes (Zhukovska et al., 2008; Frisch and Slavin, 2013). A further implication is that searches for presolar interstellar grains in meteorites led by isotopic anomalies are likely to miss a population that recondensed in the ISM. This result is because the destruction and recondensation of solids would erase isotopic anomalies. Hence, a considerable fraction of ISD could reside in meteorites and cometary material awaiting its discovery (see also Bradley et al., 1999; Keller and Messenger, 2011).

Westphal et al. (2014) presented evidence for the interstellar origin of seven dust particles collected by the Stardust Interstellar Dust Collector. Dust particles sent to the Earth for laboratory analysis have shown features consistent with a source in the contemporary interstellar dust stream. More than 50 spacecraft debris particles were also identified. The interstellar dust candidates are readily distinguished from debris impacts based on elemental composition and impact trajectory. The seven candidate interstellar particles are diverse in elemental composition, crystal structure and size. The presence of crystalline grains and multiple iron-bearing phases, including sulfide, in some particles indicates that individual interstellar particles diverge from any one representative model of interstellar dust inferred from astronomical observations and theory. Table 2.4 shows the summary of the ISD candidate's properties.

References

Aab, A., Abreu, P., Aglietta, M., et al. (The Pierre Auger Collaboration). 2020. Features of the energy spectrum of cosmic rays above 2.5×10^{18} eV using the pierre auger observatory. *Physical Review Letters* 125(12):121106 (p. 10).

Aartsen, M. G., Abbasi, R., Abdou, Y., et al. (IceCube collaboration). 2013, Jul. 8. First Observation of PeV-energy neutrinos with icecube. *Physical Review Letters* 111:021103.

Aguilar, M., Alcaraz, J., Allaby, J., et al. (AMS-01 Collaboration). 2010. Relative composition and energy spectra of light nuclei in cosmic rays: Results from AMS-01. *Astrophysical Journal* 724:329–340.

Aguilar, M., Ali Cavasonza, L., Allen, M. S., et al. 2021. Properties of iron cosmic rays: Results from the alpha magnetic spectrometer. *Physical Review Letters* 126:041104 (8 p).

Aguilar, M., Cavasonza, L. A., Alpat, B., et al. (AMS Collaboration). 2017. Observation of the identical rigidity dependence of He, C, and O cosmic rays at high rigidities by the alpha magnetic spectrometer on the international space station. *Physical Review Letters* 119:251101, (8 p).

Aguilar, M., Cavasonza, L. A., Ambrosi, G., et al. (AMS Collaboration). 2020. Properties of neon, magnesium, and silicon primary cosmic rays results from the alpha magnetic spectrometer. *Physical Review Letters* 124:211102, (8 p).

Aharonian, F. A. 2013. Gamma rays from supernova remnants. *Astroparticle Physics* 43:71–80.

Ahn, H., Allison, P. S., Bagliesi, M. G., et al. 2008. Measurements of cosmic-ray secondary nuclei at high energies with the first flight of the CREAM balloon-borne experiment. *Astroparticle Physics* 30:133–141.

ALMA. 2020. ALMA is an international partnership of the European Southern Observatory (ESO), the U.S. National Science Foundation (NSF) and the National Institutes of Natural Sciences (NINS) of Japan, together with NRC (Canada), MOST and ASIAA (Taiwan), and KASI (Republic of Korea), in cooperation with the Republic of Chile.

Altobelli, N., Postberg, F., Fiege, K., et al. 2014. Flux and composition of interstellar dust at Saturn from Cassini's cosmic dust analyzer. *Science* 352(6283):312–318.

Amato, E. 2014. The origin of galactic cosmic rays. *International Journal of Modern Physics D* 23(07):1430013. https://doi.org/10.1142/S0218271814300134.

Andersson, B.-G., Lazarian, A. and Vaillancourt, J. E. 2015. Interstellar dust grain alignment. *Annual Review of Astronomy and Astrophysics* 53:501–540.

Armstrong, J. W., Rickett, B. J. and Spangler, S. R. 1995. Electron density power spectrum in the local interstellar medium. *Astrophysical Journal* 443(1):209–221.

Aulanier, G., Démoulin, P., Schrijver, C. J., Janvier, M., Pariat, E., and Schmieder, B. 2013. The standard flare model in three dimensions II. Upper limit on solar flare energy. *Astronomy & Astrophysics* 549:A66 (7 pp).

Bamford, S. P., Nichol, R. C., Baldry, I. K., et al. 2008. Galaxy Zoo: The dependence of morphology and colour on environment. arXiv:0805.2612v2 [astro-ph].

Bauswein, A., Bastian, N.-U. F., Blaschke, D. B., Chatziioannou, K., Clark, J. A., Fischer, T. and Oertel, M. 2019, Feb. 12. Identifying a first-order phase transition in neutron-star mergers through gravitational waves. *Physical Review Letters* 122:061102.

Beck, R. 2007. Galactic magnetic fields. *Scholarpedia* 2(8):2411. http://www.scolarpedia.org/article/Galactic_magnetic_fields/

Beck, R. 2015. Magnetic field in the nearby spiral galaxy IC 342: A multi-frequency radio polarization study. *Astronomy & Astrophysics* 578:A93(27 p). Doi: 10.1051/0004-6361/201425572.

Beck, R. and Hoernes, P. 1996. Magnetic spiral arms in the galaxy NGC 6946. *Nature* 379:47–49.

Beck, R. and Wielebinski, R. 2013 *Magnetic Fields in Galaxies, in Planets, Stars and Stellar Systems*. Springer, Dordrecht, Vol. 5, pp. 641–724 (update of Sept. 2015 on http://arXiv.org/abs/1302.5663.

Behroozi, P. S., Wechsler, R. H. and Conroy, C. 2013. The average star formation histories of galaxies in dark matter halos from z=0–8. *The Astrophysical Journal*, 770(1), article id. 57, 36 p.

Bell, A. R., Schure, K. M., Reville, B. and Giacinti, G. 2013. Cosmic ray acceleration and escape from supernova remnants. arXiv:1301.7264v1 [astro-ph.HE].

Berezinskii, V. S., Bulanov, S. V., Dogiel, V. A., Ginzburg, V. L., and Ptuskin, V. S. 1990. Astrophysics of Cosmic Rays. In Ginzburg, V. L. (Ed.) pp. 534. North-Holland Publishing Co., Amsterdam, North-Holland.

Biermann, P. L. and Sigl, G. 2000. Introduction to Cosmic Rays. In Lemoine, M. and Sigl, G. (Eds.), Introductory chapter to *Physics and Astrophysics of Ultra-High-Energy Cosmic Rays*, Lecture Notes in Physics vol. 576 based on UHECR2000 (Meudon, June 26–29, 2000). See also arXiv:astro-ph/0202425v1.

Binney, J., Dehnen, W. and Bertelli, G. 2000. The age of the solar neighbourhood. *Monthly Notices of the Royal Astronomical Society* 318:658–664.

Blasi, P. 2013. The origin of galactic cosmic rays. arXiv:1311.7346v2 [astro-ph.HE].

Bloemen, H., Wijnands, R., Bennett, K., et al. 1994. COMPTEL observations of the Orion complex: Evidence for cosmic-ray induced gamma-ray lines. *Astronomy & Astrophysics.* 281:L5–L8.

Boezio, M. and Mocchiutti, E. 2012. Chemical composition of galactic cosmic rays with space experiments. arXiv:1208.1406v1 [astro-ph.HE].

Bradley, J. P. Keller, L. P., Snow, T., Flynn, G. J., Gezo, J., Brownlee, D. E., Hanner, M. S. and Bowey, J. 1999. An infrared spectral match between GEMS and interstellar grains. *Science* 285:1716–1718.

Brandenburg, A. and Subramanian, K. 2005. Astrophysical magnetic fields and nonlinear dynamo theory. *Physics Reports* 417:1–209.

Brodwin, M., McDonald, M., Gonzales, A. H., Stanford, S. A., Eisenhardt, P. R., Stern, D. and Zeimann, G. 2016. IDCS J1426.5+3508: The most massive galaxy cluster at z>1.5. *227th Meeting of American Astronomical Society*, Kissimmee, FL, 4–8 January 2016. Abstract 439.05.

Buder, S., Asplund, M., Duong, L., et al. 2018. The GALAH survey: Second data release. *Monthly Notices of the Royal Astronomical Society* 478:4513–4552.

Burlaga, L. Ness, N. F. and Stone, E. C. 2013. Magnetic field observations as voyager 1 entered the heliosheath depletion region. *Science* 341:147–150.

Bykov, A. 2014. Nonthermal particles and photons in starburst regions and superbubbles. *Astronomy and Astrophysics Review* 22: article id.77, 54 p.

Cameron, R. and Schüssler, M. 2015. The crucial role of surface magnetic fields for the solar dynamo. *Science* 347(6228):1333–1335.

Cardamone, C. N., Schawinski, K., Sarzi, M. et al. 2009. Galaxy zoo green peas: Discovery of a class of compact extremely star-forming galaxies. arXiv:0907.4155v1 [astro-ph.CO].

Carrington, R. C. 1859. Description of a singular appearance seen in the Sun on September 1, 1859. *Monthly Notices of the Royal Astronomical Society* 20:13–15.

Cassé, M., Lehoucq, R. and Vangloni-Flam, E. 1995. Production and evolution of light elements in active star-forming regions. *Nature* 373:318–319.

Cautun, M., Deason, A. J., Frenk, C. S. and McAlpine, S. 2019. The aftermath of the great collision between our galaxy and the large magellanic cloud. *Monthly Notices of the Royal Astronomical Society* 483:2185–2196.

Charbonneau, P. 2010. Living Reviews in Solar. *Physics* 7:3. www.livingreviews.org/-lrsp-2010-3 (accessed 04 April 2016).

Chiti, A., Frebel, A., Simon, J.D., et al. 2021. An extended halo around an ancient dwarf galaxy. *Nature Astronomy Letters*. Doi: 10.1038/s41550-020-01285-w.

Cho, J. and Vishniac, E. T. 2000. The generation of magnetic fields through driven turbulence. *The Astrophysical Journal* 538:217–225.

Choksi, N., Gnedin, O. Y. and Li, H. 2018. Formation of globular cluster systems: From dwarf galaxies to giants. *Monthly Notices of the Royal Astronomical Society* 480:2343–2356.

Chu, J. 2016, Jan. 7. Most distant massive galaxy cluster identified. *MIT News*.

Clavel, M. Terrier, R., Goldwurm, A., Morris, M. R, Ponti, G, Soldi, S. and Trap, G. 2013. Echoes of multiple outbursts of Sagittarius A* revealed by Chandra. *Astronomy & Astrophysics* 558:A32.

Cliver, E. W. and Dietrich, W. F. 2013. The 1859 space weather event revisited: Limits of extreme activity. *Journal of Space Weather and Space Climate* 3:AA31.

Cliver, E. W., Tylka, A. J., Dietrich, W. F. and Ling, A. G. 2014. On a solar origin for the cosmogenic nuclide event of 775A.D. *The Astrophysical Journal* 781:32.

Cole, S., Helly, J., Frenk, C. S. and Parkinson, H. 2008. The statistical properties of Λ cold dark matter halo formation. *Monthly Notices of the Royal Astronomical Society* 383:546–556.

Comerford, J. M. 2016. Merger-triggered AGN activity as traced by dual and offset AGN. *227th Meeting of American Astronomical Society*, Kissimmee, FL, 4–8 Jan. 2016. Abstract 119.03.

Comerford, J. M., Barrows, R. S., Müller-Sánchez, F., Nevin, R., Greene, J. E., Pooley, D., Stern, D. and Harrison, F. A. 2017. An active galactic nucleus caught in the act of turning off and on. arXiv:1710.00825v1 [astro-ph. GA] 02 Oct 2017.

Cormier, D., Madden, S. C., Lebouteiller, V., et al. 2015. The Herschel Dwarf Galaxy Survey. I. Properties of the low-metallicity ISM from PACS spectroscopy. *Astronomy & Astrophysics* 578:A53, (53 pp).

Cresci, G., Mannucci, F., Maiolino, R., Marconi, A., Gnerucci, A. and Magrini, L. 2010. Gas accretion as the origin of chemical abundance gradients in distant galaxies. *Nature* 467:811–813.

Cristofari, P., Gabici, S., Casanova, S., Terrier, R. and Parizot, E. 2013. Acceleration of cosmic rays and gamma-ray emission from supernova remnants in the Galaxy. arXiv:1302,2150v1 [astro-ph.HE].

Crocker, R. M., Jones, D., Melia, F., Ott, J. and Protheroe, R. J. 2010. A lower limit of 50 microgauss for the magnetic field near the Galactic Centre. *Nature* 468(7277):65. arXiv:1001.1275v1 [astro-ph.GA].

Croom, S. M., Owers, M. S., Scott, N. and Poetrodjojo, H. 2021. The SAMI galaxy survey: The third and final data release. Downloaded from https://academic.oup.com/mnras/advance-article/doi/10.1093/mnras/stab229/6123881.

Crutcher, R. M. 2003. Observations of magnetic fields in molecular clouds. In Uyanaker, B., Reich, W. and Wielebinski, R. (Eds.) *The Magnetized Interstellar Medium*, 8–12 September 2003, Antalya, Turkey, Copernicus GmbH, Katlenburg-Lindau, pp. 123–132.

Del Peloso, E. F. 2005. The age of the galactic thin disk from Th/Eu nucleocosmochronology. *Astronomy & Astrophysics* 440:1153–1159.

Del Peloso, E. F., da Silva, L. and Arany-Prado, L. I. 2005. The age of the Galactic thin disk from Th/Eu nucleocosmochronology. *Astronomy & Astrophysics* 434:301–308.

Dermer, C. D. and Razzaque, S. 2010. Acceleration of ultra-high-energy cosmic rays in the colliding shells of blazars and gamma-ray bursts: constraints from the Fermi Gamma-ray Space Telescope. *The Astrophysical Journal* 724:1366–1372.

Díaz-Santos, T., Assef, R. J., Blain, A. W., Tsai, C.-W., Aravena, M., Eisenhardt, P., Wu, J., Stern, D. and Bridge, C. 2015. The strikingly uniform, highly turbulent interstellar medium of the most luminous galaxy in the Universe. *The Astrophysical Journal Letters* 816(1):L6.

Dobbs, C. 2008. GMC formation by agglomeration and self gravity. *Monthly Notices of the Royal Astronomical Society* 391:844–858.

Dogiel, V. A., Gurevich, A. V., Istomin, I. N. and Zybin, K. P. 1987. On relativistic particle acceleration in molecular clouds. *Monthly Notices of the Royal Astronomical Society* 228:843–868.

Darg, D. W., Kaviraj, S., Lintott, C. J., et al. 2010. Galaxy Zoo: the properties of merging galaxies in the nearby Universe - local environments, colours, masses, star formation rates and AGN activity. *Monthly Notices of the Royal Astronomical Society* 401:1552–1563.

Eilek, J. 1999. Magnetic fields in clusters: Theory vs. observations. arXiv:astro-ph/9906485.

Engelmann, J., Ferrando, P., Soutoul, A., et al. 1990. Charge composition and energy spectra of cosmic-ray nuclei for elements from Be to Ni. Results from HEAO-3-C2. *Astronomy & Astrophysics* 233:96–111.

Erb D. K. 2008. A model for star formation, gas flows, and chemical evolution in galaxies at high redshifts. *The Astrophysical Journal* 674:151–156.

Erb D. K., Shapley A. E., PettiniM., Steidel C. C., Reddy N. A. and Adelberger K. L. 2006. The mass-metallicity relation at z>~2. *The Astrophysical Journal* 644:813–828.

ESO. 2020. 3D map of stellar system in the solar neighbourhood. https://www.eso.org/public/images/eso0303c/

Feretti, L., Dallacasa, D., Giovannini, G. and Tagliani, A. 1995. The radio galaxies and the magnetic field in Abell 119. *Astronomy & Astrophysics* 302:680.

Fermi, E. 1949. On the origin of the cosmic radiation. *Physical Review* 75(8):1169–1174.

Fermi, E. 1954. Galactic magnetic fields and the origin of cosmic radiation. *The Astrophysical Journal* 119:1–6.

Ferraro, F. R., Pallanca, C., Lanzoni, B., et al. 2020. A new class of fossil fragments from the hierarchical assembly of the Galactic bulge. *Nature Astronomy Articles*. Doi: 10.1038/s41550-020-01267-y.

Förster Schreiber, N. M., Genzel, R., Bouché, N., et al. 2009. The SINS survey: SINFONI integral field spectroscopy of z ~ 2 star-forming galaxies. *The Astrophysical Journal* 706:1364–1428.

Frebel, A. 2007. Discovery of HE 1523–0901, a strongly r-process-enhanced metal-poor star with detected uranium. *The Astrophysical Journal* 660:L117–L120.

Frisch, P. C. and Slavin, J. D. 2013. Interstellar dust close to the Sun. *Earth Planets Space* 65:175–182.

Fukuda, Y., Hayakawa, T., Ichihara, E., et al. 1998. Evidence for oscillation of atmospheric neutrinos. *Physical Review Letters* 81(8):1562–1567.

Gaia Collaboration. 2016a. The Gaia mission. *Astronomy & Astrophysics* 595: A1.

Gaia Collaboration. 2016b. Gaia data release 1. *Astronomy & Astrophysics* 595: A2 (pp 23).

Gaia Collaboration. 2018. Gaia data release 2. *Astronomy & Astrophysics* 616: A10 (pp29).

Gaia Collaboration. 2020. Gaia early data release 3. *Astronomy & Astrophysics* 649: A1 A special issue of the journal Astronomy & Astrophysics on Gaia data release #1 was published in November 2016, while a special issue of Astronomy & Astrophysics on Gaia data release #2 posted on April 2018.

Gaisser, T. and Halzen, F. 2014. IceCube. *Annual Review of Nuclear and Particle Science.* 64:101–123.

Gilmore, G., Wyse, R., Kuijken, K., et al. 1989. Kinematics, chemistry, and structure of the galaxy. *Annual Review of Astronomy and Astrophysics* 27:555–627.

Grasso, D. and Rubinstein, R. 2001. Magnetic fields in the early universe. arXiv:astro-ph/0009061v2.

Gregori, G. Ravasio, A., Murphy, C. D., et al. 2012. Generation of scaled protogalactic seed magnetic fields in laser-produced shock waves. *Nature* 481:480–483.

Greisen, K. 1966. End to the cosmic-ray spectrum? *Physical Review Letters* 16(17):748–750.

Guepin, C. and Kotera, K. 2017. Can we observe neutrino flares in coincidence with explosive transients?. *Astronomy & Astrophysics* 603:A76, p.18.

Gurri, P., Taylor, E. N. and Fluke, C. J. 2020. The first shear measurements from precision weak lensing. *Monthly Notices of the Royal Astronomical Society* 499:4591–4604.

Hagen, L. M. Z., Seibert, M., Hagen, A., Nyland, K., Neill, J. D., Treyer, M., Young, L. M., Rich, J. A. and Madore, B. F. 2016. On the classification of UGC 1382 as a giant low surface brightness galaxy. *The Astrophysical Journal* 826: 210 (16 p).

Halzen, F. and Klein, S. 2008. Astronomy and astrophysics with neutrinos. *Physics Today*, May 2008, pp. 29–35.

Halzen, F. and Zas, E. 1997. Neutrino fluxes from active galaxies: A model-independent estimate. *Astrophysical Journal* 488:669–674.

Hartmann, L., Ballesteros-Paredes, J. and Bergin, E. 2001. Rapid formation of molecular clouds and stars in the solar neighborhood. *The Astrophysical Journal* 562:852–868.

Hashimoto, T., Inoue, A. K., Mawatari, K., et al. 2019. "Big Three Dragons": a z=7.15 lyman break galaxy detected in $[O_{III}]$ 88 μm, $[C_{II}]$ 158 μm, and dust continuum with ALMA. arXiv:1806.00486v2.[astro-ph.GA] 8 Apr 2019.

Hawley, D. L. and Peebles, P. J. E. 1975. Distribution of observed orientations of galaxies. *The Astronomical Journal* 80:477–491.

HESS Collaboration. 2016. Acceleration of petaelectronvolt protons in the galactic centre. *Nature* 531: 476–479.

H.E.S.S. Collaboration, Aharonian, F. A., Akhperjanian, A.G., Bazer-Bachi, A.R., et al. 2006. Discovery of very-high-energy gamma-rays from the galactic centre ridge. *Nature* 439:695–698.

Hirschauer, A. S., Salzer, J. J., Skillman, E. D. et al. 2016. ALFAALFA discovery of the most metal-poor gas-rich galaxy known: AGC 198691. *Astrophysical Journal* 822(2):108.

Hodgson, R. 1859. On a curious appearance seen in the Sun. *Monthly Notices of the Royal Astronomical Society* 20:15–16.

Hodges-Kluck, E. J., Miller, M. J. and Bregman, J. N. 2016. The rotation of the hot gas around the milky way. *The Astrophysical Journal* 822(1):21.

IceCube Collaboration. 2018. Neutrino emission from the direction of the blazar TXS 0506+056 prior to the IceCube-170922A alert. *Science* 361:147–151.

IceCube Collaboration et al. 2018. Multimessenger observations of a flaring blazar coincident with high-energy neutrino IceCube-170922A. *Science* 361(6398):eaat1378.

Istomin, Ya. N. 2013. On the origin of galactic cosmic rays. arXiv:1110.5436v1 [astro-ph.GA].

Jäger, M. 2018, Jun. 2. Hubble views cosmic collision between spiral galaxies. HUBBLE space telescope.

Jiang, L., Kashikawa, N., Wang, S., et al. Evidence for GN-z11 as a luminous galaxy at redshift 10.957. *Nature Astronomy Letters*. Doi: 10.1038/s41550-020-01275-y.

Jull, A. J. T., Panyushkina, I.P., Lange, T.E., et al. 2014. Excursions in the ^{14}C record at A.D. 774–775 in tree rings from Russia and America. *Geophysical Research Letters* 41:3004–3010.

Karoff, C., Knudsen, M. F., De Cat, P., et al. 2016. Observational evidence for enhanced magnetic activity of superflare stars. *Nature Communications* 7:11058.

Kaviraj, S., Ting, Y.-S., Bureau, M., et al. 2012. Galaxy Zoo: dust and molecular gas in early-type galaxies with prominent dust lanes. arXiv:1107.5306v2 [astro-ph.CO].

Keel, W. C., Chojnowski, S. D., Bennert, V. N., et al. 2011. The Galaxy Zoo survey for giant AGN-ionized clouds: past and present black-hole accretion events. arXiv:1110.6921v2 [astro-ph.CO].

Keller, L. P. S. and Messenger, S. 2011. On the origins of GEMS grains. *Geochimica et Cosmochimica Acta* 75:5336–5365.

Keller, L. P., Messenger, S. and Bradley, J. P. 2000. Analysis of a deuterium-rich interplanetary dust particle (IDP) and implications for presolar material in IDPs. *Journal of Geophysical Research: Space Physics* 105(A5):10397–10402.

Kim, K. T., Kronberg, P. P. and Tribble, P. C. 1991. Detection of excess rotation measure due to intracluster magnetic fields in clusters of galaxies. *The Astrophysical Journal* 379:80–88.

Klein, U. 2013. *Galactic and Intergalactic Magnetic Fields*. Course astro 848, Argelander-Institut főr Astronomie, Bonn.

Kolokolova, L., Koenders, C., Rosenbush, V., et al. 2015. Dust particles alignment in the solar magnetic field: A possible cause of the cometary circular polarization. American Geophysical Union, Fall Meeting 2015. Abstract id: P41D–2089.

Kolmogorov, A. 1991. The local structure of turbulence in incompressible viscous fluid for very large Reynolds numbers. *Proceedings of the Royal Society of London. Series A* 434:9–13.

Kovaltsov, G. A. and Usoskin, I. G. 2014. Occurrence probability of large solar energetic particle events: assessment from data on cosmogenic radionuclides in lunar rocks. *Solar Physics* 289:211–220.

Kronberg, P. P. 1994. Extragalactic Magnetic Fields. *Reports on Progress in Physics* 57:325–382.

Kruijssen, J. M. D. 2019. The minimum metallicity of globular clusters and its physical origin – implications for the galaxy mass-metallicity relation and observations of proto-globular clusters at high redshift. *Monthly Notices of the Royal Astronomical Society* 486:L20–L25.

Kruijssen, J. M. D., Pfeffer, J. L., Chevance, M., Bonaca, A., Trujillo-Gomez, S., Bastian, N., reina-Campos, M., Crain, R. A. and Hughes, M. E. 2020. Kraken reveals itself – the merger history of the Milky Way reconstructed with the E-MOSAICS simulations. *Monthly Notices of the Royal Astronomical Society* 498:2472–2491.

Kruijssen, J. M. D., Pfeffer, J. L., Crain, R. A. and Bastian, N. 2019a. The E-MOSAICS project: Tracing galaxy formation and assembly with the age-metallicity distribution of globular clusters. arXiv.org>astro-ph>arXiv:1904.04261.

Kruijssen, J. M. D., Pfeffer, J. L., Reina-Campos, M., Crain, R. A. and Bastian, N. 2019. The formation and assembly history of the Milky Way revealed by its globular cluster population. *Monthly Notices of the Royal Astronomical Society* 86(3):3180–3202.

Kulsrud, R. M., Cen, R., Ostriker, J. P. and Ryu, D. 1997. The protogalactic origin for cosmic magnetic fields. *The Astrophysical Journal* 480:481–491.

Lal, D. and Venkatavaradan, V. S. 1967. Activation of cosmic dust by cosmic-ray particles. *Earth and Planetary Letters* 3:299–310.

Larsen, S. S., Romanowsky, A. J., Brodie, J. P. and Wasserman, A. 2020. An extremely metal-deficient globular cluster in the Andromeda Galaxy. *Science* 370:970–973.

Le Fèvre, O., Béthermin, M., Faisst, A., et al. 2020. The ALPINE-ALMA [CII] survey. Survey strategy, observations, and sample properties of 118 star-forming galaxies at 4 < z > 6. *Astronomy and Astrophysics* 643:A1 (19 p).

Leroy, A., Bolatto, A. D., Simon, J. D. and Blitz, L. 2005. The molecular interstellar medium of dwarf galaxies on kiloparsec scales: a new survey for CO in Northern, IRAS-detected dwarf galaxies. *The Astrophysical Journal* 625:763–784.

Li, H.-B., Blundell, R., Hedden, A., et al. 2011. Evidence for dynamically important magnetic fields in molecular clouds. *Monthly Notices of the Royal Astronomical Society* 411:2067–2075.

Li, H.-B. and Henning, T. 2011. The alignment of molecular cloud magnetic fields with the spiral arms in M33. *Nature* 479:499–501.

Lintott, C., Schawinski, K., Kell, W., et al. 2009. Galaxy Zoo : 'Hanny's Voorwerp', a quasar light echo?. arXiv:0906.5304v1 [astro-ph.CO].

Loeb, A. and Mao, S. 1994. Evidence from gravitational lensing for a non-thermal pressure support in the cluster of galaxies A2218. *The Astronomical Journal* 435:L109–L112.

Luck, R. E. and Lambert, D. L. 2011. The distribution of the elements in the galactic disk. III. A reconsideration of cepheids from l=30° to 250°. *The Astronomical Journal* 142:136–152.

Maiolino, R., Nagao, T., Grazian, A., et al. 2008. AMAZE I. The evolution of the mass-metallicity relation at z>3. *Astronomy and Astrophysics* 488:463–479.

Malkov, M. A. and Drury, L. O. 2001. Nonlinear theory of diffusive acceleration of particles by shock waves. *Reports on Progress in Physics* 64:429–481.
Mannheim, K. 1995. High-energy neutrinos from extragalactic jets. *Astroparticle Physics* 3:295–302.
Mannucci, F., Cresci, G., Maiolino, R., Marconi, A. and Gnerucci, A. 2010. A fundamental relation between mass, star formation rate and metallicity in local and high-redshift galaxies. *Monthly Notices of the Royal Astronomical Society* 408:2115–2127.
Mannucci, F., Cresci, G., Maiolino, R., et al. 2009. LSD: Lyman-break galaxies Stellar populations and Dynamics - I. Mass, metallicity and gas at z ~ 3.1. *Monthly Notices of the Royal Astronomical Society* 398(4):1915–1931.
Marrone, D. P., Spilker, J. S. Hayward, C. C., et al. 2018. Galaxy growth in a massive halo in the first billion years of cosmic history. *Nature* 553(7686): 51–54.
Masters, K. L., Nichol, R. C., Haynes, M. P., et al. 2012. Galaxy zoo and ALFALFA: Atomic gas and the regulation of star formation in barred disc galaxies. arXiv: 1205.5271v2 [astro-ph.CO].
Masters, K. L., Nichol, R. C., Hoyle, B., et al. 2011. Galaxy zoo: Bars in disk galaxies. arXiv:1003.0449v2 [astro-ph.CO].
Matsunaga, N., Feast, M. W., Bono, G., Kobayashi, N., Inno, L., Nagayama, T., Nishiyama, S., Matsuoka, Y. and Nagata, T. 2016. A lack of classical Cepheids in the inner part of the Galactic disc. *Monthly Notices of the Royal Astronomical Society* 462:414–420.
Mayerson, D. R. 2020, Oct. 27. Fuzzballs and Observations. arXiv:2010.09736v2 [hep-th].
McClure-Griffiths, N. M., Denes, H., Dickey, J. M., et al. 2018, November. Cold gas outflows from the small magellanic cloud traced with ASKAP. *Nature Astronomy* 2:901–906.
McGill Magnetar Catalog. www.physics.mcgill.ca/~pulsar/magnetar/v160321/main.html.This site is maintained by the McGill Pulsar Group, Department of Physics, McGill University, Montreal, Canada.
Melott, A. L. and Thomas, B. C. 2012. Causes of an AD 774–775 ^{14}C increase. *Nature* 491:E1–E2.
Mészáros, P. 2017. Astrophysical sources of high energy neutrinos in the icecube era. *Annual Review of Nuclear and Particle Science* 67:45–67.
Mészáros, P. 2018, Oct. Astrophysical high-energy neutrinos. *Physics Today*, pp. 36–42.
Mezcua, M., Hlavacek-Larrondo, J., Lucey, J. R., Hogan, M. T., Edge, A. C. and McNamara, B. R. 2015. The most massive black holes on the fundamental plane of black hole accretion. *Monthly Notices of the Royal Astronomical Society*: 1–16. arXiv:1710.10268v2 [astro-ph.GA] 6 Nov 2017.
Miller-Jones, J. C. A., Bahramian, A., Orosz, J. A., et al. 2021. Cygnus X-1 contains a 21-solar mass black hole – Implications for massive star winds. *Science* Doi: 10.1126/science abb3363.
Miniati, F. and Bell, A. R. 2011. Resistive magnetic field generation at cosmic dawn. *The Astrophysical Journal* 729:73 (9 p).
Miniati, F., Ryu, D., Kang, H., et al. 2000. Properties of cosmic shock waves in large-scale structure formation. *The Astrophysical Journal* 542:608–621.
Miyake, F., Nagaya, K., Masuda, K. and Nakamura, T. 2012. A signature of cosmic-ray increase in AD 774–775 from tree rings in Japan. *Nature* 486:240–242.

Miyake, F., Suzuki, A., Masuda, K. Horiuchi, K., Motoyama, H. Matsuzaki, H., Motizuki, Y., Takahashi, K. and Nakai, Y. 2015. Cosmic ray event of A.D. 774–775 shown in quasi-annual^{10}Be data from the Antarctic Dome Fuji ice core. *Geophysical Research Letter* 42:84–89.

Morfill, G. E., Forman, M. and Bignami, G. 1984. On the nature of the galactic gamma-ray sources. *The Astronomical Journal* 284:856–868.

Moskalenko, I. V. and Strong, A. W. 1998. Production and propagation of cosmic-ray positrons and electrons. *The Astrophysical Journal* 493:694–707.

Moss, D., Stepanov, R., Krause, M., Beck, R. and Sokoloff, D. 2015. The formation of regular interarm magnetic fields in spiral galaxies. *Astronomy & Astrophysics* 578: A94. Doi: 10.1051/0004-6361/201526145.

Most, E. R., Papenfort, L. J., Dexheimer, V., Hanauske, M., Schramm, S., Stöcker, H. and Rezzolla, L. 2019, Feb. 12. Signatures of Quark-Hadron phase transition in general-relativistic neutron-star merger. *Physical Review Letters* 122:061101.

Mott, A., Spitoni, E. and Matteucci, F. 2013. Abundance gradients in spiral discs: Is the gradient inversion at high redshift real? *Monthly Notices of the Royal Astronomical Society* 435:2918–2930.

Mucke, A., Protheroe, R. J., Engel, R., Rachen, J. P. and Stanev, T. 2003. BL Lac objects in the synchrotron proton blazar model. *Astroparticle Physics* 18:593–613.

Muller, C. 2014. The Carrington solar flares of 1859: Consequences on life. *Origins of Life and Evolution of Biospheres* 44:185–195.

Murase, K. 2017. *Neutrino Astronomy - Current Status, Future Prospects in Neutrino Astronomy*. Gaisser, T. and Karle, A. (Eds.) World Scientific, Singapore, pp. 15–31.

Nagano, M. and Watson, A. A. 2000. Observations and implications of the ultrahigh-energy cosmic rays. *Reviews of Modern Physics* 72:689–732.

Naoz, S. and Narayan, R. 2013. Generation of primordial magnetic fields on linear over-density scales. *Physical Review Letters* 111:051303-1/5.

Ness, M., Hogg, D. W., Rix, H.-W., Ho, A. Y. Q. and Zasowski, G. 2015. The cannon: A data-driven approach to stellar label determination. *The Astronomical Journal* 808:16. arXiv:1501.07604v2 [astro-ph.SR].

Ness, M., Hogg, D. W., Rix, H.-W., Martig, M. and Ho, A. 2016. Spectroscopic determination of masses (and implied ages) for red giants. *227th Meeting of American Astronomical Society*, Kissimmee, FL, 4–8 January 2016. Abstract 425.04.

NRAO, National Radio Astronomy Observatory. 2021. Quasar discovery sets new distance record. Public.nrao.edu/news/quasar-new-distance-record/. News Release: January 12. 2021 at 12:25 pm EST. To be published in Astrophysical Journal Letters.

Obermeier, A., Ave, M., Boyle, P., Höppner, Ch., Hörandel, J. and Müller, D. 2011. Energy spectra of primary and secondary cosmic-ray nuclei measured with TRACER. *The Astrophysical Journal*, 742:14 (11 p).

Oesch, P. A., Brammer, G. van Dokkum, P. G., et al. 2016. A remarkably luminous galaxy at z=11.1 measured with hubble space telescope grism spectroscopy. *The Astrophysical Journal* 819(2): article id. 129, 11 p.

Oka, T., Hasegawa, T., Sato, F., Tsuboi, M. and Miyazaki, A. 1998. A large-scale CO survey of the galactic center. *The Astrophysical Journal Supplement Series* 118:455–515.

Pacucci, F., Ferrara, A., Grazian, A., Fiore, F., Giallongo, E. and Puccetti, S. 2016. First identification of direct collapse black hole candidates in the early Universe in CANDELS/GOODS-S. *Monthly Notices of the Royal Astronomical Society* 459(2):1432–1439.

Panov, A. Sokolskaya, N. V., Adams, J. H., et al. 2007. Relative abundances of cosmic ray nuclei B-C-N-O in the energy region from 10GeV/n to 300 GeV/n. Results from ATIC-2 (the science flight of ATIC). In *30th International Cosmic Ray Conference*, Merida, Yucatan, Mexico, 3–11 July 2007.

Parker, E. N. 1955. Hydromagnetic dynamo models. *The Astrophysical Journal* 122:293–314.

Passot, T., Vazquez-Semadeni, E. and Pouquet, A. 1995. A turbulent model for the interstellar medium. II. Magnetic fields and rotation. *The Astrophysical Journal* 455:536–555.

Pearlman, A. B., Majid, W. A., Prince, T. A., Kocz, J. and Horiuchi, S. 2018, Oct. 20. Pulse morphology of the galactic center magnetar PSR J1745–2900. *The Astrophysical Journal* 866:160 (17p).

Penrose, R. 1965. Gravitational collapse and space-time singularities. *Physical Review Letters* 14(3):57–59.

Petersen, M. S. and Peñarrubial, J. 2020. Detection of the Milky Way reflex motion due to the Large Magellanic Cloud infall. *Nature Astronomy.* Doi: 10.1038/s41550-020-01254-3.

Petroff, E. 2019. Finding the location of a fast radio burst. *Science* 365(6453):546–547.

Petropoulou, M. and Dermer, C. 2016, Jul. 1. Properties of blazar jets defined by an economy of power. *Astrophysical Journal Letters* 825: L11 (5 p).

Picozza, P., Galper, A.M., Castellini, G., et al. 2007. PAMELA - A payload for antimatter matter exploration and light-nuclei astrophysics. *Astroparticle Physics* 27:296–315.

Pierre Auger Collaboration. 2020. Measurement of the cosmic-ray energy spectrum above 2.5×10^{18} eV using the Pierre Auger Observatory. *Physical Review D* 102:062005 (p. 27).

Pilipenko, S. V. 2013, Mar. 24. Paper-and-pencil cosmological calculator. arXiv:1303.5961 [astro-ph.CO].

Planck collaboration. 2013. Planck 2013 results. XVI. *Astronomy & Astrophysics.* arXiv:astro-ph/1303.5076.

Planck Collaboration. 2015a. Planck 2015 results. XIX. Constraints on primordial magnetic fields. *Astronomy & Astrophysics* 594: A19 (27pp).

Planck Collaboration. 2015b. Planck 2015 results. I. Overview of products and results. *Astronomy & Astrophysics* (2016).

Price, D. and Bate, M. 2008. The effect of magnetic fields on star cluster formation. *Monthly Notices of the Royal Astronomical Society* 385:1820–1834.

Ramaty, R., Kozlovsky, B., and Lingenfelter, R. E. 1979. Nuclear gamma-rays from energetic particle interactions. *Astrophysical Journal Supplement Series* 40:487–526.

Ratra, B. 1992. Cosmological 'seed' magnetic field from inflation. *The Astrophysical Journal* 391(1):L1–L4.

Reeves, H. 1994. On the origin of the light elements ($Z<6$). *Reviews of Modern Physics* 66:193–216.

Remington, B. A., Arnett, D., Drake, R. P. and Takabe, H. 1999. Modelling astrophysical phenomena in the laboratory with intense lasers. *Science* 284:1488–1493.

Remington, B. A., Drake, R. P. and Ryutov, D. D. 2006. Experimental astrophysics with high power lasers and Z pinches. *Reviews of Modern Physics* 78:755–807.

Riechers, D. A., Bradford, C. M., Clements, D. L., et al. 2013. A dust-obscured massive maximum-starburst galaxy at a redshift of 6.34. *Nature* 496: 329–333.

Riechers, D. A., T. K. Daisy Leung, T. K., Rob J., et al. 2018. Rise of the titans: A dusty, hyper-luminous "870μm Riser" galaxy at z~6. *The Astrophysical Journal* 850:1 (8 p), 2017 November 20.

Ryu, O., Kang, H., Cho, J. and Das, S. 2008. Turbulence and magnetic fields in the large-scale structure of the Universe. *Science* 320:909–912.

Ryutov, D. D., Drake, R. P., Kane, J., et al. 1999. Similarity criteria for the laboratory simulation of supernova hydrodynamics. *The Astrophysical Journal* 518:821–832.

Ryutov, D. D., Drake, R. P. and Remington, B. A. 2000. Criteria for scaled laboratory simulations of astrophysical MHD phenomena. *The Astrophysical Journal Supplement Series* 127:465–468.

Schaefer, B. E., King, J. R. and Deliyannis, C. P. 2000. Superflares on ordinary solar-type stars. *The Astrophysical Journal* 529:1026–1030.

Schawinski, K., Lintott, C., Thomas, D., et al. 2009. Galaxy Zoo: A sample of blue early-type galaxies at low redshift. arXiv:0903.3415v1 [astro-ph.CO].

Schawinskily, K., Urry, C. M., Simmons, B. D., et al. 2013. The green valley is a red herring: Galaxy zoo reveals two evolutionary pathways towards quenching of star formation in early- and late-type galaxies? arXiv:1402.4814v1 [astro-ph.GA].

Schilbach, E., Roeser, S. and Scholz, R.D. 2009. Trigonometric parallaxes of ten ultracool subdwarfs. *Astronomy & Astrophysics* 493:L27–L30.

Schlegel, E. M., Jones, C., Machacek, M. E. and Vega, L. D. 2016. NGC 5195 in M51: Feedback "Burps" after a Massive Meal? *227th Meeting of American Astronomical Society*, Kissimmee, FL, 4–8 January 2016. Abstract 118.04.

Schlickeiser, R. and Shukla, P. K. 2003. Cosmological magnetic field generation by the Weibel instability. *The Astrophysical Journal* 599:L57–L60.

Schrijver, C. J., Beer, J., Baltensperger, U., et al. 2012. Estimating the frequency of extremely energetic solar events based on solar, stellar, lunar, and terrestrial records. *Journal of Geophysical Research* 117:A08103.

Shetty, R. and Ostriker, E. 2006. Global modelling of spur formation in spiral galaxies. *The Astrophysical Journal* 647:997–1017.

Spitoni, E. and Matteucci, F. 2015. The effect of radial gas flows on the chemical evolution of the Milky Way and M31. arXiv:1502.01836 [astro-ph.GA].

Springel, V., White, S.D., Jenkins, A., et al. 2005. Simulations of the formation, evolution and clustering of galaxies and quasars. *Nature* 435:629–636.

Spruit, H. C. 2011. *The Sun, the Solar Wind, and the Heliosphere*. Miralles, M. P., and Sánchez Almeida, J. (Eds.) Springer, Berlin. p. 39.

Squire, J. and Bhattacharjee, A. 2015. Generation of large-scale magnetic fields by small-scale dynamo in shear flows. *Physical Review Letters* 115:175003.

Strandet, M. L., Weiss, A., De Breuck, C., et al. 2017. ISM properties of a massive dusty star-forming galaxy discovered at $z \sim 7$. *The Astrophysical Journal* 842:L15.

Su, M., Slatyer, T. R. and Finkbeiner, D. P. 2010. Giant gamma-ray bubbles from Fermi-LAT: active galactic nucleus activity or bipolar galactic wind? *The Astrophysical Journal* 724:1044–1082.

Sun, W. 2015. *Precision Measurement of the Boron to Carbon Ratio in Cosmic Rays with AMS-02.* Ph.D. Thesis, Massachusetts Institute of Technology, Cambridge, MA 02139, USA.

Sur, S., Schleicher, D. R. G., Banerjee, R., Federrath, C. and Klessen, R. S. 2010. The generation of strong magnetic fields during the formation of the first stars. *The Astrophysical Journal.* 721:L134–L138.

Stahler, W. and Palla, F. 2004. *The Formation of Stars.* Whiley-VCH, Weinheim, 865 p.

Staveley-Smith, L., Kraan-Korteweg, R. C., Schröder, A. C., Henning, P. A., Koribalski, B. S., Stewart, I. M. and Heald, G. 2016. The parkes HI zone of avoidance survey. *The Astronomical Journal* 151:52 (28 p).

Strong, A. W. and Moskalenko, I. V. 1998. Propagation of cosmic-ray nucleons in the galaxy. *The Astrophysical Journal* 509:212–228.

Strong, A. W. and Moskalenko, I. V. 1999. The GALPROP program for cosmic-ray propagation: new developments. *Proc. 26th ICRC (Salt Lake City, 1999),* Paper OG 3.2.18.

Swordy S. P., Müller D., Meyer P., L'Heureux J. and Grunsfeld J. M. 1990. Relative abundances of secondary and primary cosmic rays at high energies. *The Astrophysical Journal* 349:625–633.

Tacconi, L. J., Genzel, R., Neri, R., et al. 2010. High molecular gas fractions in normal massive star forming galaxies in the young Universe. *Nature* 463:781–784.

Tadaki, K., Iono, D., Yun, S. et al. 2018. Gravitationally unstable gas disk of a starburst galaxy 12 billion years ago. *Nature* 560: 613–616.

Tatischeff, V. and Kiener, J. 2004. Gamma-ray lines from cosmic-ray interactions with interstellar dust grains. *New Astronomy Reviews* 48:99–103.

Taylor, A. R. and Jagannathan, P. 2016. Alignments of radio galaxies in deep radio imaging of ELAIS N1. *Monthly Notices of the Royal Astronomical Society Letters* 459(1):L36–L40.

Taylor, G. B. and Perley, R. A. 1993. Magnetic fields in the hydra A cluster. *The Astrophysical Journal.* 416:554.

The CHIME/FRB Collaboration. 2020. A bright millisecond-duration radio burst from a Galactic magnetar. Nature 587: 05 November 2020. Doi: 10.1038/s41586-020-2863-y.

The Event Horizon Telescope Collaboration. 2019a, Apr. 10. First M87 event horizon telescope results. I. The shadow of the supermassive black hole. *The Astrophysical Journal Letters* 875:L1 (17 p).

The Event Horizon Telescope Collaboration. 2019b, Apr. 10. First M87 event horizon telescope results. II. Array and instrumentation. *The Astrophysical Journal Letters* 875:L2 (28 p).

The Event Horizon Telescope Collaboration. 2019c, Apr. 10. First M87 event horizon telescope results. III. Data processing and calibration. *The Astrophysical Journal Letters* 875:L3 (32 p).

The Event Horizon Telescope Collaboration. 2019d, Apr. 10. First M87 event horizon telescope results. IV. Imaging the central supermassive black hole. *The Astrophysical Journal Letters* 875:L4 (52 p).

The Event Horizon Telescope Collaboration. 2019e, Apr. 10. First M87 event horizon telescope results. V. Physical origin of the asymmetric ring. *The Astrophysical Journal Letters* 875:L5 (31 p).

The Event Horizon Telescope Collaboration. 2019f, Apr. 10. First M87 event horizon telescope results. VI. The shadow and mass of the central black hole. *The Astrophysical Journal Letters* 875:L6 (44 p).

Toyouchi, D. and Chiba, M. 2014, May 2. On the chemical and structural evolution of the galactic disk. arXiv:1405.0405v1 [astro-ph.GA].

Tremblay, G. R., Combes, F., Oonk, J. B. R., et al. 2018, Sep. 20. A galaxy-scale fountain of cold molecular gas pumped by a black hole. *The Astrophysical Journal* 865:13 (24 p).

Tsuboi, M., Handa, T. and Ukita N. 1999. Dense molecular clouds in the galactic center region. I. Observations and data. *The Astrophysical Journal Supplement Series* 120:1–39.

Tully, R. B., Courtois, H., Hoffman, Y. and Pomarède, D. 2014. The Laniakea supercluster of galaxies. *Nature* 513:71–73.

Ueda, J., Umehata, H., Wilson, G. W., Michiyama, T., M. Ando, M. and Kamieneski, P. 2018, Aug. 30. The gravitationally unstable gas disk of a starburst galaxy 12 billion years ago. *Nature* 560:613–616.

Urry, C. M. and Padovani, P. 1995. Unified schemes for radio-loud active galactic nuclei. *Publications of the Astronomical Society of the Pacific* 107(715):803–845..

Usoskin, I. G. and Kovaltsov, G. A. 2012. Occurrence of extreme solar particle events: assessment from historical proxy data. *The Astrophysical Journal* 757:92.

Usoskin, I. G., Kromer, B., Ludlow, F., Beer, J., Friedrich, M., Kovaltsov, G. A., Solanki, S. K. and Wacker, L. 2013. The AD775 cosmic event revisited: the Sun is to blame. *Astronomy & Astrophysics* 552:L3.

van der Marel, R. P., Fardal, M. A., Sohn, S. T., Patel, E., Besla, G., del Pino, A., Sahlmann, J. and Watkins, L. L. 2019, Feb. 10. First gaia dynamics of the andromeda system: DR2 proper motions, orbits, and rotation of M31 and M33. *The Astrophysical Journal* 872:24 (14 p).

van Weeren, R. J., Andrade-Santos, F., Dawson, W. A., et al. 2017. The case for electron re-acceleration at galaxy cluster shocks. *Nature Astronomy* 1: article number 0005.

Vayner, A., Wright, S. A., Murray, N., Armus, L., Larkin, J. E. and Etsuko, M. 2017, Dec. 20. Galactic-scale feedback observed in the 3C 298 quasar host galaxy. *The Astrophysical Journal* 851:126 (18p).

Vieira, J. D., Marrone, D.P., Chapman, S.C., et al. 2013. Dusty starburst galaxies in the early Universe as revealed by gravitational lensing. *Nature* 495:344–347.

Webster, G. 2018, Jan. 11 Jet propulsion laboratory, Pasadena, Calif. News media contact. NASA's Great Observatories Team Up to Find Magnified and Stretched Image of Distant Galaxy. News.

Wei, J.-J., Wu, X.-F., Melia, F., et al. 2015. The age-redshift relationship of old passive galaxies. *The Astronomical Journal* 150:35 (13 p).

Westphal, A. J., Stroud, R. M., Bechtel, H. A., et al. 2014. Evidence for interstellar origin of seven dust particles collected by the stardust spacecraft. *Science* 345(6198):786–791.

Widrow, L. M. 2002. Origin of galactic and extragalactic magnetic fields. *Reviews in Modern Physics* 74:775–823.

Xu, H., O'Shea, B. W., Collins, D. C., et al. 2008. The Biermann battery in cosmological MHD simulations of population III star formation. *The Astrophysical Journal* 688:L57–L60.

Yang, G., Brandt, W. N., Vito, F., et al. 2017, May 25. Linking black-hole growth with host galaxies: The accretion-stellar mass relation and its cosmic evolution. arXiv:1704.06658v2 [astro-ph.HE].

Yoshikawa, K., Tanaka, S., Yoshida, N. and Saito, S. 2020. Cosmological vlasov-poisson simulations of structure formation with relic neutrinos: Nonlinear clustering and neutrino mass. *The Astrophysical Journal* 904:159 (16 p).

Zatsepin, G. T. and Kuz'min, V. A. 1966. Upper limit of the spectrum of cosmic rays. *Journal of Experimental and Theoretical Physics Letters* 4:78–80.

Zhukovska, S., Gail, H.-P. and Trieloff, M. 2008. Evolution of interstellar dust and stardust in the solar neighborhood. *Astronomy & Astrophysics* 479:453–480.

Zweibel, E. G. and Heiles, C. 1997. Magnetic fields in galaxies and beyond. *Nature* 385:131–136.

Additional reading

Abeysekara, A. U., Albert, A., Alfaro, R., et al. 2021. HAWC observations of the acceleration of very-high-energy cosmic rays in the Cygnus Cocoon. *Nature Astronomy Letters*. Doi: 10.1038/s41550-021-01318-y.

Aguilar, M., Cavasonza, L.A., Ambrosi, G., et al. (AMS Collaboration). 2020. Properties of neon, magnesium, and silicon primary cosmic rays results from the alpha magnetic spectrometer. *Physical Review Letters* 124:211102 (pp. 1–8).

Barrow, K. S. S., Aykutalp, A. and Wise, J. H. 2018, Sep. 10. Observational signatures of massive black hole formation in the early Universe. *Nature Astronomy*, Articles, Doi: 10.1038/s41550-018-0569-y. www.nature.com/natureastronomy.

Bianchi, S., Antonucci, R., Capetti, A., et al. 2019. HST unveils a compact mildly relativistic broad-line region in the candidate true type 2 NGC 3147. *Monthly Notices of the Royal Astronomical Society: Letters* 488(1):L1–L5.

Beck, R., Fletcher, A., Shukurov, A., et al. 1996. Magnetic fields in spiral galaxies. *Annual Review of Astronomy and Astrophysics* 34:155–206.

Bergemann, M., Sesar, B., Cohen, J. G., et al. 2018. Two chemically similar stellar overdensities on opposite sides of the plane of the Galactic disk. *Nature* 555:334–337.

Caplan, M. E., Schneider, A. S. and Horowitz, C. J. 2018. The elasticity of nuclear pasta. *Physical Review Letters*. 121, 132701, 5 p.

Chen, X., Wang, S., Deng, L., de Grijs, R., Liu, C. and Tian, H. 2019. An intuitive 3D map of the Galactic warp's precession traced by classical Cepheids. *Nature Astronomy, Letters*. Doi: 10.1038/s41550-018-0686-7. Published online: 04 February 2018.

Cheng, X., Anguiano, A., Majewski, S. R., et al. 2020. Exploring the galactic warp through asymmetries in the kinematics of the galactic disk. *The Astrophysical Journal* 905:49 (15 p).

Choi, S. K., Hasselfield, M., Ho, S.P.P., et al. 2020. The atacama cosmology telescope: A measurement of the cosmic microwave background power spectra at 98 and 150 GHz. *Journal of Cosmology and Astroparticle Physics* 12:045.

Cooke, K. C., Kirkpatrick, A., Estrada, M., et al. 2020. Dying of the light: An X-ray fading cold quasar at z ~0.405. *The Astronomical Journal* 903:106 (9 p).

Cromartie, H. T., Fonseca, E., Ransom, S. M., et al. 2019. Relativistic Shapiro delay measurements of an extremely massive millisecond pulsar. *Nature Astronomy Letters* Doi: 10.1038/s41550-019-0880-2.

Delhaize, J., Heywood, I., Prescott, M., et al. 2021. MIGHTEE: are giant radio galaxies more common than we thought?. *Monthly Notices of the Royal Astronomical Society* 501:3833–3845.

Dessauges-Zavadsky, M., Richard, J., Combes, F., et al. 2019. Molecular clouds in the Cosmic Snake normal star-forming galaxy 8 billion years ago. *Nature Astronomy Letters* Doi: 10.1038/s41550-019-0874-0.

Di Teodoro, E. M., McClure-Griffiths, N. M., Lockman, F. J. and Armillotta, L. 2020. Cold gas in the Milky Way's nuclear wind. *Nature* 584:364–367.

Do, T., Witzel, G., Gautam, A. K., et al. 2019. Unprecedented Near-infrared Brightness and Variability of Sgr A*. *The Astrophysical Journal Letters* 882:L27 (7 p).

Donati, J.-F., Moutou, C., Farès, R., Bohlender, D., Catala, C., Deleuil, M., Shkolnik, E., Cameron, A. C., Jardine, M. M., and Walker, G. A. H. 2008. Magnetic cycles of the planet-hosting star τ Bootis. *Monthly Notices of the Royal Astronomical Society* 385:1179–1185.

Donlon, T., Newberg, H. J., Sanderson, R. and Widrow, L. M. 2020. The Milky Way's Shell Structure reveals the time of radial collision. *The Astrophysical Journal* 902:119 (27 p).

Drlica-Wagner, A., Bechtol, K., Mau. S., et al. 2020. Milky way satellite census. I. The observational selection function for milky way satellites in DES Y3 and pan-STARRS DR1. *The Astrophysical Journal* 893(1):47.

Farihi, J., van Lieshout, R., Cauley, P. W., et al. 2018. Dust production and depletion in evolved planetary systems. *Monthly Notices of the Royal Astronomical Society* 481(2):2601–2611.

Farrell, S. A., Servillat, M., Pforr, J., et al. 2012, Jan. 10. A young massive stellar population around the intermediate mass black hole ESO 243-49 HLX-1. arXiv:1110.6510v3 [astro-ph.CO].

Fattahi, A., Belokurov, V., Deason, A. J., Frenk, C. S., Gómez, F. A., Grand, R. J. J., Marinacci, F., Pakmor, R. and Springel, V. 2019. The origin of galactic metal-rich stellar halo components with highly eccentric orbits. *Monthly Notices of the Royal Astronomical Society* 484(4):4471–4483.

Frederick, S., Gezari, S., Graha, M. J., et al. 2019. A new class of changing-look LINERs. *The Astrophysical Journal* 883:31 (30 p), Doi: 10.3847/1538-4357/ab3a38.

Fu, R. R., Weiss, B. P., Lima, E. A., et al. 2014. Solar nebula magnetic fields recorded in the Semarkona meteorite. *Science* 346(6213):1089–1092, Doi: 10.1126/science.1258022.

Gao, F., Wang, L., Pearson, W. J., Gordon, Y. A., Holwerda, B. W., Hopkins, A. M., Brown, M. J. I., Bland-Hawthorn, J. and Owers, M. S. 2020. Mergers trigger active galactic nuclei out to z ~0.6. *Astronomy & Astrophysics* 637:A94-16 p. Doi: 10.1051/0004-6361/201937178.

Harkane, Y., Ouchi, M., Ono, Y., et al. 2019. SILVERRUSH VIII. Spectroscopic identification of early large-scale structures with protoclusters over 200 Mpc at z~6–7: Strong associations of dusty star-forming galaxies. *The Astrophysical Journal* 883:142 (16 p).

Helmi, A., Babusiaux, C., Koppelman, H. H., Massari, D., Veljanoski, J. and Brown, A. G. A. 2018. The merger that led to the formation of the Milky Way's inner stellar halo and thick disk. *Nature* 563:85–88.

Heywood, I., Camilo, F., Cotton, W. D., et al. 2019. Inflation of 430-parsec bipolar radio bubbles in the Galactic Centre by an energetic event. *Nature* 573:235–237.

Holoien, T. W.-S., Vallely, P. J., Auchettl, K., et al. 2019. Discovery and early evolution of ASASSN-19bt, the first TDE detected by TESS *The Astrophysical Journal* 883:111 (17 p), Doi:10.3847/1538-4357/ab3c66.

Horvat, R., Trampetic, J. and You, J. 2017. Spacetime deformation effect on the early Universe and the PTOLEMY experiment. *Physics Letters B* 772:130–135.

Horvat, R., Trampetic, J. and You, J. 2018. Inferring type and scale of noncommutativity from the PTOLEMY experiment. *European Physical Journal C* 78:572 (6 p).

Impellizzeri, C. M. V. 2019. Counter-rotation and high-velocity outflow in the parsec-scale molecular torus of NGC 1068. *The Astrophysical Journal Letters* 884:L28 (6 p).

Ito, H., Matsumoto, J., Nagataki, S., Warren, D. C., Barkov, M. V. and Yonetoku, D. 2019. The photospheric origin of the Yonetoku relation in gamma-ray bursts. *Nature Communications* 10, Article number: 1504.

Kruijssen, J. M. D., Dale, J. E., Longmore, S. N., et al. 2019. The dynamical evolution of molecular clouds near the Galactic Centre – II. Spatial structure and kinematics of simulated clouds. *Monthly Notices of the Royal Astronomical Society* 484(4):5734–5754.

Lee, J.-E., Lee, S., Baek, G., Aikawa, Y. Cieza, L. Yoon, S.-Y., Herczeg, G., Johnstone, D. and Casassus, S. 2019. The ice composition in the disk around V883 Ori revealed by its stellar outburst. *Nature Astronomy, Letters* Doi: 10.1038/s41550-018-0680-0.

Lehner, N. 2020. Project AMIGA: The Circumgalactic Medium of Andromeda. *The Astrophysical Journal* 900:9 (44 p).

Levesque, E. M., Kewley, L. J., Graham, J. F. and Fruchter, A. S. 2013. A high-metallicity host environment for the long-duration GRB 020819. arXiv:1001.0970v2 [astro-ph. HE]4 Feb 2010.

Liu, J., Zhang, H., Howard, A.W., et al. 2019. A wide star-black-hole binary system from radial velocity measurements. *Nature* 575:618–621.

López Maroto, A. and Beltrán Jiménez, J. 2010. Dark energy, non-minimal couplings and the origin of cosmic magnetic fields. *Journal of Cosmology and Astroparticle Physics* (12). ISSN 1475-7516.

Lucchini, S., D'Onghia, E., Fox, A. J., Bustard, C., Bland-Hawthorn, J. and Zweibel, E. 2020. The magellanic corona as the key to the formation of the magellanic stream. *Nature* 585:203–206.

Martin, L., Ho, S. H., Kacprzak, G. G. and Churchill, C. W. 2019. Kinematics of circumgalactic gas: Feeding galaxies and feedback crystal. *The Astrophysical Journal*, 878:84 (28 p), Doi:10.3847/1538-4357/ab18ac.

Matsuoka, Y., Onoune, M., Kashikawa, N., et al. 2019, Jan 29. Discovery of the first low-luminosity quasar at z >7. arXiv:1901.10487v1 [astro-ph.GA].

McConnell, D., Hale, C. L., Lenc, E., et al. 2020. The rapid ASKAP continuum survey I: Design and first result. *Publications of the Astronomical Society of Australia* 37:e048 (18 p).

Meingast, S., Alves, J. and Fürnkranz, V. 2019. Extended stellar systems in the solar neighborhood. *Letter to the Editor*, A/A 622, L13 pp. 1–6.

Miki, Y., Mori, M. and Kawaguchi, T. 2021. Destruction of the central black hole gas reservoir through head-on galaxy collision. *Nature Astronomy, Letters*. Nature Astronomy | www./nature.com/natureastronomy.

Miller-Jones, J. C. A., Bahramian, A., Orosz, J. A., et al. 2021, Feb 18. Cygnus X-1 contains a 21-solar mass black hole – implications for massive star winds. *Science*. Doi: 10.1126/science.abb3363.

Mor, R., Robin, A. C., Figueras, F., Roca-Fàbrega, S. and Luri, X. 2019. Gaia DR2 reveals a star formation burst in the disc 2–3 Gyr ago. *Astronomy & Astrophysics* 624: L1 (Article Number). Letters to the Editor, Published online. Doi: 10.1051/0004-6361/201935105.

Mróz, P., Poleski, R., Gould, A., et al. 2020, Oct 20. A terrestrial-mass rogue planet candidate detected in the shortest-timescale microlensing event. arXiv:2009.12377v2 [astro-ph.EP].

Nadler, E. O., Wechsler, R. H., Bechtol, K., et al. 2020. Milky way satellite census. II. Galaxy–halo connection constraints including the impact of the large magellanic cloud. *The Astrophysical Journal* 893(1):48.

Neeleman, M., Prochaska, J. X., Kanekar, N. and Rafelski, M. 2020. A cold, massive, rotating disk galaxy 1.5 billion years after the Big Bang. *Nature* 581:269–272.

Neijssel, C. J., Vinciguerra, S., Vigna-Gómez, A., et al. 2021. Wind mass-loss rates of stripped stars inferred from Cygnus X-1. *The Astrophysical Journal* 908:118 (9 p).

Nokhrina, E. E., Gurvits, L. I., Beskin, V. S., Nakamura, M., Asada, K., and Hada, K. 2019. M87 black hole mass and spin estimate through the position of the jet boundary shape break. *Monthly Notices of the Royal Astronomical Society* 489(1):1197–1205.

O'Brien, T., Tarduno, J. A. Anand, A., et al. 2020. Arrival and magnetization of carbonaceous chondrites in the asteroid belt before 4562 million years ago. *Communications Earth & Environment* 1:54. Doi: 10.1038/s4347-020-00055-w.

Pasham, D. R., Strohmayer, T. E. and Mushotzky, R. F. 2014. A 400-solar-mass black hole in the galaxy M82. *Nature* 513:74–76. Doi: 10.1038/nature/13710.

Prochaska, J. X.., Macquart, J.-P. and McQuinn, M. 2019. The low density and magnetization of a massive galaxy halo exposed by a fast radio burst. *Science* Doi: 10.1126/science.aay0073. Supplementary materials science.sciencemag.org/cgi/content/full/science.aay0073/DC1 Materials and Methods.

Rich, R. M. 1990. Kinematics and abundances of K giants in the nuclear bulge of the galaxy. *The Astronomical Journal* 362:604–629.

Riley, J., Mandel, I., Marchant, P., et al. 2020. Chemically homogeneous evolution: A rapid population synthesis approach. arXiv:2010.00002v1 [astro-ph.SR] 29 Sep 2020.

Schellenberger, G., David, L., O'Sullivan, E., et al. 2019. Forming one of the most massive objects in the universe: The quadruple merger in abell 1758. *The Astrophysical Journal* 882:59 (19 p).

Snios, B., Siemiginowska, A., Sobolewska, M., et al. 2020. X-Ray properties of young radio quasars at z>4.5. *The Astrophysical Journal* 899:127 (10 p).

Spingola, C., Dallacasa, D., Belladitta, S., et al. 2020. Parsec-scale properties of the radio brightest jetted AGN at z>&. *Astronomy & Astrophysics* 643:L12 (8 p).

Teague, R., Bae, J. and Bergin, E. 2019. Meridional flows in the disk around a yang star. *Nature* 574:378–381.

The Tibet Asγ Collaboration. 2021. Potential PeVatron supernova remnant G106.3+2.7 seen in the highest-energy gamma rays. *Nature Astronomy Letters.* Doi: 10.1038/s41550-020-01294-9.

Tremblay, P.-E., Fontaine, G., Fusillo, N. P. G., et al. 2019. Core crystallization and pile-up in the cooling sequence of evolving white dwarfs. *Nature* 565:202–205.

Turatto, M. 2003. Classification of supernovae. arXiv:astro-ph/0301107.

VERA collaboration. 2020. The First VERA astrometry catalog. arXiv:2002.03089v1.

Vitral, E. and Mamon, G. A. 2021. Does NGC 6397 contain an intermediate-mass black hole or a more diffuse inner subcluster?. *Astronomy and Astrophysics* 646:A63 (p. 27).

Wang, F., Yang, J., Fan, X., et al. 2021. A luminous quasar at redshift 7.642. *The Astrophysical Journal Letters* 907:L1 (7 p).

Watkins, W. W., Van der Marel, R. P., Sohn, S. T. and Evans, N. W. 2019, Feb 8. Evidence for an intermediate-mass milky way from GAIA DR2 halo globular cluster motion. arXiv:1804.11348v3 [astro-ph.GA].

Wei, J.-J., Wu, X.-F., Fulvio Melia, F., et al. 2015. The age–redshift relationship of old passive galaxies. *The Astronomical Journal* 150: 35 (13 p).

Xing, Q.-F., Zhao, G., Aoki, W., Honda, S., Li, H.-N., Ishigaki, M. N. and Matsuno, T. 2019, Apr. 29. Evidence for the accretion origin of halo stars with an extreme r-process enhancement. *Nature Astronomy Letters* Doi: 10.1038/s41550-019-0764-5.

Xue, Y. Q., Zheng, X. C., Li, Y., et al. 2019. A magnetar-powered X-ray transient as the aftermath of a binary neutron-star merger. *Nature* 568:198–201. See also arXiv:1904.05368 [astro-ph.HE].

Yang, Z., Bethge, C., Tian, H., et al. 2020. Global maps of the magnetic field in the solar corona. *Science* 369(6504):694–697.

Zhao, X., Gou, L., Dong, Y., et al. 2021. Reestimating the Spin Parameter of the Black Hole in Cygnus X-1. *The Astrophysical Journal* 908:117 (11 p).

3

Galactic Chemical Evolution

3.1 Age–Metallicity Relation

Metallicity is a measure of the proportion of "heavy elements", which are called metals by astronomers and include all elements heavier than helium, that a particular star contains. Usually, the metallicity of an object is a measure of the relative amount of iron (Fe) and hydrogen (H) compared with the solar value of this ratio as determined from the analysis of their absorption lines. Metallicity is usually denoted as [Fe/H] and is given by

$$[\mathrm{Fe/H}] = \log_{10}(\mathrm{Fe/H})_{\mathrm{star}} / (\mathrm{Fe/H})_{\mathrm{sun}} = \log_{10}(\mathrm{Fe/H})_{\mathrm{star}} - \log_{10}(\mathrm{Fe/H})_{\mathrm{sun}}$$

For example, if the metallicity [Fe/H] of a star is 0, then it has the same iron abundance as the Sun. If [Fe/H] of a star is –1, it is one-tenth of the abundance of heavy elements found in the Sun. If [Fe/H] of a star is +1, its heavy metal abundance is ten times the solar value.

All elements heavier than helium, "metals", are denoted by Z. For the Sun, the total amount of metals by mass is about $Z_{\odot}=0.02$, or 2% of the mass in the Sun is not hydrogen or helium.

It is, in principle, possible to find the chemical composition of a galaxy as a function of position and time by measuring the element abundances of stars with different birthplaces and ages. This approach is valid, provided that their atmospheres represent the composition of the gas from which they were formed. Such studies may give valuable information about the chemical evolution of galaxies and even about the structure of the matter in the very early phases of the Universe.

Edmunds and Phillipps (1997) examined the metallicity distribution within and between galaxies and hence determined their present-day mean abundance. Adding in components for intergalactic gas, they arrived at an estimate of the mean metal abundance of the Universe. The result was that the overall mean abundance is close to solar abundance. The time evolution of this quantity, using simplified but general models of galactic chemical evolution, can be discussed. The variation of the Universe's total metal content with epoch is constrained within fairly well-defined limits for plausible differences of the mean global star formation rate. What one would observe at any given redshift depends basically on the choice of galaxies regions that are being sampled and on

DOI: 10.1201/9781003181330-3

the formation histories of these particular regions. Edmunds and Phillipps (1997) also investigated the constraints provided jointly by observations of the metallicity evolution and Universe gas content, as measured by quasar (QSO) damped Lyman-α absorbers. They noted a generic inconsistency in global models and introduced a more realistic model with different types of evolution for different types of galaxies. Current chemical data do not require that the global average star formation rate in the Universe should have decreased by a significant factor since galaxy formation began.

The mean abundance is defined as the total mass density of metals divided by the overall baryon density. This is the most fundamental measure of the integrated chemical evolution which has occurred to date in the Universe, although it may not correspond most directly to observational measurements. "Mean metallicity" of a gas-rich galaxy is the ratio of its metallicity and its cross-sectional area. This area-weighted gas-phase mean metallicity and its evolution were considered by Phillipps and Edmunds (1996).

Figure 3.1 shows an estimate of star formation rate (M_{Sun}/year/Mpc^3) and metallicity (Fe/H) with time passed since the Big Bang or redshift (z). It is customary to use the ratio of iron abundance to hydrogen (Fe/H) as a representation of the metallicity of stars and the Universe in general. The bottom part of Figure 3.1 shows an estimate of the evolution of the metallicity of the Universe (adapted from Chopra and Lineweaver, 2018). Curves of this type for other life-essential elements should be constructed as functions of time from the observations of the oldest, low-metallicity stars. Such a procedure would yield a more precise estimate for the time of origin of life in the Universe.

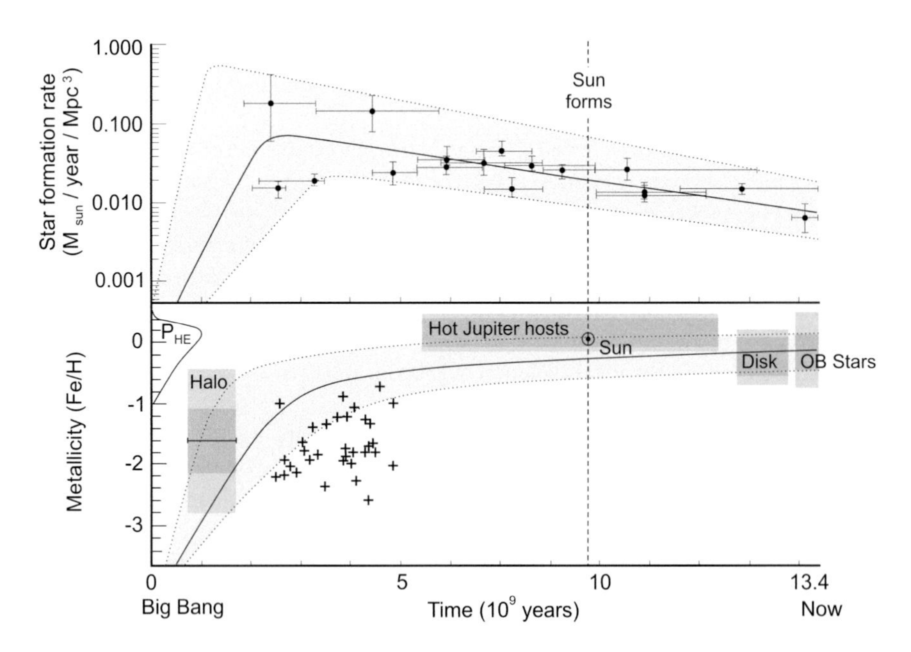

FIGURE 3.1 A rough estimate of the star formation rate in the Universe is shown in the top. The integral of the top curve produces the monotonically increasing metallicity curve (shown on the bottom of the figure), following Chopra and Lineweaver (2018).

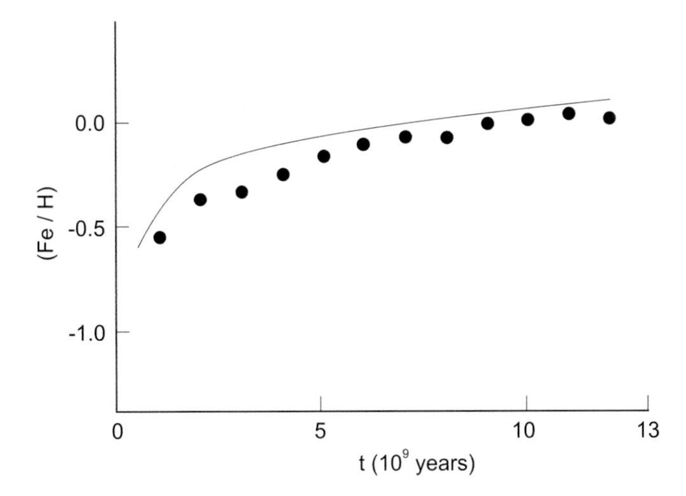

FIGURE 3.2 The theoretical age–metallicity ([Fe/H] is a logarithmic measure of the iron abundance relative to the Sun) relationship compared with the observational data of Twarog (1980) for solar neighborhood stars (full dots). The time is in units of 10^9years. (Adopted from Matteucci, 2001.)

The chemical evolution model used by Matteucci (2001) follows the evolution of the element's mass because of stellar nucleosynthesis, mass ejection and the infall of gas of primordial chemical composition. The model also takes into account the temporal delay in the chemical enrichment due to stellar lifetimes, which is essential to correctly account for the contribution of long-living stars. The predicted iron abundance as a function of time is found to be in good agreement with the age–metallicity relationship for solar neighborhood stars (Twarog, 1980), as shown in Figure 3.2. This fact indicates that the iron produced by intermediate-mass stars, in addition to the large ones, does not lead to an overproduction of this element when the corresponding yield is a decreasing function of time.

The existence of an age–metallicity relation (AMR) in the disk is an essential issue for developing chemical evolution models. However, there is some uncertainty in the published literature that whether an AMR exists at all (Friel, 1995).

H and He were present very early on in the Universe, while all metals having $A \geq 12$ (except for a tiny fraction of Li) were produced through nucleosynthesis in stars. The fraction by mass of heavy elements is denoted by Z.

- The Sun's abundance Z_{Sun} is ~0.02.
- Most of the metal-poor stars in the Milky Way have $Z \sim 10^{-5}$–10^{-4} Z_{Sun}.

Each star burns H and He in its nucleus and produces heavy elements. These elements are partially returned into the interstellar gas at the end of the star's life through winds and supernovae explosions. Some fraction of the metals is locked into the remnant of the star. This fact implies that the chemical abundance of the gas in a star-forming galaxy should evolve. The metal abundance of the gas and subsequent generations of stars should increase with time. The evolution of chemical element abundances in a galaxy provides a clock for galactic aging.

The study of the evolution of overall metallicity and mass–metallicity (M–Z) relation as a function of redshift is critical for the understanding of different redshift samples. The most significant problems in M–Z relation studies at $z > 1$ (Yuan et al., 2012) are the following:

1. The metallicity is not based on the same diagnostic method.
2. The stellar mass is not derived using the same manner.
3. The samples are selected differently, and the effects of selection on mass and metallicity are not understood.

Yuan et al. (2012) attempted to minimize these problems by recalculating the stellar mass and metallicity, and by expanding the lens-selected sample at $z > 1$. The samples used in their work include the following:

1. The Sloan Digital Sky Survey (SDSS) sample ($z \sim 0.07$); see Zahid et al. (2011).
2. The Deep Extragalactic Evolutionary Probe 2 (DEEP2) sample ($z \sim 0.8$); see Zahid et al. (2011).
3. The UV-selected sample ($z = 2$) for which the data are presented by Erb et al. (2006).
4. The lensed sample ($1 < z < 3$).

The authors presented an observational picture of the metallicity evolution of star-forming galaxies as a function of stellar mass between $0 < z < 3$. Their findings are summarized as follows (also presented in Figure 3.3): There exists a relation between the mean and median metallicities of star-forming galaxies and the redshift value. The mean metallicity falls to ~0.18 dex for redshift values from 0 to 1 and then falls further to ~0.16 dex for redshift values from 1 to 2. (The term dex is a conventional notation for the decimal exponent, i.e., dex $(2.35) = 10^{2.35}$). A faster evolution is present for $z \sim 1 \rightarrow 3$ than for $z \sim 0 \rightarrow 1$ in the case of very high-mass galaxies ($10^{9.5}$ $M_\odot$ $< M\star <$ 10^{11} $M_\odot$). There is almost twice as much enrichment between $z \sim 1 \rightarrow 3$ and $z \sim 1 \rightarrow 0$.

Figure 3.3 shows dZ/dz in the unit of Δdex per redshift. Yuan et al. (2012) derived dZ/dz for the SDSS to the DEEP2 ($z \sim 0$–0.8), and the DEEP2 to the lensed ($z \sim 0.8$–2.0) samples from Zahid et al. (2011) (black squares/lines). For comparison, they also derived dZ/dz from $z \sim 0.8$ to 2.0 using the DEEP2 samples and the samples from Erb et al. (2006) (full circles/dashed lines). Filled and empty squares are results from the mean and median quantities. The dash-dotted line shows the model prediction from the simulation of Davé et al. (2011). The dZ/dz in different mass ranges are shown for different redshifts and cosmic time frames. A negative value of dZ/dz means a positive metal enrichment from high redshift to the local Universe. The negative slope of dZ/dz versus cosmic time indicates a deceleration in the metal enrichment from high z to low z (after Yuan et al., 2012).

The smaller increase in metal enrichment from $z \sim 2 \rightarrow 0.8$ to $z \sim 0.8 \rightarrow 0$ is significant in the high-mass galaxies ($10^{9.5} M_\odot < M_\star < 10^{11} M_\odot$), consistent with a mass-dependent chemical enrichment. The authors compared the metallicity evolution of star-forming galaxies from $z = 0 \rightarrow 3$ with the most recent simulations. For lower mass, the metallicity model

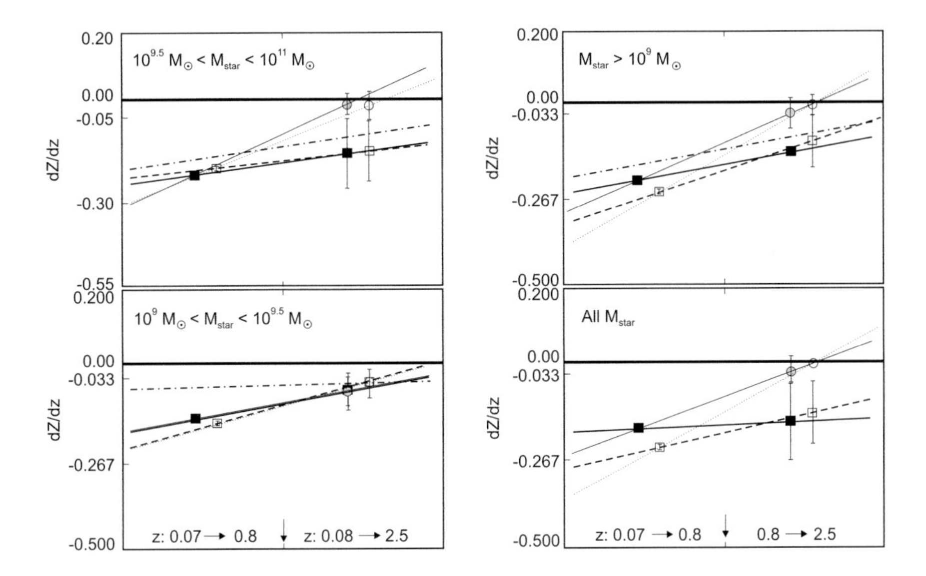

FIGURE 3.3 Cosmic metal enrichment rate (dZ/dz) in two epochs of redshift (cosmic time). dZ/dz is defined as the slope of the Z–z relation, Yuan et al. (2012).

is consistent with the observed metallicity within the observational error. However, for higher-mass bins, the model over-predicts the metallicity at all redshifts. The over-prediction is most significant in the most top mass interval of 10^{10-11} $M_\odot$. Additional theoretical investigation into the metallicity of the highest mass galaxies is needed to determine the cause of this discrepancy. The median of metallicity values for the lensed sample is 0.35±0.06 dex lower than for local SDSS galaxies and is 0.28±0.06 dex lower than for the z ~ 0.8 DEEP2 galaxies.

Konami et al. (2013) successfully derived the spatial metal abundance patterns in the hot ISM of NGC 3079, using both Chandra and Suzaku for the first time. The combination of these two missions enabled the optimal utilization of the advantages of both proper spatial and excellent spectral resolution at high sensitivity. Figure 3.4 shows the abundance patterns in the 0.5′ circle, 0.5′–1′ ring, 1′–2′ ring and 1′ circle regions. The abundance patterns of NGC 3079 can be compared with the solar abundance table (Lodders, 2003), and those expected from the supernovae type II (SNeII) and SNeIa yields. The SNeII yields computed by Nomoto et al. (2006) are used here. These refer to an average over the Salpeter initial mass function for stellar masses 10–50 $M_\odot$, with a progenitor metallicity of Z=0.02 (Konami et al. 2013)". The SNeIa yields from the model of Iwamoto et al. (1999) were adopted. Within a 4.5 kpc region (i.e., a circle with a radius of 1′), the abundance pattern is consistent with that of SNeII within the uncertainties. Konami et al. (2013) inferred this region to be significantly enriched by SNeII metal yields, undoubtedly related to the starburst activity in this system. In contrast, the abundance pattern is not SNeII-like beyond 4.5 kpc, presumably because the starburst has had little impact here. This emission comes from the intrinsic galactic halo and is not

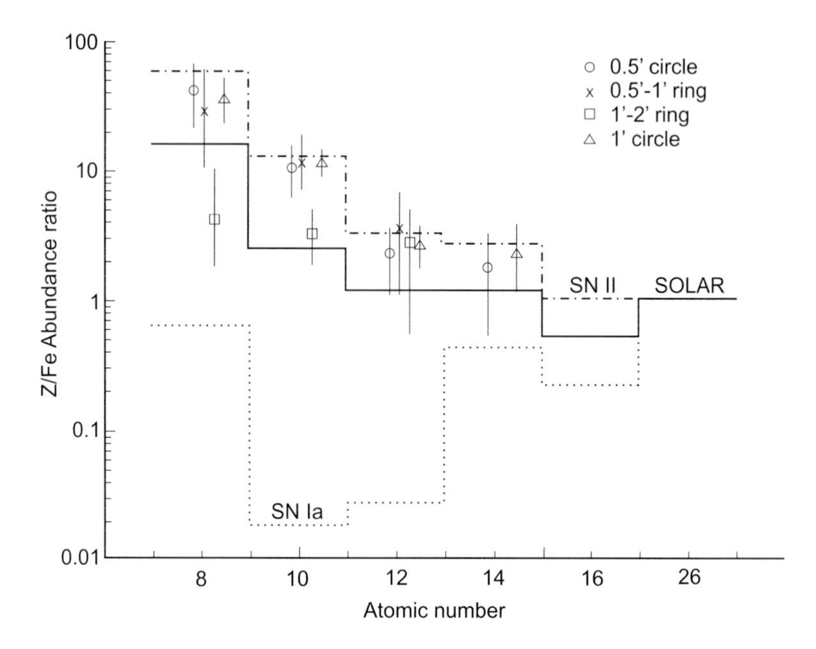

FIGURE 3.4 Abundance ratios of O, Ne, Mg and Si to Fe for the best-fit model of NGC 3079 derived by simultaneously fitting the Chandra and Suzaku spectra. Open circle, cross, open square and open triangle show the abundance patterns of 0.5′, 0.5′–1′, 1′–2′ and 1′, respectively. Solid, dash-dotted and dotted lines represent the number ratios of metals to Fe for solar abundance, for SNeII and SNeIa products (Lodders, 2003; Iwamoto et al., 1999; Nomoto et al., 2006), respectively. (After Konami et al., 2013.)

enriched by the starburst outflow yet. Using AKARI, Yamagishi et al. (2010) detected extended emission from hot dust (~30 K) in the center regions (4 kpc) of NGC 3079, possibly associated with starburst activity. The emission ratios of ionized to neutral polycyclic aromatic hydrocarbons are higher in the center (~2 kpc) than in the disk region. Both facts indicate that star formation is relatively active in the center region.

Hollek et al. (2011) reported the abundance values or upper limits for 20 chemical elements, including Li, C, Mg, Al, Si, Ca, Sc, Ti, Cr, Mn, Fe, Co, Ni, Zn, Sr, Y, Zr, Ba, La and Eu in 16 new stars and four standard stars. These data were obtained from high-resolution, high-S/N Magellan Inamori Kyocera Echelle (MIKE) spectra via traditional manual analysis methods using the MOOG code. They found that, except for Mg, abundance values are in agreement with the values of other halo stars reported in Cayrel et al. (2004).

In the pilot sample, the authors find that stars are grouped in several distinct chemical groups, indicating different enrichment mechanisms for each of these groups. However, the exact mechanism is still uncertain. They found four new stars with [Fe/H] <-3.6, where the metallicity distribution function severely drops. They also reported that CS 22891-200 has a lower [Fe/H] value than the one published by McWilliam et al. (1995), resulting in the total number of stars with [Fe/H] <-3.6 being five. All of these stars have an enhanced [C/Fe] abundance ratios relative to solar values. They have four carbon-enhanced metal-poor (CEMP) stars in their sample. Two of these are extreme extremely metal-poor (EEMP) CEMP-no stars. The trend of an increasing enhancement of C toward the lowest metallicities is confirmed. This behavior is present without invoking the contribution of AGB stars at the weakest metallicities. This fact may suggest that massive stars released C from their atmospheres and enriched the local ISM and that all the sample stars did not form in the same region. They detected La and Eu in five stars. Of these, four are rich in neutron-capture elements with their [Eu/Fe] ratios ranging from 0.02 to 0.79, indicating r-process signatures. The star with the highest [Eu/Fe] ratio is a mildly r-process-enhanced star. For the remaining stars, they found scaled solar system abundance ratios with very small scatter in the abundances. This fact may indicate the presence of core-collapse supernovae in the early Universe (Hollek et al., 2011).

The majority of metal-poor stars, with [Fe/H] <–1, show abundance patterns similar to the solar system when scaled down by metallicity. There are two main differences: There is an enhancement in the α-elements (e.g., [Mg/Fe]) and depletion in some of the Fe-peak elements (e.g., [Mn/Fe]) compared to the solar abundance ratios; see Figure 3.5. This situation could be a consequence of the enrichment by previous core-collapse supernovae (see, for example, Heger and Woosley, 2010). The 10% of stars considered chemical outliers among stars, with [Fe/H] < –2.0, show great diversity in their abundance patterns. Many stars have over-abundances of selected groups of elements, e.g., the rapid (r) neutron-capture process elements (Sneden et al., 2008) and the slow (s) neutron-capture process elements. The number of chemically unusual stars increases with decreasing metallicity, with stars often belonging to multiple chemical outlier groups. Not included

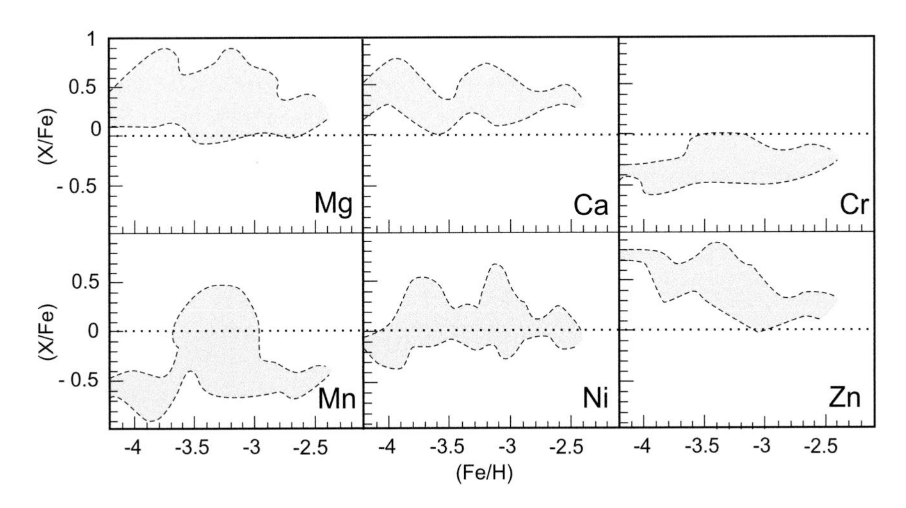

FIGURE 3.5 [X/Fe] abundance ratios for life-essential elements vs [Fe/H] for each of the elements measured using three sets of data. Only the abundance regions, not individual data, are shown. The dotted line represents the solar abundance ratio. Some of Ni abundances data shown are preliminary. (After Hollek et al., 2011.)

in this estimate of chemically peculiar stars are the CEMP stars (with [C/Fe] >0.7); this is at least ~15% of stars with [Fe/H] <–2.0. At very low metallicities, the frequency of CEMP stars also increases (Carollo et al., 2011, and references therein). All three [Fe/H] <–4.5 stars in a sample of 14 extremely metal-poor stars studied by Hollek et al. (2011) are CEMP stars.

Of our interest are, in particular, Fe-peak elements. According to Hollek et al. (2011), the [Cr/Fe] and [Mn/Fe] abundance ratios for all stars in their sample are found to be deficient relative to the solar abundance by −0.23 and −0.49 dex, respectively. Keep in mind that the [Cr/Fe] ratios are determined based on only the Cr I abundances, and there is a ~0.35 dex difference between derived Cr I and Cr II abundances. Figure 3.6 shows the [Co/Fe], [Ni/Fe] and [Zn/Fe] ratios in the sample stars being generally enhanced relative to the solar abundance ratios by 0.42, 0.05 and 0.25 dex, respectively.

The Fe-peak elements are produced in various late burning stages (see Woosley and Weaver 1995), as well as in supernovae. The abundances of Fe-peak elements follow those of other halo star samples and show an increase in these elements over time (e.g., McWilliam, 1997). The elemental abundances of Fe-peak elements in the large CASH sample can be used to define yields of the nucleosynthesis processes on the progenitor stars. For example, the measured Zn abundances, which can be measured in the high-resolution spectrograph (HRS) spectra, are dependent on the explosion energy of supernovae (Nomoto et al., 2006; Heger and Woosley, 2010).

The study of neutron-capture elements allows for searching of different sites of nucleosynthesis, in addition to proton and α-capture. The neutron-capture process occurs in two locations: in the envelopes of highly evolved AGB stars (s-process) and during some explosive events (r-process), such as a core-collapse supernova. The contributions of an individual process to the total abundance of neutron-capture elements in a given star can be determined by evaluating the common neutron-capture abundance patterns. For example, the s-process contributes ~80% of the Ba abundance to the solar system abundances, with a contribution of ~20% from the r-process. In contrast, Eu is made almost

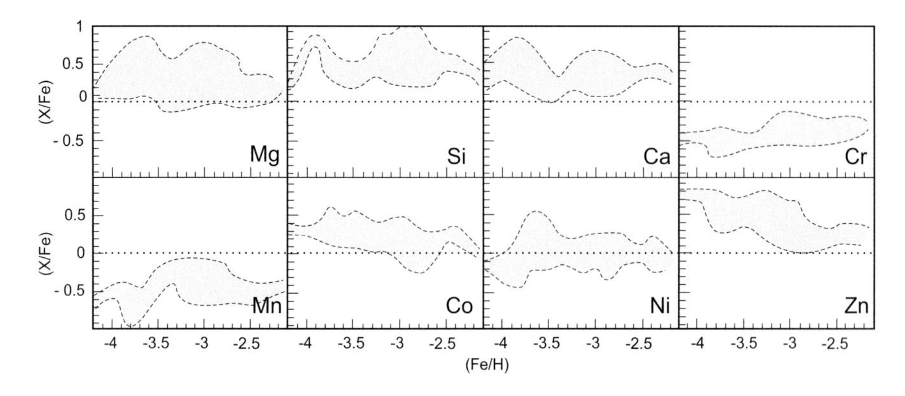

FIGURE 3.6 [X/Fe] abundance ratios for life-essential elements vs [Fe/H] for the elements up to Zn for two data sets only. The abundance region, not individual data, are shown. The dotted line represents the solar abundance ratio. (After Hollek et al., 2011.)

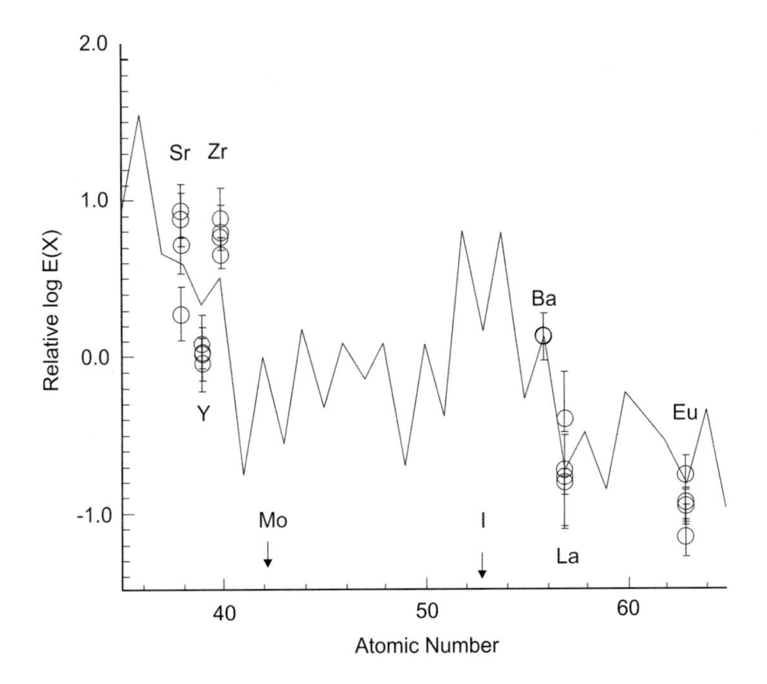

FIGURE 3.7 Relative log ε(X) abundances for all r-process-enhanced stars in the Magellan Inamori Kyocera Echelle (MIKE) sample. Abundances are adjusted to the Ba abundance to fit a scaled solar system r-process curve (solid line). (After Hollek et al., 2011.)

entirely by the r-process (Sneden et al., 2008). However, these ratios may differ in the early Universe. There are not enough neutron-capture elements detectable in the HRS snapshot spectra to determine whether the abundances of neutron-capture elements in a given star have an s- or r-process origin (Hollek et al., 2011). The r-process contributions to the abundances of some heavy elements are shown in Figure 3.7.

Mannucci et al. (2010) demonstrated that the M–Z functional relationship observed in the local Universe is due to a more general relationship between stellar mass M_*, gas-phase composition and star formation rate (SFR). Data from local galaxies define a surface in this 3D space, called fundamental metallicity relation (FMR), with a small residual dispersion of only ~ 0.05 dex. For low stellar mass, metallicity decreases sharply with increasing SFR. For high stellar mass, metallicity does not depend on SFR.

Galaxies with high redshift value (up to z ~2.5) are following the same FMR defined by local SDSS galaxies, without any indication of evolution. Therefore, it looks like the FMR defines the properties of metal enrichment of galaxies in the last 80% of cosmic time. Mannucci et al. (2010) observed the evolution of the M–Z relation up to z=2.5. In their work, galaxies with progressively higher SFRs and, therefore, lower metallicities are chosen at increasing redshifts, representing different parts of the same FMR (Mannucci et al., 2010). With the introduction of the new quantity $\mu_\alpha=\log(M_*)-\alpha\log(SFR)$, with $\alpha=0.32$, Mannucci et al. (2010) defined a new projection for the FMR minimizing the metallicity value spread for local galaxies. This procedure also cancels out any redshift change up to $z \sim 2.5$; i.e., all galaxies have the same functional relation between $\mu_{0.32}$ and metallicity and have the same range of values as $\mu_{0.32}$. At $z>2.5$, the evolution of about 0.6 dex off the FMR is found, with high-redshift galaxies having lower metallicities.

The existence of the FMR is explained by the influence of two processes: the infall of natural gas and the outflow of enriched material. This effect is responsible for the dependence of metallicity with SFR. It is the dominant effect at high redshift and introduces the dependence on stellar mass. The combination of these two effects explains the shape of FMR and the role of $\mu_{0.32}$. The small-metallicity scatter around the FMR supports the smooth infall scenario of gas accretion in the local Universe (Mannucci et al., 2010).

Mannucci et al. (2010) studied several samples of galaxies at different redshifts whose metallicity, M_* and SFR have been measured to test the hypothesis of correlation in the present Universe and at high redshifts. Three sets of data samples were used for this study: $z=0$ (SDSS), $z=0.5$–2.5 and $z=3$–4.

$z=0$ (SDSS): The local galaxies are well measured in the SDSS project (Abazajian et al., 2009). The MPA/JHU catalogue of emission-line fluxes and stellar masses from SDSS DR7 available at http://www.mpa-garching.mpg.de/SDSS and described in Kauffmann et al. (2003), Brinchmann et al. (2004) and Salim et al. (2007) was used. The catalogue includes 927,552 galaxies whose spectroscopic properties, such as emission-line fluxes and spectroscopic indexes, have been measured. Emission-line galaxies in redshift interval 0.07–0.30 (47% of the total sample) were selected. After additional reductions, their final galaxy sample contained 141,825 galaxies.

$z=0.5$–2.5: Many galaxies have been observed at high redshift, and they provided data to study the evolution of metallicity for the other properties of galaxies. Mannucci et al. (2010) extracted from the literature data a total of 182 objects grouped in three samples of galaxies at intermediate redshifts and published values of emission-line fluxes, M_* and dust extinction: $0.5<z<0.9$, $1.0<z<1.6$ and $2.0<z<2.5$.

$z=3$–4: A sample of 16 galaxies at redshift between 3 and 4 was observed by Maiolino et al. (2008) and Mannucci et al. (2009) on the implementation of the LSD and AMAZE projects. Published values of stellar masses, line fluxes and metallicities are available for these galaxies, which can be compared with lower-redshift data.

Based on the above data samples, Mannucci et al. (2010) reported the study of the dependence of gas-phase metallicity 12+log(O/H) on stellar mass M_* and SFR in a few samples of galaxies from $z=0$ to 3.3. In the local Universe, they found that metallicity is tightly related to both M_* and SFR, and this defines the FMR. The observed residual metallicity dispersion of local SDSS galaxies around this FMR is about 0.05 dex, i.e., about 12%. Of importance are the popular M–Z relation, luminosity–metallicity relation and velocity–metallicity relation. Neglecting the dependence of metallicity on SFR results in doubling the observed dispersion.

When high-redshift galaxies are compared to the FMR defined locally, there is no evolution up to $z=2.5$; i.e., high-redshift galaxies follow the same FMR defined by SDSS galaxies even if they have higher SFRs. This fact implies the same physical processes being in place in the local Universe and at high redshifts. The observed functional dependence of the M–Z relation is due to the increase in the average SFR with redshift. This situation is the result of different sampling parts of the same FMR at different redshifts. At an even higher redshift, $z \sim 3.3$, the evolution of ~0.6 dex concerning the FMR is found, although several observational effects and selection biases may affect the size of this evolution. This fact is an indication of different mechanisms of domination. Even if the nature of the FMR is three-dimensional in M_*, SFR and 12+log(O/H), metallicity is found to be correlated with $\mu_{0.32}=\log(M_*)-0.32\log(SFR)$. Galaxies at any redshifts up to $z=2.5$ follow the same $\mu_{0.32}$–metallicity relation and have the same range of values as $\mu_{0.32}$. Metallicity in galaxies of any mass is reported to have the same dependence of SFR. For galaxies above, the threshold of $SFR=10^{-10}/y$ is showing rapidly decreasing metallicities with increasing SFR (Mannucci et al., 2010).

Following these authors, the interpretation of the existence of the FMR, its dependence on SFR, and the role of $\mu_{0.32}$ depends on the relevant timescales. If dynamical times are shorter than timescales for chemical enrichment, the reliance of metallicity on SFR can be explained by dilution by infall. In this case, this effect dominates the metallicity evolution of galaxies at high redshift when they grow due to massive accretions of metal-poor gas and produce large SFR. The galaxies at $z=3.3$ can also fit into this scheme of large masses of infalling gas, in agreement with other recent independent measurements (Mannucci et al., 2009; Cresci, 2010). In the local Universe, galaxies with large SFR are rare and often associated with merging events, and other effects become dominant, which relate metallicity mainly to M_*. The outflow is a possibility, although downsizing could also work. When infall and SFR evolve on timescales much longer than the chemical enrichment, a sort of steady-state situation is created. Continuous infall of metal-poor gas appears sustaining SFR and diluting metallicity and outflow of metal-rich gas in galactic winds. In this case, the flows must depend on both the mass and SFR.

The small residual scatters around the FMR in the local Universe support the smooth accretion scenario, where galaxy growth is dominated by continuous accretion of cold gas. Interacting and merging galaxies are showing a more extensive spread around the FMR, in agreement with what is observed. Galaxies with intermediate and high redshifts show larger metallicity dispersions, which could be due to uncertainties in the measurements, or intrinsic dispersion, or both. This effect prevents the study of the evolution of residual scatters with redshift. Nevertheless, the absence of significant biases in metallicity or SFR up to $z=2.5$ points toward the existence of the same physical effects and the dominance of smooth accretion even at intermediate redshift (Mannucci et al., 2010).

Gas dynamics and star formation rate regulate the gas-phase oxygen abundance (metallicity) of star-forming galaxies. In this case, oxygen in the Universe is formed in the late-stage evolution of massive stars. Oxygen thrown into the interstellar medium by supernovae explosions and stellar winds increases the metallicity of galaxies, increasing their stellar mass. In most star-forming galaxies ($z \leq 2$), the increase in stellar mass is generated by cosmological inflows of gas from the medium; see, for example, Noeske et al. (2007) and Whitaker et al. (2012). However, the outflows of gas are observed in star-forming galaxies up to $z \sim 3$ by Weiner et al. (2009), Chen et al. (2010) and Steidel et al. (2010). Because metallicity is determined by the interplay of gas flows and star formation, observations of the chemical evolution of galaxies provide essential constraints for these physical processes in models of galaxy evolution; see, for example, Davé et al. (2011), Zahid et al. (2012) and Møller et al. (2013).

Zahid et al. (2013) calculated the stellar M–Z relation at five epochs up to the range of z ~ 2.3. They quantified evolution in the shape of the M–Z curve as a function of redshift. The M–Z relation flattens at late times. There is an upper limit to the gas-phase oxygen abundance in star-forming galaxies that is independent of redshift. From an examination of the M–Z relation and its observed scatter, they show that the shape of the curve, flattening at late times, is a consequence of evolution in the stellar mass. At these times, galaxies reach this empirical upper metallicity limit. There is also further development of the fraction of galaxies at a fixed stellar mass that reaches this limit. The stellar mass corresponding to the start of metallicities saturation is ~0.7 dex smaller in the local Universe than in z ~ 0.8 galaxies. The M–Z relation is a measure of the average gas-phase oxygen abundance as a function of stellar mass.

Figure 3.8 shows the M–Z relations at five epochs, as reported by Zahid et al. (2013). They determined the M–Z relation for the SDSS, the Smithsonian Hectospec Lensing Survey (SHELS) and the DEEP2 survey samples by binning the data. The galaxies were sorted into equally populated bins of stellar mass, and the median stellar mass and metallicity for each bin were plotted.

> The metallicity of stars in our galaxy ranges from [Fe/H] = −4 to +0.5 dex. The solar iron abundance value is ε(Fe) = (7.51 ± 0.01) dex. The average values of [Fe/H] are -0.2. -1.6, -0.2 dex for the solar neighborhood, the halo, and galactic bulge, respectively.
>
> *McWilliam (1997)*

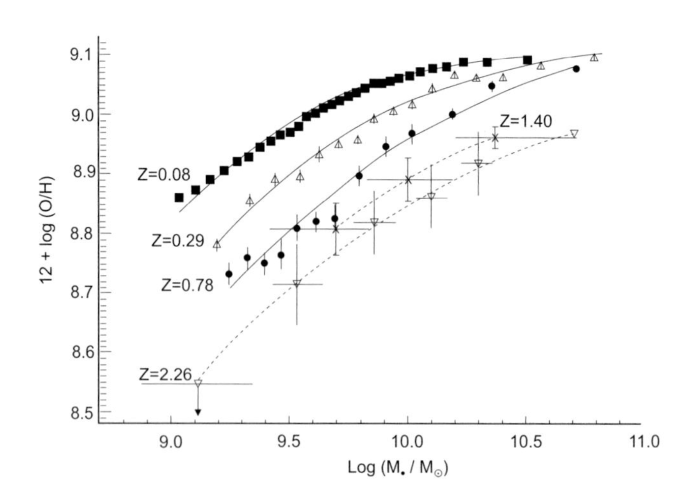

FIGURE 3.8 M–Z relation at five epochs ranging from z = 0.08 to z = 2.26. The solid curves fit the data by Zahid et al. (2013), while the bottom two (dashed) curves indicate metallicities converted using the formulae of Kewley and Ellison (2008). (Adapted from Zahid et al., 2013.)

The galactic disk, halo and bulge are characterized by unique abundance patterns of O, Mg, Si, Ca and Ti, and neutron-capture elements. These signatures show that the environment plays an essential role in chemical evolution. Also, supernovae come in many flavors with a range of element yields.

> The 300-fold dispersion in heavy element abundances of the most metal-poor stars suggests incomplete mixing of ejecta from the individual supernova, with vastly different yields, in clouds of $\sim 10^6$ $M_{\odot}$. The composition of Orion association stars indicates that star-forming regions are very self-enriched on timescales of 80×10^6 years. The rapid self-enrichment and inhomogeneous chemical evolution models are needed to match observed abundance trends and the dispersion in the age-metallicity relation.
>
> *McWilliam (1997)*

Globular clusters (GCs) are roughly spherical agglomerations of thousands to millions of stars, bound by their mutual gravity. Their central densities can exceed 10^6 solar masses per cubic parsec. GCs formed early in the history of the Universe and therefore recorded the early stages of galaxy formation and evolution. Larsen et al. (2020) reported an extremely metal-deficient GC in the Andromeda Galaxy. GCs in the Andromeda Galaxy align spatially and kinematically with stars in the outer parts of the galaxy. The authors investigated the globular cluster RBC EXT8 at a projected distance of 27 kpc from the galaxy center. The authors concluded that EXT8 is about 0.5 dex more metal-poor than the value of [Fe/H] = −2.39 found for the Andromeda Galaxy. Their model fitting yielded abundances for several additional elements (Mg, Si, Ca and Ti). Metal-poor GCs are expected to have formed in the early Universe in low-mass galaxies that merged to form larger galaxies. The correlation between the mass and metallicity of galaxies therefore imprints a maximum mass for a GC that could form with a given metallicity. At [Fe/H] = −2.9, the maximum mass is expected to be about $10^5 \times$ solar mass (Choksi et al., 2018, Kruijssen, 2019).

Roediger et al. (2013) presented an extensive literature overview of age, metallicity and chemical abundance pattern information describing the 41 galactic globular clusters (GGCs) studied by Schiavon et al. (2005). Their overview constitutes a notable improvement over previous similar work, particularly in terms of chemical abundances. It enables detailed evaluations of and refinements to stellar population synthesis models designed to recover the information needed for unresolved stellar systems based on their integrated spectra. However, since the Schiavon et al. (2005) sample spans a wide range of the known GGC parameter space, their compilation also benefits other investigations studying the formation of the Milky Way, the chemical evolution of GGCs and stellar evolution. For instance, they confirm with their compiled data that the GGC system has a bimodal metallicity distribution, and at the same time, it is uniformly enhanced in the α-elements.

When paired with the ages of studied clusters, Roediger et al. (2013) found evidence supporting a scenario in which the Milky Way obtained its GCs through two channels: in situ formation and accretion of satellite galaxies. The abundance distributions of C,

N, O and Na and the dispersions thereof per cluster corroborate the fact that all GGCs studied so far concerning multiple stellar populations have been found to harbor them. Finally, using data on individual stars, they also confirm that the atmospheres of stars become progressively polluted by C–N–(O)-processed material after they leave the main sequence. They also show evidence that suggests the α-elements, Mg and Ca originate from more than one nucleosynthetic production site. Their compilation incorporates all relevant analyses from the literature up to mid-2012. To help the investigators in the fields named above, they provide online the detailed electronic tables of the data upon which their work is based; see Roediger et al. (2013).

Some of the conclusions presented by Roediger et al. (2013) are discussed next. From the ages, as reported in Marin-Franch et al. (2009), it is tempting to conclude that the GGCs, from Schiavon et al. (2005), originated from a two-component star formation history. This star formation history could be described as consisting of either a sharp burst superimposed upon a comparatively steady background (lasting $\geq 4\times 10^9$ years), or a vigorous early episode that quickly peaked and was then reduced down to a more sustainable level. By extending the overall distribution to more extreme ages, the additional literature data seem to agree better with the first scenario. These data would strengthen this scenario by spreading the burst component over a longer timescale of $\sim(1.0–1.5)\times 10^9$ years. The metallicity distribution for the GGCs, listed in Schiavon et al. (2005), also supports the idea that these clusters arose from at least two distinct processes, given their definite bimodal shape, with a peak-to-peak separation of ~1.0 dex.

When the available ages (Marin-Franch et al., 2009) of the clusters comprising each metallicity subgroup are studied, there is no strong correlation between the two parameters (Roediger et al., 2013). The metal-poor and metal-rich GGCs of their sample, separated at [Fe/H]=–1.0 dex, have age and rms of $(12.0\pm 1.1)\times 10^9$ and $(12.8\pm 0.7)\times 10^9$ years, respectively; the distinction of these two age-groups worsens when we consider all of the adopted values.

The situation, as shown in Figure 3.9, can be attributed to the presence of several old GGCs (age $>12\times 10^9$ years) in the metal-poor subsample, whereas only one of the metal-rich GGCs has an age of $<12\times 10^9$ years (after Komatsu et al., 2011). Moreover, the metal-rich subsample harbors a high incidence of ancient clusters in which three of the six oldest objects are present. If the metal-rich/metal-poor boundary lays slightly lower, i.e., at Fe/H=–1.05 dex, four objects will be included. Since the sample is representative of the entire GGC system, these findings have some implications for the formation of the Milky Way and its halo.

For instance, the parameter spread in Figure 3.9 would be hard to explain within a scenario in which all of the GGCs, listed by Schiavon et al. (2005), formed in situ since one would expect the metal-poor GGCs to be older than the metal-rich ones. This spread is consistent with a possibility that the GGC system either arose in situ or was accreted from satellite galaxies. Although the possible correlation of these two possibilities with metallicity remains unclear, Roediger et al. (2013) interpreted the older, metal-rich clusters as the descendants of the former and the younger, metal-poor clusters as the descendants of the latter. Other research groups have reached similar conclusions by analyses of the GGC age–metallicity relation using much larger samples (see, for example,

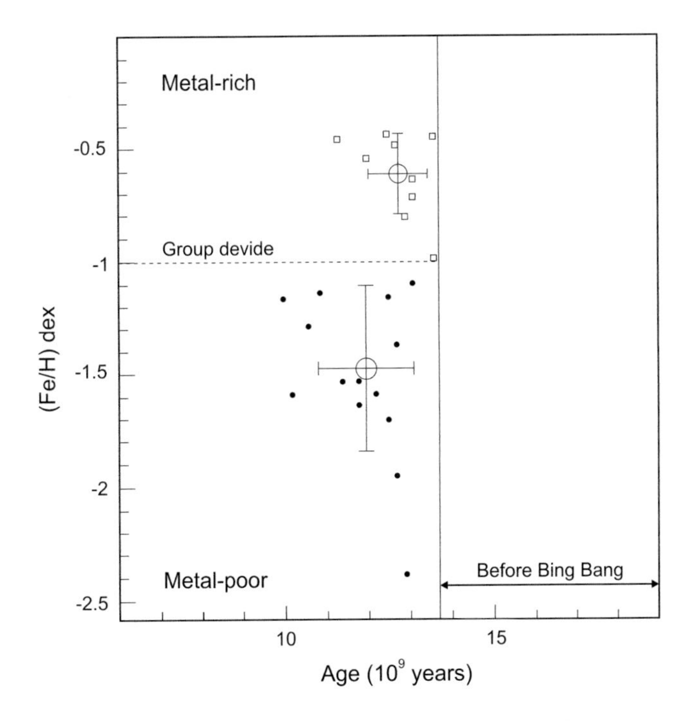

FIGURE 3.9 Metallicity versus age for the galactic globular clusters. The mean ages and metallicities (with corresponding rms uncertainties) of the metal-poor (solid dots) and metal-rich (squares) GGCs in Komatsu et al.'s (2011) sample are denoted by the open circles with error bars, while the dashed line at [Fe/H] = –1.0 dex roughly marks the transition between these two groups. The region corresponding to ages higher than that of the Universe is also indicated. (After Komatsu et al., 2011; Roediger et al., 2013.)

Marin-Franch et al., 2009; Forbes and Bridges, 2010; Dotter et al., 2011). Additional insight into this topic might be achieved by searching for correlations between the above parameters and those from the Ha10 catalogue, mainly velocity information.

More information about the stellar populations that comprise GGCs can be obtained from distributions of their mean chemical abundance ratios, as shown in Figure 3.10. These, in turn, provide further information on the formation of the Milky Way, as well as certain aspects related to stellar evolution. Figure 3.10 shows the distributions of the mean abundances of the α-elements (Mg, Ca, Si and Ti) of these systems show sharp peaks corresponding to super-solar values. The median values of the distributions are [Mg/Fe] = +0.38 dex, [Ca/Fe] = +0.30 dex, [Si/Fe] = +0.36 dex and [Ti/Fe] = +0.24 dex. These values imply that the GGC system was formed over rapid timescales. Slightly broader distributions are present for mean abundances of carbon, nitrogen and oxygen among the GGCs. Apart from possible undersampling effects, the authors interpret the broader distributions of these elements as being a consequence of the combined and well-known phenomena of atmospheric mixing and multiple stellar populations within these clusters.

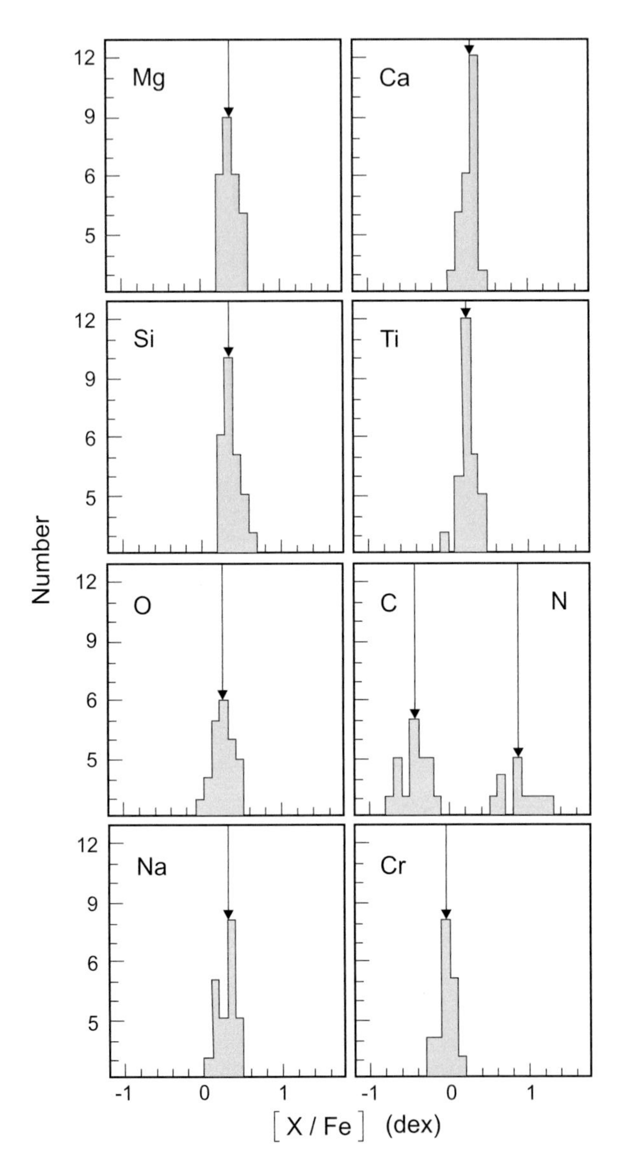

FIGURE 3.10 The specific abundance ratios for individual elements in the stellar populations that comprise galactic globular clusters. The arrow indicates the median value of each distribution.

One of the best tools for chemical tagging of GC stars is the Na–O anti-correlation, which represents the unique characteristics of GCs (see Figure 3.11, which summarizes the results of the Fibre Large Array Multi-Element Spectrograph (FLAMES) survey with ~2500 red giant stars in 25 GCs following Carretta (2015a)). The huge spreads in Na and O tell us that GCs are not pure stellar populations (by definition, coeval stars with

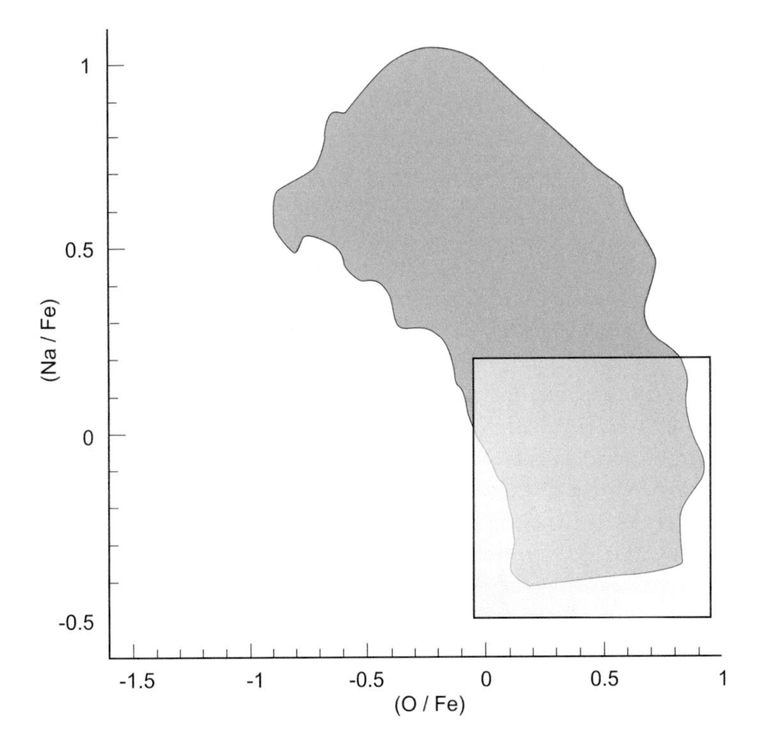

FIGURE 3.11 Na/Fe as a function of O/Fe for GC (data from Carretta et al. 2009a, 2009b and references in Carretta, 2015a) and field stars (Gratton et al., 2000, 2003) is distributed within the shaded region. The box on the bottom encompasses the field stars. (After Carretta, 2015a.)

the same initial chemical composition) because GC stars have very different Na and O contents.

FLAMES is located at 2,635 m above sea level in the Atacama Desert of Chile, at ESO's Paranal Observatory. It is one of the very best astronomical observing sites in the world and is the flagship facility for European ground-based astronomy. In addition to the FLAMES, it also houses the intermediate- and high-resolution spectrographs installed on the second Unit Telescope (UT2) at the Very Large Telescope (VLT). With FLAMES, one can access targets over a field of view of diameter of 25 arcmin. FLAMES also feeds two different spectrographs: GIRAFFE and UVES, covering the whole visual spectral range. GIRAFFE can observe up to 130 objects at a time or do integral field spectroscopy, with intermediate or high resolution (typically from R ~ 5,000 to R ~ 30,000). UVES provides higher spectral resolution (R ~ 47,000), but can access only up to 8 objects at a time (Pasquini et al., 2002).

The Na–O anti-correlation, discovered by the Lick-Texas group (Kraft, 1994), joined other features such as the anti-correlation of C and N and of other elements such as Al, Mg and even K by increasing the Coulomb barrier (Cohen and Kirby, 2012, Carretta, 2014, Carretta, 2015b).

Boesgaard et al. (2005) reported the observations of five almost identical stars at the turnoff of the metal-rich globular cluster M71 with the Keck I telescope, W. M. Keck Observatory, Mauna Kea, Hawaiian Islands, and HIRES (High Resolution Echelle Spectrometer) at a resolution of ~45,000. They derive stellar parameters and abundances of several elements. The mean Fe abundance, Fe/H=−0.80±0.02, is in excellent agreement with previous determinations from both giants and near-turnoff stars. There is no evidence for any star-to-star abundance differences or correlations in the sample. Abundance ratios of the Fe-group elements Cr and Ni are similar to Fe. The turnoff stars in M71 galaxy have consistent enhancements of 0.2–0.3 dex in Si/Fe, Ca/Fe and Ti /Fe, similar to the red giants. The Mg/Fe ratio is somewhat lower than that suggested by other studies.

The authors compare their mean abundances for the five M71 stars with field stars of similar metallicity Fe/H (8 with halo properties and 17 with disk properties). The abundances of the α-fusion elements (Mg, Si, Ca and Ti) agree with both samples, but seem to have a closer match to the disk stars. The Mg abundance of M71 is at the lower edge of the disk and halo samples. The elements Y and Ba, made by neutron capture, are enhanced relative to solar in the M71 turnoff stars. The ratio Ba/Fe is similar to that of halo field stars; however, it is above that for the disk field stars (factor of 2). The Ba/Y ratio is significantly lower than the M71 giant values. This value may be a consequence of the cluster material being exposed to a neutron flux higher than the disk stars or self-enrichment that has occurred after cluster star formation. The content of Na in the M71 turnoff stars is very similar to that in the disk field stars; however, it is more than a factor of 2 higher than that of the halo field star sample. The authors found Na/Fe=+ 0.14±0.04, with a spread smaller than half of that found in the red giants of M71. The lack of α-element variations suggests that the material needed to explain the abundance patterns in M71 did not arise from explosive nucleosynthesis. It was formed in a standard s-process environment, similar to that provided by AGB stars. The determination of light s-peak abundances should indicate whether this pollution occurred before or after cluster formation.

Edvardsson et al. (1993) derived abundances of 13 elements in 189 field F and G dwarf stars in the galactic disk. Some of these stars have Fe/H values similar to that of globular cluster M71. Boesgaard et al. (2005) separated those stars with [Fe/H] within ±0.20 dex of −0.80 to compare with the M71 turnoff mean abundances; there are 17 such disk stars. Figure 3.12 shows the mean Fe-normalized M71 abundances (open circles with a dot) compared to halo field stars of similar metallicity from Stephens and Boesgaard (2002) (black squares) for elements essential for life. Figure 3.13 shows the mean Fe-normalized M71 abundances (open circles with a dot) compared to disk field stars of similar metallicity from Edvardsson et al. (1993) (black squares). This presentation is similar to Figure 3.12 except that the plot for [Na/Fe] is included instead of Cr/Fe because Edvardsson et al. (1993) did not determine Cr abundances. Figure 3.14 is a similar plot to Figure 3.13 with the disk stars indicated by filled circles.

The M71 means of the α-elements appear to match the abundances of the disk dwarfs of comparable metallicity better than those of the halo field stars. The value of Mg/Fe ratio is at the bottom of the range for the disk stars and the halo stars, while Si/Fe, Ca/Fe and Ti/Fe ratios are somewhat higher in M71 than in the halo stars. The abundances well

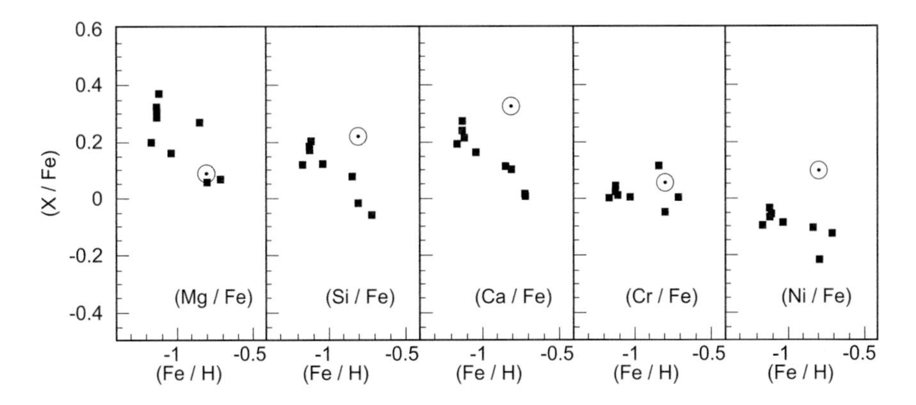

FIGURE 3.12 Mean Fe-normalized globular cluster M71 abundances (open circles) compared to halo field stars of similar metallicity from Stephens and Boesgaard (2002) (black squares). (After Boesgaard et al., 2005.)

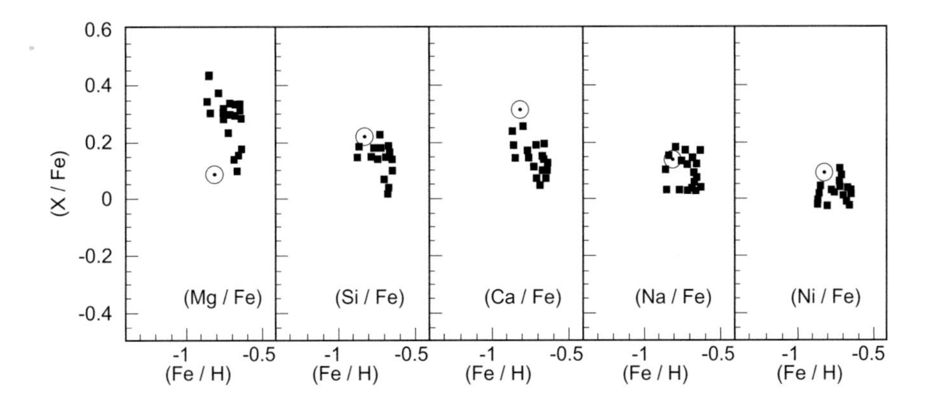

FIGURE 3.13 Mean Fe-normalized globular cluster M71 abundances (dot in open circles) compared to disk field stars of similar metallicity (black squares). This figure is similar to Figure 3.12 except that the plot for Na/Fe is included instead of Cr/Fe because Edvardsson et al. (1993) did not determine Cr abundances after Boesgaard et al. (2005).

match the abundance of the neutron-capture element, Ba, in M71 halo stars; however, the mean [Ba/Fe] value is more than a factor of 2 higher than in the disk stars (Boesgaard et al., 2005).

Figure 3.14 shows the elemental abundances normalized to Fe as a function of the atomic number. It includes values for globular cluster M71 turnoff stars, 7 of the 10 M71 red giants reported in Sneden et al. (1994) and the halo field stars. The giants and turnoff stars are in agreement with Na and Ni abundances. The α-element Ca is high in the turnoff stars, but would perfectly fit at +0.14 to the Grevesse and Sauval (1998) data for solar Ca. The α-elements Ti and Si are a bit lower in the turnoff stars than in the giants.

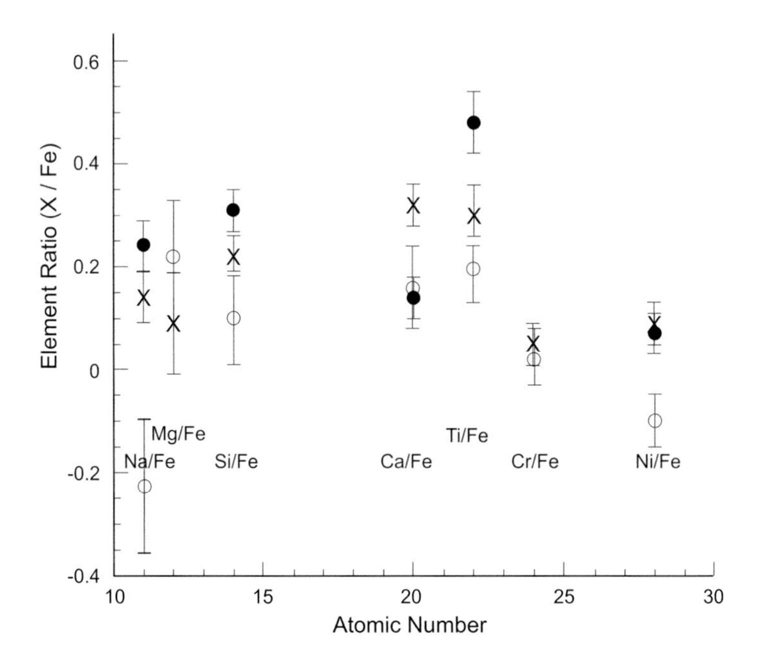

FIGURE 3.14 Abundance comparisons of mean abundances for the six M71 turnoff stars (marked X) with those for red giants (marked black dots) in M71 (Sneden et al., 1994) and those for halo field stars (marked open circles) (Stephens and Boesgaard, 2002) of similar Fe/H to M71. (After Boesgaard et al., 2005.)

Here, Ni abundance obtained by Boesgaard et al. (2005) is in excellent agreement with the values from the giants reported by Sneden et al. (1994).

Wanajo and Janka (2011) examined the r-process nucleosynthesis in the neutrino-driven wind from the thick accretion disk (or "torus") around a black hole. Such systems are expected as the remnants of a binary neutron star (NS–NS) or a neutron star–black hole (NS–BH) merger. They consider a simplified, analytic, time-dependent evolution model of a $3M_\odot$ central black hole surrounded by a neutrino-emitting accretion torus with a radius of 90 km, which serves as the basis for computing spherically symmetric neutrino-driven wind solutions. They find mass outflow domination by ejecta with modest entropies and reasonable expansion timescales (~100 ms). The mass-integrated abundances are in good agreement with the solar system r-process abundance distribution if a minimum in the electron fraction at the charged-particle freeze-out, $Y_{e,\,min} \sim 0.2$, is achieved. In the case of $Y_{e,\,min} \sim 0.3$, the production of r-elements beyond A ~ 130 does not reach the third peak, but could still be necessary for an explanation of the abundance signatures of r-process-deficient stars in the early galaxy. The overall mass of the ejected r-process nuclei is estimated to be $\sim 1 \times 10^{-3} M_\odot$. If their model was representative, this demands a galactic event rate of $\sim 2 \times 10^{-4}$/y for black-hole-torus winds from merger remnants to be the most crucial source of the r-process elements. Their result thus suggests

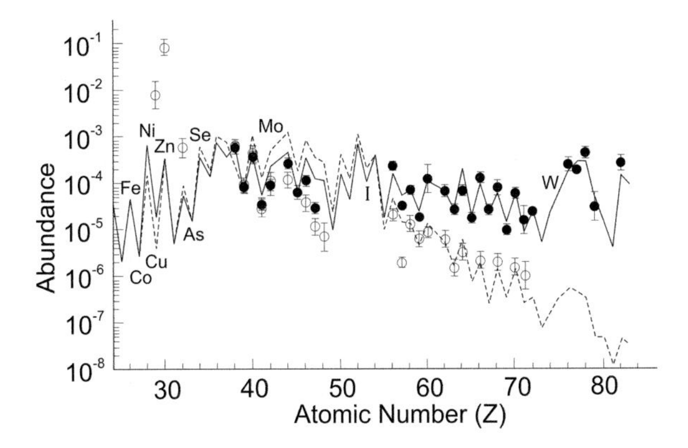

FIGURE 3.15 Time-integrated nucleosynthetic abundances for the entire torus for case 1 (solid line) and case 2 (dashed line). Examples 1 and 2 are compared with the spectroscopic abundances of r-enhanced (CS 31082-001, filled circles; Sneden et al., 2003) and r-deficient (HD 122563, open circles; Honda et al., 2006) galactic halo stars, respectively. For HD 122563, the Cd and Lu values are from Roederer et al. (2010) and the Ge value is from Cowan et al. (2005). For both stars, the abundances are vertically shifted to match the calculated Eu abundances (Wanajo and Janka, 2011).

that black-hole-torus winds from compact binary mergers have the potential to be the significant, but probably not the dominant, production site of r-process elements (Wanajo and Janka, 2011).

Some of the results presented by Wanajo and Janka (2011) are plotted in Figure 3.15 showing the nucleosynthetic abundances for two cases (variants) of their model. Example 1 (solid line; as a function of atomic number) is compared with r-process elements enhanced in star CS 22892-052 with [Fe/H] ~ −3.1 (filled circles; scaled to match the calculated Eu abundance, Sneden et al. (2003)). They find quite a good agreement with the stellar abundances distribution of CS 22892-052. Even if the production of r-elements heavier than A ~ 130 is small, as found in Example 2, their model could be a possible explanation for the abundance distribution found in star HD 122563 with [Fe/H] ~ −2.7 (open circles, scaled to match the calculated Eu abundance, Honda et al. (2006)). This fact suggests that the NS–NS and NS–BH mergers could at least be the origin of some trans-iron elements up to Z ~ 50 (A ~ 120). Although their calculations span the whole periodic table of elements, of relevance to our considerations are only the life-essential elements (up to W).

Nissen (2016) has recently presented results on high-precision abundances of several elements in solar twins with the accent on trends of elemental ratios with stellar age. His findings show that in stars younger than 6×10^9 years, ratios [Sc/Fe], [Mn/Fe], [Cu/Fe] and [Ba/Fe] are strongly correlated with stellar age, which is also the case for the other elements previously studied (see Figure 3.16). Linear relations between [X/Fe] ratios and age have $\chi_{red}{}^2 \sim 1$, and for most stars, the residuals do not depend on the condensation temperature of elements. For ages $6–9\times10^9$ years, the [X/Fe] ratio–age correlations break down, and the stars split up into two groups according to [X/Fe] values for the odd-Z

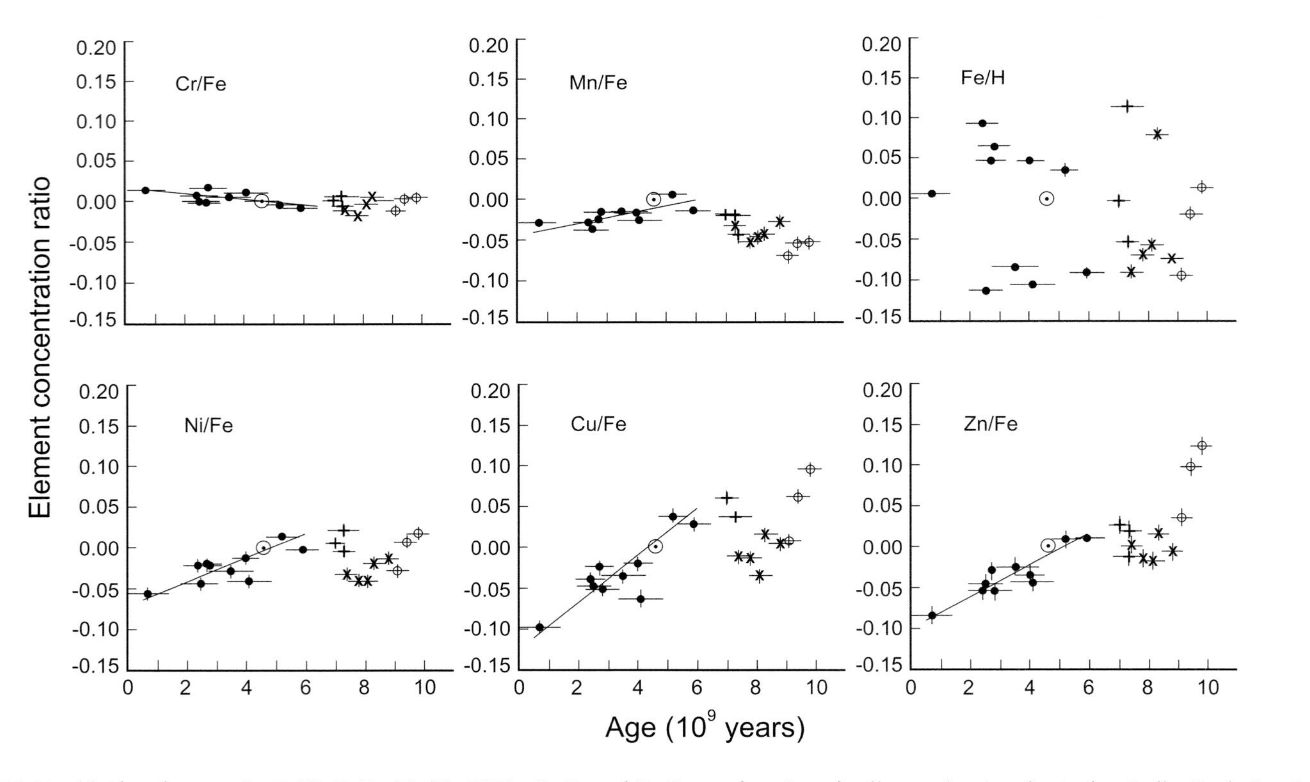

FIGURE 3.16 (a) Abundance ratios Fe/H, Cr/Fe, Mn/Fe, Ni/Fe, Cu/Fe and Zn/Fe as a function of stellar age (as given by Aarhus Stellar Evolution Code, ASTEC); see Christensen-Dalsgaard (2008). For stars younger than 6×10^9years (black filled circles), the lines show linear fits to the data. The Sun demonstrated with the $\odot$ symbol was not included in the fits. Mark "x" shows five old stars with low Na/Fe ratio relative to three stars with similar age shown with mark "+". Three [α/Fe]-enhanced stars are shown with open circles. (After Nissen, 2016). (b) Abundance ratios C/Fe, O/Fe, Na/Fe, Mg/Fe, Al/Fe, Si/Fe, S/Fe and Ca/Fe as a function of stellar age; The symbols used are the same as in Figure 3.16a.

(Continued)

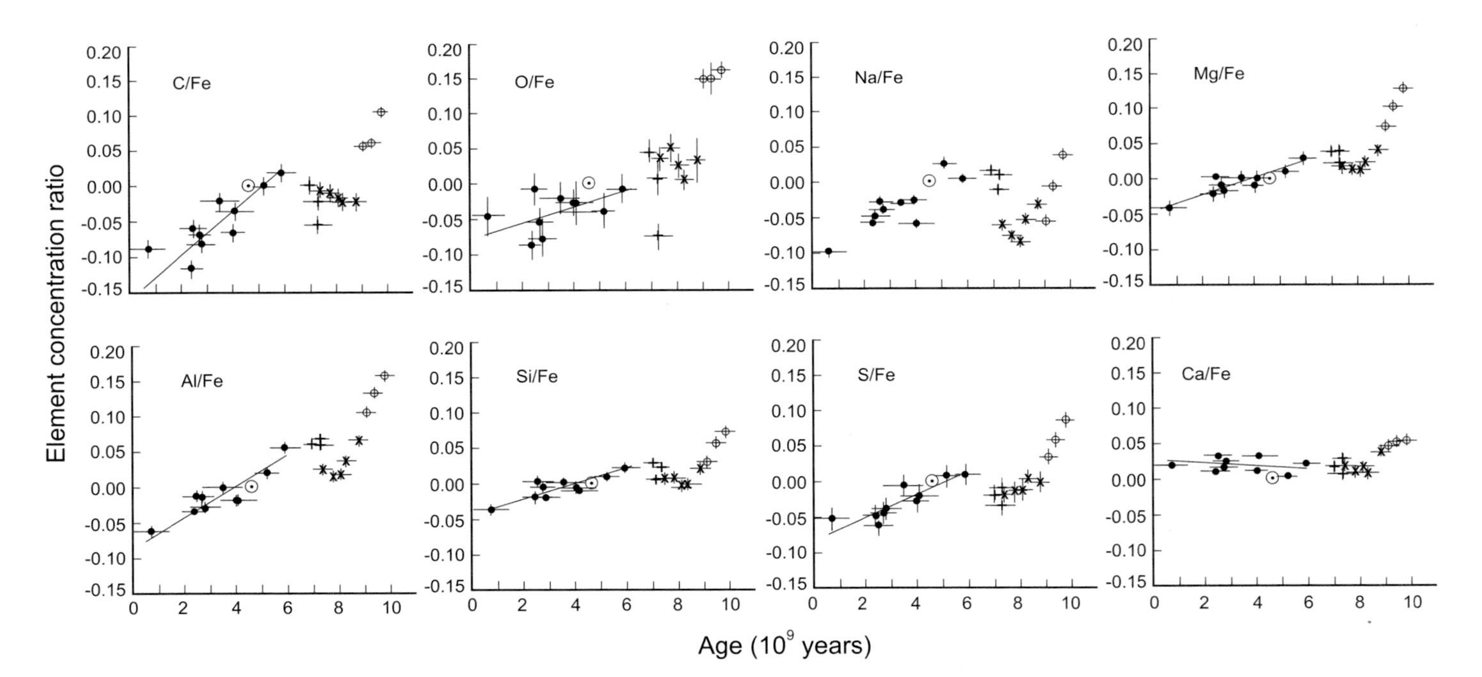

FIGURE 3.16 (*Continued*) (a) Abundance ratios Fe/H, Cr/Fe, Mn/Fe, Ni/Fe, Cu/Fe and Zn/Fe as a function of stellar age (as given by Aarhus Stellar Evolution Code, ASTEC); see Christensen-Dalsgaard (2008). For stars younger than 6×10^9years (black filled circles), the lines show linear fits to the data. The Sun demonstrated with the $\odot$ symbol was not included in the fits. Mark "x" shows five old stars with low Na/Fe ratio relative to three stars with similar age shown with mark "+". Three [α/Fe]-enhanced stars are shown with open circles. (After Nissen, 2016). (b) Abundance ratios C/Fe, O/Fe, Na/Fe, Mg/Fe, Al/Fe, Si/Fe, S/Fe and Ca/Fe as a function of stellar age; The symbols used are the same as in Figure 3.16a.

elements Na, Al, Sc and Cu. According to Nissen (2016), stars in the solar neighborhood younger than $\sim 6 \times 10^9$ years were formed from interstellar gas with a smooth chemical evolution. The older stars have probably originated from regions enriched by supernovae with different neutron excesses. Correlations between elemental abundance ratios and stellar age suggest that:

i. the SNeII make Sc along with the α-capture elements;
ii. the yield ratio of SNeII to SNeIa for Mn and Fe is approximately the same;
iii. the weak *s*-process mainly makes Cu in massive stars;
iv. the Ba/Y ratio for asymptotic giant branch stars is increasing with decreasing stellar mass;
v. the ratios Y/Mg and Y/Al can be used as chemical clocks when determining the ages of solar-metallicity stars.

It was Nissen (2015) who discussed the so-called chemical clock, referring to a strong linear relation between [Y/Mg] and the age of the solar twins. The author shows that both the high- and low-[Na/Fe] stars, and the [α/Fe]-enhanced stars fit the relation very well. A maximum likelihood fit (including all stars and the Sun) with errors in both coordinates taken into account gives

$$\mathrm{Y/Mg} = (0.170\ -\ 0.0371)\ \times\ \mathrm{Age}\left[10^9 \mathrm{years}\right]$$

with $\chi_{\mathrm{red}}{}^2 = 1.0$ and an rms scatter of 0.024 dex in [Y/Mg] corresponding to a scatter of 0.6×10^9 years in age. An excellent correlation between [Y/Al] and stellar age has also been observed, and it could be fitted with the formula:

$$\mathrm{Y/Al} = (0.196 \pm 0.0427)\ \times\ \mathrm{Age}\ \left[\mathrm{in}\, 10^9 \mathrm{years}\right]$$

In this case, $\chi_{\mathrm{red}}{}^2$ is close to one and the scatter in [Y/Al] is only 0.025 dex. The fact that both [Y/Mg] and [Y/Al] have a strong linear relationship with age means that [Al/Mg] is also closely correlated with age. A linear fit to the data provides the formula:

$$\mathrm{Al/Mg}\ =\ (0.0057\ \pm\ 0.0008)\ +\ (-0.027\ \pm\ 0.005)\ \times\ \mathrm{Age}\left[\mathrm{in}\ 10^9 \mathrm{years}\right]$$

with a scatter of 0.011 dex in Al/Mg. The data of three α-enhanced stars fit the above relation very well, showing that Al behaves like an α-element with a little higher amplitude than Mg.

This subject has also been discussed by da Silva et al. (2012), Ramírez at al. (2014) and recently by Tucci Maia et al. (2016).

Considering the origin of life issue, it is of interest to identify the region in the Universe where the ratios of elements essential for life is similar to the one in living matter. Among others, the Cu/Na ratio, as shown in Figure 3.17, must satisfy this criterion for the origin of life candidate location.

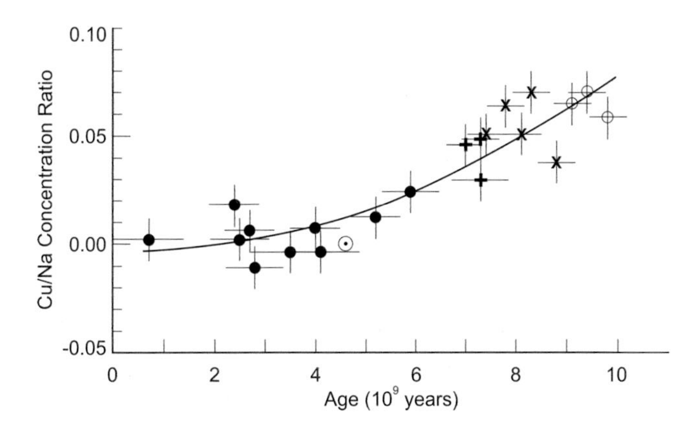

FIGURE 3.17 The Cu/Na concentration ratio versus the stellar age with the same symbols as in Figure 3.16a. (After Nissen, 2016.)

3.2 Galactic Chemical Evolution Models

By now, it is well established that different chemical elements are produced by different sources in the Universe. They are different masses of stars, supernovae, binary systems, etc. The relative contribution of each source depends on time and environment (mass and type of galaxy). This generates a need for the construction of a galactic chemical evolution (GCE) model. However, the issue is complicated by the agalactic acquisitions done in galaxy history; this is best illustrated by GCs. Namely, GCs are dense, gravitationally bound systems of thousands to millions of stars. They are preferentially associated with the oldest components of galaxies, so measurements of their composition can constrain the buildup of chemical elements in galaxies during the early Universe.

The GC systems in most galaxies are characterized by low abundances of elements heavier than H and He ("metals"), relative to the composition of the Sun. The most metal-poor GCs in the Milky Way have iron abundances of [Fe/H] ~−2.5, where square brackets denote the abundance ratio of the elements relative to the solar photospheric composition on a logarithmic scale. The number of iron atoms per hydrogen atom in the most metal-poor GCs in the Milky Way is about 1/300 that in the Sun. The value of [Fe/H] ~−2.5 is considered to be a metallicity floor for GCs formed in the early Universe with sufficient mass to survive until the present day. A metallicity floor for GCs has implications for clusters and star formation and for the buildup of metals in galaxies during the early Universe

This point is challenged by the observation of an extremely metal-deficient GC in the Andromeda Galaxy, as reported by Larsen et al. (2020). A massive GC, named RBC EXT8, has an iron abundance about 1/800 that of the Sun and about 1/3 that of the most iron-poor GCs previously known. Chemical element Mg is also strongly depleted in RBC EXT8. Their measurements do not support the proposal for the existence of a metallicity floor for GCs, nor the theoretical expectations that the massive GCs could not be formed at such a low metallicity.

GCE models must be based on the fact that different chemical elements are produced by different astronomical sources: stars of different masses, supernovae and binary systems. The relative contribution of each source depends on time and environment, i.e. mass and type of galaxy. Since the next generation of stars is formed from dust clouds that include heavy elements from the previous stellar generation, stars in the present-day galaxy are fossils that retain the information on the properties of stars from the past.

Elements with $A \geq 12$ are formed inside stars starting by the triple-α reaction, and ^{12}C excited state. The so-called α-elements (O, Mg, Si, S and Ca) are mostly produced in massive stars before being ejected by a core-collapse supernova. About half of the iron-peak elements: Cr, Mn, Fe, Ni, Co, Cu and Zn, are produced by SNeIa, which are produced by binary systems of white dwarfs. The elements beyond Fe are synthetized by the two extreme cases of neutron-capture processes: the slow (s, $\rho_N \sim 10^7$ n/cm^3) and rapid (r, $\rho_N \geq 10^{20}$ n/cm^3) processes depending on the neutron density. Neutron stars mergers provide suitable conditions for the r-process.

Kobayashi et al. (2020) constructed GCE models for all stable elements (from C to U) using the published observations of elemental abundances in the Milky Way together with the latest results of stellar astrophysics. Their models are calculated from the first principles by addressing the origin of all of the elements within the same framework. The authors present the time–metallicity evolution of neutron-capture elements for the solar neighborhood, halo, bulge and the thick disk of the Milky Way galaxy. Their models took into account all of the chemical enrichment sources: AGB stars, core-collapse supernovae, neutrino-driven winds, neutron star mergers and SNeIa.

The basic equations of chemical evolution are described in Kobayashi et al. (2000). The code follows the time evolution of elemental and isotopic abundances in a system where the ISM is instantaneously well mixed – one-zone model. The gas fraction and the metallicity of the system evolve as a function of time as a consequence of star formation and inflow and outflow of matter to/from the outside of the system. According to the model, the solar ratio [Fe/H]$=0$ at the time of the Sun's formation 4.6×10^9 years ago. The consequence of this is that the Sun is slightly more metal-rich than the average ISM of the solar neighborhood.

The GCE model constructed by Kobayashi et al. (2020) presents the evolution of elemental abundance ratios [X/Fe] against [Fe/H] in the solar neighborhood, where elements X range from C to Zn. Slightly less impressive are models' predictions for s-process' elements (Ga–U). The uncertainties of the yields from core-collapse supernovae due to the lack of physical explosion models could largely change the iron-peak element abundances relative to α-elements. The authors summarized the origin of stable elements in the solar system as follows:

- H, He and Li are produced in the Big Bang, with some uncertainties on Li production.
- Be and B are produced by cosmic rays.
- C (49%), F (51%) and N (74%) are produced by AGB stars.
- Elements O, Ne, Mg. Si, S, Ar and Ca are mainly produced by core-collapse supernovae and SNeIa.

- Cr, Mn, Fe and Ni are produced by SNeIa; Co, Ca, Zn, Ga and Ge are largely produced by hypernovae (HNe).
- Sr (32%), Y (22%) and Zr (44%) are produced from electron capture supernovae (ECSNe).
- For heavier neutron-capture elements, contributions from both NS–NS/NS–BH mergers and magnetorotational supernovae (MRSNe) are required.

In order to improve the models, more experimental data are required from astrophysical observations, nuclear reactions and cross-sections measurements at the energies of interest.

Investigating heavy element abundances of distant objects over a range of redshifts is essential for understanding the chemical history of the Universe. As already suggested, SNeII from massive stars mainly supply α-elements such as O, Ne and Mg up to Fe. SNeIa from binary systems mainly supply Fe. It is estimated that the lifetime of SNeIa is one to two orders of magnitude longer than that of SNeII, resulting in the delay in Fe enrichment compared to α-elements. This has a consequence that the [α/Fe] as a function of time is not a smooth curve, but rather, it has a break. Sameshima et al. (2017) performed an analysis of Mg II λ2798 and Fe II UV emission lines from the archival SDSS quasars in order to evaluate Mn/Fe abundance ratio. Their sample consisted of 17,432 quasars selected from the SDSS DR7 with a redshift range of $0.72 < z < 1.63$. Their calculations suggested that Mg II and Fe II emission lines are created at different regions of the photoionized cloud. Their results also indicate that Mg II/Fe II flux ratio depends largely on the cloud gas density. Note that the Mg II/Fe II flux ratio is used as a first-order proxy for the Mg/Fe abundance ratio in chemical evolution studies with quasar emission lines.

By comparison of the derived Mg/Fe abundance ratios with that of chemical evolution models, the authors suggested that the α-enrichment by mass loss from metal-poor intermediate-mass stars occurred at $z \sim 2$ or earlier (Sameshima et al., 2017). For their sample of SDSS quasars at $z < 2$, they have derived the average values for [Mg/Fe] ~ -0.2 and [Fe/H] $\sim +0.5$, which is in fair agreement with chemical evolution models taking into account the mass loss effect from metal-poor intermediate-mass stars.

Sameshima et al. (2020) extended these studies to quasars at high redshifts by performing near-infrared spectroscopy of six luminous quasars at $z \sim 2.7$. Their WINERED instrument, a near-IR spectrograph for a wide wavelength range of 0.9–1.35 μm with the resolving power of 28,000, was mounted on the New Technology Telescope at the La Silla Observatory. The measured Mg II/Fe II flux ratio was very close to the previously published data for $0.72 < z < 1.63$ (Sameshima et al., 2017), suggesting that there is no evolution over a long period of time. The authors also discuss the star formation history through a direct comparison with chemical evolution models.

3.3 Damped Lyman-Alpha Systems

Damped Lyman-alpha (DLA) systems or damped Lyman-alpha absorption systems are concentrations of neutral hydrogen gas detected in the spectra of quasars. They are defined as the systems where the column density (density projected along the line of sight to the quasar) of hydrogen is more significant than 2×10^{20} atoms/cm^2 (Coles and Lucchin, 2002), the highest of them having N~10^{22}. Standard lines in the Lyman-alpha forest are typically ~10^{13}–10^{15}. They are detected in the spectra of a background quasar in the Lyman-alpha region. This absorption occurs when the electron of a hydrogen atom goes from n=1 to n=2. A photon of appropriate energy/wavelength is then "removed" from the continuum, thus creating a spectral line. The measured spectra consist of neutral hydrogen Lyman-alpha absorption lines, which are broadened by radiation damping, as shown in Figure 3.18.

These systems can be observed at relatively high redshifts of 2–4 when they contain most of the neutral hydrogen in the Universe. The systems are believed to be associated with the early stages of galaxy formation, as the high neutral hydrogen densities of DLA systems are also typical of sightlines in the Milky Way and other nearby galaxies. They are observed in absorption spectra rather than by their individual stars as opposite to the galaxy. Therefore, they offer the opportunity of studying the dynamics of the gas in the early galaxies.

In their publication, Rafelski et al. (2012) showed spectra measured with Echellette Spectrograph and Imager (ESI; Sheinis et al., 2002). The 25 quasar spectra with 32 confirmed DLA systems having z>4, in order of increasing value of the redshift of DLA system for each spectrum, are presented. These spectra contain other lower-redshift DLA systems; however, only those with z>4 are shown. In addition to the 25 quasars with z>4 DLA systems, the authors also analyzed ESI observations and HIRES follow-up from 2008, which were obtained in a similar unbiased fashion. All the data were analyzed with the "XIDL" package developed in IDL by J. X. Prochaska (http://www.ucolick.org/~xavier/IDL/). Notably, the ESI observations were analyzed with "ESIRedux" (Prochaska et al., 2003b), and the HIRES observations were analyzed with "HIRedux". Both software are publicly available. The data were extracted in an optimal manner, and the data were fit using the "x_continuum" routine within XIDL (Rafelski et al., 2012).

DLA systems are interesting to probe at large redshifts. These objects are not luminous enough to be observed otherwise. They dominate the mass density of neutral gas in the Universe, containing at redshift z ~ 3–5 a comoving mass density of neutral gas comparable to the mass density of a star in the present-day galaxies. DLA systems are tracers

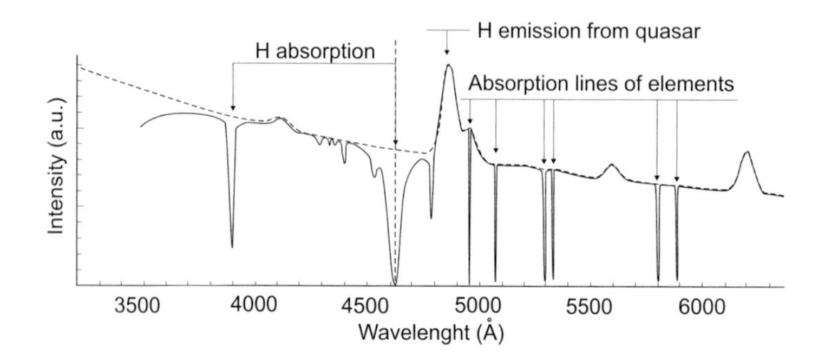

FIGURE 3.18 A schematic presentation of the measured spectra.

of the available material at high redshift for star formation. It is suggested that this gas is gradually converted into stars in the present-day galaxies.

DLA systems are predominantly neutral, making it easier to measure heavy elements abundances. They are neutral because "self-shielding" prevents the ionization of the gas. UV radiation ionizes the gas, but at high column density, ionized material quickly recombines, becoming neutral again. Gas-phase heavy elements abundances (with low ionization potential) are dominant spectral features. It is enough to divide the column density of the heavy element species by the neutral hydrogen density to get the abundance.

The study of high-redshift galaxies and access to their physical properties can be done using either emission- or absorption-line spectroscopy. Absorption-line spectroscopy is a potent technique. It has several advantages, and more information can be obtained compared to emission-line spectroscopy. According to Fan et al. (2001), it allows one to detect objects up to very high redshift; QSOs detected up to redshifts of 6.2 can be used as background searchlights for obtaining useful probes of the intervening Universe up to lookback times of 95% of the age of the Universe. Also, through the analysis of QSO absorption lines, one can study the spatial distribution, motion, chemical enrichment and ionization histories of gaseous structures, ranging from the intergalactic medium to high-column-density absorption systems associated with galaxies. Finally, since the detection of material intercepting a line of sight to a given QSO depends on the gas column density and the luminosity of the QSO, this is a unique technique for probing the chemical composition and physical conditions in the interstellar medium of various types of galaxies over a broad range of lookback times. The interstellar medium is detected independently of its distance, luminosity, star formation history and morphology.

The metallicity of DLA systems is discussed by Pettini (2003) in terms of [Fe/H] and [Zn/H] values and is shown in Figure 3.19 as a function of redshift.

The presence of dust in DLA systems is deduced from comparing the gas-phase abundances of two elements, which are depleted by differing amounts in local interstellar clouds. The Cr/Zn ratio is the most suitable of such pairs. From the earliest abundance measurements in DLA systems, it became apparent that this ratio is generally sub-solar since a fraction of the Cr has been incorporated into dust grains. Figure 3.20 indicates the end of decreasing Cr depletion with decreasing metallicity.

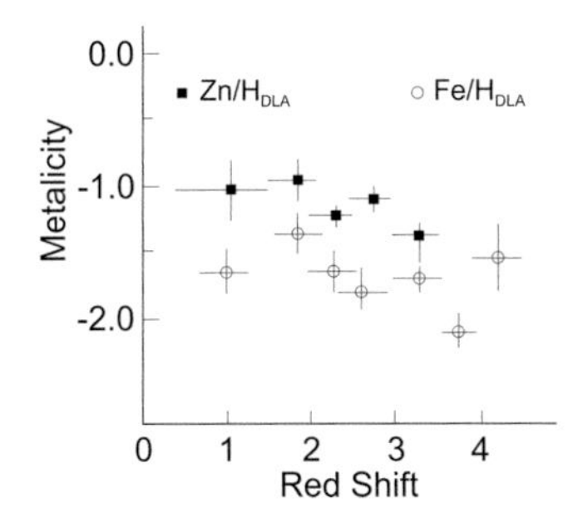

FIGURE 3.19 Column-density-weighted metallicities of DLA systems in different redshift intervals. (After Pettini, 2003.)

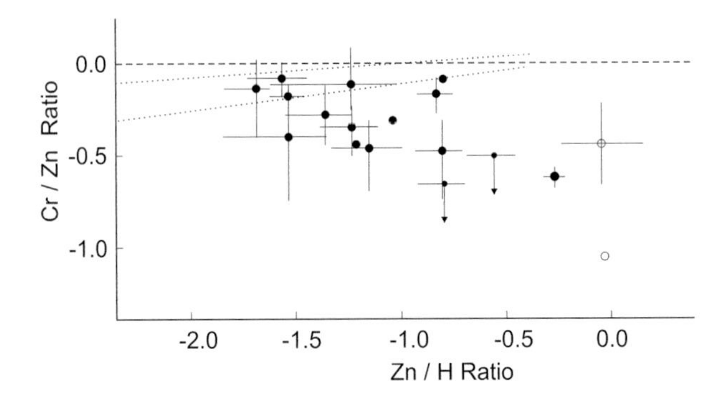

FIGURE 3.20 Cr abundance relative to Zn in 18 DLA systems (filled symbols). The region between dashed lines indicates how the Cr/Fe ratio varies in the galactic stars in this metallicity regime. The open circles show the typical Cr/Fe ratios measured in interstellar clouds in the disk and halo of our galaxy, where the under-abundance of Cr relative to Zn is ascribed to dust depletion. (After Pettini, 2003.)

Here we should mention the work performed by Rafelski et al. (2014) showing the rapid decline in the metallicity of DLA systems at $z \sim 5$. They presented chemical abundance measurements for 47 DLA systems, 30 at $z > 4$, observed with the ESI, and the HIRES on the Keck telescopes. H_I DLA column densities are measured by fitting the Ly-α profiles to Voigt profile, and a higher number of false DLA identifications with SDSS at $z > 4$ were found due to the increased density of the Lyman-α forest. Ionic column densities are determined by the apparent optical depth method. They combine their new measurements with previous surveys to determine the evolution of the cosmic metallicity of neutral gas. They found that the metallicity of DLA systems is decreasing with increasing redshift. The significance of the trend is improved by going to higher redshifts (from $z = 0.09$ to $z = 5.06$), with a linear fit of (-0.22 ± 0.03) dex per unit redshift. The "floor" of $\sim 1/600$ solar value continues out to $z \sim 5$, despite finding DLA systems with much lower metallicities. However, this value is not statistically different from a steep tail to the distribution (see Figures 3.21 and 3.22). They also find that the intrinsic scatter of metallicity among DLA systems of ~ 0.5 dex continues out to $z \sim 5$. Metallicity distribution and α/Fe ratios of $z > 2$ DLA systems are consistent with the same parent population as those of halo stars. Therefore, it is possible that the halo stars in the Milky Way formed out of gas that commonly exhibits DLA absorption at $z > 2$ (Rafelski et al., 2012, 2014).

Prochaska et al. (2003a) reported the observation of over 25 elements in a galaxy at redshift $z = 2.626$ (age of 2.5×10^9 years). The elements until Fe on the periodic table are produced during the nuclear reactions that produce energy in stars or during the collapse of the stellar core and the resulting explosion defined as a supernova (Burbidge et al., 1957). Regarding supernovae, the current thinking states that the supernovae of massive stars synthesize even-atomic-number (even-Z) nuclei – including the "α-elements" O, Ne, Mg, Si, S, Ar and Ca. In contrast, a substantial amount of Fe-peak elements is produced on significantly longer timescales in supernovae with lower mass progenitors. As

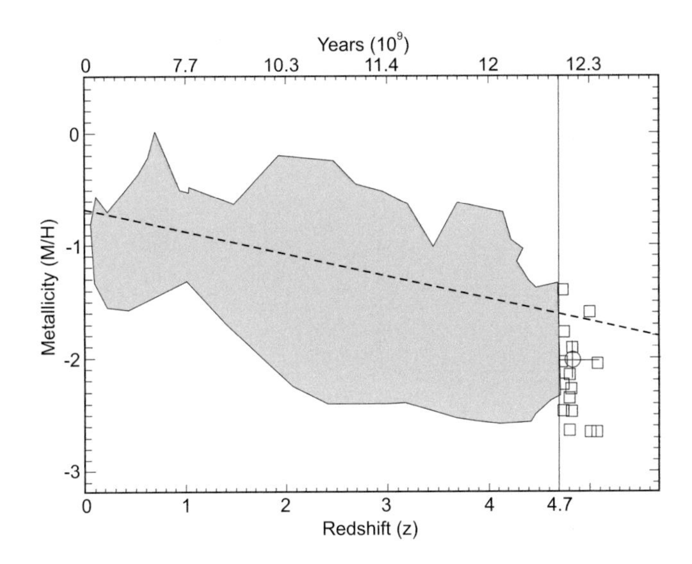

FIGURE 3.21 DLA metallicity vs redshift, showing a sharp decrease in metallicity at z>4.7. The metallicities of DLA systems inhabit the shaded region at z<4.7, and the squares are DLA systems at z>4.7. The dotted line represents a linear fit to the (z) data points in redshift space for DLA systems at z<4.7. The open circle is (z) deduced from DLA systems at z>4.7 and is significantly below the linear fit. (After Rafelski et al., 2014.)

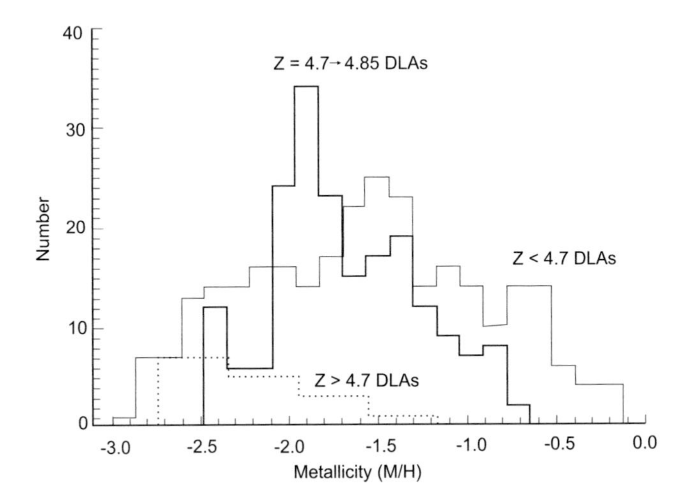

FIGURE 3.22 Histogram of DLA metallicities, showing a very different metallicity distribution of z>4.7 DLA systems. The thin solid line region represents the z<4.7 DLA systems, the thick solid line region corresponds to z<4.7 DLA systems evolved to z=4.85, and the dotted line part is for z>4.7 DLA systems. (After Rafelski et al., 2014.)

TABLE 3.1 Abundances of Some Chemical Elements which Might Have Played a Role in the Origin of Life for the Galaxy at z = 2.626

Element	Element/H	Δ_{dust}	Element/O
B	−0.57	0.1	−0.03
N	>−2.24	0.0	>−1.80
O	−0.54	0.1	0.00
Mg	−0.78	0.3	−0.04
Al	>−2.00	>0.5	>−1.06
Si	−0.91	0.3	−0.17
P	<−1.06	<0.3	<−0.32
S	−0.87	0.1	−0.33
Cl	−1.55	>0.0	>−1.11
Ti	−1.87	>0.7	>−0.73
Cr	−1.61	>0.7	>−0.47
Mn	<−1.85	0.7	<−0.71
Fe	−1.69	>0.7	>−0.55
Co	<−1.48	>0.7	>−0.34
Ni	−1.73	>0.7	>−0.59
Cu	<−1.11	>0.7	>0.03
Zn	−0.91	0.2	−0.27
As	<0.26	0.0	<0.70
Sn	<−0.27	0.0	<0.17
Pb	<−0.19	0.0	<0.34

Source: Modified from Prochaska et al. (2003a).

a result, comparisons between the α-elements and the Fe-peak elements clarify the star formation history and age of the galaxy (Tinsley, 1979).

Table 3.1 shows gas-phase abundances relative to solar on a logarithmic scale; for example, [X/H] = −1 implies 1/10 solar abundance. Throughout the analysis, the authors have adopted $N(H_I) = 10^{21.35}$ cm^{-2} measured from a fit to the DLA profile.

Dust corrections are estimated from the depletion patterns of galactic gas with comparable depletion levels (Savage and Sembach, 1996). The correction values are added to the gas-phase abundances to result in the underlying nucleosynthetic model (very conservative corrections).

The life-essential elements of this galaxy are shown in Figure 3.23. The solid line represents the solar abundance pattern scaled to the oxygen of the galaxy. To some extent, the galaxy's enrichment pattern resembles the Sun, indicating that their nucleosynthetic enrichment histories are similar to the case presented in the paper by Prochaska et al. (2003a).

In their paper, Dessauges-Zavadsky et al. (2004) presented a useful set of chemical element abundances in the DLA system. It is assumed that the DLA systems probably represent some early stages in the evolution of the galaxies. At that time, most of the galaxy mass still resided in the interstellar medium rather than in the newly formed stars. These objects

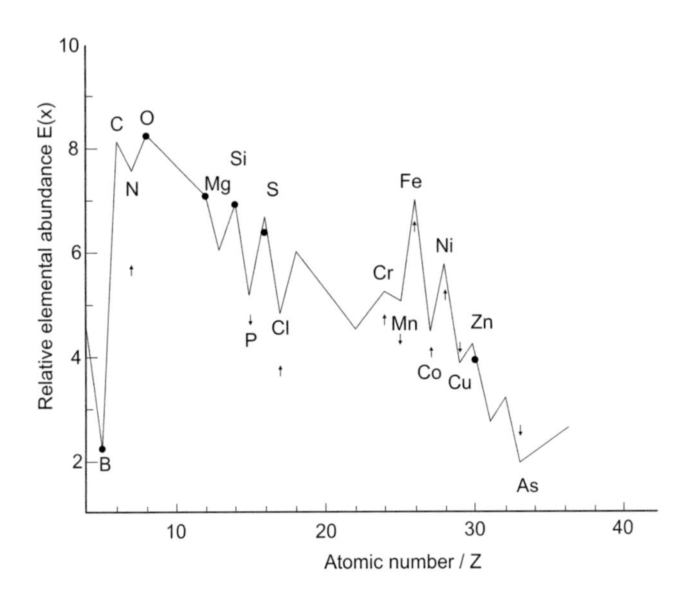

FIGURE 3.23 Dust-corrected elemental abundances of life-essential elements in the protogalaxy at z ~ 2.626 toward FJ081240.6.320808 on a logarithmic scale with hydrogen defined to have an abundance number Є(X)=12. The adjustments were performed following the depletion patterns observed in the local Universe. This fact explains the lower limits for the Fe, Cr, Al, Co and Ni abundances and similarly the upper limit for Mn. Typical statistical errors are <0.1 dex, that is, the size of the plot symbols. The solid line represents the solar abundance pattern normalized to the observed oxygen abundance ([O/H]=−0.44, after dust correction). To zero order, the pattern of this high-redshift galaxy resembles that of the Sun (Prochaska et al., 2003a).

give us the best opportunity to study the galaxies at high redshift and in their early stages of evolution. They also give us the opportunity for tracking the galactic chemical evolution through the cosmic ages. Dessauges-Zavadsky et al. (2004) presented the results of column density measurements for 21 ions of 15 elements (N, O, Mg, Al, Si, P, S, Cl, Ar, Ti, Cr, Mn, Fe, Ni and Zn) in 4 DLA systems in the redshift interval z_{abs}=1.7–2.5. This is in contrast with the data for the majority of DLA systems for which only a handful of elements: usually Si and Fe, and occasionally Cr, Zn and Ni, are detected (for example, Lu et al., 1996; Prochaska and Wolfe, 1999; Prochaska et al., 2001). The small amount of information on individual systems has severely limited the interpretation of the DLA abundance forms.

Figure 3.24a–c presents the nucleosynthetic patterns for life-essential elements of the DLA systems toward Q1331+17, z_{abs}=1.776; Q2231−00, z_{abs}=2.066; and Q0100+13, z_{abs}=2.309, respectively. At first glance, the abundance patterns of these high-redshift galaxies resemble that of the solar neighborhood, indicating that their nucleosynthetic enrichment histories are not too dissimilar from our galaxy. However, at careful inspection, one can see several important differences.

Figure 3.24a shows the nucleosynthetic abundance form of the DLA at z_{abs}=1.776 toward Q1331+17. The dust-corrected elemental abundances are shown on a logarithmic

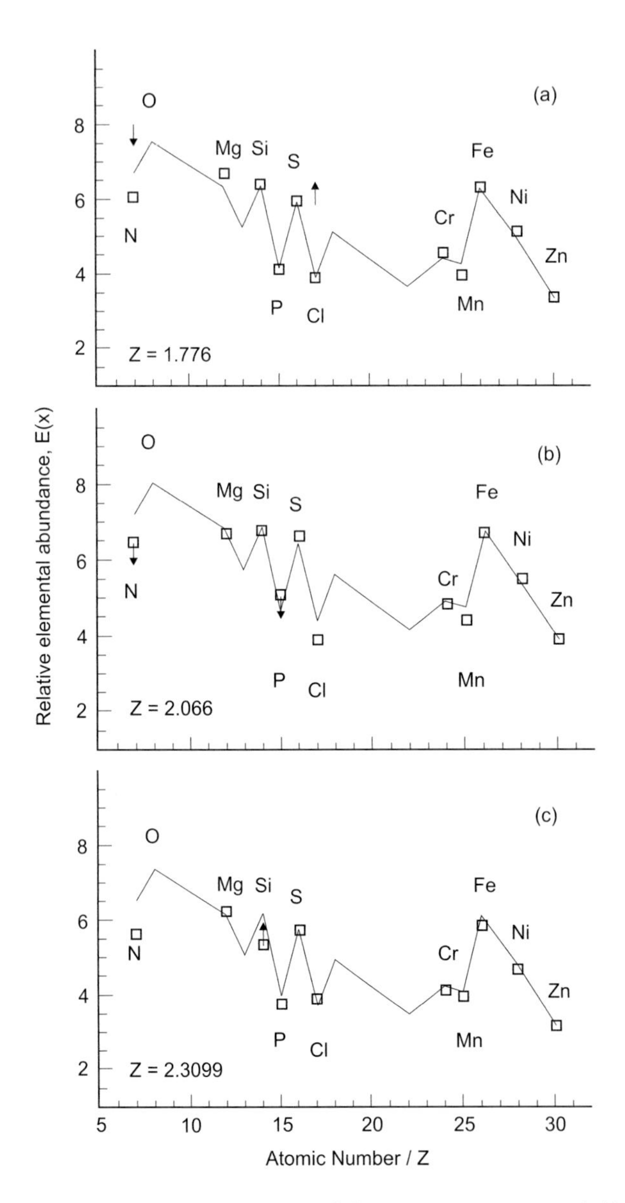

FIGURE 3.24 The nucleosynthetic patterns of the DLA systems toward (a) Q1331+17, (b) Q2231–00 and (c) Q0100+13. (a) The abundance pattern of the DLA at z_{abs}=1.776. The solar values (solid line) were scaled to [S/H]=–1.26. (b) The abundance pattern of the DLA at z_{abs}=2.066. The solar values were scaled to match the dust-corrected silicon metallicity, $[Si/H]_{cor}$=–0.78. (c) The abundance pattern of the DLA at z_{abs}=2.309 following the data presented by Dessauges-Zavadsky et al. (2004).

scale, where hydrogen is defined to have $\varepsilon(H)=12$ and generally, for any element X, $\varepsilon(X)=\log(X/H)+12$. The DLA system toward Q1331+17 has almost a solar abundance pattern. Actually, few differences could be observed. The Fe-peak elements (Fe, Ni and Cr) follow each other; the two α-elements S and Si, and the ratios of α-elements to Fe-peak elements (Si/Fe, S/Zn and S/Fe) are all solar. It should be noted that the Mg/Fe ratio is slightly oversolar.

Figure 3.24b shows the nucleosynthetic abundance pattern of the DLA system toward Q2231-00 at $z_{abs}=2.066$. The solar values were scaled to the dust-corrected silicon metallicity, [Si/H]cor=−0.78. The DLA system toward Q2231-00 shows a solar [Si/Fe] ratio, but enhanced [S/Zn, Fe] and [Ti/Zn, Fe] ratios.

Figure 3.24c shows the nucleosynthetic abundance pattern of the DLA at $z_{abs}=2.309$ toward Q0100+13. The solar values were scaled to the observed sulfur metallicity, [S/H]=−1.44, of the DLA system.

The DLA system toward Q0100+13 shows a slight enhancement of the α-elements S and Mg abundances relative to the Zn and Fe abundances, namely [S/Zn]=+0.19±0.09 and $[Mg/Fe]_{cor}=+0.29\pm0.11$. This α/Fe-peak enhancement suggests the enrichment by massive stars. The α-elements are produced within 2×10^7 years by SNeII resulting from massive stars, while the Fe-peak elements are mainly produced by SNeIa on longer timescales, between 10^8 and 10^9 years. The Fe-peak elements Fe, Ni and Cr have solar values one relative to the other. In this system, one can observe the odd–even effect, which corresponds to an under-abundance of odd-Z elements compared to the even-Z elements of the same nucleosynthetic origin. Finally, the DLA shows an under-solar [N/S] ratio of −0.83±0.14 at [S/H]=−1.48±0.11. Nitrogen is an element of particular interest since it has a complex nucleosynthetic origin.

A chemical evolution model by Calura et al. (2003) contains the development of element abundances, including the processes in the stars and restoration into the ISM by the supernova explosions and stellar winds. The DLA systems offer the best possibility to study the high-redshift galaxies. Accurate chemical abundances of these systems can be obtained over a large interval of cosmic time, and they thus offer the best opportunity to track the chemical evolution of galaxies in the Universe.

3.4 Multi-messenger Astronomy

Multi-messenger astronomy stands for the coordinated observation of different classes of signals that originate from the same astrophysical event. In other words, different probes are used to observe the same astrophysical process. In principle, it could provide a wealth of information about the studied astrophysical process (Abbott et al., 2017). They reported the global-level effort that led to the first joint detection of gravitational and electromagnetic radiation from a single source. An ~100 s long gravitational wave signal (GW170817) was followed by short gamma-ray bursts (sGRBs, GRB 170817A) and an optical transient (SSS 17a/AT 2017gfo) found in the host galaxy NGC 4993. The source was detected across the electromagnetic spectrum – in the X-ray, ultraviolet, optical, infrared and radio bands – over hours, days and weeks. These observations supported the hypothesis that GW170817 was produced by the merger of two neutron stars in NGC

4993, followed by an sGRB and a kilonova powered by the radioactive decay of *r*-process nuclei synthesized in the ejecta.

The paper by Abbott et al. (2017) showed the simultaneous observation of gravitational and electromagnetic waves from a single source for the first time. The gravitational wave observation of a binary neutron star merger is also the first of its kind. Three observations have been made:

i. a prompt sGRB that demonstrates that BNS mergers are the progenitor of at least a fraction of such bursts;
ii. an ultraviolet, optical and infrared transient (kilonova), which allows for the identification of the host galaxy and is associated with the aftermath of the BNS merger; and
iii. delayed X-ray and radio counterparts that provide information on the environment of the binary.

All of these observations are described in detail in a number of articles cited in Abbott et al. (2017). They offer a comprehensive, sequential description of the physical processes related to the merger of a binary neutron star. The results of this campaign demonstrate the importance of collaborative gravitational wave, electromagnetic and neutrino observations of the same object and mark a new era in multi-messenger, time-domain astronomy. However, such a multi-messenger astronomy correlates signals from known fundamental forces and standard model particles.

Many of the open problems of modern physics suggest the existence of fields with light quanta with mass $\ll 1$ eV/c^2 (see Dailey et al., 2020). They showed that networks of precision quantum sensors that are shielded from or are insensitive to conventional standard model physics signals can be a powerful tool for multi-messenger astronomy. The authors considered the case in which high-energy astrophysical events produce intense bursts of exotic low-mass fields (ELFs). They estimated ELF signal amplitudes, delays, rates and distances of gravitational wave sources to which global networks of atomic magnetometers (Budker and Kimball, 2013; Afach et al., 2018) and atomic clocks (Derevianko and Pospelov, 2014; Roberts et al., 2017) could be sensitive. Bursts of ELFs could be generated by cataclysmic astrophysical events, such as black hole or neutron star mergers or supernovae. Ultralight bosons are among the objects considered as possible ELFs. They have a small mass, and consequently, a high-energy event is not required for their production.

As an example of successful multi-messenger approach, we mention here the work by Dietrich et al. (2020) who reported the observation of binary neutron stars mergers. Their aim was to elucidate the properties of matter under extreme conditions and to determine the expansion rate of Universe as quantified by the Hubble constant. The authors performed a joint analysis of the gravitational wave event GW170817 with its electromagnetic counterparts AT2017gfo and GRB 170817A, and the gravitational wave event GW190425, both originating from neutron star mergers. In order to constrain the neutron star equation of state, they combined this analysis with previous measurements of pulsars using x-ray and radio observations, and nuclear theory computations using chiral effective field theory. They reported the following results:

- The radius of a neutron star with mass $M_{ns}=1.4\ M_{sun}$ is equal to $R=\left(11.75^{+0.86}{}_{-0.81}\right)$ km at the level of confidence of 90%.
- The Hubble constant at 1σ uncertainty is $H=\left(66.2^{+4.4}{}_{-4.2}\right)$ km/Mpc/s.

References

Abazajian, K. N., Adelman-McCarthy, J. K., Agueros, M. A., et al. 2009. The seventh data release of the Sloan digital sky survey. *The Astrophysical Journal Supplement Series* 182:543–558.

Abbott, B. P., Abbott, R., Abbott, T. D., et al. 2017. Multi-messenger observations of a binary star merger. *Astrophysics Journal Letters* 848:L12.

Afach, S., Budker, D., DeCamp, G., et al. 2018. Characterization of the global network of optical magnetometers to search for exotic physics (GNOME). *Physics of the Dark Universe* 22:162–180.

Boesgaard, A. M., King, J. R., Cody, A. M., Stephens, A. and Deliyannis, C. P. 2005. Chemical composition in the globular cluster M71 from keck hires spectra of turn-off stars. *The Astrophysical Journal* 629:832–848.

Brinchmann, J., Charlot, S., White, S. D. M., Tremonti, C., Kauffmann, G. and Heckman, T. 2004. The physical properties of star-forming galaxies in the low-redshift Universe. *Monthly Notices of the Royal Astronomical Society* 351:1151–1179.

Budker, D. and Kimball, D. E., (Eds). 2013. *Optical Magnetometry.* Cambridge University Press, Cambridge, UK.

Burbidge, E. M., Burbidge, G. R., Fowler, W. A. and Hoyle, F. 1957. Synthesis of elements in stars. *Reviews of Modern Physics* 29(4):547–654.

Calura, F., Matteucci, F. and Vladilo, G. 2003. Chemical evolution and nature of damped Lyman α systems. *Monthly Notices of the Royal Astronomical Society* 340(1):59–72.

Carretta, E. 2015a. Globular clusters and their contribution to the formation of the Galactic halo. In Bragaglia, A., Arnaboldi, M., Rejkuba, M. and Romano, D. (Eds.) *The General Assembly of the Galaxy Halos: Structure, Origin, and Evolution. Proceedings IAU Symposium.* No. 317. arXiv:1510.00507 [astro-ph.GA].

Carretta, E. 2015b, Five groups of red giants with distinct chemical composition in the globular cluster NGC 2808. *The Astrophysical Journal* 810:148.

Carretta, E., Bragaglia, A., Gratton, R., D'Orazi, V. and Lucatello, S. 2009a. Intrinsic iron spread and a new metallicity scale for globular clusters. *Astronomy and Astrophysics* 508(2):695–706.

Carretta, E., Bragaglia, A., Gratton, R. and Lucatello, S. 2009b. Na-O anticorrelation and HB. VIII. Proton-capture elements and metallicities in 17 globular clusters from UVES spectra *Astronomy and Astrophysics* 505(1):139–155.

Carollo, D., Beers, T. C., Bovy, J., et al. 2011. Carbon-enhanced metal-poor stars in the inner and outer halo components of the Milky Way. arXiv:1103.3067 [astro-ph.GA].

Cayrel, R., Depagne, E., Spite, M., et al. 2004. First stars V - Abundance patterns from C to Zn and supernova yields in the early Galaxy. *Astronomy and Astrophysics* 416:1117–1138.

Chen, Y.-M., Tremonti, C. A., Heckman, T. M., et al. 2010. Absorption-line probes of the prevalence and properties of outflows in present-day star-forming galaxies. *The Astronomical Journal* 140:445–461.

Choksi, N., Gnedin, O. Y., Li H. 2018. Formation of Globular Cluster Systems: From Dwarf Galaxies to Giants. *Monthly Notices of the Royal Astronomical Society* 480:2343–2356.

Chopra, A. and Lineweaver, C. 2018. The cosmic evolution of biochemistry. In Gordon, R. and Sharov, A. (Eds.) *Habitability of the Universe Before Earth*, Volume 1 in the series: Astrobiology: Exploring Life on Earth and Beyond, series editors: Rampelotto, Seckbach, P. H. J. & Gordon. pp. 75–87. R. ISSN 2468–6352. Doi: 10.1016/B978-0-12-811940-2.00004-6©2018 Elsevier Inc. Available from: https://www.researchgate.net/publication/323624615_The_Cosmic_Evolution_of_Biochemistry [accessed May 13 2019].

Christensen-Dalsgaard, J. 2008. ASTEC -- the Aarhus stellar evolution code. *Astrophysics and Space Science* 316:13–24. See also arXiv:0710.3114 [astro-ph].

Cohen, B.A., Swindle, T. D. and Kring, D. A. 2000. Support for the lunar cataclysm hypothesis from lunar meteorite impact melt ages. *Science* 290(5497):1754–1755.

Coles, P. and Lucchin, F. 2002. *Cosmology: The Origin and Evolution of Cosmic Structure*, 2nd edition. John Wiley & Sons, England, pp. 430–432.

Cowan, J. J., Sneden, C., Beers, T. C., Lawler, J. E., Simmerer, J., Truran, J. W., Primas, F., Collier, J. and Burles, S. 2005. Hubble space telescope observations of heavy elements in metal-poor galactic halo stars. *The Astrophysical Journal* 627:238–250.

Cresci G., Mannucci F., Maiolino R., Marconi A., Gnerucci A. and Magrini L. 2010. Gas accretion as the origin of chemical abundance gradients in distant galaxies. *Nature* 467:811–813.

Dailey, C., Bradley, C., Kimball, D. F. J., et al. 2020. Quantum sensor networks as exotic field telescopes for multi-messenger astronomy. *Nature Astronomy Letters*, Doi: 10.1038/s41550-020-01242-7.

Da Silva, R., Porto de Mello, G. F., Milone, A. C., et al. 2012. Accurate and homogeneous abundance patterns in solar-type stars of the solar neighborhood: A chemo-chronological analysis. *Astronomy and Astrophysics* 542:A8 (26 p).

Davé, R., Finlator, K. and Oppenheimer, B. D. 2011. Galaxy evolution in cosmological simulations with outflows–II. Metallicities and gas fractions. *Monthly Notices of the Royal Astronomical Society (MNRAS)* 416:1354–1376.

Derevianko, A. and Pospelov, M. 2014. Hunting for topological dark matter with atomic clock. *Nature Physics* 10:933–936.

Dessauges-Zavadsky, M., Calura, F., Prochaska, J. X., D'Odorico, S. and Matteucci, F. 2004. A comprehensive set of elemental abundances in damped Lyα systems: Revealing the nature of these high-redshift galaxies. *Astronomy & Astrophysics* 416:79–110.

Dietrich, T., Coughlin, M. W. and Pang, P. T. H. 2020. Multimessenger constraints on the neutron star equation of state and the Hubble constant. *Science* 370:1450–1453.

Dotter, A., Sarajedini, A. and Anderson, J. 2011. Globular clusters in the outer galactic halo: New hubble space telescope/advanced camera for surveys imaging of six globular clusters and the galactic globular cluster age-metallicity relation. *The Astrophysical Journal* 738(1):article id. 74 (11 p).

Edmunds, M. G. and Phillipps, S. 1997. Global chemical evolution - II. The mean metal abundance of the Universe. *Monthly Notices of the Royal Astronomical Society* 292:733–747.

Edvardsson, B., Andersen, J., Gustafsson, B., Lambert, D. L., Nissen, P. E. and Tomkin, J. 1993. The chemical evolution of the galactic disk - part one - analysis and results. *Astronomy and Astrophysics* 275:101–152.

Erb, D. K., Shapley, A. E., Pettini, M., Steidel, C. C., Reddy, N. A. and Adelberger, K. L. 2006. H_α observations of a large sample of galaxies at z ~ 2: implications for star formation in high-redshift galaxies. *The Astrophysical Journal* 647:128–139.

Fan, X., Narayanan, V. K., Lupton, R. H., et al. 2001, Survey of z >5.8 Quasars in the Sloan digital sky survey I: Discovery of three new quasars and the spatial density of luminous quasars at z~6. *The Astrophysical Journal* 122:2833–2849.

Fedonkin, M. A. 2008. Ancient biosphere: The origin, trends and events. *Russian Journal of Earth Sciences* 10:ES1006, pp. 1–9. Doi: 10.2205/2007ES000252.

Forbes, D. A. and Bridges, T. 2010. Accreted versus in situ Milky Way globular clusters. *Monthly Notices of the Royal Astronomical Society* 404:1203–1214.

Friel, E. D. 1995. The old open clusters of the milky way. *Annual Review of Astronomy and Astrophysics* 33:381–414.

Gratton, R.G., Carretta, E., Claudi, R., Lucatello, S. and Barbieri, M. 2003. Abundances for metal-poor stars with accurate parallaxes. I. Basic data. *Astronomy and Astrophysics* 404:187–210.

Gratton, R.G., Sneden, C., Carretta, E. and Bragaglia, A. 2000. Mixing along the red giant branch in metal-poor field stars. *Astronomy and Astrophysics* 354:169–187.

Greenberg, J. M. 1971. Interstellar grain temperatures effects of grain materials and radiation fields. *Astronomy & Astrophysics* 12:240–249.

Grevesse, N. and Sauval, A. J. 1998. Standard solar composition. *Space Science Reviews* 85(1/2):161–174.

Heger, A. and Woosley, S. E. 2010. Nucleosynthesis and evolution of massive metal-free stars. *The Astrophysical Journal* 724(1):341–373.

Hollek, J. K., Frebel, A., Roederer, I. U., Sneden, C., Shetrone, Timothy, M., Beers, C., Kang, S.-J. and Thom̀, C. 2011. The chemical abundances of stars in the halo (cash) project. II. A sample of 14 extremely metal-poor stars. *The Astrophysical Journal* 742(1):54 (19 p).

Honda, S., Aoki, W., Ishimaru, Y., Wanajo, S. and Ryan, S. G. 2006. Neutron-capture elements in the very metal poor star HD 122563. *The Astrophysical Journal* 643:1180–1189.

Iwamoto, K., Brachwitz, F., Nomoto, K., Kishimoto, N., Umeda, H., Hix, W. R. and Thielemann, F.-K. 1999. *The Astrophysical Journal Supplement Series (ApJS)* 125:439–462.

Kauffmann G., Heckman, T. M., Tremonti, C., et al. 2003. The host galaxies of active galactic nuclei. *Monthly Notices of the Royal Astronomical Society* 346:1055–1077.

Kewley, L. J. and Ellison, S. L. 2008, Metallicity calibrations and the mass-metallicity relation for star-forming galaxies. *The Astrophysical Journal* 681(2):1183–1204.

Kobayashi, C., Karakas, A. I. and Lugaro, M. 2020. The origin of elements from carbon to uranium. *The Astrophysical Journal* 900:179 (33 p).

Kobayashi, C., Tsujimoto, T. and Nomoto, K. 2000. The history of the cosmic supernova rate derived from the evolution of the host galaxies. *Astrophysical Journal* 539:26–38.

Komatsu, E., Smith, K. M., Dunkley, J., et al. 2011. Seven-year wilkinson microwave anisotropy probe (WMAP) observations: Cosmological interpretation. *The Astrophysical Journal Supplement Series* 192:18 [arXiv:0803.0547].

Konami, S., Matsushita, K., Gandhi, P. and Tamagawa, T. 2013. Spatial distribution of abundance patterns in the starburst galaxy NGC 3079 revealed with Chandra and Suzaku. *Publications of the Astronomical Society of Japan (PASJ)* 1: arXiv:1205.6005v1 [astro-ph.GA 27 May 2012].

Kraft, R.P. 1994. Abundance differences among globular-cluster giants: Primordial versus evolutionary scenarios. *Astronomical Society of the Pacific, Publications* 106(700):553–565.

Krüger, H. and Grün, E. 2009. Interstellar dust inside and outside the heliosphere. *Space Science Reviews* 143:347–356.

Kruijssen J. M. D. 2019. The minimum metallicity of globular clusters and its physical origin - implications for the galaxy mass-metallicity relation and observations of proto-globular clusters at high redshift. *Monthly Notices of the Royal Astronomical Society* 486:L20–L25.

Larsen, S. S., Romanowsky, A. J., Brodie, J. P. and Wasserman, A. 2020. An extremely metal-deficient globular cluster in the Andromeda Galaxy. *Science* 370:970–973.

Li, A. and Draine, B.T. 2001. Infrared emission from interstellar dust. II. The diffuse interstellar medium. *The Astrophysical Journal* 554:778–802.

Li, A. and Greenberg, J. M. 1997. A unified model of interstellar dust. *Astronomy & Astrophysics* 323:566–584.

Lodders, K. 2003. Solar system abundances and condensation temperatures of the elements. *The Astrophysical Journal* 591:1220–1247.

Lu, L., Sargent, W. L. W., Barlow, T. A., Churchill, C. W. and Vogt, S. S. 1996. Abundances at high redshifts: The chemical enrichment history of damped Lyalpha galaxies. *Astrophysical Journal Supplement* 107:475–519.

Maiolino, R., Nagao, T., Grazian, A., et al. 2008, AMAZE. I. The evolution of the mass-metallicity relation at z>3. *Astronomy and Astrophysics* 488(2):463–479.

Mannucci, F., Cresci, G., Maiolino, R., Marconi, A. and Gnerucci, A. 2010. A fundamental relation between mass, star formation rate and metallicity in local and high-redshift galaxies. *Monthly Notices of the Royal Astronomical Society* 408:2115–2127.

Mannucci, F., Cresci, G., Maiolino, R., et al. 2009. LSD: Lyman-break galaxies stellar populations and dynamics – 1. Mass, metallicity and gas at z~3.1. *Monthly Notices of the Royal Astronomical Society* 398:1915–1931.

Marin-Franch, A., Aparicio, A., Piotto, G., et al. 2009. The ACS survey of galactic globular clusters. VII. Relative ages. *The Astrophysical Journal* 694(2):1498–1516.

Matteucci, F. 2001. *The Chemical Enrichment of Galaxies*. Kluwer Academic Publ, Dordrecht, Netherlands (headquarters). p. 293.

McWilliam, A. 1997. Abundance ratios and galactic chemical evolution. *Annual Review of Astronomy and Astrophysics* 35:503–556.

McWilliam, A., Preston, G. W., Sneden, C. and Searle, L. 1995. Spectroscopic analysis of 33 of the most metal poor stars. II. *Astronomical Journal* 109:2757–2799.

Møller, P., Fynbo, J. P. U., Ledoux, C. and Nilsson, K. K. 2013. Mass–metallicity relation from z=5 to the present: evidence for a transition in the mode of galaxy growth at z=2.6 due to the end of sustained primordial gas infall. *Monthly Notices of the Royal Astronomical Society* 430:2680–2687.

Müller, H. S. P., Thorwirth, S., Roth, D. A. and Winnewisser, G. 2001. The cologne database for molecular spectroscopy, CDMS. *Astronomy & Astrophysics* 370:L49–L52.

Nissen, P. E. 2015. High-precision abundances of elements in solar twin stars Trends with stellar age and elemental condensation temperature. *Astronomy and Astrophysics* 579:A52 (15 p).

Nissen, P. E. 2016. High-precision abundances of Sc, Mn, Cu, and Ba in solar twins. Trends of element ratios with stellar age. *Astronomy and Astrophysics* 593:A65 (12 p).

Noeske, K. G., Weiner, B. J., Faber, S. M., et al. 2007. Star formation in AEGIS field galaxies since z=1.1: The dominance of gradually declining star formation, and the main sequence of star-forming galaxies. *The Astrophysical Journal* 660(1):L43–L46.

Nomoto, K., Tominaga, N., Umeda, H., Kobayashi, C. and Maeda, K. 2006. Nucleosynthesis yields of core-collapse supernovae and hypernovae, and galactic chemical evolution. *Nuclear Physics A* 777:424–458.

Pasquini, L., Avila, G., Blecha, A., et al. 2002, Dec. Installation and commissioning of FLAMES, the VLT Multifibre Facility. *The Messenger* 110:1–9 (ISSN 0722-6691).

Pettini, M. 2003. Element abundances through the cosmic ages. Lectures given at the XIII Canary Islands winter school of astrophysics "Cosmochemistry: The Melting Pot of Elements". http://www.ast.cam.ac.uk/pettini/canaries13. arXiv:astro-ph/0303272.

Phillipps, S. and Edmunds, M. G. 1996. Global chemical evolution – I. QSO absorbers and the chemical evolution of galaxy discs. *Monthly Notices of the Royal Astronomical Society* 281:362–368.

Prochaska, J. X., Gawiser, E., Wolfe, A. M., Castro, S. and Djorgovski, S. G. 2003b. The age-metallicity relation of the Universe in neutral gas: the first 100 damped Lyα systems. *Astrophysical Journal* 595:L9–L12.

Prochaska, J. X., Howk, J. C. and Wolfe, A. M. 2003a. The elemental abundance pattern in a galaxy at z=2.626. *Nature* 423:57–59.

Prochaska, J. X. and Wolfe, A. M. 1999. Chemical abundances of the damped Lyα systems at z>1.5. *Astrophysical Journal Supplement* 121:369–415.

Prochaska, J. X., Wolfe, A. M., Tytler, D., et al. 2001. The UCSD HIRES/Keck I damped Lyα abundance database. I. The data. *Astrophysical Journal Supplement* 137:21–73.

Ramírez, I., Meléndez, J., Bean, J., et al. 2014, The solar twin planet search I. Fundamental parameters of the stellar sample. *Astronomy and Astrophysics* 572:A48.

Rana, N. C. 1991. Chemical evolution of the galaxy. *Annual Review of Astronomy and Astrophysics* 29:129–162.

Rafelski, M., Neeleman, M., Fumagalli, M., Wolfe, A. M. and Prochaska, J. X. 2014. The rapid decline in metallicity of damped Lyα systems at z~5. *The Astrophysical Journal Letters* 782:L29 (6p).

Rafelski, M., Wolfe, A. M., Prochaska, J. X., Neeleman, M. and Mendez, A. J. 2012. Metallicity evolution of damped Ly-α systems out to z ~ 5. *The Astrophysical Journal* 755:89 (21 p).

Roberts, B. M., Blewitt, G., Dailey, C., et al. 2017. Search for domain wall dark matter with atomic clocks on bord global positioning system satellites. *Nature Communications* 8:L195.

Roederer, I. U., Sneden, C., Lawler, J. E. and Cowan, J. J. 2010. New abundance determinations of cadmium, lutetium, and osmium in the r-process enriched star BD +17 3248. *The Astrophysical Journal Letters* 714:L123–L127.

Roediger, J. C., Courteau, S., Graves, G. and Schiavon, R. P. 2013. Constraining stellar population models - I. Age, metallicity and abundance pattern compilation for galactic globular clusters. arXiv:1310.3275v1 [astro-ph.GA] 11 Oct 2013.

Salim, S., Rich, R. M., Charlot, S., et al., 2007. UV star formation rates in the local universe. *The Astrophysical Journal Supplement Series* 173(2):267–292.

Sameshima, H., Yoshii, Y. and Kawara, K. 2017. Chemical evolution of the universe at 0.7<z<1.6 derived from abundance diagnostics of the broad-line region of quasars. *The Astrophysical Journal* 834:203 (18 p).

Sameshima, H., Yoshii, Y. and Matsunaga, N., et al. 2020. Mg II and Fe II fluxes of luminous quasars at z ~ 2.7 and evolution of the Baldwin effect in the flux-to-abundance conversion method for quasars. arXiv:2010.10548v2 [astro-ph.GA 23 Nov 2020.]

Savage, B. D. and Sembach, K. R. 1996. Interstellar abundances from absorption-line observations with the Hubble Space Telescope. *Annual Review of Astronomy and Astrophysics* 34:279–330.

Schiavon, R. P., Rose, J. A., Courteau, S. and MacArthur, L. A. 2005. A library of integrated spectra of galactic globular clusters. *The Astrophysical Journal Supplement Series* 160:163–175.

Sheinis, A. I., Bolte, M., Epps, H. W., Kibrick, R. I., Miller, J. S., Radovan, M. V., Bigelow, B. C. and Sutin, B. M. 2002. ESI, a new keck observatory Echellette spectrograph and imager. *The Publications of the Astronomical Society of the Pacific* 114(798):851–865.

Sneden, C., Cowan, J.J. and Gallino, R. 2008. Neutron-capture elements in the early galaxy. *Annual Review of Astronomy and Astrophysics* 46:241–288.

Sneden, C., Cowan, J. J., Lawler, J. E., et al. 2003. The extremely metal-poor, neutron capture–rich star CS 22892-052: a comprehensive abundance analysis. *The Astrophysical Journal* 591:936–953.

Sneden, C., Kraft, R. P., Langer, G. E., Prosser, C. F. and Shetrone, M. D. 1994. Oxygen abundances in halo giants. V. Giants in the fairly metal-rich Globular cluster M71. *The Astrophysical Journal* 107:1773.

Steidel, C. C., Erb, D. K., Shapley, A. E., et al. 2010. The structure and kinematics of the circumgalactic medium from Far-ultraviolet spectra of $\tilde{z}$= 2–3 galaxies. *The Astrophysical Journal* 717:289–322.

Stephens, A. and Boesgaard, A. M. 2002. Abundances from high-resolution spectra of kinematically interesting halo stars. *The Astronomical Journal* 123:1647–1700.

Tinsley, B. M. 1979. Stellar lifetimes and abundance ratios in chemical evolution. *The Astrophysical Journal* 229:1046–1056.

Tucci Maia, M., Ramírez, I., Meléndez, J., et al. 2016. The solar twin planet search III. The [Y/Mg] clock: Estimating stellar ages of solar type stars. *Astronomy and Astrophysics* 590:A3.

Twarog, B. A. 1980. The chemical evolution of the solar neighborhood. II - The age-metallicity relation and the history of star formation in the galactic disk. *Astrophysical Journal, Part 1* 242:242–259.

Wanajo, S. and Janka, H. T. 2011, Dec. 7. The r-process in the neutrino-driven wind from a black-hole torus. arXiv:1106.6142v2 [astro-ph.SR].

Weiner, B. J., Coil, A. L., Prochaska, J. X., et al. 2009. Ubiquitous outflows in DEEP2 spectra of star-forming galaxies at z=1.4. *The Astrophysical Journal* 692:187–211.

Whitaker, K. E., van Dokkum, P. G., Brammer, G. and Franx, M. 2012. The Star-formation Mass Sequence out to z=2.5. *The Astrophysical Journal Letters* 754:L29 (6 p).

Woosley, S. E. and Weaver, T. A. 1995. The evolution and explosion of massive stars. II. Explosive hydrodynamics and nucleosynthesis. *Astrophysical Journal Supplement* 101:181–235.

Yuan, T. T., Kewley, L. J. and Richard, J. 2012, Nov 27. The metallicity evolution of star-forming galaxies from redshift 0 to 3: combining magnitude limited survey with gravitational lensing. arXiv:1211.6423v1 [astro-ph.CO].

Zahid, H. J., Dima, G. I., Kewley, L. J., Erb, D. K. and Davé, R. 2012. A census of oxygen in star-forming galaxies: an empirical model linking metallicities, star formation rates, and outflows. *The Astrophysical Journal* 757:54 (22p).

Zahid, H. J., Geller, M. J., Kewley, L. J., Hwang, H. S., Fabricant, D. G. and Kurtz, M. J. 2013. The chemical evolution of star-forming galaxies over the last 11 billion years. *The Astrophysical Journal Letters* 771:L19 (6 p).

Zahid, H. J., Kewley, L. J. and Bresolin, F. 2011. The mass-metallicity and luminosity-metallicity relations from DEEP2 at z ~ 0.8 *The Astrophysical Journal* 730(2): article id. 137 (15 p).

Additional Reading

Adams, W. S. 1949. Observations of interstellar H and K, molecular lines, and radial velocities in the spectra of 300 O and B stars. *Astrophysical Journal* 109:354.

Balestra, I., Tozzi, P., Ettori, S., et al. 2007. Tracing the evolution in the iron content of the intra-cluster medium. *Astronomy & Astrophysics* 462:429–442.

Bisbas, T. G., Tan, J.C., Csengeri, T., et al. 2018. The inception of star cluster formation revealed by [C II] emission around an infrared dark cloud. *Monthly Notices of the Royal Astronomical Society: Letters* 478(1):L54–L59.

Boissier, S. and Prantzos, N. 1999. Chemo-spectrophotometric evolution of spiral galaxies—I. The model and the Milky Way. *Monthly Notices of the Royal Astronomical Society* 307:857–876.

Bose, S., Deason, A. J. and Frenk, C. S. 2018. The imprint of cosmic reionisation on the luminosity function of galaxies. arXiv:1802.10096v2 [astro-ph.GA], 18 pages.

Carretta, E., Bragaglia, A., Gratton, R. G., et al. 2010. Properties of stellar generations in globular clusters and relations with global parameters. *Astronomy & Astrophysics* 516:A55 (29 p).

Compiègne, M., Verstraete, L., Jones, A., Bernard, J.-P., Boulanger, F., Flagey, N., Le Bourlot, J., Paradis, D. and Ysard, N. 2011. The global dust SED: Tracing the nature and evolution of dust with DustEM. *Astronomy & Astrophysics* 525:A103 (14 p).

De Barros, S., Oesch, P.A., Labbé, I., et al. 2019, Apr. 04. The GREATS Hβ+[O III]luminosity function and galaxy properties at z~8~: Walking the way of JWST. *Monthly Notices of the Royal Astronomical Society*, stz940, Doi: 10.1093/mnras/stz940.

Del Peloso, E. F., da Silva, L., Porto de Mello, G. F. and Arany-Prado, L. I. 2005. The age of the galactic thin disk from Th/Eu nucleocosmochronology. III. Extended sample. *Astronomy & Astrophysics* 440:1153–1159.

Dessauges-Zavadsky, M., Calura, F., Prochaska, J. X., D'Odorico, S. and Matteucci, F. 2007. A new comprehensive set of elemental abundances in DLAs III. Star formation histories. *Astronomy and Astrophysics* 470:431–448.

Dessauges-Zavadsky, M., Prochaska, J.X. D'Odorico, S. Calura, F. and Matteucci, F. 2006. A new comprehensive set of elemental abundances in DLAs. II. Data analysis and chemical variation studies. *Astronomy and Astrophysics* 445(1):93–113.

Fischer, D. A. and Valenti, J. 2005. The planet-metallicity correlation. *The Astrophysical Journal* 622(2):1102–1117.

Ginsburg, A., McGuire, B., Plambeck, R., et al. 2019. Orion SrcI's disk is salty. arXiv:1901.04489v1 [astro-ph.GA].

Hou, J., Prantzos, N. and Boissier, S. 2000. Abundance gradients and their evolution in the Milky Way disk. *Astronomy and Astrophysics* 362:921–936.

Masseron, T., García-Hernández, D. A., Santoveña, R., et al. 2020. Phosphorus-rich stars with unusual abundances are challenging theoretical predictions. *Nature communications* 11:3759. Doi: 10.1038/s41467-020-17649-9.

McClure-Griffiths, N. M., Dénes, H., Dickey, J. M., et al. 2018. Cold gas outflows from the small magellanic cloud traced with ASKAP, *Nature Astronomy* 2:901–906.

Meiksin, A. A. 2009. The physics of the intergalactic medium. *Reviews of Modern Physics* 81:1405–1469.

Pentericci, L., Risaliti, G., Salvati, M. and Silva, L. 2008. AMAZE. I. The evolution of the mass-metallicity relation at z>3. *Astronomy and Astrophysics* 488:463–479.

Prochaska, J. X., Chen, H.-W., Dessauges-Zavadsky, M. and Bloom, J. S. 2007, Sep. 1. Probing the interstellar medium near star-forming regions with gamma-ray burst afterglow spectroscopy: gas, metals, and dust. *The Astrophysical Journal* 666:267Y280.

Spitooni, E., Romano, D., Matteucci, F. and Ciotti, L. 2015, Feb 9. The effect of stellar migration on galactic chemical evolution: A heuristic approach. arXiv:1407.5797v3 [astro-ph.GA].

Toyouchi, D. and Chiba, M. 2014, May 2. On the chemical and structural evolution of the galactic disk. arXiv:1405.0405v1 [astro-ph.GA].

Troja, E., Castro-Tirado, A.J., Becerra González, J., et al. 2019. The afterglow and kilonova of the short GRB 160821B. *Monthly Notices of the Royal Astronomical Society*, stz2255. Doi: 10.1093/mnras/stz2255.

Veilleux, S. 2008, Jul 29. AGN host galaxies. arXiv:0807.3904v2 [astro-ph].

Werner, N., Urban, O., Simionescu, A. and Allen, S. W. 2013. A uniform metal distribution in the intergalactic medium of the Perseus cluster of galaxies. *Nature* 502:656–658.

Zheng, W., Postman, M., Zitrin, A., et al. 2012. A magnified young galaxy from about 500 million years after the Big Bang. *Nature* 489(7416):406–408.

Zitrin, A., Labbe, I., Belli, S., Bouwens, R. Ellis, R. S., Roberts-Borsani, G., Stark, D. P., Oesch, P. A. and Smit, R. 2015. Ly alpha emission from a luminous z=8.68 galaxy: Implications for galaxies as tracers of cosmic reionization. *The Astrophysical Journal Letters*. 810:L12.

Zubko, V., Dwek, E. and Arendt, R. G. 2004. Interstellar dust models consistent with extinction, emission, and abundance constraints. *The Astrophysical Journal Supplement Series* 152:211–249.

4

Living Matter

4.1 Introduction

Here we start with a list of some properties of living matter that separate it from non-living matter.

- Living matter is based on organic molecules organized into complex structures.
- Living matter grows and develops.
- Living matter has the capability to reproduce and pass on genetic material as a blueprint for growth and subsequent reproduction.
- Living matter collects matter and energy from the external environment and converts it into different forms.
- Living matter evolves.
- Living matter can take the required chemical elements against the concentration gradient.

Also, the autopoiesis (referring to a system capable of reproducing and maintaining itself) is a necessary condition shared by any living entity. It is, therefore, universal (Lucantoni and Luisi, 2012; Maturana and Varela, 1980).

Living organisms are grouped into one or more of a few major categories, a discipline known as taxonomy. Modern taxonomy divides all living organisms into three domains: Bacteria (includes the kingdom Eubacteria), Archaea (consists of the kingdom Archaebacteria) and Eukarya (consists of the kingdoms Protista, Fungi, Plantae and Animalia). The hierarchical system of classification was originally conceived by Linnaeus (1758), and it includes seven levels or taxa. These seven levels representing general and specific organism attributes are kingdom, phylum, class, order, family, genus and species. Species names of organisms always contain two words, consisting of a genus and species designation.

All organisms are made of cells. They can be either single-celled or multicellular. The shape of the cell is helpful in the determination of its function in multicellular organisms. Here are two examples: (i) Red blood cells are doughnut-shaped to be able to exchange oxygen easily and to pass through narrow blood vessels freely. (ii) Nerve cells are long, so by connections to other nerve cells, they can span long distances in the organism. According to modern cell theory, cells are the basic units of organisms, and

DOI: 10.1201/9781003181330-4

cells arise only by the division of a previously existing cell. All eukaryote cells have three significant parts: (i) a nucleoid or nucleus (central portion of the cell containing genetic material), (ii) cytoplasm (semi-fluid or gel, which fills the interior of the cell) and (iii) plasma membrane (bilayer with proteins surrounding the cell).

Prokaryotes are the simplest organisms. They are grouped into two main types of prokaryotes: archaea and bacteria. Most of the prokaryotes' cells have a strong wall outside of the plasma membrane. The internal organization of prokaryotic cells is straightforward, having only a few internal compartments and no subunits. Prokaryotic cells do not possess a nucleus surrounded by a membrane. Their genetic material is in the form of a simple circle of DNA. Eukaryotes, members of the domain Eukarya, have organelles bounded by a membrane. They have a nucleus surrounded by a double membrane (nuclear envelope). DNA inside the nucleus is organized into chromosomes.

The cell is the basic unit of all forms of life. Modern cells use lipid membranes to control what kind of molecules may enter and exit the cell selectively. Many multicellular organisms contain several levels of cell organization. Cells are sorted in the hierarchical levels of the organization. The lowest level of organization is the cell; next in the hierarchy are tissues, organs and organ systems. A tissue is made of a group of similar cells working together to perform a particular function (e.g., nerve or muscle tissue). An organ is made of a collection of different tissues that carry out a specific purpose (e.g., liver). An organ system is made of a group of organs that function to carry out a particular task in the organism (e.g., digestive system). At each level of the organization, the structure helps determine the purpose. The highest level of organization in multicellular life is the whole organism.

By organization of the body's cells into specialized tissues, organs and organ systems, a division of labor in the body is created, which makes multicellular life possible. All living organisms must have a source of energy for survival. All known species of life conserve energy in a peculiar form of ion gradients across cell membranes. Adenine triphosphate (ATP) is the molecule whose function is to store and release energy to enable chemical reactions in each cell. Plants use the energy of the sunlight to make chemical energy (which is stored as chemical bonds of glucose) and, ultimately, structural components for the body of the organism, as well as energy for work. Producing chemical energy from light (electromagnetic radiation) is called photosynthesis.

On the other hand, heterotrophs collect energy by ingesting food, including plants or other animals. Consumers feeding only on plant material are called herbivores; consumers that prey on other animals are called carnivores; consumers of both plant and animal material are called omnivores; the organisms that absorb chemical energy from the environment are called saprobe or saprotroph. The heterotrophs use chemical energy and nutrients making molecular building blocks obtained from these food sources to build new body structures or to convert it to energy for work. Metabolism is a name for all of the chemical reactions in an organism that occur to manage its material and energy resources.

All living organisms existing on the Earth are made up of chemical compounds based mostly on the element carbon. Chemical element carbon can form covalent bonds with up to four atoms. This characteristic allows carbon to compose many diverse molecules.

The majority of biological molecules consist of carbon atoms bonded to other carbon atoms or atoms of oxygen, nitrogen, sulfur or hydrogen. Carbon-containing molecules can make structures such as chains, branches or rings. The biological molecules, such as sugars, are relatively small. Some others are large and complex and are called macromolecules.

In most cases, the macromolecules are polymers, which have long chains of similar, linked subunits. The starch, proteins and nucleic acids are all polymers of rather complicated carbohydrates, a large group of molecules containing carbon, hydrogen and oxygen. They store energy and provide building materials. This group includes (i) sugars such as glucose, (ii) double sugars such as sucrose and lactose, and (iii) starches (polysaccharides). Lipids (fats and oils) make up membranes and store energy.

Lipids do not dissolve in water. They have long sections of non-polar carbon–hydrogen bonds. When placed in water, lipid molecules form clusters with polar parts facing toward the exterior and non-polar portions toward the interior (away from water). Lipids include (i) triglycerides (dense energy storage built in fat cells), (ii) phospholipids (structural component of the cell membrane phospholipid bilayer), (iii) steroids (e.g., cholesterol) and (iv) waxes (for waterproofing). Proteins perform the chemistry of cells. Proteins are made of chains of amino acids that may interact and fold over each other to form different shapes and structures. Protein functions include structural support, enzymes for reactions, transport of other molecules, storage, signaling, movement in the organism and immune defense. Nucleic acids store and transfer genetic information. Nucleic acids are composed of sugar, a phosphate group and a nitrogen base. The two major types of nucleic acids are deoxyribonucleic acid (DNA) and ribonucleic acid (RNA). DNA and RNA sequences are a code for the inherited traits of an organism.

Recently, mathematicians and physicists have an increased interest in biochemistry. Some of these efforts are described in the book by Stcherbic and Buchatsky (2015). Their book provides a review of biochemistry as an algebra of molecules of living matter. It utilizes Clifford algebras to discuss the underlying biochemical processes of DNA replication, DNA transcription, RNA splicing and translation. Specifically, viral carcinogenesis is defined by the structural features that are used to identify a particular Clifford algebra. Useful examples of the transformation of genetic information into Clifford algebras are provided.

4.2 Elemental Composition of Living Matter

The organic cycle includes eleven elements: H, C, N, O, Na, Mg, P, S, Cl, K and Ca, which form the bulk of the living matter. They all have very low atomic weights, and they belong to the lowest 20 elements of the periodic table. These elements are also the most abundant in the Universe. The only nonessential light elements, with the exclusion of noble gases, are Li, Be, B (boron is essential for some plants) and maybe Al. Also, some elements have been recognized as essential trace elements. For example, essential trace elements for warm-blooded animals are F, Si, V, Cr, Mo, Fe, Co, Ni, Cu, Zn, Se, Mo, Sn, I and W. For the most part, the essential trace elements are transition metals with unfilled d-orbital. None of the 39 elements beyond tungsten (Z=74) has ever been shown to have any physiological significance.

The so-called bulk elements oxygen, carbon, hydrogen and nitrogen make up (96.8%±0.1%) of the mass of living matter, while phosphorus and sulfur make up (1.0%±0.3%). The amount of (2.2%±0.2%) is dominated by potassium, sodium, calcium, magnesium and chlorine, while the remaining mass of (0.03%±0.01%) is attributed to "trace" elements such as iron, copper and cobalt. These values, reported in Chopra and Lineweaver (2009), were derived from the averaged elemental abundances in humans and bacteria, as presented in Snyder et al. (1975) and Porter (1946). The uncertainties largely reflect the range of abundances among humans and bacterial species rather than instrumental uncertainties.

The significant elemental abundance differences between life, the oceans and the Sun are discussed by Chopra and Lineweaver (2009). They presented the correlation between the elemental composition of humans and bacteria, the Sun and seawater. Several authors, including Davies and Koch (1991) and Bowen (1979), presented the relative proportions of the chemical elements in organisms such as humans and bacteria and noted the similarity in their compositions. Bowen (1979) also pointed out that the elemental abundances in these species show a strong correlation with the elemental abundances of seawater. Life forms show many common aspects, but also differ in their composition and usage of elements because of the different environments they live in. Any life form can only be made of the chemical elements available to it from its environment. Figure 4.1 shows the comparison of elemental abundance in humans as compared to solar abundances.

The distribution of important dietary elements for mammals is shown in Table 4.1. The chemical elements are as follows: (i) four essential elements: H, C, N and O; (ii) essential bulk elements: Na, Mg, P, S, Cl, K and Ca; (iii) essential trace elements: Cr, Mn, Fe, Co, Ni, Cu, Zn, Se, Mo, I and W; (iv) elements with no confirmed functions in humans, but in plants, animals, etc.: Li, B, F, Si, V and As; and (v) the rest: environmental pollutants (see also Williams and Frausto da Silva, 2007).

Trace elements are being used by all organisms and make the components of proteins having unique coordination, catalytic and electron transfer properties. Although many trace element-containing proteins are characterized, little is known about the general

TABLE 4.1 Dietary Elements

H																		He
Li	Be												B	C	N	O	F	Ne
Na	Mg												Al	Si	P	S	Cl	Ar
K	Ca		Sc	Ti	V	Cr	Mn	Fe	Co	Ni	Cu	Zn	Ga	Ge	As	Se	Br	Kr
Rb	Sr		Y	Zr	Nb	Mo	Tc	Ru	Rh	Pd	Ag	Cd	In	Sn	Sb	Te	I	Xe
Cs	Ba	a	Lu	Hf	Ta	W	Re	Os	Ir	Pt	Au	Hg	Tl	Pb	Bi	Po	At	Rn
Fr	Ra	b	Lr	Rf	Dd	Sg	Bh	Hs	Mt	Ds	Rg	Cn	U3	Fl	U5	Lv	U7	U8

U3→Ununtrium, Z=113, temporary symbol Uut. U5→Ununpentium, Z=115, temporary symbol Uup. U7→Ununseptium, Z=117, temporary symbol Uus. U8→Ununoctium, Z= 118, temporary symbol Uuo.

[a] Lanthanides are La, Ce, Pr, Nd, Pm, Sm, Eu, Gd, Tb, Dy, Ho, Er, Tm and Yb.

[b] Actinides are Ac, Th, Pa, U, Np, Pu, Am, Cm, Bk, Cf, Es, Fm, Md and No.

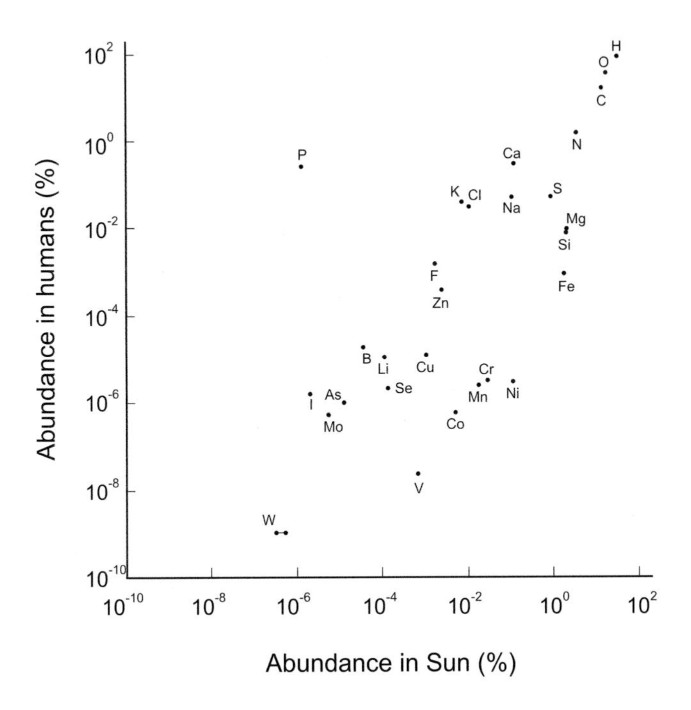

FIGURE 4.1 The comparison of elemental abundance in humans with solar abundances. (After Chopra and Lineweaver, 2009.)

trends in trace element utilization. Zhang and Gladyshev (2010) performed genomic analyses of copper, molybdenum, nickel, cobalt (vitamin B12) and selenium (selenocysteine) in 747 sequenced organisms for the elements serving:

i. in transporters and transport-related proteins,
ii. as cofactor biosynthesis traits, and
iii. in trace element-dependent proteins.

Only a few organisms were found to utilize all five trace elements, whereas many symbionts, parasites and yeasts used only one or none. The investigation of metalloproteomes and selenoproteomes showed examples of increased use of proteins with copper in land plants, with cobalt in *Dehalococcoides* and *Dictyostelium* and with selenium in fish and algae. In contrast, nematodes were found to have a broad diversity of copper transporters.

These considerations also characterized trace element metabolism in the usual model organisms and suggested the use of new model organisms for experimental studies of separate trace elements. Discrepancies in the occurrence of user proteins and corresponding transport systems revealed deficiencies in the understanding of trace element biology. Biological interactions among some of the trace elements were observed; however, such interactions were limited, and it appeared those trace elements had unique utilization patterns. Finally, it should be mentioned that environmental factors, such as oxygen requirement and habitat, are correlated with the utilization of some trace elements.

4.2.1 Essential Trace Elements for Life

Trace elements are vital components of all living systems and are required as a carrier of metabolic and structural functions. Among these activities is a catalytic role in a vast array of cellular reactions, including energy generation in the first place. A cell's elemental contents are defined as its metallome. It includes all the major elements required for all life, plus bio-essential chemical elements, most of them being transition metals that include Fe, Mn, Co, Mo, Zn, Ni, Cr, Cu and V. Some elements, such as Fe and Zn, have a universal utility in most organisms, while others perform specific functions only in individual organisms (Frausto da Silva and Williams, 2001).

Trace elements function in different ways. Some are essential parts of enzymes that are used to interact directly with components and often facilitate their conversion to final products. Some essential trace elements donate or accept electrons in reactions of reduction and oxidation; some structurally stabilize biological molecules. Some control biological processes by promoting the binding of molecules to receptor sites on cell membranes (Mertz, 1998).

Approximately one-third of all proteins possess a bound metal. An examination of the Protein Data Bank (PDB, http://www.rcsb.org) reveals that Mg and Zn are the most abundant, while Ca, Mn, Fe and Ni are also frequently observed (Shi et al., 2005).

Differences in metallomes may be explained by different microbial requirements and the availability of elements that have varied in the environment through time (Williams and Frausto Da Silva, 2003; Anbar and Knoll, 2002; Dupont et al., 2006). These variations, produced by changing environmental conditions on the Earth, probably have directly impacted the evolution of life. The evidence for this is probably documented in differences in element utilization by groups of organisms that evolved at various periods of Earth's history. For example, Ni and W are fundamental requirements for the growth of some anaerobic species. The species considered are archaeal methanogens and hyperthermophiles, respectively. They operate with metabolisms that are generally thought to be quite primitive (Frausto da Silva and Williams, 2001; Kietzin and Adams, 1996). However, the use of Mo for nitrogen fixation and nitrate reduction was most likely initiated after the oxygenation of the Earth at ~2.2×10^9years (Cloud, 1972; Canfield, 1998). With the increase in oxygen in the atmosphere, Mo oxides with relatively high solubility became a more biologically available species (Stiefel, 2002; Williams and Frausto da Silva, 2002).

Hyperthermophiles grow best at temperatures >80°C and are predominant members of the archaeal domain of life. There are two main archaeal groups: the largely hyperthermophilic Crenarchaeota and the methanogen-containing Euryarchaeota. Hyperthermophiles often occupy the deepest and shortest lineages in both prokaryotic domains (Woese et al., 1990; Stetter, 1996) and are found in unique high-temperature environments such as deep-ocean hydrothermal vents or terrestrial geothermal sites (Stetter, 1996; Stetter et al., 1990). These environments typically contain high, and in some cases, possibly toxic concentrations of metals that surprisingly sustain the growth of microorganisms. For example, the seawater concentrations of Fe^{2+} is generally <1 nM, but it usually exceeds values of 10 mM in hydrothermal vent fluids (Kelley et al., 2002; Von Damm et al., 1985a, b). However, all organisms, including hyperthermophiles, have established strategies and mechanisms for coping with environmental stressors. For instance, some microorganisms are able to produce metal chelators. Their function is to both secure metals for the cell and, conversely, bind and remove harmful species (see, for example, Bruland, 1989; Edgcomb et al., 2004; Holden and Adams, 2003; Liermann et al., 2005; Neilands, 1995; Sander et al., 2007; Williams, 2001).

Our understanding of how the concentration of essential trace elements has changed in the environment over time has increased significantly. However, the same is not valid for our knowledge of the majority of archaeal prokaryotes or their extreme living habitats. In the same way, the utilization of trace elements by different microbial groups is a function of many factors such as availability, cellular requirements, environmental constraints, and substitution and competition between elements (Morel et al., 2003). Many research groups have attempted to quantify the intracellular elemental quotas of various primarily marine organisms. In all such investigations, it was difficult to prove whether the reported metal quotas are representative of the true metallome of the studied organisms or merely the media composition for specific growth conditions. Also, it is not known whether eukaryotic element stoichiometries are also representative of prokaryotic quotas. For prokaryotes, these problems are exacerbated by the fact that microorganisms requiring different elements for different metabolisms also need various growth media.

An examination of the correlations between trace element utilization, microbial metabolisms and the microorganisms themselves could be used to conclude environmental changes through time and could be exploited as potential bio-signatures (Anbar and Rouxel, 2007; Archer and Vance, 2006; Beard et al., 1999; Zerkle et al., 2005). The capability of ancient microorganisms to adapt and take advantage of varying elemental conditions would have been favorable in that it could have allowed the optimization of metabolic processes as well as the adaptation of microorganisms to new environments.

No experimental data have been reported yet for hyperthermophilic microorganisms' metallomes. Their element requirements for growth are known to be unique. Cameron et al. (2012) conducted measurements to determine the following:

i. The cellular trace element concentrations of the hyperthermophilic archaea *Methanococcus jannaschii* and *Pyrococcus furiosus*.
ii. An estimate of the metallome for these hyperthermophilic species via ICP-MS.

The elemental contents of these cells were compared to simultaneous experiments with the mesothermophilic bacterium *Escherichia coli* grown under both aerobic and anaerobic conditions. Fe and Zn were usually the most abundant metals in cells. The element concentrations of *E. coli* grown aerobically decreased in the order Fe>Zn>Cu>Mo>Ni>W>Co. Contrary to that, *M. jannaschii* and *P. furiosus* showed almost the reverse pattern with elevated Ni, Co and W concentrations. A bio-signature is potentially demonstrated only for the methanogen *M. jannaschii*. That may be partly related to the requirements of methanogenesis. The bioavailability of trace elements has varied over time. If hyperthermophiles are very ancient, then the observed trace element patterns may provide some insights into the Earth's earliest cells and early Earth chemistry. Their data show that element quotas are regulated over the range of experimental trace element concentrations and at least partly determined by cellular requirements and not dictated entirely by the growth media.

Evolutionary and physiological considerations argue that the study of hyperthermophilic archaea should reveal new molecular aspects of DNA stabilization and repair. So far, these unusual prokaryotes have yielded several genes and enzymatic activities with established mechanisms of excision repair, photo-reversal and trans-lesion synthesis. DNA enzymes of hyperthermophilic archaea show biochemical properties that may be related to DNA stability or repair at extremely high temperatures, but they remain difficult to be evaluated rigorously in vivo. The most interesting feature of the hyperthermophilic archaea is that all of them lack genes of the nucleotide excision and DNA mismatch repair pathways, which are otherwise often present in biology. Although the growth properties of these microorganisms interfere with experimentation, there is evidence that some systems of excision repair and mutation avoidance operate in *Sulfolobus* spp. According to Grogan (2004), it will be of significance to formulate and test hypotheses in *Sulfolobus* spp. and other hyperthermophilic archaea. Mechanisms and gene products involved in the repair of UV photoproducts and DNA mismatches should be tested.

The essential chemical elements for mammals are as follows: (i) four basic organic elements: H, C, N and O; (ii) essential bulk elements: Na, Mg, P, S, Cl, K and Ca; (iii) essential trace elements: Cr, Mn, Fe, Co, Ni, Cu, Zn, Se, Mo and I; and (iv) elements with no confirmed functions in humans, but in plants, animals, etc.: Li, B, F, Si, V and As (see Figure 4.2). The vast majority of the essential trace elements serve as crucial components of the enzyme system or proteins with vital functions.

Table 4.2 shows the elemental composition of crop plants (after Epstein, 1994), which contributes significantly to dietary input for many animals and humans.

4.2.1.1 Lithium

Lithium is one of the alkali metals with high chemical activity. It is relatively widespread in the Earth's environment. Although Li has not been identified to serve as a required cofactor of an enzyme or enzymatic transport system, it is increasingly considered as an essential trace element for animals and humans. However, the physiological role of Li in the human body is still not recognized. The deficiency of Li may disturb protein metabolism and reproductive ability. Elevated Li levels in drinking water have a beneficial influence on the circulatory system and can prevent cardiovascular diseases (Kabata-Pendias and Mukherjee, 2007).

Furthermore, a latest study suggests a positive role in the prevention of Alzheimer's disease (Young, 2011). Li appears to alter neurotransmission at a synaptic level in the brain. Therefore, carbonate and other Li salts have been used in psychiatry for over 50 years, mainly to treat moderate mood swings (Mukherjee, 2007; Young, 2011; Schrauzer, 2002; Aral and Vecchio-Sadus, 2008). The deficiency of Li in humans is unlikely to reach a degree of severity, as observed in experimental animals. Still, if any symptoms of Li insufficiency in humans had occurred, they would be expected to be mild and would manifest themselves by behavioral rather than physiological abnormalities (Schrauzer, 2002).

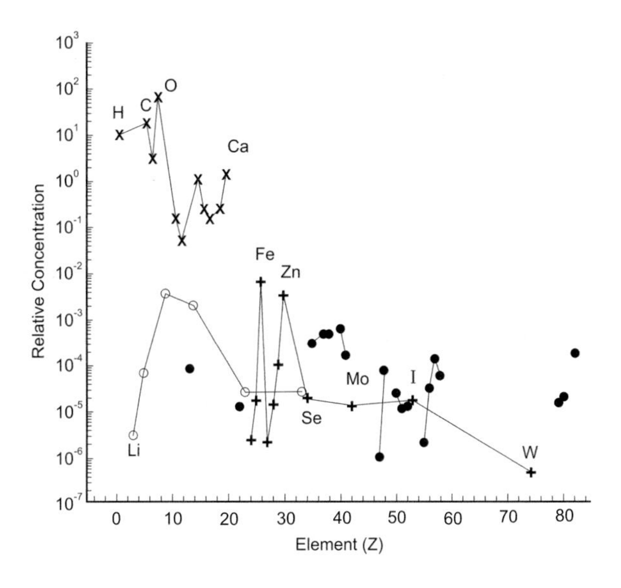

FIGURE 4.2 Living matter – concentrations of chemical elements essential for mammals are as follows: (i) four basic organic elements: H, C, N and O; (ii) essential bulk elements: Na, Mg, P, S, Cl, K and Ca – symbol X; (iii) essential trace elements: Cr, Mn, Fe, Co, Ni, Cu, Zn, Se, Mo, I and W – symbol +; (iv) no confirmed functions in humans, but in plants, animals, etc.: Li, B, F, Si, V and As – open circles; and (v) the rest: environmental pollutants. The concentrations of some of the elements are shown with black circles.

TABLE 4.2 Essential Elements in Crop Plants (After Epstein, 1994)

Chemical Element	Concentration (Dry Weight, %)	Comment
N	0.5–6	Macronutrients
P	0.15–0.5	
S	0.1–1.5	
K	0.8–8	
Ca	0.1–6	
Mg	0.05–1	
Na	0.001–8	
Si	0.1–10	
Fe	20–600	Essential trace element
Mn	10–600	
Zn	10–250	
Cu	2–50	
Ni	0.05–5	
B	0.2–800	
Cl	10–80,000	
Mo	0.1–10	
Co	0.05–10	
Al	0.1–500	Not confirmed

Today, lithium carbonate is successfully used for controlling wild mood swings from depression to delight characteristic of manic depressive illness (also known as bipolar disorder). It produces a dramatic therapeutic improvement of any drug used in psychiatry. Read more at:http://www.discoveriesinmedicine.com/Hu-or/Lithium.html#ixzz45nGsUDwU.

Despite clearly detrimental effects of Li on plants, there are indications that this element at low (subtoxic) concentrations stimulates the growth and development of plants. Therefore, in several countries, the beneficial effects of Li are used in agriculture in the form of Li-containing fertilizers to increase starch levels and biomass in potatoes. However, it is still not proven that Li is an essential nutrient or an active substance for plants because they do not need this element to complete their life cycle. Thus, the essentiality of Li for plants has to be elucidated (Schäfer, 2012).

4.2.1.2 Boron

Boron abundances on the Earth are estimated to be as follows: crustal abundance – 10.0 mg/kg; oceanic abundance – 4.44 mg/kg. Boron has served essential functions since the early evolution of life. This is evidenced by the role of the element in the heterocystous cyanobacteria. They were predominant organisms during the Middle Precambrian Period (Bonilla et al., 1990). The chemical element boron is essential for at least some phylogenetic organisms. The first natural biomolecules found to contain element boron were antibiotics produced by species in Eubacteria (Bonilla et al., 1990; Chen et al., 1981). Specific brown algae and diatoms in Stramenopila and all higher plants in Viridiplantae require boron (Lovatt and Dugger, 1984). Specific fungi have a demonstrated physiological response to boron (Bennett et al., 1999), an important finding because species in the kingdom Fungi are thought to share a common ancestor with animals exclusive of plants (Carney and Bowen, 2004).

Boron satisfies the essentiality criteria in humans and most of the higher animals (Expert Consultation WHO/FAO/IAEA 1996; Frieden, 1984; Mertz, 1970; Underwood and Mertz, 1987):

1. It reacts with biological material or forms chelates.
2. It is present in healthy tissues of different animals at comparable concentrations.
3. Toxicity results only at relatively high intakes.
4. Tissue concentrations during short-term variations in the input are constant because of homeostatic mechanisms.
5. Its depletion prevents growth and completion of the life cycle.
6. Its depletion consistently results in a reduction in a physiologically relevant function.
7. And when it is an integral part of an organic structure, its depletion causes a reduction in the performance of a vital task.

The importance of element boron in the living world is strongly related to its implications in the prebiotic origins of genetic material (Scorei, 2012). It is possible that during the evolution of life, the most crucial role of boron has been to provide thermal and chemical stability to cells in hostile environments. The complexation of boric acid and borates with organic cis-diols is the most probable chemical mechanism for the role of boron in the evolution of the living world. The boron is essential for heterocystous cyanobacteria, principally Nostocaceae. They were predominant organisms during the Middle Precambrian Period (some 2×10^9years ago). This finding indicates that boron was a crucial element during the early evolution of life.

Boron and compounds containing boron have a variety of known biological functions. For example, boron is present in bacterial antibiotics, in the bacterial quorum-sensing molecule and in other important molecules in bacteria. Transporters that are responsible for the efficient uptake of boron by roots, xylem loading of boron and distribution of boron among leaves have been described by Scorei (2012). Experimental data from plants and animals show that boron plays a critical role in membranes and extracellular matrix. The importance of borate as a cross-linking molecule is revealed by the discovery of a variety of borate-dependent molecules.

Boron has been shown to possess the following abilities in large parts of biology:

i. to be a cell signaling molecule,
ii. to be a cofactor of the enzymes it regulates,
iii. to be a non-enzymatic cofactor,
iv. to play both structural and functional roles, including electron transfer, redox sensing and structural modules, and
v. to play a role in the cytoskeleton structure.

Although boron has not yet been convincingly shown to be an essential nutrient in animal cells, more data will probably support such a role in the future.

In addition to its role in plants (Bonilla et al., 2009), other boron functions have been identified. Boron's essentiality for the growth of many organisms has been linked to the stability of cell walls and envelopes. Diatoms (Smyth and Dugger, 1981), some cyanobacteria (Bonilla et al., 1990) and bacteria of the genus *Frankia* (Bolaños et al., 2002) are some examples.

The complexation of borate with organic cis-diols remains the most probable chemical mechanism for the involvement of this chemical element in the evolution of the living world. Animal and plant cells actively transport borate anions when the boron concentration in the environment is low; boric acid can diffuse directly through cell membranes at high environmental levels.

Boron has another attractive characteristic as the basis for a living organism. Also, the bonds between boron and nitrogen, which is located to the right of carbon in the periodic table, can mimic carbon–carbon bonds when present in the same molecule. The boron–nitrogen compounds can form aromatic structures, enabling the possibility of DNA analogs (Schulze-Makuch and Irwin, 2008). Numerous gaseous, liquid and solid hydrocarbon species contain alkyl groups, and this will make possible the encounter between such groups and boron so that a hindered dialkyl or trialkyl borane is produced (Abbas and Schulze-Makuch, 2007).

The trace elements boron and copper may be of nutritional significance similar to selenium. When the diets of animals and humans are manipulated to cause possible changes in cellular integrity or hormone responsiveness, a large number of responses to dietary boron occur. The findings indicate that boron is essential for optimal calcium and, thus, bone metabolism (Nielsen, 1990).

Cyanobacteria require boron for the formation of nitrogen-fixing heterocysts. Boron may also be beneficial to animals. Boron deficiency in plants produces various symptoms: Many functions have been postulated. Deficiency symptoms first appear on growing points, within hours in root tips and within minutes or seconds in pollen tube tips. Cell wall abnormalities characterize them. Boron-deficient tissues are brittle or fragile, while plants grown on high-boron levels may have unusually flexible or resilient tissues. Borate forms cyclic diesters with appropriate diols or polyols. The most stable is formed with cis-diols on a furanoid ring. Two compounds in the plant cell wall have this structure: ribose in ribonucleotides and RNA, and apiose. Germanium can be a substitute for boron in some carrot cell cultures. Both boron and germanium are localized primarily in the cell wall. It is assumed that borate–apiofuranose ester cross-links are the auxin-sensitive acid-growth link in vascular plants. It is also assumed that the cyanobacterial envelope depends on the borate cross-linking of some molecules. Another assumption is that boron in diatoms forms ester cross-links in the polysaccharide cell wall matrix instead of boron–silicon interactions. The process of complexing of ribonucleotides is probably a factor in boron toxicity.

The effect of boron on dinitrogen-fixing cyanobacteria was examined in the work by Bonilla et al. (1990). According to them, the absence of boron inhibited growth and nitrogenase activity in *Nodularia* sp., *Chlorogloeopsis* sp. and *Nostoc* sp. cultures. The examinations of boron-deficient cultures showed changes in heterocyst morphology. This does not apply to the cultures of cyanobacteria: *Gloeothece* sp. and *Plectonema* sp., grown in the absence of boron; they did not show any alteration in growth or nitrogenase activity. These results suggest that boron is required only by heterocystous cyanobacteria. A possible role for boron in the early evolution of photosynthetic organisms is proposed.

Bonilla et al. (1990) suggested a possible role for B in the stabilization of the inner layer of heterocysts by interacting with their -OH groups. Their suggestion is based on the fact that boric acid forms esters with cis-diols. This mechanism has been proposed for higher plant cell membranes (Parr and Loughman, 1983; Pollard et al., 1977). Boron deficiency could lead to an alteration in the heterocyst envelope. This alteration would facilitate O_2 diffusion and result in an inhibition of nitrogenase activity. This hypothesis is consistent with the inhibitory effect that B deficiency has only on heterocystous cyanobacteria. Furthermore, a drastic alteration in the protecting O_2 diffusion envelope of the heterocysts in Nodularia sp. cells has also been shown.

Together, these results suggest the following:

i. The essential function of B in cyanobacteria is restricted to heterocystous species.
ii. Boron is involved in heterocyst function.

A role for this element in the stabilization of heterocyst structure could explain its requirement only in heterocystous cyanobacteria. Given their biological antiquity, this fact might suggest that B was necessary in the early history of life. Finally, the findings show that cyanobacteria are adequate models for the study of mineral nutrient requirements concerning the origin of life (Bonilla et al., 1990).

In the work reported by Fleischer et al. (1998), it was found that *Chenopodium album* L. cells are capable of continuous, long-term growth in a boron-deficient medium. In comparison with cultures grown in boron-containing media, these cultures contained more enlarged and detached cells, had increased turbidity due to the rupture of a small number of cells, and contained cells with an increased cell wall pore size. The observed characteristics were reversed by the addition of boric acid (≥7 μM) to the boron-deficient cells. C. album cells were grown in the presence of 100 μM boric acid, which entered the stationary phase when they were not subcultured, and remained viable for at least 3 weeks. The transition between the growth phase and the stationary one was accompanied by a decrease in the wall pore size. Cells grown without boric acid or with 7 μM boric acid were not capable of reducing their wall pore size at the transition to the stationary phase. These cells could not stay viable in the stationary phase because they continued to expand and died as a result of wall rupture. The addition of 100 μM boric acid prevented wall rupture, and the wall pore size was reduced to typical values. One can conclude that boron is required to maintain the standard pore structure of the wall matrix and to stabilize the wall at growth termination mechanically.

For its role in humans, see an interesting article by Pizzorno (2015). According to the author, boron has been proven to be an essential trace mineral because:

1. it is necessary for the growth and maintenance of bone;
2. it greatly improves wound healing;
3. it beneficially influences the body's use of estrogen, testosterone and vitamin D;
4. it boosts magnesium absorption;
5. it reduces levels of some biomarkers, such as high-sensitivity C-reactive protein (hs-CRP) and tumor necrosis factor alpha (TNF-α);
6. it raises levels of antioxidant enzymes superoxide dismutase (SOD), catalase and glutathione peroxidase;
7. it protects against pesticide-induced oxidative stress and heavy metal toxicity;
8. it improves the brain's electrical activity, cognitive performance and short-term memory for elders;
9. it influences the formation and activity of vital biomolecules such as S-adenosyl methionine (SAM-e) and nicotinamide adenine dinucleotide (NAD+);
10. it demonstrated preventive and therapeutic effects in several cancers such as prostate, cervical and lung cancers, and multiple and non-Hodgkin's lymphomas; and
11. it may help improve the adverse effects of some traditional chemotherapeutic agents (after Pizzorno, 2015).

4.2.1.3 Fluorine

Biologically synthesized organofluorines are found in microorganisms and plants (Gribble, 2002), but not in animals (Murphy et al., 2003). The most common example is fluoroacetate, which is an active poison molecule identical to commercial "1080". It is used as a defense agent against herbivores by at least 40 green plants in Australia, Brazil and Africa (Proudfoot et al., 2006). Other organofluorines which are biologically synthesized include ω-fluoro fatty acids, fluoroacetate and 2-fluorocitrate (Murphy et al.,

2003). In bacteria, the enzyme adenosyl-fluoride synthase with the carbon–fluorine bonding has been isolated. The discovery is assumed to possibly lead to biological routes for organofluorine synthesis (O'Hagan et al., 2002). Fluoride is not considered an essential element for mammals and humans. Small quantities of fluoride may be beneficial for bone strength, but this is an issue only in the formulation of artificial diets (Nielsen, 2009).

Interesting speculation has been elaborated by Budisa et al. (2014). For the life that uses some fluorinated building blocks as monomers of choice for self-assembling its catalytic polymers, fluorine might be an element of choice in polar aprotic solvents. Organofluorine compounds are scarce in the chemistry of life as we know it. Biomolecules, such as peptides or proteins, when fluorinated, exhibit a "fluorous effect". They became fluorophilic (neither hydrophilic nor lipophilic). This type of polymers, capable of creating self-sorting assemblies, resist denaturation by organic solvents by the exclusion of fluorocarbon chains from the organic phase. Fluorous cores are made of a compact interior shielded from the surrounding solution. One can anticipate that fluorine-containing "Teflon"-like or "non-sticking" building blocks are probably the monomers needed for the synthesis of organized polymeric structures in fluoride-rich planetary environments. Although no fluorine-rich terrestrial environment is known, theoretical considerations might help us define chemistries that might support life in such situations. One possible scenario is that all molecular oxygen may be used up by oxidation reactions on a planetary surface. In this case, fluorine gas released from F-rich magma later in the history of a planetary body would result in a fluorine-rich terrestrial environment.

4.2.1.4 Silicon

Silicon is an essential trace element in bone formation and metabolism, and its decrease in the mammalian diet leads to abnormal bone formation. Silicon is a component of many biomaterials enhancing bone generation around implants. Despite this, the mechanism of action has not yet been elucidated, and a therapeutic dosage has not been determined. Explaining the biological role of silicon and designing a delivery system to enhance early bone mineralization has been described by Gurpreet (2014).

The soil water (soil solution) contains silicon, mainly as silicic acid, H_4SiO_4, at 0.1–0.6 mM – concentrations on the order of those of potassium, calcium and other major plant nutrients, and well over those of phosphate. Terrestrial plants readily absorb silicon; they contain it in appreciable concentrations, from 1% of the dry matter to several percent, and in some cases to 10% or even higher. In spite of this prevalence of silicon as a mineral constituent of plants, it is not considered to be among the elements defined as "essential" for terrestrial higher plants except for the members of the Equisetaceae. Because of that, it is not part of any of the commonly used nutrient solutions. The plant physiologist's solution-cultured plants contain only what silicon is derived as a contaminant of their environment. Ample evidence is available that silicon, when available to plants, plays a significant role in their growth, mechanical strength and resistance to fungal diseases. Plants grown in conventional nutrient solutions present an experimental artifact. The withdrawal of silicon from solution cultures leads to wrong results of experiments on the growth and development of plants and responses to environmental stress (Epstein, 1994).

It has been shown by Currie and Perry (2007) that the mechanisms by which silicification proceeds contain the following parts: an energy-dependent Si transporter; Si as a biologically active element triggering natural defense mechanisms; and how abiotic toxicities are alleviated by silica. A complete understanding of silica formation in vivo still requires an understanding of the role of the environment in which silica formation occurs.

4.2.1.5 Vanadium

Vanadium compounds have been blamed for various implications in the pathogenesis of some human diseases and also in maintaining normal body functions. It is known that salts of vanadium can interfere with several enzymatic systems, including different ATPases, protein kinases, ribonucleases and phosphatases. Vanadium deficiency is blamed for several physiological malfunctions, including thyroid, glucose and lipid metabolism. Also, several genes are regulated by vanadium or by its compounds, including genes for TNF-α, interleukin-8 (IL-8), activator protein 1 (AP-1), mitogen-activated protein kinase (MAPK), p53 and nuclear factor-kappa B. Therefore, vanadium is not far from recognition as an element of pharmacological and nutritional significance, which is revealed through its increasing therapeutic uses in diabetes (Mukherjee et al., 2004).

Vanadium has many functions; the main functions, as summarized by Rehder (1992), include the regulation of phosphate-metabolizing enzymes; the halogenation of organic compounds by vanadate-dependent non-heme peroxidases; the reductive protonation of nitrogen (nitrogen fixation) by alternative, i.e., vanadium-containing, nitrogenases from N2-fixing bacteria; vanadium sequestering by sea squirts (ascidians); and amavadin, a low-molecular-weight complex of V(IV) accumulated in the fly agaric and related toadstools.

The function of vanadium, while still elusive in ascidians and toadstools, begins to be clarified in vanadium–enzyme interaction. Investigations of the structure and function of model compounds play an increasingly important role in elucidating the biological significance of vanadium.

Crans et al. (2004) reviewed the chemistry and biochemistry of vanadium and the biological activities exerted by vanadium compounds. Vanadium compounds and vanadium-containing proteins have shown a wide range of properties and reactivity; some of them are small-molecule-based, while others provide the needed structure or function in a biomolecule. Areas identified in their review include the vanadium-binding proteins, vanabins, isolated from the tunicates, which are proposed to be the first vanadium transport proteins and represent only the third class of proteins (VHPO, VNase and vanabins) that bind vanadium naturally. Since many enzymes are found to have dual functions in biological systems, the fact that the addition of vanadium can change the action of a protein may have far-reaching implications for how vanadium affects biological systems.

Of particular relevance is the observation that V-compounds have been reported to exhibit increasing numbers of enzyme activities. The combined range of vanadium bioactivity is vast. The vanadate–phosphate analogy places vanadate in a central position to interact effectively with phosphorylases. Although vanadate may not be a perfect transition state analog and may not be able to provide the entire binding affinity anticipated for an ideal analog, very few alternatives exist, and indeed not any with similar high binding affinities. However, other useful properties of V-complexes are beginning to be recognized, such as the observed photochemically induced cleavage of myosin and F1 ATPase at one specific amino acid residue. Other activities, such as the peroxidase and haloperoxidase activities exhibited by amavadin, provide additional important information. If amavadin is truly kinetically inert, these activities would imply outer sphere reactions and would thus serve as a new prototype of reactivity exerted by V-complexes. The ability of V-complexes to act as peroxidases, catalases and nitrogenases is critical. This suggests that many reactions of V-compounds include redox processes (Crans et al., 2004).

4.2.1.6 Chromium

The interest in chromium (Cr) has been around since the end of the 19th century. At that time, the carcinogenic effects of hexavalent Cr were discovered. The essentiality of trivalent Cr was demonstrated as early as 1959. In the 1970s, Cr^{3+} has been studied in humans and laboratory animals, and in the 1990s, Cr has been studied in livestock animals. Trivalent chromium is vital to healthy carbohydrate, lipid and protein metabolism. Cr is biologically active as part of an oligopeptide – chromodulin – increasing the effectiveness of insulin by facilitating its binding to receptors at the cell surface. Since chromium is acting as a cofactor of insulin, Cr activity in the organism is parallel to insulin functions. Absorption of chromium is low, ranging between 0.4% and 2.0% for inorganic compounds, while for organic compounds, it is more than ten times higher. Absorbed Cr circulates in the blood bound to the β-globulin plasma fraction. It is transported to tissues being bound to transferrin. Absorbed Cr is excreted in urine by glomerular filtration. A small amount is excreted through sweat, bile and milk. The demand for Cr is increasing as a result of factors commonly referred to as stressors, especially during different forms of nutritional, metabolic and physical strain (Pechova and Pavlata, 2007).

Although chromium may occur in all oxidation states from –2 to +6, it is most often found in 0, +2, +3 and +6. The elemental chromium (0) is not naturally present in the Earth's crust and is biologically inert. Trivalent chrome, Cr^{3+}, is the most stable oxidation state in which Cr is found in living organisms. It cannot cross cell membranes easily (Mertz, 1992) and has low reactivity, which is the most significant biological feature distinguishing it from Cr^{6+}. Trivalent Cr forms several coordination complexes, hexadentate ligands being the basic form. Hexavalent chromium (Cr^{6+-}) is the second most stable form of Cr and a potent oxidizing agent, especially in acidic media. Hexavalent chromium can be bound to oxygen as chromate $\left(CrO_4^{2-}\right)$ or dichromate $\left(Cr_2O_7^{2-}\right)$ with strong oxidative capacity. This form of Cr passes biological membranes quickly. It reacts with protein components and nucleic acids inside the cell by deoxygenation to Cr^{3+}. The reactions with genetic matter provide for the carcinogenic properties of Cr^{6+}.

On the request from the European Commission, the Panel on Dietetic Products, Nutrition and Allergies (EFSA-NDA, 2014) considered the evidence for setting Dietary

Reference Values for chromium. Trivalent chromium, Cr^{3+}, has been assumed to be necessary for the efficiency of insulin in regulating the metabolism of carbohydrates, lipids and proteins. By considering the criteria for the essentiality of a trace element, it was noted that attempts to create chromium deficiency in animal models had resulted in inconsistent results. Also, there was no evidence of essentiality of Cr^{3+} in animal nutrition. By evaluating the possibility of Cr^{3+} as an essential element for humans, the most convincing was considered to be the evidence from reported improvements associated with chromium supplementation in patients on total parenteral nutrition. Still, overall data do not provide sufficient information on the reversibility of the possible deficiencies and the nature of any dose–response curve to identify a dietary requirement for humans. The Panel concluded that no Average Requirement and no Population Reference Intake for chromium could be defined. Several studies evaluated the effect of chromium supplementation on glucose and lipid metabolism. In the only study which had information on total chromium intake, parameters of glucose metabolism of normoglycemic subjects for the placebo and chromium-supplemented periods were the same. The Panel also concluded that the setting of an Adequate Intake for chromium is not appropriate; for details, see the report EFSA-NDA (2014) and references therein.

4.2.1.7 Manganese

The chemical element manganese (Mn) is an essential element for humans, animals and plants and is required for their growth, development and maintenance of health. It is present in most tissues of all living organisms, and it is present in rocks, soil, water and food. High-dose oral, parenteral or inhalation exposures result in increased tissue Mn levels that may lead to the development of adverse neurological, reproductive or respiratory effects. Manganese-induced clinical neurotoxicity is associated with motor dysfunction syndrome, usually referred to as manganese (Peres et al., 2016). Mn is an essential element with homeostatically regulated absorption and excretion; therefore, a dose–response curve shown in Figure 4.3 should apply (see Section 4.2.2).

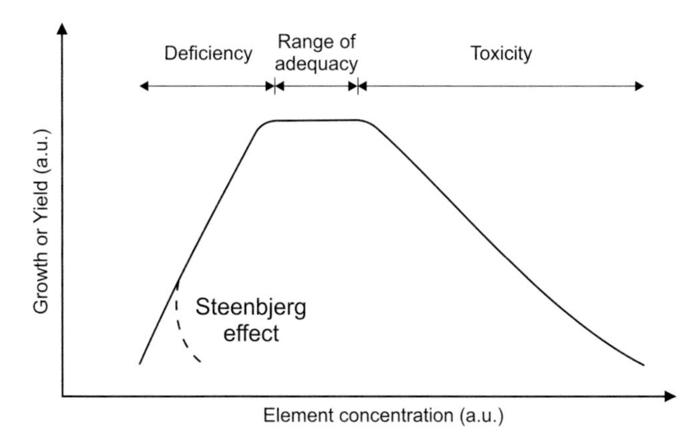

FIGURE 4.3 Dose–response curve for an essential element. This curve can be considered as a definition of element essentiality.

In their work, Santamaria and Sulsky (2010) derived the lowest-observed-adverse-effect levels (LOAELs), no-observed-adverse-effect levels (NOAELs) and benchmark dose levels (BMDs). These values have been derived from studies that were conducted to evaluate subclinical neurotoxicity in human occupational cohorts exposed to Mn. These studies showed that rodents and primates maintain stable tissue Mn concentrations. This situation is a result of homeostatic mechanisms that regulate absorption and excretion of Mn and limit tissue concentrations at low to moderate levels.

4.2.1.8 Iron

Iron was far more abundant and available than any other transition metals in the environment where life originated and evolved. This fact explains its almost universal role in the biochemistry of living cells. Because of its abundance in the Universe, iron–sulfur proteins occur in all organisms and perform diverse functions such as electron transfer and catalytic action. They maintain structural integrity (Rouault and Klausner, 1996) and the bio-sensing of oxidative stress. Also, they perform redox-dependent gene regulation in prokaryotes and, potentially, eukaryotes (Zheng and Storz, 2000). Of particular importance are the [2Fe-2S], [3Fe-4S] and [4Fe-4S] clusters in the active centers of the proteins that catalyze the redox reactions (Fontecave, 2006). These multinuclear combinations of iron and sulfur atoms inside the protein molecules are universal cofactors of enzymes and are typical for all living organisms. The Fe-proteins seem to be the oldest on the Earth (Fedonkin, 2008). That [2Fe-2S] and [3Fe-4S] clusters take part in the catalysis of the reaction involving the transfer of one electron, while more complex [4Fe-4S] clusters catalyze much more full diversity of the biochemical reactions is interesting. This difference may gain a geohistorical interpretation. But so far, it seems inevitable that iron as the most abundant transition metal could play a vital role in the initiation of life. The primary environment of life could be related to the places of the metal sulfide precipitation. The overlap of the potential redox span of iron respiration and the redox potential of the biochemical reactions involving H_2, H_2S, CH_4 and NH_4^+ (Gaidos et al., 1999) could be of interest. This coincidence may indicate the primary role of iron in starting the hydrogen metabolism.

Fe and Mn are both components of a wide range of proteins, close homologs of which may bind other metals in individual organisms or under certain conditions. An acidophilic archaeon, *Ferroplasma acidiphilum*, has recently been found to possess a large number of Fe-binding proteins, about 86% of the entire protein repertoire. The list included many metalloproteins that bind different metals (Zn and Mn, in particular) in organisms and proteins (Ferrer et al., 2007).

Iron is a component of molecules that transport oxygen in the blood. Numerous enzymatic reactions involving oxidation and reduction use iron as the agent for adding oxygen, removing hydrogen or transferring electrons. The types of enzymes dependent on iron for activity include the following:

- the oxidoreductases (xanthine oxidase/dehydrogenase);
- monooxygenases (amino acid oxidases and cytochrome P450);

- dioxygenases (amino acid dioxygenases, lipoxygenases, peroxidases, fatty acid desaturases and nitric oxide synthases); and
- different enzymes such as aconitase.

4.2.1.9 Cobalt

Cobalt mainly comes bound to the corrin ring of vitamin B12, a cofactor involved in methyl group transfer and rearrangement reactions (Banerjee and Ragsdale, 2003). Also, Co is found in several non-corrin Co-containing enzymes. These enzymes are nitrile hydratase (NHase) and methionine aminopeptidase (Kobayashi and Shimizu, 1999).

Cobalt plays several crucial roles in many biological functions in the form of vitamin B12. However, recent studies have provided information on the biochemistry and bioinorganic chemistry of several proteins containing cobalt in a form other than that in the corrin ring of vitamin B12. To date, eight non-corrin cobalt-containing enzymes have been isolated and characterized. They are:

- methionine aminopeptidase,
- prolidase,
- nitrile hydratase,
- glucose isomerase,
- methylmalonyl-CoA carboxytransferase,
- aldehyde decarbonylase,
- lysine-2,3-aminomutase and
- bromoperoxidase.

A cobalt transporter is involved in the biosynthesis of nitrile hydratase, the host cobalt-containing enzyme. Understanding the differences between cobalt and nickel transporters might lead to drug development for gastritis and peptic ulceration (Kobayashi and Shimizu, 1999). Comparative genomics studies have shown that all B12-dependent proteins depend on this coenzyme. This fact is a result of the specific binding site of B12 (Zhang et al., 2009; Rodionov et al., 2006).

4.2.1.10 Nickel

Among the known nickel-dependent enzymes are the following:

- urease (Burne and Chen, 2000),
- [NiFe] hydrogenase, carbon monoxide dehydrogenase (Ni-CODH) (Kerby et al., 1997),
- acetyl-coenzyme A decarbonylase/synthase (Doukov et al., 2002),
- superoxide dismutase (sodN) (Eitinger, 2004),
- methyl-coenzyme M reductase (Diekert et al., 1981) and
- glyoxalase I (Sukdeo et al., 2004).

Nickel occupies the active site of six diverse microbial enzymes that catalyze some critical chemical reactions.

Ni-containing proteins are strictly dependent on this element (Zhang et al., 2009). The authors reported the comparative genomic analyses to examine the occurrence and evolutionary dynamics of the use of Ni and Co in metalloproteases and transport systems. Ni and Co are widely used in bacteria and archaea; Cbi/NikMNQO is the most common prokaryotic Ni/Co transporter. The authors also provided the list of the most widespread metalloproteins for both Ni and Co, which include Ni-dependent urease, NiFe hydrogenase, B12-dependent methionine synthase (MetH) and methylmalonyl-CoA mutase. The occurrence of other metalloenzymes showed a mosaic distribution, and a new B12-dependent protein family was predicted. *Deltaproteobacteria* and *Methanosarcina* generally have more significant Ni- and Co-dependent proteomes.

On the other hand, the utilization of these two metals is limited to eukaryotes and only some of them use both elements. The Ni-utilizing eukaryotes are mostly fungi and plants, while animals need Co in the form of B12. The NiCoT transporter family is the most common eukaryotic Ni transporter, the best known being eukaryotic urease, a Ni-dependent enzyme, and MetH, a B12-dependent enzyme. Finally, the investigation of environmental conditions and sorts of organisms that are dependent on Ni or Co showed that host-associated organisms (particularly obligate intracellular parasites and endosymbionts) have the tendency to lose Ni/Co utilization (Zhang et al., 2009).

4.2.1.11 Copper

Several proteins and enzymes use copper. The list includes the following (Sakurai and Kataoka, 2007):

- cytochrome c oxidase,
- Cu–Zn superoxide dismutase,
- tyrosinase,
- a large number of blue Cu proteins,
- multicopper oxidases (MCOs),
- His, Cys and methionine.

Many studies have focused on Cu transport, homeostasis, storage and Cu-binding proteins (see Sakurai and Kataoka (2007), Balamurugan and Schaffner (2006), Harris (2000), MacPherson and Murphy (2007)). However, the evolution of Cu utilization, especially the Cu dependence of various cuproprotein families, has thus far received less attention.

In humans, the transport and cellular metabolism of Cu depend on a series of membrane proteins and smaller soluble peptides that make a functionally integrated system for maintaining cellular Cu homeostasis. The passage of Cu through the membrane is a function of integral membrane proteins forming the channels that select Cu ions for passage. Two Cu-transporting ATPase enzymes (ATP7A and ATP7B) bound to the membrane catalyze an ATP-dependent transfer of Cu to intracellular compartments or deport Cu from the cell. The ATP7A and ATP7B work together with a series of smaller peptides. This action includes the copper chaperones and Cu, Zn superoxide dismutase compounds. Copper chaperones are a class of proteins assuring safe handling and delivery of potentially harmful copper ions to an essential copper protein. They exchange Cu at the ATPase sites or incorporate the Cu directly into the structure of Cu-dependent enzymes such as cytochrome c oxidase. These mechanisms are active for a high intake of Cu or for healthy Cu metabolism (Harris, 2000).

Nearly all organisms utilize copper as an essential trace element. Biological roles of copper include the following:

- electron transfer,
- oxidation of organic substrates and metal ions,
- dismutation of superoxide,
- monooxygenation,
- transport of dioxygen and iron and
- reduction of dioxygen, nitrite, nitrous oxide, etc. (Frausto da Silva and Williams, 2001).

These functions of copper parallel those of iron due to their shared redox-active properties. Regarding the oxidation state, copper(I) and copper(II) states are utilized in biological systems, while the involvement of the copper(III) state has been erroneously suggested for galactose oxidase.

Copper in proteins has been classified into three groups based on their spectroscopic and magnetic properties: type I copper, type II copper and type III copper. Type I copper shows intense absorption around 600 nm and narrow hyperfine splittings in electron paramagnetic resonance (EPR) spectroscopy. Type II copper does not give strong absorption at 600–700 nm and shows hyperfine splittings of the typical magnitude in the EPR spectrum. Unlike type I copper and type II copper, type III copper is not detected in the EPR spectrum because of strong antiferromagnetic interaction. This classification of copper has been based on coppers contained in lacquer laccase, the prototype MCO. On the other hand, copper in proteins containing only one type of copper has been classified into blue copper, non-blue copper and EPR non-detectable copper. These coppers have also been extensively called type I copper, type II copper and type III copper, respectively (Sakurai and Kataoka, 2007).

Blue copper protein is a class of copper proteins containing only a blue copper center in a relatively small-sized protein molecule of 9–23 kDa that functions in electron transfer (Solomon et al., 2004; Vila and Fernandez, 2001; Messerschmidt et al., 2001). The subgroup called phytocyanin in higher plants is considered to function as a radical scavenger. Red copper protein (Nitrosocyanin) is a variant of blue copper protein with a modified ligand set (Lieberman, et al., 2001). The same copper center, type I copper, is also present in MCOs containing four copper centers: a type I copper center and a trinuclear copper center comprised of a type II copper center and a pair of type III copper centers (Messerschmidt, 1997; Solomon et al., 1996; Sakurai and Kataoka, 2007). The function of both the blue copper center and the type I copper center is electron transfer. Since MCO is an oxidase, the electron transfer pathways to and from type I copper are different: Electrons withdrawn from the substrate are transferred to the trinuclear copper center via type I copper and are used to convert dioxygen to water.

On the other hand, blue copper proteins shuttle an electron between exogenous redox partners through the coordinating His imidazole group. The directional electron transfer occurs in plant plastocyanin located on the thylakoid membrane surface (Guss and Freemank, 1983; Mussiani et al., 2005). Copper-containing nitrite reductase (NIR) involved in the anaerobic respiration system is a subgroup of MCOs, which contains a type I copper and a type II copper (Nakamura and Go, 2005). The type I copper in NIR is used for the entry of electrons from a blue copper protein or cytochrome c. The type II copper in NIR is used as a binding and reducing site for substrates. Nitrous oxide reductase (N2OR) and cytochrome c oxidase (COX) have a binuclear copper center, CuA, which is a variant of type I copper and iron–sulfur center. They function to facilitate the entry of electrons toward the reduction centers of nitrous oxide (CuZ) and dioxygen (heme *a*3-CuB), respectively (Kroneck, 2001). Sakurai and Kataoka (2007) reviewed the structure and function of type I copper in MCOs and their variants in comparison with those of blue copper proteins and their modifications.

Several comparative genomics studies have been performed with the aim to analyze the utilization of Cu in cuproproteins and cuproproteomes (Andreini et al., 2004; Andreini et al., 2008; Ridge et al., 2008; Zhang and Gladyshev, 2010). These studies provided the first information on Cu utilization in the three domains of life and provided a new resource for studying the Cu dependence of cuproprotein families. The majority of known cuproprotein families are dependent on Cu, including Cu–Zn SOD, blue Cu proteins and MCOs. The loss of Cu ligands has been observed in members of several cuproprotein families, which came together with changes in the function of proteins.

Cytochrome oxidase is one of the best studied cuproprotein families. It acts as a terminal enzyme in the respiratory process in aerobic organisms. There are two major subgroups of this family: COX and quinol oxidase (QO) (Musser and Chan, 1998).

Tyrosinases (catechol/polyphenol oxidases) contain a type III Cu center and are distributed in all three domains of life. These Cu-containing proteins are essential for pigmentation and are essential factors in wound healing and primary immune defense (Claus and Decker, 2006; Schallreuter et al., 2008).

4.2.1.12 Zinc

Zn is thought to be essential for all organisms and to have a significant role in the origin of the life process (Mulkidjanian, 2009). It is present in a great variety of enzymes, structural and proteins, transcription factors, ribosomal proteins, etc. (Prasad, 1995). The uptake of Zn, its storage, homeostasis and user proteins have thoroughly been discussed in many articles and reviews, for example, Murakami and Hirano (2008), Hantke (2005), Gaither and Eide (2001) and Dupont et al. (2010).

Zn plays a vital role in nucleic acid metabolism, cell replication, and tissue repair and growth. Zn deficiency is connected with a range of pathological conditions, including impaired immunity, retarded growth, brain development disorders and slow wound healing. Many reports have suggested that Zn is involved in cancer development. The levels of Zn in serum and in malignant tissues of patients with various types of cancer have been found to be abnormal. Zn may affect tumor cells by regulating gene expression profiles and cell viability, both being mediated, in part, by tumor-induced changes in Zn transporter expression.

Also, Zn could affect the immune response through the interaction with tumor cells with the influence on the processes within the cancer microenvironment. The functions and activity levels of immune cells that attack tumor cells are influenced by the intracellular Zn concentrations within those cells. In both cases, Zn contributes to intracellular element homeostasis and signal transduction in tumor and immune cells. The article by Murakami and Hirano (2008) presents the summary of the current understanding of the roles of Zn homeostasis and signaling primarily in immune cells and their contributions to oncogenesis.

In humans, Zn deficiency is widespread throughout the world. It is more prevalent in areas where the population subsists on cereal proteins. Conditioned Zn deficiency is seen in many disease states. Its deficiency during growth periods results in growth failure and a lack of gonadal development in males. Other effects of Zn deficiency include skin changes, poor appetite, mental lethargy, delayed wound healing, neurosensory disorders and cell-mediated immune disorders. Severe Zn deficiency, as seen in acrodermatitis enteropathica (a genetic disorder), is fatal if Zn is not administered to these patients. A clinical diagnosis of marginal Zn deficiency in humans remains problematic. Assays of Zn in granulocytes and lymphocytes provide better diagnostic criteria for marginal Zn deficiency than plasma Zn. Approximately 300 enzymes are known to require Zn for their activities. DNA synthesis, cell division and protein synthesis all require Zn (Prasad, 1995).

According to Hantke (2005), many bacteria use the so-called ATP-binding cassette (ABC) transporter for high-affinity uptake of zinc with a cluster 9 solute-binding protein. Other members of this cluster transport manganese. There are difficulties in the differentiation of zinc-specific and manganese-specific transporters based on sequence analysis. In bacteria, low-affinity ZIP-type zinc transporters have also been identified. Most high-affinity zinc uptake systems are controlled by Zur proteins, which form at least three separate parts of the Fur protein family (regulators of iron transport). Transport of zinc out of the periplasmic region makes a problem for the cell because zinc is a cofactor of some periplasmic enzymes. Specific zinc-binding proteins in the periplasm might function as chaperones to supply these enzymes with zinc.

Two families of transporters, called ZIP (Zrt-, Irt-like proteins) and CDF (Cation Diffusion Facilitator), play essential roles in zinc transport in eukaryotic organisms. These are ancient gene families found in all phylogenetic levels. The characterized members of each group have been involved in the transport of metal ions, in particular zinc, across lipid bilayer membranes. This remarkable conservation of function suggests that other, as yet uncharacterized, members of the family should also be involved in metal ion transport. Many of the ZIP family transporters are involved in cellular zinc uptake, and at least one of them, the Zrt3 transporter of *S. cerevisiae*, transports stored zinc out of an intracellular compartment during adaptation to zinc deficiency. Contrary to that, CDF family members mediate zinc efflux out of cells or facilitate zinc transport into intracellular departments for detoxification and storage. The response to zinc regulates the activity of many of these transporters through transcriptional and post-transcriptional processes. These processes maintain zinc homeostasis at both cellular and organismal levels.

4.2.1.13 Arsenic

Many millions of people worldwide are exposed to inorganic arsenic mainly through drinking water contaminated by known mineral deposits (Smedley and Kinniburgh,

2002). Humans metabolize inorganic arsenic to methylated compounds that are largely excreted through urine together with unchanged inorganic arsenic. Exposure to inorganic arsenic is an established cause of cancers of the bladder, lung and skin. Increasing evidence indicates that inorganic arsenic may also cause cancers of the kidney, prostate and liver; cardiovascular disease; diabetes; and developmental and reproductive effects (Navas-Acien and Guallar, 2008 and references therein).

Although arsenic compounds are best known historically for their toxicity, their pharmacological action is also well documented. Slightly less documented is the increasing evidence of the essential function fulfilled by small dietary arsenic intakes, as found in four species of experimental animals. The biological effects of arsenic depend markedly on the chemical form in which the element is taken, inorganic compounds being more toxic than organic ones. Most living organisms convert the inorganic arsenic by methylation into a large variety of less toxic organoarsenic compounds, which are then excreted. Cacodylic acid and methanearsonic acid are typical urinary excretion products in humans. Organic compounds containing arsenobetaine, dimethylarsenoribosides or arsenolipids are metabolic products of aquatic organisms. These compounds contribute substantial amounts of arsenic to human intake by diets containing fish and other seafood.

For a detailed discussion of the arsenic forms and toxicity in biological systems, see the paper by Akter et al. (2005).

4.2.1.14 Selenium

Selenium is an essential micronutrient for many organisms (including humans) and has antioxidant, redox-regulatory, anti-inflammatory, chemopreventive and other properties (Rayman, 2000). Selenium is the essential trace element, and it is of fundamental importance to human health. As a constituent of selenoproteins, selenium has structural and enzymic roles, being best known for its role as an antioxidant and catalyst for the production of active thyroid hormone. Selenium is required for the proper functioning of the immune system and appears to be the main nutrient in counteracting the development of virulence and inhibiting HIV progression to AIDS. It is needed for sperm motility and may reduce the risk of miscarriage. Deficiency has been connected to adverse mood states. Findings have been equivocal in linking selenium to the risk of cardiovascular disease, although higher selenium status is beneficial to conditions such as oxidative stress and inflammation. An elevated selenium intake may be associated with reduced cancer risk. In the context of these health effects, low or diminishing selenium levels in some parts of the world, including some European countries, are giving cause for concern (Rayman, 2000).

Selenium is an essential trace element for the functioning of many organisms by serving critical catalytic roles in the form of the 21st co-translationally incorporated amino acid selenocysteine (Sec). It is mostly found in redox-active proteins in species of all three domains of life, and analysis of a large number of genome sequences has facilitated the identification of the encoded selenoproteins. Data available from the variety of reports indicate that gram-positive bacteria and enterobacteria synthesize and incorporate Sec via the same pathway. However, recent in vivo studies suggest that Sec decoding is much less stringent in gram-positive bacteria than in Escherichia coli. For years, the knowledge about the path of Sec synthesis in Archaea and Eukarya has only been incomplete. Still, genetic and biochemical studies guided by analysis of genome sequences of Sec-encoding archaea have led to the characterization of the pathways, also showing that they are principally identical. Stock and Rother (2009) summarized the current knowledge about the metabolic pathways of Archaea and gram-positive bacteria where selenium is involved. Their review refers to the known selenoproteins and the paths employed in selenoprotein synthesis.

Sec is the 21st amino acid. It exists in all kingdoms of life as a key defining the selenoproteins. Sec is a cysteine (Cys) residue analog with a selenium-containing selenol group in place of the sulfur-containing thiol group in Cys. The selenium atom gives Sec quite different properties from Cys. The most obvious difference is the reduced pK(a) of Sec, and Sec is also a stronger nucleophile than Cys. Proteins containing Sec are often enzymes, employing the reactivity of the Sec residue during the catalytic cycle, and because of that, Sec is ordinarily essential for their catalytic efficiencies. Other unique features of Sec, not shared by any of the other 20 standard amino acids, is the consequence of the atomic weight and chemical properties of selenium and the particular occurrence and features of its stable and radioactive isotopes. Sec is also incorporated into proteins by an expansion of the genetic code as the translation of selenoproteins involves the decoding of a UGA codon, otherwise being a termination codon. In contrast to other elements, Se is co-translationally inserted into proteins in the form of Sec residue (Böck et al., 1991). Biosynthesis of Sec and insertion into selenoproteins involve sophisticated molecular machinery modifying UGA codons (usually functioning as stop signals) to act as Sec codons (Böck et al., 1991; Papp et al., 2007; Zhang and Gladyshev, 2009; Hatfield et al., 2006).

Recently, progress has been made in the discovery of new selenoproteins, most of which are identified by bioinformatics approaches and further verified by experiments (see, for example, Kryukov et al., 1999; Leacure et al., 1999; Zhang and Gladyshev, 2005; Kryukov and Gladyshev, 2004; Kryukov et al., 2003; Zhang et al., 2005; Zhang and Gladyshev, 2008b). Let us mention that 25 selenoprotein genes were found in humans and 24 in mice and rats (Kryukov et al., 2003). An exciting feature of these proteins is that almost all selenoproteins have widespread homologs in which Sec is replaced with Cys. Thus, Se utilization can be examined by analyzing the evolutionary dynamics of Se-dependent and Se-independent, Cys-containing forms of various selenoprotein families. Sec can significantly increase the catalytic activity of proteins compared to their Cys-containing homologs (Johansson et al., 2005; Kim and Gladyshev, 2005).

4.2.1.15 Molybdenum

Molybdenum enzymes have been isolated from organisms throughout the entire kingdom of life, from simple bacteria to humans. It is believed that Mo enzymes are the chemical descendants of tungsten enzymes. This hypothesis is based on the discovery of tungsten enzymes in archaebacteria, where these W-enzymes have structures and functions very similar to those of the Mo enzymes found in higher animals, including humans.

In living matter, molybdenum is mainly used to form Mo cofactor (Moco) in molybdoenzymes (Mendel and Bittner, 2006; Zhang and Gladyshev, 2009). Comparative genomics studies have shown that molybdoenzymes are strictly dependent on this cofactor (Zhang and Gladyshev, 2008a, 2010).

Transition chemical element molybdenum is of essential importance for nearly all biological systems as it is required by enzymes catalyzing diverse critical reactions in the global carbon, sulfur and nitrogen metabolism. The metal itself is biologically inactive unless a particular cofactor complexes it. Except for bacterial nitrogenase, where Mo is a constituent of the FeMo cofactor, Mo is bound to a pterin, thus forming the molybdenum cofactor, which is the active compound at the catalytic site of all Mo-enzymes. In eukaryotes, the most important Mo-enzymes are as follows:

1. sulfite oxidase, which catalyzes the final step in the degradation of sulfur-containing amino acids and is involved in detoxifying excess sulfite,
2. xanthine dehydrogenase, which is included in purine catabolism and reactive oxygen production,
3. aldehyde oxidase, which is involved in oxidizing a variety of aldehydes and is essential for the biosynthesis of the phytohormone abscisic acid, and
4. nitrate reductase, which catalyzes the crucial step in inorganic nitrogen assimilation.

All Mo-enzymes, except plant sulfite oxidase, need at least one or more redox-active center, many of them involving iron for electron transfer. In the biosynthesis of Moco, which is the complex interaction of six proteins in a process with four steps, iron and copper are also included in an indispensable way. Moco, as released after synthesis, is probably distributed to the apoproteins of Mo-enzymes by putative Moco-carrier proteins. In order to become active, the xanthine dehydrogenase and aldehyde oxidase require the sulfuration of their Mo-sites. This final step is catalyzed by a Moco sulfurase enzyme, which activates sulfur from L-cysteine in a pyridoxal phosphate-dependent manner, typical for cysteine desulfurases (Mendel and Bittner, 2006).

4.2.1.16 Iodine

Thyroid hormones (THs: thyroxine and triiodothyronine) are essential cell signaling molecules in vertebrates. The function of THs is to regulate and coordinate physiology within and between cells, tissues and whole organisms, in addition to the control of embryonic growth and development, via dose-dependent regulatory effects on essential genes. Although invertebrates and plants do not possess thyroid glands, many utilize THs for development, while some other store iodine as TH derivatives or TH precursor molecules (iodotyrosines) – or produce similar hormones acting in analogous ways. Such typical developmental roles for iodotyrosines across kingdoms indicate that a common endocrine signaling mechanism may be responsible for coordinated evolutionary change in all multicellular organisms (Crockford, 2009). Iodine acts as a powerful antioxidant. It is essential for life in many unicellular organisms, including ancient cyanobacteria. Starting from an initial role as membrane antioxidant and biochemical catalyst, the spontaneous coupling of iodine with tyrosine has created a versatile, highly reactive and mobile molecule, which, over time, became integrated into the process of energy production, gene function and DNA replication in mitochondria.

Although it is not clear precisely what iodine is used for biochemically, plants take up iodine from water and store it. In addition, iodine is coupled to various carbohydrates, polyphenols and proteins. Debris of vegetation absorbs and fixes even more iodine, about ten times more. Therefore, soils rich in organic matter, such as peat, are particularly rich in stored iodine. Marine phytoplanktonic species and aerobic soil bacteria are capable of reducing inorganic iodate to iodide at or within their cell walls. While some iodine is transferred into the cytoplasm during this process, much of it is released as iodide and organic iodinated compounds, including methyl iodide, CH_3I. Macroalgae also release iodide; they are known for their ability to accumulate large quantities of iodine, but the function of these stored reserves has not been determined.

In some single-celled organisms, including most anaerobic bacteria, fungi, viruses and yeasts, I_2 is damaging the cell membrane. However, some of the choanoflagellates and most α-proteobacteria are resistant to I_2, as are all animals. Although I_2 kills some organisms, it does not mean they do not use iodine at all, only that they must get it in another form. Certain ferric iron and sulfate-reducing anaerobic bacteria can reduce iodate to iodide. Some facultative anaerobes can also do this. Based on that, we can assume that the earliest anaerobic cells might have had this ability, too. The fact that virtually all species can convert iodate to iodide and that iodide enters cells in many of these suggests that iodine always enters cell cytoplasm and is a participant in essential element biochemistry (Crockford, 2009).

The only physiologic role of iodine in the human body is the synthesis of thyroid hormones in the thyroid gland. The dietary requirement of iodine is defined by normal thyroxine (T_4) production by the thyroid gland without affecting the thyroid iodide trapping mechanism or raising thyroid-stimulating hormone (TSH) levels. Iodine uptake by the diet is absorbed throughout the gastrointestinal tract. The iodide ion is bioavailable and absorbed from food and water. Dietary iodine is converted to the iodide ion before it is absorbed. This conversion does not happen for the iodine within thyroid hormones ingested for therapeutic purposes. Iodine enters the circulation as plasma inorganic iodide, which is cleared from circulation by the thyroid and kidney. The thyroid gland performs the synthesis of thyroid hormones by the use of iodide, while the kidney excretes iodine with urine. The excretion of iodine in the urine is a good measure of iodine intake. In a healthy population having no evidence of clinical iodine deficiency in the forms of endemic goiter and endemic cretinism, urinary iodine excretion reflects the average daily iodine requirement. Therefore, for determining the iodine requirements, the critical indexes are serum T_4 and TSH levels (indicating normal thyroid status) and urinary iodine excretion (for details, see FAO, 2001).

4.2.1.17 Tungsten

The chemical nature of tungsten compared to molybdenum is probably responsible for its role in the living systems. Together with chromium and molybdenum, tungsten belongs to group VI of the periodic table of elements, similar to molybdenum in its chemical properties. The vital role of the latter in biological processes is well known (Johnson et al., 1996). W and Mo have the same atomic (1.40 Å) and ionic (0.68 Å) radii and similar electronegativity (1.4 for W and 1.3 for Mo); some other coordination characteristics are also very close (Kletzin and Adams, 1996). Both elements can be in various oxidation states (from +2 to +6) and can form polynucleotide complexes; however, only the oxidation states +4, +5 and +6 and mononucleotide systems are biologically relevant (Pope, 1987; McCleverty, 1994).

Tungsten and molybdenum are rare in occurrence, and they rank 54th and 53rd in natural abundance, respectively (Greenwood and Earnshaw, 1984). The solubility of tungsten salts is smaller than that of molybdenum salts. Its concentration in freshwater rarely exceeds 20 nM and is usually less than 0.5 nM, whereas the level of molybdenum is two or more orders of magnitude higher than this value. The concentration of tungsten in seawater is very low, 5×10^5 times lower than that of molybdenum.

The biological importance of tungsten has been proved by the isolation of several tungsten-containing enzymes (W-enzymes) from hyperthermophilic archaea. W was previously considered only as an antagonist of Mo because the replacement of Mo by W leads to the inactivation of Mo-containing enzymes (Mo-enzymes). In addition to the "true W-enzymes" in which tungsten cannot be replaced by molybdenum, recently, some enzymes have been isolated, which can use either molybdenum or tungsten in the catalytic process. The review by L'vov et al. (2002) briefly summarizes data on the participation of tungsten in catalysis by some enzymes and the structure of the active sites of W-enzymes.

Since for many years W has been considered to be the biological antagonist of Mo, it is often used for the study of the properties and functions of Mo in Mo-enzymes. This is because tungsten can replace molybdenum in Mo-enzymes, forming catalytically inactive (or possessing very low activity) analogs. All organisms can be divided into at least three groups based on the way they use molybdenum and tungsten (L'vov et al., 2002):

1. Organisms preferring molybdenum to tungsten.
2. Organisms preferring tungsten to molybdenum.
3. Organisms, to some extent, able to use both metals.

This spectacular progress in the knowledge of the biological role of tungsten was achieved after the discovery of hyperthermophilic archaea in the early 1990s. Tungsten enzymes isolated from hyperthermophilic archaea cells are exceptions from the group of W-enzymes because the dependence of their catalytic activity on tungsten is obligatory. This element cannot be replaced by vanadium or molybdenum (Mukund and Adams, 1996). Some hyperthermophilic archaea, such as *Pyrococcus furiosus*, synthesize three W-enzymes called aldehyde ferredoxin oxidoreductase, formaldehyde ferredoxin oxidoreductase and glyceraldehyde-3-phosphate dehydrogenase (Mukund and Adams, 1996). An essential difference between W-enzymes and their Mo-analogs is that the former is expressed constitutively in many studied microorganisms, while Mo-analogs are induced only in the presence of Mo (Hochheimer et al., 1996, 1998). In addition, W-enzymes catalyze oxidation–reduction reactions at much lower redox potentials than their Mo-analogs.

4.2.2 Concentration Factors

The most important characteristics of living cells is their ability to take up chemical elements from the environment against their concentration gradient. This ability is most obvious, for example, in marine microorganisms that obtain their nutrients directly from the seawater. The concentration factor is then defined as follows:

$$\text{Concentration factor} = \frac{\text{Element concentration in organism / cell}}{\text{Element concentration in the cell environment}}$$

Concentration factors for some elements in plankton and algae can be as high as 60,000 (Bowen, 1966). The organisms concentrate all elements present in their environment.

The concentration factor, C_F, is a function of the number of physical, chemical and biological parameters. Figure 4.3 shows the relation of the uptake of essential elements to yield or growth and may be considered as a definition of essentiality. There is a rather narrow range of adequacy of element concentration in the organisms. Smaller levels result in different abnormalities induced by deficiencies, which are accompanied by pertinent specific biochemical changes. Higher concentrations result in toxicity. In plants, it is possible to have, under severe deficiency conditions, a decrease in the level of an element, which results in a small increase in growth (Steenbjerg, 1951).

The ability of the cell to maintain a specific trace element within a certain homeostatic range is mainly dependent on the processes of uptake, storage and excretion. The relative importance of these processes varies among trace elements and organisms. High-affinity uptake systems for some trace elements are found in both prokaryotes and eukaryotes. Among them, the ABC transporters are the most frequently used uptake systems for metals (Van Gossum and Neve, 1998). Besides high-affinity transporters, metal ions could be transported via general cation influx systems, although the efficiency of such processes may be low (Maguire, 2007). For some elements, element-specific transporters have not been identified.

An example is Se; the uptake (in the form of selenate or selenite) is thought to be maintained up by the sulfate transport system. On the other hand, excessive uptake of certain metals may result in metal overload and toxicity. The storage of trace elements in inactive sites (or forms) and excretion/export systems are critical mechanisms that prevent inappropriate amounts of trace elements in the cell (see, for example, Gold et al., 2008). Also, the release of a trace element from a storage site may be essential to avoid deficiency.

The utilization of trace elements in cells is complex and not completely understood. Most metals are directly used as cofactors inserted into proteins involved in various metabolic pathways. In contrast, Mo and Co are mainly used in the form of cofactor (Moco) and cobalamin (vitamin B12), respectively. The number of metalloprotein families also varies from less than 10 (Ni-binding proteins) to more than 300 (Zn-binding proteins). The use of Se is quite different from other trace elements. It is principally used in the form of Sec, the 21st amino acid, found in selenoproteins of the three domains of life (Böck et al., 1991; Hatfield and Gladyshev, 2002).

Organisms inhabiting the metal-polluted environments develop resistance mechanisms that enable efficient detoxification and transformation of toxic forms to nontoxic

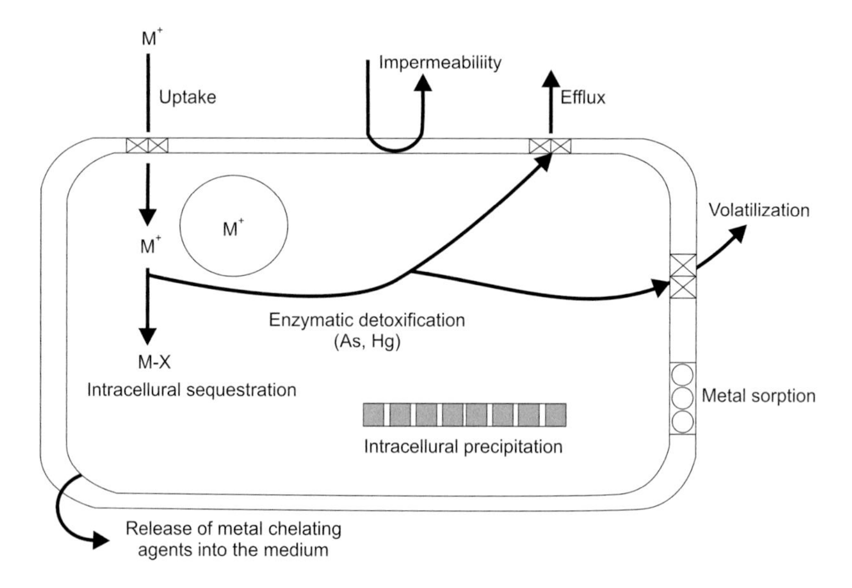

FIGURE 4.4 General mechanisms adapted by bacteria, eukaryotes and archaea for metal resistance. (Modified from Srivastava and Kowshik, 2013.)

forms. The majority of Bacteria and Eukarya tolerate metals by reducing influx and increasing output (Gadd, 1993; Blaudez et al., 2000), or sometimes by the enzymatic detoxification followed by volatilization (Nies, 1999; Silver and Phung, 2009). Intracellular compartmentalization is observed only in eukaryotes (Gadd, 1993). Figure 4.4 shows the various mechanisms of metal resistance exhibited by all three domains of life. All three domains exhibit the following tools for metal resistance:

sorption of metals,
volatilization,
release of metal-chelating compounds in the medium,
enhanced efflux,
impermeability,
decreased uptake,
enzymatic detoxification, and
intracellular chelation.

Organellar compartmentalization is observed only in eukaryotes, except for magnetosomes in magnetotactic bacteria (Srivastava and Kowshik, 2013).

4.2.2.1 Possible Effects of Magnetic Field on Concentration Factor

Living cells possess electric charges produced by ions or free radicals, which act as endogenous magnets. These endogenous magnets can be affected by the exogenous magnetic fields, which can orient unpaired electrons. External magnetic fields can influence both the activation of ions and polarization of dipoles in living cells (Goodman et al., 1995;

Belyavskaya, 2004). Treatments with magnetic field are supposed to enhance seed energy by influencing the biochemical processes controlled by free radicals and by stimulating the activity of proteins and enzymes (Dhawi et al., 2009). Fesenko et al. (2010) studied the effect of the "zero" magnetic field (0.2 μT) on early embryogenesis in mice and found several abnormalities. "Zero" magnetic field effects have also been studied in pathogen bacteria (Creanga et al., 2004).

Roughly one-third of all living species became extinct at the close of the Cretaceous, a period marked by the resumption of polarity reversals of Earth's magnetic field following a very long period of regular magnetic activity. There is mounting evidence that a correlation exists between major faunal extinctions and geomagnetic polarity reversals. The validity of this correlation in recent geological time seems to have been well established by studies of fossil species of single-celled marine microorganisms (Hays, 1971). Several mechanisms linking changes in the geomagnetic field with effects on living organisms have been proposed. Most are based on the assumption that during polarity reversals, the dipole component of the geomagnetic field weakens or disappears for periods of a few thousand years, allowing a much higher flux of both solar protons and galactic cosmic rays to bombard the Earth's surface. Other mechanisms include climate change and the effects of a significant reduction in the content of ozone in the atmosphere, which would increase exposure to ultraviolet radiation.

Direct magnetic field effects on growth have been discussed (Crain, 1971; Bowen, 1966). A mechanism through which the geomagnetic field might have influenced the extinction of the species has been proposed (Valkovic, 1977). As an effect of the magnetic field on living organisms, he introduced the concentration factor dependence on the magnetic field intensity. Geomagnetic field intensity decreases by an order of magnitude during polarity reversals. During this time, species are affected by low magnetic field intensity H_1. If they were living in an environment that had provided them with elements, they would be in the range of adequacy when Earth's magnetic field was H_0. Because of the assumed dependence of concentration factor on magnetic field intensity, they will be in the range of deficiency or toxicity when the magnetic field is H_1 (depending on the slope of the assumed functional dependence). Prolonged deficiency or toxicity for several generations may lead to disastrous effects, including the species' disappearance. This effect can happen for one or more essential elements. Magnetic field dependence of concentration factors can bring living organisms into the range of deficiency or toxicity without changing trace element availability in the environment. One can imagine a species living in a situation such that the supply of a given essential trace element is near the edge of the range of adequacy. In this case, only small disturbances are needed to have an inadequate supply of the essential trace element and consequent development of abnormalities.

As described by Valkovic (1978), two experiments were performed. The organism used was a respiratory-deficient mutant of *M. bacilliformis*, which also had lost the ability to grow as mycelium. Instead, it exists as spherical cells, which reproduce only by budding at the expense of alcoholic fermentation. The mutant is sensitive to temperature (dies at 30°C) and is also susceptible to vigorous mechanical shaking. It grows relatively slowly in a defined medium.

In the first experiment, the growth of the culture took place at room temperature in round-bottom flasks containing 14 mL of medium. Each container was inoculated to have 10^3 cells/mL; six vials were placed in solenoids whose magnetic fields were extremely uniform over the active region of growth, and six flasks were used as controls. Growth (increase in cell number) was monitored daily by measurements of the turbidity of the cell suspensions. After 7–10 days, when the cell number reached saturation, the cells were harvested. Trace elements were analyzed by proton-induced X-ray emission spectroscopy (PIXE). Concentrations of all essential elements may be measured simultaneously. Measured data for Mn/Zn and Cu/Zn ratios indicate that the concentration ratios for these elements within the microorganisms might be dependent on magnetic field intensity. Both Mn/Zn and Cu/Zn ratios decreased with the increase in the magnetic field (Mn/Zn for 5%; Cu/Zn for 13%). However, this decrease was still within the experimental error.

The above results were checked by performing the second measurement. Although absolute values for Cu/Zn concentration ratios are slightly higher than in previous experiments, the Cu/Zn ratio seems to follow the trend observed in the first experiment: It decreases with the increase in magnetic field intensity. Similar, but not certain, conclusions are also valid for Mn/Zn and Fe/Zn concentration ratios. The Ni/Zn concentration ratio does not show a tendency toward decrease, while the magnetic field intensity increases (Valkovic, 1978).

4.3 Origin of Trace Element Requirements

Research on the role of trace elements in chemical evolution should be reviewed by considering the following three points:

i. The origin of the essential requirements for trace elements in the present biological systems.
ii. The roles of trace elements in chemical evolution and origin of life.
iii. The origin of enzymatic activity by metal ions, i.e., the origin of metalloenzymes.

The metals play a fundamental role in biocatalysis. Most of the known enzymes contain transition metal ions as a cofactor of their active sites. The metalloenzymes lose their catalytic activity when the metal ions are being removed from the protein molecule. These facts may indicate the primary role of the metals in the origin of biocatalysis. The taxonomic distribution of the metalloproteins gives a hint on the biogenesis as well. For example, the tungsten enzymes have been discovered so far in prokaryotes only. However, obligatory dependence on tungsten is documented merely for hyperthermophilic archaea. Their basal position on the molecular tree of life points to the W-rich hydrothermal systems as a cradle of life (Fedonkin, 2008). The relatively recent discovery of the essential biological roles of nickel and tungsten (Cammack, 1988; L'vov et al., 2002; Ragsdale, 1998) illustrates that the list of heavy metal cofactors of active centers of enzymes may expand. Reconstruction of the evolution of metabolism could be done by the synthesis of geochemistry data of early Earth and the taxonomic distribution data. The role trace elements play in the chemical evolution of the present biological systems is considered by Kobayashi and Ponnamperuma (1985).

The origin of the trace element requirement by living matter is discussed by Valkovic (1990). McClendon (1976) proposed that the relative abundance of an element should be a decisive factor in the origin of its nutritional essentiality. The distribution of essential elements in the periodic table of elements might contain answers to questions on the origin of element requirements and possibly about the origin of life in prebiotic time. The assumption that vital body chemistry should bear similarity to the primordial chemical environment is a reasonable one. An organism would not make itself dependent on a rare element for its existence provided a more abundant element could play the same role. The presence of an element is a necessary prerequisite for the development of an essential metabolism based on that element (Valkovic, 1990).

If an element has not been selected, this may be because its abundance in the available environment was too low, either on the absolute scale or in comparison with some other element that can play the same role. This approach can explain why a particular element has been selected for an essential biochemical role.

There are four groups of elements required for life:

1. Unique requirements dating from the origin of life: H, C, N, O, K, Mg, P, S and Fe.
2. Individual needs acquired later: B, Se and I.
3. The primordial requirements, which could be satisfied by several elements. Evolutionary adoption was made to the more abundant member: K *vs* Rb, Mg *vs* Be, S *vs* Se, Cl *vs* Br, and H *vs* F.
4. The same as under 3, but at a later time: Ca *vs* Sr, Na *vs* Li, and Si *vs* Ge.

Although it is often said that the composition of organisms resembles that of the ocean, one might get different conclusions from the following discussion. Rather high concentrations of nonessentials: Br, Sr, Rb and some others, in the ocean waters are evident. Metals that serve as essential trace elements are strongly depleted in the ocean compared to plant or animal requirements. Also, the content of both carbon and nitrogen is low compared to living matter concentrations and cosmic abundances.

The question we can ask is as follows: Which environment should be compared to the composition of living organisms? The answer is the one with which the organism has an intimate contact. Contrary to all expectations, the universal (cosmic) abundance curve of chemical elements is in best agreement with the distribution of essential life elements within the periodic table. Essential elements are the most abundant elements. Essential trace elements are almost all grouped in the second peak (around Fe). This region is characterized by a maximum in nucleon binding energy for the nucleus. This fact is responsible for the peak in the universal abundance curve.

The laws of the chemistry of trace elements dictate the molecular speciation and reactivity both within cells and the environment at large. Using protein structure and comparative genomics, Dupont et al. (2010) explained several major influences chemistry has had upon biology. All life species exhibit the same proteome size-dependent scaling for the number of metal-binding proteins within a proteome. This evolutionary constant shows that the selection of one element occurs at the exclusion of another. An exception is Fe for Zn and Ca, which is a defining feature of eukaryotic proteomes. Early life did not have structures required for the control of intracellular metal concentrations and that of the metal-binding proteins used in the catalyses of electron transport and redox transformations. The recent development of protein structures used in metal homeostasis happened simultaneously with the emergence of metal-specific structures, which predominantly bound metals (after Dupont et al., 2010). This event potentially promoted the diversification of emerging lineages of Archaea and Bacteria through the establishment of biogeochemical cycles.

Structures binding Cu and Zn evolved later, a fact that is considered to provide evidence that availability in the environment influenced the element selection process. The late evolution of Zn-binding proteins is fundamental to eukaryotic cellular biology. Therefore, Zn bioavailability may have been a limiting factor in eukaryotic evolution. The results presented by Dupont et al. (2010) provide an evolutionary timeline based on genomic characteristics, and alternative geochemical methods can test critical hypotheses.

The arrival of large-scale whole-genome sequencing can help elucidate how trace metal chemistry influenced the evolution of protein macromolecules and how the different kingdoms of life incorporated this machinery. Specific amino acid residues and an appropriate arrangement of those residues are both required for proper metal binding by a protein, meaning that an analytical approach using protein structure is more relevant than a purely sequence-based methodology. A recent survey of Fe-, Zn-, Mn- and Co-binding structures in over 300 genome-encoded proteomes showed that the abundance of proteins binding a specific element within a proteome relates to proteome dimensions obeying power law with given slopes (α) for each kingdom and metal (Dupont et al., 2006). Accurately, the proteomes of eukaryotes, and organisms in kingdoms Archaea and Bacteria without a nucleus were described by power-law function with values of $\alpha>1$ for Fe-, Mn- and Co-binding structures. In contrast, the eukaryotic proteomes have values of $\alpha>1$ for Zn-binding structures. The different power-law slopes for Fe, Mn, Zn and Co parallel the hypothetical chemical environment in which each kingdom supposedly evolved, providing full-genome evidence for the influence of trace metal geochemistry on biological evolution.

The inadequate knowledge of the order of metalloenzyme evolution precluded any conclusions on the impact the different trace metals had on the emergence of diversified life. Although the power-law relation revealed a consistent evolutionary trajectory, it lacked direction, impeding interpretation over evolutionary history. In the work by Dupont et al. (2010), the relative timing for the evolution of metal-binding protein structures was determined using a phylogeny of protein architectures that have been developed and refined previously (Caetano-Anolles and Caetano-Anolles, 2003; Wang et al., 2007). The abundances and occurrences of all of the diverse protein families binding Fe, Zn, Mn, Co, Ni, Cu, Mo and Ca in 313 genomes were also determined. By examining the intersections of these data sets, Dupont et al. (2010) identified several genomic fossils that illuminate how trace metal chemistry has influenced evolution. Environmental concentrations of trace metals influenced what type of metal-binding proteins would first evolve. The discovery of the protein machinery needed to solve the equilibrium puzzle of the Irving–Williams Series (Irving and Williams, 1953) coincided with the birth of metal-binding electron transfer proteins. By using proteome content, protein domain evolution and cell biology, it is speculated that low Zn, Mo and Cu environmental concentrations prevented the widespread emergence and diversification of the eukaryotic kingdom until the advent of a planet-wide shift in redox state (Dupont et al., 2010).

Bánfalvi (2011) has redefined the term heavy metals as those trace elements that have ≥3 g/cm^3 densities and may cause harmful biological effects. He has arrived at this definition by clarifying first the light elements based on their electronic configurations and compatibility with group of elements (C, H, N, O, P, S) in constructing biomolecules. As for compatibility criteria, the chemical bond formations between *s*–*p* electrons and *p*–*p* electrons were taken, allowing the tetrahedral three-dimensional construction of biological compounds with four bonding partners. The compatibility range ended at $1s^2 2s^2 2p^6 3s^2 3p^6 4s^2$ electronic configuration corresponding to calcium, which is the 20th element in the periodic table. High reactivity, rich coordination chemistry and complex formation of transition metals are due to the outer *d* and *f* electron subshells and explain their crucial catalytic role in enzyme reactions and toxicity at higher cellular concentrations.

4.4 Present Life on the Earth

Until the 20th century, it was considered all living things can be classified as either a plant or an animal. However, in the 1950s and 1960s, most biologists realized that this system failed to accommodate the fungi, protists and bacteria. By the 1970s, a network of five kingdoms was accepted as the model by which all living things could be classified. At a fundamental level, a distinction was made between the prokaryotic bacteria and the four eukaryotic kingdoms (plants, animals, fungi and protists). The difference recognizes the common traits that eukaryotic organisms share, such as nuclei, cytoskeletons and internal membranes.

In the 1970s, Carl Woese and his team began investigating the sequences of bacteria to develop a better picture of bacterial relationships (see, for example, Woese and Fox, 1977). In addition to the bacteria and eukaryotes, the analysis showed that there must exist an additional group of methane-producing microbes. These methanogens were already known to be strange in the microbial world, since they could be killed by oxygen, they produced unusual enzymes, and they had cell walls different from all known bacteria.

Figure 4.5 shows a basic phylogenetic tree with the present-day accepted relationships between these domains. Each line denotes an evolutionary "distance" between the groups shown, which reflects the time since they split from a common ancestor. There is no particular significance to the groups explained. They merely illustrate some examples. It can be seen that the tree is "rooted" in a single ancestor. This point is generally called the last universal common ancestor or LUCA. Additionally, in this present-day accepted thinking, the first eukaryotic common ancestor (FECA) is positioned where and when Eukarya separated from Archaea. Since FECA already possessed complex, eukaryote-typical features inherited from the LUCA, we could assume these to be transferred to the last eukaryotic common ancestor (LECA) by FECA in subsequent evolution. However, there are other possibilities, as discussed by Mariscal and Doolittle (2015),

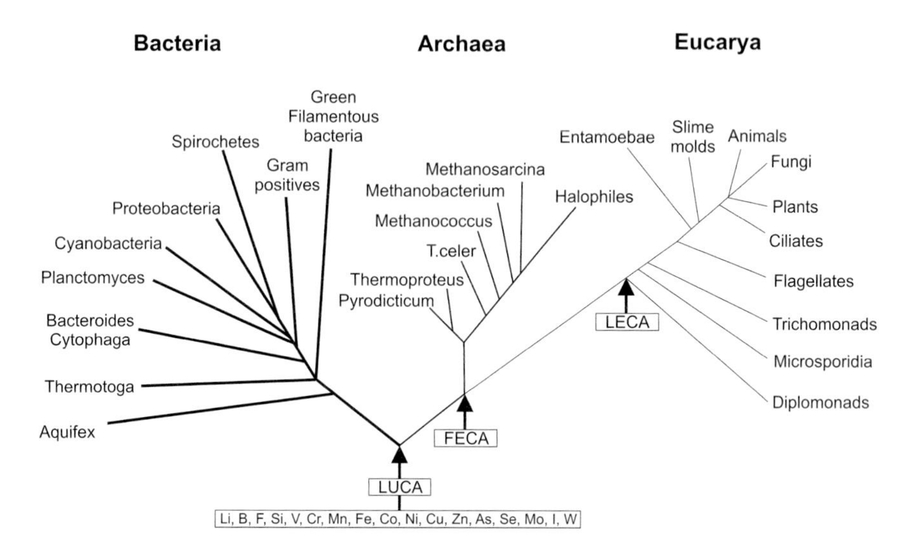

FIGURE 4.5 The present-day accepted the version of the phylogenetic tree of life.

where all three domains are directly rooted in the LUCA and Bacteria, and Archaea jointly separated from Eukarya.

There are different types of phylogenetic trees. For some trees, the length of a branch is reflecting the number of genetic changes that happened in a particular DNA sequence. In some other trees, branch length can represent chronological time and branching points can be determined from the fossil record. These trees are designed to convey a different type of information or combined information. The choice of the tree will depend upon the knowledge that a researcher is interested in communicating.

One can also find trees drawn in different ways; some use diagonal branches or curved branches. Many of these convey the same information as trees drawn with horizontal branches and vertical lines at the nodes; they are just different methods of presentation.

All organisms presented on the tree of life share a common ancestor. One of the most critical challenges has been determining the nature of the LUCA. The first attempt to do this was based on an analysis of paralogue protein couples. These are proteins encoded by genes that have duplicated. If the gene duplication is found in all three domains, this implies that the duplication occurred before the three domains diverged. Thus, by examining the distribution of a variety of paralogous genes, one can infer when domains separated. This fact led many scientists to assume that the tree of life is rooted in bacteria with a later divergence between archaea and eukaryotes. A more recent examination using sequenced whole genomes to identify genes found in all three domains of life revealed approximately 100 common genes that were then used to construct a phylogenetic tree, with similar results.

In contrast to this view is the one which places the LUCA nearer the eukaryotes with a later divergence between bacteria and archaea. This thinking follows from the idea that many complex processes involving RNA are found in eukaryotes, including the splicing of RNA, which has been hypothesized to be a relic of an ancient pre-biological RNA world. A concern with making these deep inferences is that biases can be introduced into tree reconstruction by varying rates of evolutionary change in different species. For example, the evolutionary rates of ribosomal RNA changes vary by up to 100 times in foraminifera.

When a universal tree of life is constructed, making special efforts to use only slowly evolving genes, the root of life does not so clearly sit within the Bacteria and instead suggests that the rooting in the bacterial lineage could be an artifact. Part of the problem in defining the real evolutionary root of the tree is the low number of genes being considered. As a more significant amount of organism genomes are sequenced, the quantity of information that can be used to determine the root will increase and so the accuracy of the inferences.

A question of particular interest is what the functional capabilities of the LUCA were. Some biochemical pathways, such as the TCA cycle and components of electron transport chains, are very widely distributed and present in all three domains of life, suggesting that they are ancient and could have been present in the LUCA. An early analysis of the tree of life indicated that the deepest branching organisms were heat-loving ones. The result included *thermophiles* and *hyperthermophiles*. This fact led to numerous hypotheses to explain the observations.

The LUCA had a more straightforward metabolism compared to modern forms, with a partially full complement of essential amino acids and perhaps a few non-specific enzymes. Keep in mind that enzymes are biochemical catalysts that are molecularly larger and more specific than inorganic catalysts. Aromatic amino acids (phenylalanine, tyrosine and tryptophan) appear to have arisen sometime after the LUCA arose, as did more complex and specific enzymes (Ahmad and Jensen, 1988). While the actual incorporation of iodine into essential biochemistry appears to have come after the rise of the LUCA, it was suggested that its influence probably started with the LUCA itself (Barrinton, 1962; Crockford, 2009).

For example, intense impact bombardment in the early history of the Earth was suggested to have created an evolutionary bottleneck for selecting high-temperature-tolerant organisms. After bombardment, when impacts were more infrequent and the Earth cooler, life could have entered into a greater diversity of microorganisms adapted to growing at lower temperatures (Cockell, 2015).

However, the hot environment ancestry of life has been questioned. The heat-loving bacteria have acquired up to 24% of their genes from the archaea, suggesting considerable horizontal gene transfer. The DNA reverse gyrase gene, which encodes for an enzyme that introduces a supercoil to DNA, making it more difficult to unwind and conferring resistance to high temperatures, is an example of a gene that is likely to have been passed by horizontal gene transfer from archaea. It is found in bacteria such as *Thermotoga* and *Aquifex* that live in high-temperature environments. If the root of life is bacterial, then the existing data might argue against a hot origin of life if they acquired these genes from elsewhere (Cockell, 2015).

There are several complexities in unravelling the functional capabilities of the LUCA. First, the open question of where the tree of life is rooted must be resolved as only then can it be determined whether high-temperature adaptation is a primary inherited trait. Second, even if the tree of life does not originate in organisms adapted to living at high temperatures, this does not rule out a hot origin of life. This reasoning is based on the fact that a considerable time (and extinct organisms) might exist between the origin of life and the root of the tree as defined by extant organisms.

All these considerations are done by assuming that the three existing domains of life define the LUCA's capabilities and characteristics. We cannot be sure that there were not domains that went extinct early in the history of life and took with them some crucial information (Cockell, 2015).

According to Lineweaver and Schwartzman (2005), the deepest roots of the 16S rRNA phylogenetic tree are hyperthermophilic (see also Wong et al., 2007). Some of the extant organisms are hyperthermophiles able to tolerate temperatures above 90°C. These organisms – Aquifex, Thermotoga, Nanoarchaeota and Korarchaeota – seem to be the best representatives of the LUCA. The existence of hyperthermophilic organisms close to the root suggests that the LUCA could be hyperthermophilic. By extension of this thinking, the origin of life on the Earth was hyperthermophilic (Lineweaver and Chopra, 2012).

It is known that all living organisms require a source of energy. The organisms that use radiant light are called phototrophs; the ones that use an organic form of carbon are called heterotrophs. Others that oxidize inorganic compounds are called lithotrophs.

The carbon needs of all organisms must be met by organic carbon or by CO_2. Organisms that use organic carbon are heterotrophs, while those that use CO_2 as a sole source of carbon for growth are called autotrophs.

4.4.1 Domain Bacteria

Domain Bacteria also includes the kingdom Eubacteria. Bacteria must find in their environment all of the substances required for energy generation and cellular biosynthesis. Chemicals and elements of this environment utilized for their growth are referred to as nutrients or nutritional requirements. Many bacteria can be grown in the laboratory in culture media, which are designed to provide all the essential nutrients in a solution for bacterial growth. Bacteria are symbionts or obligate intracellular parasites of other cells. It is usually difficult to grow them outside of their natural host cells. The microbe could be a mutualist or parasite, but in both cases, the host cell must provide the nutritional requirements of its residents. Many bacteria are identified in the environment by inspection using genetic techniques. The attempts to isolate them and grow in artificial culture have been unsuccessful; this is especially thought for prokaryotes.

Differences in the composition of the membrane lipids in Bacteria and Eukarya are the following: Membrane lipids in Bacteria are composed of ester linkages, while membrane lipids in Archaea have ether linkages (see, for example, Berg et al., 2002; Jain et al., 2014), shown in Figure 4.6.

The nutritional requirements of bacteria are revealed by the cell's elemental composition, which consists mainly of chemical elements C, H, O, N, S. P, K, Mg, Fe, Ca and Mn, with traces of Zn, Co, Cu and Mo. These elements are found as inorganic ions, small

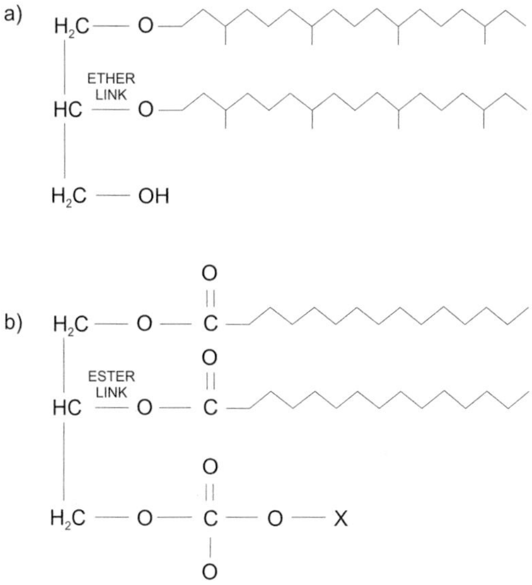

FIGURE 4.6 Differences in the composition of the membrane lipids in (a) Archaea – top figure; and (b) Bacteria and Eukarya – bottom figure. Membrane lipids in bacteria are composed of ester linkages, while membrane lipids in Archaea have ether linkages.

molecules and macromolecules which serve either a structural or functional role in the cells.

Sometimes bacteria are referred to as group based on the patterns of growth under various chemical (nutritional) or physical conditions. Here are some examples:

Phototrophs are organisms using light as an energy source.
Anaerobes are organisms growing without oxygen.
Thermophiles are organisms growing at high temperatures (Todar, 2012).

Let us discuss in some detail *Aquifex*, which is a genus of bacteria, one of the few in the phylum Aquificae. The two species usually classified in *Aquifex* are *A. pyrophilus* and *A. aeolicus*. Both species are highly thermophilic, growing best in water temperature of 85°C to 95°C near underwater volcanoes or hot springs (Madigan and Martinko, 2005). They are the right bacteria, contrary to the other inhabitants of extreme environments, namely the Archaea (Reysenbach, 2001). Both the known species of *Aquifex* are rod-shaped with a length of 2–6 μm and a diameter of around 0.5 μm. They are non-spore-forming, gram-negative autotrophs. Aquifex means water maker in Latin and refers to the fact that its method of respiration creates water. *Aquifex* tends to form cell aggregates composed of up to 100 individual cells. *Aquifex* takes carbon dioxide from the environment to get all of the carbon they need. They are chemolithotrophic, meaning that they take energy for biosynthesis from inorganic chemical sources. The enzymes this organism uses for aerobic respiration are very similar to the enzymes found in other aerobic bacteria (Deckert et al., 1998). *A. aeolicus* requires oxygen to survive, but it can grow in levels of oxygen as low as 7.5 ppm. *A. pyrophilus* can even grow anaerobically by reducing nitrogen instead of oxygen. Like other thermophilic bacteria, *Aquifex* has important uses in the industrial process.

Aquifex aeolicus survives best at temperatures around 80°C and can withstand up to 95°C. Along with *Aquifex pyrophilus*, it is one of the most thermophilic bacteria known. It is an obligate chemoautotrophic bacterium, meaning it only derives energy from inorganic compounds. This gram-negative bacterium does not form spores; it performs biotin biosynthesis. Biotin is a water-soluble vitamin essential for fatty acid biosynthesis and catabolism. Biotin is a growth factor for many cells; it is also known as vitamin B7. *A. aeolicus* can grow on hydrogen, oxygen, carbon dioxide and mineral salts (Huber et al., 1992). It contains three thermostable and oxygen-tolerant hydrogenases. Two of the hydrogenases serve in energy conservation. (These two are membrane-bound periplasmic hydrogenases.) The third, soluble cytoplasmic hydrogenase, is most likely involved in the carbon dioxide fixation pathways.

The genome of *A. aeolicus* has successfully been mapped by Deckert et al. (1998). This mapping was made more accessible by the fact that the length of the genome is only a third of the length of the genome for *E. coli*. A comparison of the *A. aeolicus* genome with other organisms showed that around 16% of its genes originated from the Archaea domain. Members of this genus are assumed to be some of the earliest members of the Eubacteria domain. *A. aeolicus* was first discovered north of Sicily, while *A. pyrophilus* was first found just north of Iceland. The complex metabolic machinery required by *A. aeolicus* to function as a chemolithoautotroph is encoded within a genome. Chemolithoautotroph organism uses an inorganic carbon source for biosynthesis and an inorganic chemical energy source. Metabolic flexibility seems to be reduced because of the limited genome size. The use of oxygen as an electron acceptor is allowed by the presence of a complex respiratory apparatus. Although this organism grows at 95°C, only a few specific indications of thermophily are apparent from the genome. Deckert et al. (1998) described the complete genome sequence containing 1,551,335 base pairs of this evolutionarily new organism.

Bonch-Osmolovskaya et al. (1990) reported the finding of a new species, called *Thermoproteus uzoniensis*, a rod-shaped archaebacterium in the samples of hot springs water and soil of the Uzon Caldera (SW of Kamchatka Peninsula). Cells were rods from 1 to 20 μm in length and 0.3–0.4 μm in width, sometimes branching or with spherical protrusions on the ends. The cell wall consisted of two layers: an internal one with a distinct hexagonal structure and the outer one with a less clear structure and variable thickness. Cells were non-motile and had no flagella. The organism grew anaerobically by fermenting peptides and, at the same time, was reducing elemental sulfur to H_2S. Fermentation products were acetate, isobutyrate and isovalerate. The G+C content of the DNA was 56.5 mol. %.

The sulfur-dependent archaebacterium *Thermoproteus tenax*, which has a cylindrical cell shape variable in length, but constant in diameter, is described in the paper by Wildhaber and Baumeister (1987). Its complete surface is covered by a normal protein layer (S-layer). The characteristics of the lattice are p6 symmetry and a lattice constant of 32.8 nm. The three-dimensional reconstruction obtained from a tilt series of isolated and negatively stained S-layer shows a complex mass distribution of the protein: A prominent, pillar-shaped protrusion is located at the sixfold crystallographic axis with radiating arms connecting neighboring hexamers in the vicinity of the threefold axis. The base vectors of the S-layer lattice have a preferred orientation for the longitudinal axis of the cell. The layer can be seen as a helical structure consisting of a right-handed, two-stranded helix, with the individual chains running parallel. Supposing that the new S-layer protein is inserted at lattice faults (wedge disclinations) near the poles, the growing of the layer would then proceed by moving a disclination at the end of the helix. The constant shape of the cell, as well as the particular structure of the layer, strongly suggests that this S-layer has a shape-maintaining function. *Thermoproteus* neutrophilus, a hyperthermophilic archaeon, uses a new autotrophic CO_2 fixation pathway, the dicarboxylate–hydroxybutyrate cycle. This regulation of the central carbon metabolism was studied on the level of whole cells. Also, the enzyme activity, the proteome, transcription and gene organization were described by Hugo Ramos-Vera et al. (2010). The organism proved to be an autotroph, which prefers organic acids as carbon sources that can easily feed into the metabolite pools of this cycle.

Bacterium requires an energy source, a source of carbon and a source of other required nutrients, and an agreeable range of physical conditions such as O_2 concentration, temperature and pH to be able to grow. Table 4.3 shows major elements, their sources and functions in bacterial cells (after Todar, 2012). This table ignores the occurrence of trace

TABLE 4.3 Major Elements, Their Sources and Functions in Bacterial Cells (After Todar, 2012)

Element	% Dry Weight	Source	Function
Carbon	50	Organic compounds or CO_2	The main constituent of cellular material
Oxygen	20	H_2O, organic compounds, CO_2 and O_2	The component of cell material and cell water; O_2 is an electron acceptor in aerobic respiration
Nitrogen	14	NH_3, NO_3, organic compounds, N_2	The element of amino acids, nucleic acids nucleotides and coenzymes
Hydrogen	8	H_2O, organic compounds, H_2	The main part of organic compounds and cell water
Phosphorus	3	Inorganic phosphates (PO_4)	The component of nucleic acids, nucleotides, phospholipids, LPS, teichoic acids
Sulfur	1	SO_4, H_2S, SO, organic sulfur compounds	The element of cysteine, methionine, glutathione, several coenzymes
Potassium	1	Potassium salts	The principal cellular inorganic cation and cofactor for certain enzymes
Magnesium	0.5	Magnesium salts	Inorganic cellular cation, a cofactor for specific enzymatic reactions
Calcium	0.5	Calcium salts	Inorganic cellular cation, a cofactor for many enzymes
Iron	0.2	Iron salts	Component of cytochromes and some non-heme iron proteins and a cofactor for some enzymatic reactions

TABLE 4.4 Major Nutritional Types of Prokaryotes (after Todar, 2012)

Nutritional Type	Energy Source	Carbon Source	Examples
Photoautotrophs	Light	CO_2	Cyanobacteria, some purple and green Bacteria
Photoheterotrophs	Light	Organic compounds	Some purple and green bacteria
Chemoautotrophs or lithotrophs (lithoautotrophs)	Inorganic compounds: H_2, NH_3, NO_2, H_2S	CO_2	A few bacteria and many archaea
Chemoheterotrophs or heterotrophs	Organic compounds	Organic compounds	Most bacteria and some archaea

elements in bacterial nutrition. Bacteria require trace elements in such small amounts that they satisfy nutritional requirements by their presence as impurities in the water or media components. The trace elements act as cofactors for essential enzymatic reactions in the cell, even as metal ions. The common cations that qualify as trace elements in bacterial nutrition are Mn, Co, Zn, Cu and Mo.

Based on carbon and energy sources for growth, four major nutritional types of prokaryotes may be defined (see Table 4.4, after Todar, 2012).

Materials stimulating the growth (growth factors) are required in small amounts by cells because they fulfil specific roles in metabolism. The requirement for a growth factor results from either a blocked or missing metabolic pathway in the cells. Growth factors are classified into three categories.

i. purines and pyrimidines: required for the synthesis of nucleic acids (DNA and RNA);
ii. amino acids: needed for the synthesis of proteins; and
iii. vitamins: needed as coenzymes and functional groups of some enzymes.

Some bacteria, such as *E. coli*, do not need any growth factors: They can synthesize all essential purines, pyrimidines, amino acids and vitamins from their carbon source as part of their metabolism. Some other bacteria, such as *Lactobacillus*, require purines, pyrimidines, vitamins and several amino acids to grow. These compounds must be provided in advance to culture media that are used to produce these bacteria. The growth factors are not used directly as sources of carbon or sources of energy; instead, they are accumulated by cells to fulfil their specific role in metabolism (Todar, 2012).

Neveu et al. (2014) showed that cell–sediment separation methods can potentially enable the determination of the elemental composition of microbial communities by eliminating the sediment elemental contribution from bulk samples. They demonstrated that a separation method could be applied to determine the structure of prokaryotic cells. The technique uses chemical and physical means to extract cells from benthic sediments and mats. Recovery yields are between 5% and 40%, as determined from cell counts. The method conserves cellular element contents to within 30% or better, as concluded by comparing C, N, P, Mg, AI, Ca, Ti, Mn, Fe, Ni, Cu, Zn and Mo concentrations in *Escherichia coli*. Contamination by C, N and P from chemicals used during the procedure was negligible. Elements Na and K were not conserved, being likely exchanged through the cell membrane as cations during separation. The V, Cr and Co abundances could not be determined due to significant (>100%) measurement uncertainties. They applied this method to measure elemental contents in extremophilic communities of Yellowstone National Park hot springs. The technique was generally successful at separating cells from sediment, but failed to discriminate between cells and detrital biological or noncellular material of similar density. This procedure resulted in Al, Ti, Mn and Fe contamination, which can be tracked using proxies such as metal/Al ratios. In such a way, Neveu et al. (2014) were able to estimate the elemental abundances of a chemosynthetic community. The communities had C/N ratios typical of aquatic microorganisms and were low in P, and their metal abundances varied between hot springs by orders of magnitude.

Here, we find mentioning the work by Chopra et al. (2010) of interest, which illustrates that the relative concentrations of chemical elements in organisms such as humans and bacteria that represent two different domains of life (Eukarya and Bacteria) are similar (see Figure 4.8). However, we should mention that there is a significant fluctuation of elemental compounds of bacteria (indicated by the error bar on Fe only in Figure 4.7), as reported by Novoselov et al. (2013) and shown in Figure 4.8.

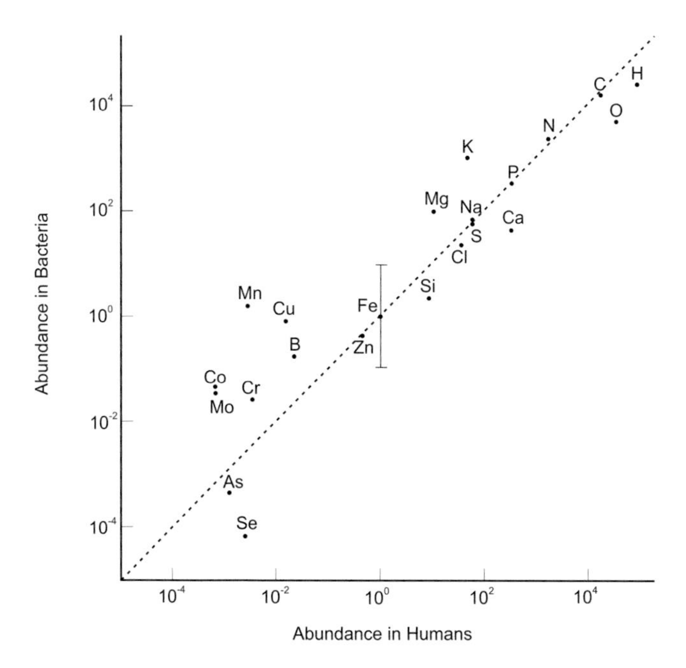

FIGURE 4.7 A positive correlation between elemental abundances (by the number of atoms) in the representatives of two domains of life: Bacteria (bacteria) and Eukarya (humans). The abundances are normalized to iron. (After Chopra et al., 2010.)

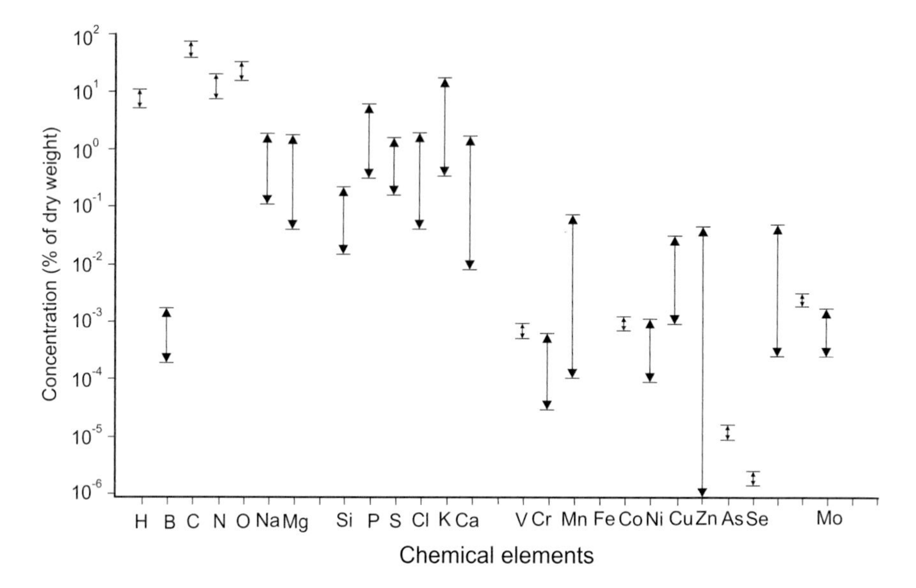

FIGURE 4.8 The elemental composition of contemporary bacteria (after Novoselov et al., 2013). Concentration (% of dry weight) range for life-essential elements is shown.

4.4.2 Domain Archaea

Archaea are believed to be the most primitive forms of life on Earth, as reflected by their name, which is derived from archaea meaning "ancient" (Woese, 1981; Brock et al., 1994). This type of bacteria is genealogically neither prokaryotes nor eukaryotes. This discovery meant there were not two lines of descent of life but three: the archaebacteria, the true bacteria and the eukaryotes.

Archaea is today recognized as the third domain of life. The present effort is directed toward understanding the molecular biology and biochemistry of Archaea and understanding the complete structure and the depth of the phylogenetic tree of life. DNA replication is one of the most critical events in living organisms, and DNA polymerase is the crucial enzyme in the molecular machinery that drives the process. DNA polymerases are classified into families A, B, C and X based on their amino acid sequences. These families are represented by *Escherichia Coli* DNA polymerase I (family A), DNA polymerase II (family B), DNA polymerase III α-subunit (family C) and others such as DNA polymerase β and terminal transferase (family X). Most of the biochemical properties of the DNA polymerases in the same family are similar. All archaeal DNA polymerases are assumed to belong to family B. All cloned products of pol genes showed amino acid sequence similarities, including three eukaryal DNA replicases and Escherichia coli DNA polymerase II. Recently, a new heterodimeric DNA polymerase from the hyperthermophilic archaeon has been found in *Pyrococcus furiosus*. The genes coding for the subunits of DNA polymerase are saved in the euryarchaeotes whose genomes have been completely sequenced. The characteristics of the novel DNA polymerase family suggest that its members play an essential role in DNA replication within euryarchaeal cells (Ishino and Cann, 1998).

Archaeans are inhabitants of some of the most extreme environments. Some of them live near rift vents in the deep sea at temperatures well over 100°C; others live in hot springs or extraordinarily alkaline or acid waters. They have been found living inside the digestive tracts of cows, termites and marine life, where they produce methane. In addition, they live in the anoxic muds of marshes and at the bottom of the ocean, and even in petroleum deposits deep underground.

Some archaeans can survive the desiccating effects of extremely saline waters. One salt-loving group of archaea is called *Halobacterium*. The light-sensitive pigment (bacteriorhodopsin) gives *Halobacterium* its purple color and provides it with the possibility to produce chemical energy. Bacteriorhodopsin pumps protons outside of the membrane. During flowback, these protons are used in the synthesis of ATP, the energy source of the cell. This compound is chemically similar to the light-detecting pigment rhodopsin found in the vertebrate retina (Woese and Fox, 1977).

Archaeans are the only organisms capable of living in extreme habitats such as thermal vents or hypersaline water. Recent research found that archaeans were also abundant in the plankton of the open sea. More needs to be learned about these microbes, but it is clear that the Archaea is a remarkably diverse and successful group of organisms.

For a description of some molecular features of Archaea, in comparison with Bacteria and Eukarya, see Table 4.5, modified from Zlatanova (1997). According to this author, if Archaea and Eukarya have a common evolutionary ancestor, it must have had some of

TABLE 4.5 Comparisons of Eubacteria, Archaea and Eukaryotes (following Zlatanova, 1997)

Characteristic	Bacteria	Archaea	Eukaryotes
Cell wall	Gram-positive or gram-negative structure	Murein absent	Plasma membrane
Multicellular (predominantly)	No	No	Yes
Nucleus, membrane-bound organelles	No	No	Yes
DNA	Circular	Circular	Linear
Ribosome	70S	70S	80S
Membrane: lipids ester-linked	Yes	No	Yes
Photosynthesis with chlorophyll	Yes	No	Yes
Growth above 80° C	Yes	Yes	No
Histone proteins present	No	Yes	Yes
Operons	Yes	Yes	No
Introns in genome	None	Some	Most
Transcription factors required	Yes	No	Yes
Methanogenesis	No	Yes	No
Nitrification	Yes	No	No
Denitrification	Yes	Yes	No
Nitrogen fixation	Yes	Yes	No
Chemolithotrophy	Yes	Yes	No
Gas vesicles present	Yes	Yes	No
Sensitive to chloramphenicol, kanamycin and streptomycin	Yes	No	No

the molecular features of the information processing system seen in a modified form in both domains.

Domain Archaea also includes the kingdom Archaebacteria. They are prokaryotes, meaning that they have no cell nucleus or any other membrane-bound organelles in their cells. Archaeal cells have some unique properties which separate them from the other two domains of life, Bacteria and Eukaryota. The Archaea are further divided into multiple phyla. Classification is difficult because the majority has not been studied in a laboratory; their nucleic acids were analyzed in samples collected in the environment.

Archaea and bacteria are somewhat similar in size and shape, although a few archaea have strange shapes, as those seen in *Haloquadratum walsbyi* (Stoeckenius, 1981). Despite this visual similarity to bacteria, archaea have genes and several metabolic pathways that are more closely related to those of eukaryotes. An example is the enzymes that are involved in transcription and translation. Also, other aspects of archaeal biochemistry are unique, such as the presence of ether lipids in their cell membranes. Archaea energy sources are different from the one used by eukaryotes. Their energy sources may be organic compounds, sugars, ammonia, metal ions or even hydrogen gas. Archaea, which is tolerant to salt (the *Haloarchaea*), use sunlight as an energy source. Other species of archaea fix carbon; however, no known species of archaea does both.

Archaea reproduce by binary fission, fragmentation or budding. However, unlike bacteria and eukaryotes, no known species forms spores.

Earlier, Archaea were viewed as extremophiles living in harsh environments, such as hot springs and salt lakes, but they have since been found in a broad range of habitats, including soils, oceans, marshlands, and the human colon, oral cavity and skin (Bang and Schmitz, 2015). Many Archaea live in the oceans. The archaea in plankton are probably one of the most abundant groups of organisms on the planet. Archaea are a significant part of the Earth's life, playing roles in both the carbon cycle and the nitrogen cycle. There are no examples of archaeal pathogens or parasites because they are often mutualists or commensals.

An example is the methanogens that inhabit human and ruminant guts, where their vast numbers help digestion. Methanogens are used for biogas production and in sewage treatment. Enzymes from extremophile archaea capable of enduring high temperatures and organic solvents are exploited in biotechnology.

The methanogenic bacteria form a large and diverse group united by three properties:

- They produce large quantities of methane as the primary product of their energy metabolism.
- They are strict anaerobes.
- They are members of the domain Archaea.

A unique model of energy metabolism characterizes the methanogenic bacteria; however, they are very diverse in their other properties.

Methanogenic bacteria obtain their energy for growth from the conversion of a limited number of substances to methane gas. The significant substrates are H_2+CO_2, formate and acetate. The list of substrates for the growth of methanogens may be divided into three groups.

First group: The energy substrate (electron donor) is H_2, formate or certain alcohols, and the electron acceptor is CO_2, which is reduced to methane.

Second group: The energy substrate is one of a variety of methyl-containing C-1 compounds, which can serve as substrates for a few taxa or methanogens. Usually, these compounds are disproportionate. Molecules of the substrate are oxidized to CO_2. The electron acceptors are the remaining methyl groups, which are reduced directly to methane.

Third group: Acetate is the primary source of methane, but the ability to catabolize this substrate is limited to only some species. Since acetate is present in many environments, methane synthesis proceeds by an acetoclastic reaction. In this reaction, the methyl carbon of acetate is reduced to methane, while the carboxyl carbon is oxidized to CO_2.

An additional distinctive feature of methanogens is their extreme sensitivity to oxygen. The methanogens are very strict anaerobes generally present in nature only in anoxic environments.

Yet another distinctive feature of methanogens is that they belong to archaebacteria (Jones et al., 1987; Woese, 1987). They are different from other archaebacteria because they are abundant in environments of moderate temperature, pH and salinity. Unlike other archaebacteria, methanogens contain large amounts of coenzymes essential for methane synthesis.

Scherer et al. (1983) determined the elemental composition of ten species using inductively coupled plasma emission spectrometry and by CHN analyzer. The ten species were representative of all three orders of the methanogens and were cultivated under defined conditions. Special emphasis was given to *Methanosarcina barkeri*, represented by five strains cultivated on various substrates. The following elements were determined: C, H, N, Na, K, S, P, Ca, Mg, Fe, Ni, Co, Mo, Zn, Cu and Mn. The range of concentration values for individual elements was as follows: (1) C (37%–44%), (2) H (5.5%–6.5%), (3) N (9.5%–12.8%), (4) Na (0.3%–4.0%), (5) K (0.13%–5.0%), (6) S (0.56%–1.2%), (7) P (0.5%–2.8%), (8) Ca (group I: 85–550 ppm; group II: 1,000–4,500 ppm), (9) Mg (0.09%–0.53%), (10) Fe (0.07%–0.28%), (11) Ni (65–180 ppm), (12) Co (10–120 ppm), (13) Mo (10–70 ppm), (14) Zn (50–630 ppm), (15) Cu (<10–160 ppm) and (16) Mn (<5–25 ppm). The highest variations were found for N and K, which both seem to have unknown important physiological functions, whether zinc and copper are essential trace elements for methanogens. All investigated species contained remarkably high zinc concentration values. In contrast, copper seemed to be present only in some species.

The primary transcription process of archaea is similar to that of eukaryotes. However, their transcriptional regulatory processes appear to be different. Although archaeal genomes encode a few homologs of eukaryote-like regulators, many potential bacterial-type transcriptional regulators have been identified (Aravind and Koonin, 1999). Control of the eukaryotic-like transcription machinery in archaea mainly proceeds via bacterial-like regulators. Several studies have revealed that most of these archaeal transcriptional regulators act as repressors of transcription (Bell, 2005) and provided evidence of positive control by these regulators (Brinkman et al., 2002; Ouhammouch et al., 2003). In addition to bacterial-like regulators, archaea appear to contain archaea-specific regulators (Gregor and Pfeifer, 2001; Hochheimer et al., 1999). A gene cluster responsive for copper resistance has been identified in multiple archaeal genomes (Ettema et al., 2006) by using a comparative genomics approach.

The prokaryotes live in nature in a broad range of physical conditions with variable O_2 concentration, hydrogen ion concentration (pH) and temperature. The exclusion limits for life on the planet, concerning environmental parameters, are always set by some microorganisms, most often a prokaryote and frequently an archaeon.

The UCSC Archaeal Genome Browser presents information on the biology of more than 100 microbial species from the domain Archaea. Basic gene annotation is derived from NCBI GenBank–RefSeq entries (see http://www.ncbi.nlm.nih.gov/genome/browse/). It contains overlays of sequence conservation across multiple species, nucleotide and protein motifs, non-coding RNA predictions, operon predictions, and more. Also, available gene expression data (microarray or high-throughput RNA sequencing) are displayed on http://archaea.ucsc.edu (see also Chan et al. (2012), Schneider et al. (2006) and Karolchik (2004)).

Archaeal organisms are currently recognized as very exciting and useful experimental materials. A significant challenge for molecular biologists studying the biology of Archaea is their DNA replication mechanism. Undoubtedly, a full understanding of DNA replication in Archaea requires the identification of all the proteins involved. In each of four wholly sequenced genomes, only one DNA polymerase was reported. This observation suggested that a single DNA polymerase might perform the task of replicating the genome and repairing the mutations, or these genomes contain other DNA polymerases that cannot be identified by amino acid sequence. Recently, a heterodimeric DNA polymerase has been discovered in the hyperthermophilic archaeon, *Pyrococcus furiosus*. The review by Cann and Ishino (1999) summarizes the 20-year-old knowledge about archaeal DNA polymerases and their relationship with accessory proteins, which were predicted from the genome sequences.

4.4.3 Domain Eukarya

The domain Eukarya starts with FECA, who has inherited eukaryotic-typical features from the LUCA and, by subsequent evolution, transferred them to the LECA. The domain Eukarya includes the kingdoms Protista, Fungi, Plantae and Animalia.

The kingdom Protista means "the very first". Protists are mostly unicellular, although some protists are multicellular (algae). They can be heterotrophic or autotrophic, and most live in water (though some live in moist soil or even the human body). All are eukaryotic (have a nucleus). A protist is any organism different from a plant, animal or fungus. Protists can be classified based on how they obtain nutrition and based on how they move. So, we have animal-like protists (protozoa), plantlike protists (algae) and fungus-like protists.

There are four phyla of animal-like protists classified based on the way they move:

i. Zooflagellates move by using one or two flagella, absorbing food across a membrane.

ii. Sarcodines (for example, ameba) move using pseudopodia, which are extensions filled with cytoplasm (false feet).
   i. Ciliates move by cilia. They have two nuclei: macronucleus and micronucleus. They gather food through the mouth pore into the gorge, which forms food vacuoles, and use anal pore for removing waste.
   ii. Sporozoans do not move on their own.

Despite their diversity, plantlike protists are often collectively referred to as algae. Most single-celled plantlike protists live near the surface of oceans, lakes and other bodies of water. They are a fundamental food source for other organisms sharing their habitat. Plantlike protists are classified into the following types:

i. *Red algae* are multicellular and have chlorophyll and red pigment that gives their color.
ii. *Brown algae* live in a colder climate, and they have chlorophyll and yellowish-brownish pigment.
iii. *Green algae* live in water and moist soil, but can be found in melting snow and inside other organisms. Some are single-cellular, and some are multicellular; they are green because chlorophyll is the main pigment.
iv. *Diatoms* are single-cellular and live in saltwater and freshwater and sometimes hold on to plants, shellfish, sea turtles and whales. They use photosynthesis.
v. *Dinoflagellates* are mainly single-cellular and live mostly in saltwater, but some are found in freshwater and snow. They have two flagella that make the protists spin. Most use photosynthesis, but some get food as consumers, decomposers or parasites; they are sometimes red and produce a potent poison.
vi. *Euglenoids* live mostly in freshwater, and they are single-cellular having plant and animal characteristics. They use photosynthesis, but when there is not enough light, they are consumers. Some do not use photosynthesis at all because they do not have chloroplast; instead, they eat other protists or take in nutrients.

Fungus-like protists share many features with fungi. They are heterotrophs; i.e., they must receive food outside themselves. They have cell walls and reproduce by forming spores. Fungus-like protists usually do not move, but a few develop movement at some point in their lives. Two major types of fungus-like protists are the following:

i. Slime molds usually measure about one or two centimeters; however, some slime molds are big up to several meters. They are of bright colors, such as vibrant yellow; others are brown or white.
ii. Water molds live mostly in water and moist soil. Often, they are parasites of plants and animals and get nutrients from these organisms and also from decaying organisms.

The kingdom Fungi includes yeast, molds, smuts, mushrooms and toadstools, distinct from the green plants. So far, approximately 70,000 species of fungi have been identified. Since they absorb nutrition from other organisms, they play the vital role of ecological decomposers. Since they have complex cells with a nucleus and organelles, the members of the kingdom Fungi are eukaryotes. Most are multicellular, except for single-celled yeast. Fungi are composed of filaments called hyphae. The hyphae group together form a conglomeration called mycelium. Fungi release enzymes that essentially digest the food to which they are attached (Maria de Lourdes and Rai, 2014). Once the organism is broken down by enzymes, the fungi can absorb the nutrients to live.

Fungi are usually classified into four divisions: the *Chytridiomycota* (chytrids), *Zygomycota* (bread molds), *Ascomycota* (yeasts and sac fungi) and *Basidiomycota* (club fungi). Divisions are characterized by the process of how the fungus reproduces sexually. The shape and internal structure of the sporangia, which produce the spores, are the most useful character for identifying these various significant groups; for more details, see https://ucmp.berkeley.edu/fungi/fungisy.html; Stephenson (2010); Petersen (2013).

The kingdom Plantae could be defined as multicellular, autotrophic eukaryotes, which conduct photosynthesis. All members of this family are made of a true nucleus and advanced membrane-bound organelles. The kingdom Plantae contains about 300,000 different species of plants. The kingdom Plantae is essential, as they are the source of food for all other living creatures present on planet Earth, which depends on plants to survive.

Various classification systems of plants have been used. The current system of classification is widely accepted. According to this, the kingdom Plantae has been divided into five major groups (see Raven, 1992; Abramoff and Thomson, 1995; Quattrocchi, 2000; Schmidt, 2006; Mabberley, 2008; Rose, 2009). They are as follows:

i. Thallophyta.
ii. Bryophyta.
iii. Pteridophyta.
iv. Gymnosperms.
v. Angiosperms.

Each group of plants has individual and unique features that belong exclusively to that group.

i. Thallophytes are the simplest of plants, and they are usually found in moist or wet places due to the absence of "true roots" and vascular tissue that is needed to transport water and minerals. They are autotrophic; most members of this group manufacture their food. Reserve food is generally starch: After photosynthesis, glucose is produced and consumed almost immediately; the remaining glucose is converted into complex compounds called starch. Thallophytes have a cell wall composed of cellulose around their cells. Unlike other plants, xylem and phloem are absent, and sex organs are simple, single-celled; there is no embryo formation after fertilization.
ii. Bryophytes are the second-largest taxonomic group in the plant kingdom. Bryophytes consist mainly of carbohydrates. Bryophytes mostly inhabit shady and humid environments. Bryophytes (Bryophyta) are the most basilar group among the land plants. They include small green plants of simple construction with unlignified conducting tissue and without roots. Bryophytes are an essential group for

the study of the early evolution of phenolic compounds (flavonoids, lignans and possibly lignin) in land plants. The earliest recognizable bryophytes in the macrofossil record include liverworts that appear to have their closest affinities with the Metzgeriales (Krassilov and Schuster, 1984; Frahm, 2001).

iii. Pteridophytes are a flowerless, seedless, spore-producing vascular plant that has successfully invaded the land. Pteridophytes represent an intermediate position between bryophytes and spermatophytes (gymnosperm and angiosperm). The pteridophytes are the first terrestrial plants to possess vascular tissues – xylem and phloem, and they are found in cold, damp, shady places. Some may also flourish well in sandy soil. The main plant body in pteridophytes is the sporophyte. It is differentiated into real root, stem and leaves, and these organs have well-differentiated vascular tissues. The leaves in Pteridophyta are small as in Selaginella or large macrophylls as in ferns. The sporophytes bear sporangia subtended by leaf-like appendages called sporophylls. Sometimes, sporophylls may form distinct compact structures called strobili or cones. The sporangia produce spores in spore mother cells by meiosis. The spores germinate and give rise to inconspicuous, small, free-living, thalloid gametophytes called prothallus.

iv. Gymnosperms are seed-bearing vascular plants with the ovules or seeds not enclosed in an ovary. Gymnosperm is named after the Greek word *gymnospermous*, meaning naked seeds. Gymnosperm seeds develop either on the surface of scale or leaf-like appendages of cones or at the end of short stems. Studies of their DNA have shown that the gymnosperms consist of four major, related groups (a–d): conifers, cycads, ginkgo and gnetophytes.

   a. Approximately 588 living species represent conifers, which is the most diverse and by far the most ecologically and economically important gymnosperm group. Most conifers are trees; they appeared in the fossil record about 290 million years ago and have been in widespread groups ever since then.
   b. Cycads are a group of slow-growing tropical and subtropical palm-like trees that have barely changed since before the time of the dinosaurs. Cycad fossils date back to the late Paleozoic Era, $290–265 \times 10^6$ years ago.
   c. Ginkgo's morphology has changed little over the 100 million years (Zhou and Zheng, 2003). There are some differences between the present-day ginkgo and Jurassic ones. The leaves of the Jurassic plants are divided into several lobes, similar to those of the chestnut tree. The new fossils show that after 50 million years, the lobes had joined up into the small fans of today's specimens. The way the seeds are formed has also changed. Today's ginkgo makes a few seeds, only one of which reaches full maturity, on a single stalk. Jurassic trees have sprays of trunks, each sporting a single seed. Ginkgo is herbal medicine's favorite tree.
   d. Gnetophytes are a group of plants divided into three different families, each having a single genus – *Gnetum*, *Ephedra* and *Welwitschia*.

      Gnetum has 28 species consisting of a few trees and shrubs accompanied by several woody vine species. The male and female reproductive structures resemble flowers and are grown on separate trees making the Gnetum plants dioecious.

v. Angiosperms are plants having a complex structure and a very well-developed vascular system and reproductive system. They are also known as "flowering plants" because flowers are an integral part of their reproductive structure. Angiosperms' use of flowers to reproduce made them more reproductively successful. While gymnosperms relied primarily on the wind to achieve sexual reproduction by transferring pollen – which contains the male reproductive cells for plants – into the ovaries of female plants, angiosperms used sweet- smelling, brightly colored flowers and sugary nectar to attract insects and other animals. Today, angiosperms contain about 80% of all plant species on the Earth.

The kingdom Animalia is also known as Metazoa. This kingdom does not contain prokaryotes. All the members of the kingdom Animalia are multicellular eukaryotes. They are heterotrophs, and they depend on other organisms directly or indirectly for their food. Most animals ingest food and digest it in an internal cavity. Most of them are motile, meaning they can move independently and spontaneously. Additional characteristics of the kingdom Animalia are as follows:

a. Most of the animals live in the seas; fewer are seen in freshwater; and even fewer are on land.
b. Around 9–10 million animal species inhabit the Earth.
c. There are 36 phyla recognized in the animals' kingdom. Each phylum shares particular properties structurally and functionally, which together separate it from other phyla.
d. The sizes of animals range from a few celled organisms such as the mesozoans to animals weighing many tons such as the blue whales.
e. Animal bodies are made of cells organized into tissues that perform specific functions. In most animals, the tissue is organized into complex organs, which form organ systems.
f. The animal cell contains the nucleus, mitochondria, ribosomes, endoplasmic reticulum, vacuoles, centrioles and cytoskeleton, called organelles.
g. Animals contain many organ systems that aid in performing specific functions that are necessary for the survival of the organism.
h. The list of organ systems includes the skeletal, muscular, digestive, respiratory, circulatory, excretory, reproductive, immune and endocrine systems.
i. Most of the animals are bilaterally symmetrical. Primitive animals are asymmetrical, while cnidarians and echinoderms are radially symmetrical.
j. Respiration is a process of exchange of gases between organisms and atmosphere, taking in oxygen and giving out carbon dioxide. This process is performed in organs of respiration, such as the lungs, gills, book gills and book lungs, and some animals' skin is also used for breathing.
k. The sensory mechanism and the coordination of the organ systems are carried out by the nervous system. In animals, the nervous system comprises nerve ganglions, or brain, spinal cords and nerves.
l. The distribution of nutrients, the exchange of gases and the removal of wastes take place by the circulatory system. This system is composed of the heart, blood vessels and blood.

m. The excretory system takes care of the removal of wastes from kidneys.
n. The skeletal system supports and protects the animal.
o. The majority of animals reproduce sexually by the fusion of cells with a single set of unpaired chromosomes, such as the eggs and the sperms.
p. Glands of the endocrine system help in the control and coordination of the body system.

4.4.3.1 Trace Elements in Human Nutrition and Health

The element content of the human body and its organs has been the subject of many studies. Here we make reference to the Report of the Task Group on Reference Man (Snyder et al., 1975). The elemental composition of the human body (reference man) is shown in Table 4.6 (after Helmenstine, 2019; see also Emsley, 1998). Life-essential elements are presented in bold. The ratio of iron to hydrogen concentration, [Fe/H] – called metallicity, a term often used in astrophysics, amounts to 6×10^{-4} for standard reference man. When element Zn is used as a measure of metallicity, [Zn/H], we obtain a value of 3.29×10^{-4}.

Conventionally, risk assessments have focused primarily on the effects of high doses of chemicals, which ultimately may induce toxicity. However, for several metals that are essential to life, harmful effects also occur at low levels of intake due to deficiency. This fact poses a challenge to basic assumptions of risk assessment, where the underlying paradigm aims at minimizing exposure as far as possible. For an essential element, an unbalanced concern about high-dose effects may lead to recommendations that lead to harm from deficiency. Therefore, the fact sheet (HERAG, 2007) was prepared by International Council on Mining & Minerals (ICMM) to guide a careful consideration

TABLE 4.6 Elements in the Human Body (After Helmenstine, 2019; see also Emsley, 1998)

Element	Mass (g)	Element	Mass (g)	Element	Mass (g)	Element	Mass (g)
O	43×10^3	Rb	0.68	**Cr**	14×10^{-3}	Te	0.7 mg
C	16×10^3	Sr	0.32	**Mn**	12×10^{-3}	Y	0.6 mg
H	7×10^3	Br	0.26	**As**	7×10^{-3}	Bi	0.5 mg
N	1.8×10^3	Pb	0.12	**Li**	7×10^{-3}	Tl	0.5 mg
Ca	1.0×10^3	**Cu**	72×10^{-3}	Cs	6×10^{-3}	In	0.4 mg
P	780 g	Al	60×10^{-3}	Hg	6×10^{-3}	Au	0.2 mg
K	140 g	Cd	50×10^{-3}	Ge	5×10^{-3}	Sc	0.2 mg
S	140 g	Ce	40×10^{-3}	**Mo**	5×10^{-3}	Ta	0.2 mg
Na	100 g	Ba	22×10^{-3}	**Co**	3×10^{-3}	**V**	0.11 mg
Cl	95 g	**I**	20×10^{-3}	Sb	2×10^{-3}	Th	0.1 mg
Mg	19 g	Sn	20×10^{-3}	Ag	2×10^{-3}	U	0.1 mg
Fe	4.2 g	Ti	20×10^{-3}	Nb	1.5×10^{-3}	Sm	50 μg
F	2.6 g	**B**	18×10^{-3}	Zr	1×10^{-3}	Be	36 μg
Zn	2.3 g	**Ni**	15×10^{-3}	La	0.8×10^{-3}	**W**	20 μg
Si	1.0 g	**Se**	15×10^{-3}	Ga	0.7×10^{-3}		

Note: Bold – life-essential elements.

of both nutritional essentiality and high-dose toxicity in the overall risk assessment and risk characterization procedures for essential metals and their compounds.

The accepted criteria for essentiality for human health state that the deficiency of the element in the diet results in either functional or structural abnormalities and that the defects are related to the specific biochemical processes that can be reversed by the intake of the essential metal (WHO, 1996). By definition, the roles fulfilled by an essential metal cannot be replaced by any other substance. Essential trace element levels are subject to homeostatic control mechanisms that may include regulation of absorption and excretion and tissue retention. These mechanisms enable adaptation to variations in nutrient intakes to ensure a safe and optimum systemic supply of essential trace elements for the performance of essential functions.

In the monograph "Principles and methods for the assessment of risk from essential trace elements" (WHO, 2002), the World Health Organization considers the following trace elements to be essential for human health: Cu, Zn, Fe, Cr, Mo, Se, Co and I. The second group of elements is named by the WHO as "probably essential for humans": Si, Mn, Ni, B and V.

The WHO concept for the evaluation of essentiality (see Figure 4.9 after WHO, 2002) is founded on the definition of boundaries between deficient and excess oral intakes of essential trace elements, finally resulting in an "acceptable range of oral intake". The acceptable range of oral intake is supposed to limit the probability of deficient and excess intakes occurring in healthy populations and is defined for different age and sex groups and physiological states such as pregnancy and lactation. As essential trace element intakes drop below A (lower limit of the acceptable range of oral intake where 2.5% of the population under consideration will be at risk of deficiency), an increasing proportion will be at risk of deficiency. At extremely low intakes, all subjects will manifest deficiency. As essential trace element intakes exceed B (where 2.5% of the population under

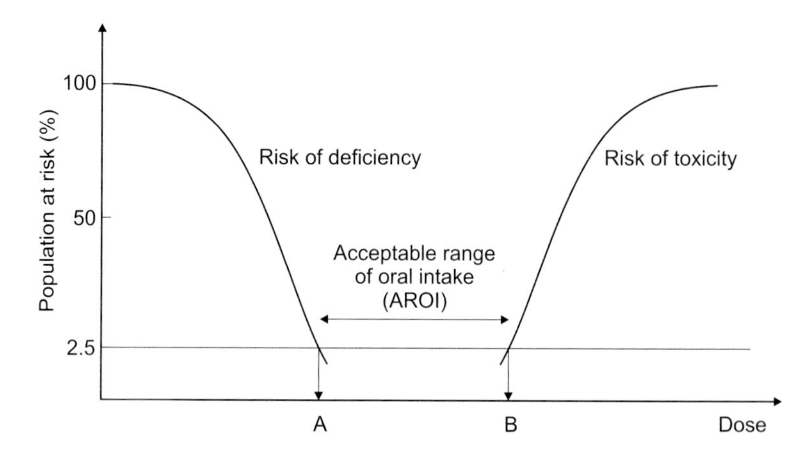

FIGURE 4.9 WHO concept on deficiency and toxicity from deficient and excess oral intakes of essential trace elements. (Adapted from WHO, 2002.)

consideration will be at risk of toxicity), a progressively more significant proportion of communities will be at risk of toxicity.

There are several reports described in the literature on the problem of assessment of the trace element status of individual humans and populations (Delves, 1985; Hambidge, 1988; Gibson, 1989; Angelova et al., 2014; and many others). These studies are usually done for a single element or a group of elements; for example, Hotz et al. (2003) studied zinc and copper.

The intake of essential trace elements outside the Adequate Range of Oral Intake (AROI) is related to an increased risk of different diseases, including cancers (Silvera and Rohan, 2007).

4.5 Chirality Phenomenon

Together with the problem of trace element essentiality, we should consider a closely related problem of chirality. Laboratory synthesis made from optically inactive starting materials yields (1:1) mixtures of L- and D-isomers. However, only L-amino acids are found in proteins in living organisms. The exception is glycine, which has two (indistinguishable) hydrogen atoms attached to its alpha carbon. L-amino acids are used exclusively for protein synthesis by all life on our planet. The function of a protein is determined by its shape. A protein with a D-amino acid will have its R group sticking out in the wrong direction (for details, see, for example, LibreTexts, 2020). Figures 4.10 and 4.11 show examples of D- and L-isomers amino acids and sugars, respectively.

Here is the obvious question: Could there have been some asymmetric factors in the environment that influenced the choice when life began? There have been several

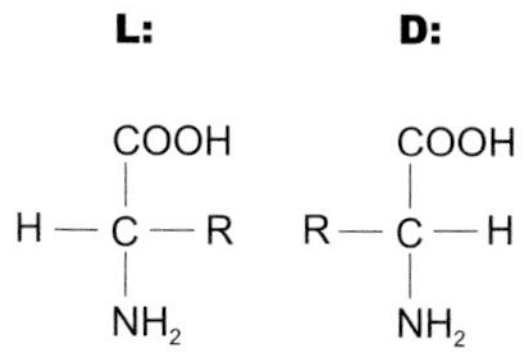

L: COOH, H—C—R, NH_2

D: COOH, R—C—H, NH_2

FIGURE 4.10 Two enantiomers (left-handed – L and right-handed – D) of generic chiral amino acid.

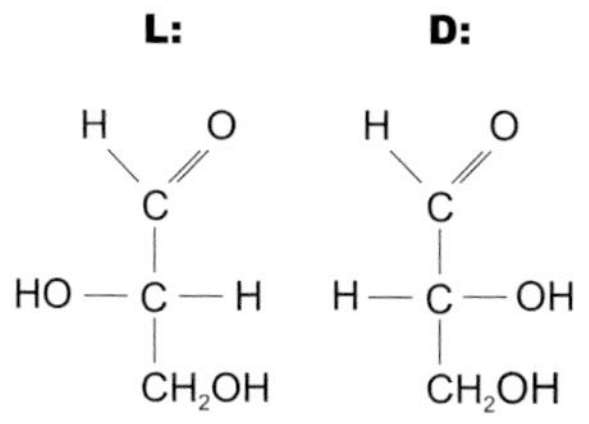

L: H, O, C, HO—C—H, CH_2OH

D: H, O, C, H—C—OH, CH_2OH

FIGURE 4.11 Glyceraldehyde has one chiral center and therefore exists as two different enantiomers with opposite optical rotation.

suggestions, including (i) polarized light, (ii) optically active quartz and (iii) natural radioactivity. The last scheme assumed the origin of life in a radioactive environment. The result of e-radioactivity is longitudinally polarized β-rays, which produce circularly polarized bremsstrahlung. This radiation was considered to have preferentially destroyed D-isomers. Some recent experiments have not supported this assumption. Assuming that amino acids could be synthesized on dust particles in interstellar space, the observed optical activity may be a result of cosmic-ray bombardment. High-energy polarized protons in cosmic rays may be able to preferentially destroy one isomer because of significant asymmetry in proton (in cosmic rays)–proton (in amino acid) scattering.

The process of demarcation of the origins of the chemistry of life starts with the attention to the molecules that might have existed on the Earth. The process extends to the possibility of potential mechanisms for the assembly of these molecules into polymers capable of self-replication and transmittance of genetic information. At some point along this pathway, the single chirality of sugars and amino acids emerges as their hallmark in biological molecules. Researchers have developed abstract mathematical theses for the origin of homochirality from a presumably racemic collection of prebiotic molecules. Before the end of the 20th century, experimental findings supported several basic features of these scenarios, although these studies considered chemical systems without direct prebiotic relevance. Currently, researchers are examining plausible conditions that couple chemical and physical processes leading to a one-chirality of sugars and amino acids followed by chemical reactions enhancing molecular complexity. While these studies have been conducted in the frame of the "RNA world" scenario, the experimental findings support more a "metabolism first" scenario of the origin of life. Hein and Blackmond (2012) incorporated both chemical and physical phenomena that allow for the amplification of a small initial imbalance of either sugar by amino acids or amino acid by sugars. They suggested that an enantio-enriched chiral pool of one type of molecule could lead to a similarly enantio-enriched pool of the other.

Interstellar dust appears to play a critical role in the formation of interstellar molecules. Molecules may be formed on or in grain surfaces. Many molecules have been observed in interstellar space. Johnson (1972) reported the observation of interstellar porphyrins (molecule $MgC_{46}H_{30}N_6$). The existence of interstellar molecules suggests the following:

1. Such molecules support or are the metabolic products of an interstellar biota.
2. Such molecules participating in planetary condensation from the interstellar medium can make a significant contribution to the origin of terrestrial life.

As early as 1984, Joyce et al. suggested that for life to emerge, something first had to crack the symmetry between left-handed and right-handed molecules, an event biochemists call "breaking the mirror". Since then, the search for the origin of life's handedness in the prebiotic worlds has mainly been focused on physics and chemistry, not biology. Joyce et al. (1984) showed that the polymerization of activated mononucleotides proceeded readily in a chiral system. However, it is severely inhibited by the presence of the opposing enantiomer. This finding poses a challenge for the spontaneous emergence of RNA-based life. It has been suggested that (i) RNA was preceded by some other genetic polymer that is not subject to chiral inhibition, or (ii) chiral symmetry was broken through chemical processes before the origin of RNA-based life.

Once an RNA enzyme arose that could catalyze the polymerization of RNA, it would be possible to distinguish between the two enantiomers, enabling the process of RNA replication and the start of RNA-based evolution. It is assumed that the earliest RNA polymerase and its substrates have been of the same handedness, but this might not be the case. Replication of D- and L-RNA molecules may have emerged together, assuming the ability of structured RNAs of one handedness to catalyze the templated polymerization of activated mononucleotides of the opposite handedness.

Sczepanski and Joyce (2014) developed such a cross-chiral RNA polymerase ribozyme, using in vitro evolution, starting from a population of random-sequence RNAs. The D-RNA enzyme, consisting of 83 nucleotides, catalyzes the joining of L-mono- or oligonucleotide substrates on a complementary L-RNA template, and similar behavior occurs for the L-enzyme with D-substrates and a D-template. Chirality inhibition is avoided because the 10^6-fold rate acceleration of the enzyme only pertains to cross-chiral substrates. The enzyme's activity can generate full-length copies of its enantiomer through the templated joining of 11 component oligonucleotides.

Recently, Yin et al. (2015) have found that any molecule, if large enough (several nanometers) and with an electrical charge, will seek their type with which to form a large assembly. The authors have shown that homochirality, or how molecules select other like molecules to create larger assemblies, may not be as mysterious as previously imagined. The remaining problem is understanding how homochirality occurred at the onset of life. If chirality emerged sometime after the origin of life, the question remains: Why did right-handed RNA win? Left- and right-handed molecules have chemically identical properties, so there is no apparent reason for one to triumph.

Yin et al. (2015) showed that chiral macro-anions demonstrate chirality recognition behavior by forming a homogeneous blackberry structure via long-range electrostatic interactions between the individual enantiomers in their racemic solutions. Adding chiral co-anions suppresses the self-assembly of one enantiomer while maintaining the assembly of the other one. This process leads to a natural chirality selection and chirality amplification process, indicating that some environmental preferences can lead to a complete chirality selection. The fact that the relatively simple inorganic macro-ions exhibit chirality recognition and selection during their assembly process indicates that the related features of bio-macromolecules might be due to their macro-ionic nature via long-range electrostatic interactions.

The existence of the single chirality of biological molecules, almost exclusively left-handed amino acids and right-handed sugars, presents us with two questions (Blackmond, 2010):

i. What served as the original template for the biased production of one enantiomer over the other in the chemically austere and racemic environment of the prebiotic world?
ii. How was this sustained and propagated to result in the biological world of single chirality that surrounds us?

A review by Blackmond (2010) focuses primarily on the second question: the plausible mechanisms for the evolution of molecular chirality as exemplified by the D-sugars and L-amino acids found in the living organisms today. "Symmetry breaking" is the term used for describing the occurrence of a difference between left and right enantiomeric molecules. This imbalance is measured in terms of the enantiomeric excess, or ee, where ee=(R–L)/(R+L). R and L stand for concentrations of the right- and left-handed molecules, respectively. Speculations for how an imbalance might have come about could be grouped as either terrestrial or extraterrestrial and further subdivided into either random or deterministic. The evidence of L-enantiomeric excess in amino acids found in chondritic meteor deposits (Pizzarello, 2006) allows the hypothesis that the initial imbalance is older than our world.

4.5.1 Origin of Chirality

The observed chirality might be a consequence of the preferential destruction of one enantiomer. This process may be a consequence of cosmic-ray bombardment and caused by asymmetry in proton–proton scattering. Protons in the cosmic rays get scattered of different targets on their path through the Universe and along with the magnetic field force including individual atoms in the molecules in molecular clouds in particular one on the dust particle surfaces. Among other processes, the process of interest is elastic proton–proton scattering, as shown in Figure 4.12.; either proton (beam or target) can be polarized; see Valkovic and Obhodas (2020).

Analyzing power $A_N(t)$ for p–p elastic scattering in the energy range of energetic cosmic-ray protons has been studied by several groups; see Akchurin et al. (1993), Bravar et al. (2005) and Okada et al. (2008). Some of this work is summarized by Bazilevsky et al. (2011), and the results are presented in Figure 4.13:

The analyzing power (A_N) can be extracted from the asymmetry between the number of scatterings on the left versus on the right corrected for the left and right detector acceptances or from the asymmetry between the number of scattering (e.g., on the left) for target polarization up and target polarization down, corrected for the integrated luminosities for the corresponding target spin states. These two approaches can be combined in the so-called sqrt formula, which cancels the contributions from different

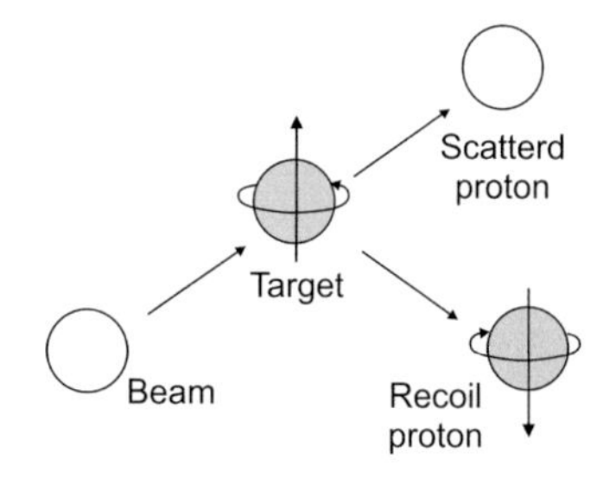

FIGURE 4.12 The elastic scattering process. A_N arises from the interference between a spin-flip (Coulomb) and spin-non-flip (nuclear) amplitude.

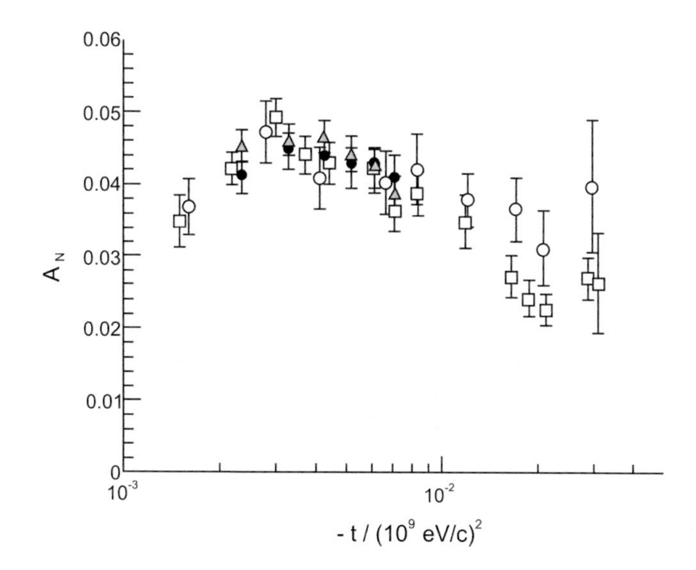

FIGURE 4.13 Analyzing power (A_N) for elastic p–p scattering: 24 and 100 GeV ($s^{1/2}$=6.8 and 13.7 GeV), 31 and 250 GeV ($s^{1/2}$=7.7 and 21.7 GeV); all data are from different Relativistic Heavy Ion Collider (RHIC) runs. The uncertainties shown are the quadratic sum of the statistical and systematic uncertainties. (After Bazilevsky et al., 2011.)

left–right detector acceptances and different luminosities in the measurements with up and down target polarization states to the asymmetry:

$$A_N = \frac{1}{P_T} \frac{\sqrt{N_L^{\uparrow} \cdot N_R^{\downarrow}} - \sqrt{N_R^{\uparrow} \cdot N_L^{\downarrow}}}{\sqrt{N_L^{\uparrow} \cdot N_R^{\downarrow}} + \sqrt{N_R^{\uparrow} \cdot N_L^{\downarrow}}},$$

where

$N_{L\,(R)}^{\uparrow\,(\downarrow)}$ is the number of recoil protons selected from p–p elastic scattering events detected on the left (right) side of the beam, P_T is the target polarization, and the arrows give the direction of the target polarization.

The A_N measurements were performed for the recoil proton kinetic energy range T_R=1–4 MeV corresponding to a momentum transfer $0.002 < -t < 0.008$ $(\text{GeV/c})^2$ ($-t = 2m_p T_R$, where m_p is the proton mass).

For DNA today, radiation increases the frequency of gene mutations; this has been known since the pioneering work of Muller (1927) that showed that the mutation rate is proportional to the radiation dose, much of it attributable to ionization by cosmic rays. Ionization by spin-polarized radiation could be enantio-selective (Zel'dovich et al., 1977). Therefore, one can argue that the mutation rate of L- and D-organisms would be different; see Globus and Blandford (2020). As there could be 10^9 or even 10^{12} generations of the earliest and simplest life forms, a small difference in the mutation rate could easily sustain one of the two early chiral choices.

On the other hand, laboratory experiments have demonstrated that it is possible to induce an enantiomeric excess of amino acids by irradiation of interstellar ice analogs with UV CPL (de Marcellus et al., 2011). However, this raises two problems. First, circular dichroism is also wavelength-, pH- and molecule-specific (D'hendecourt et al., 2019). It is hard to see how one sense of circular polarization can enforce a consistent chiral bias, given the large range of environments in which the molecules are found. Second, it is often supposed that astronomical sources supply the polarization. If we seek a universal chiral light source that consistently emits one polarization over another, then we are again drawn to the weak interaction in order to account for a universal asymmetry. One option is to invoke spin-polarized particles, which can radiate one sense of circular polarization through Cherenkov radiation or bremsstrahlung and can preferentially photolyze chiral molecules of one handedness. Another option is to invoke supernova neutrinos (Boyd et al., 2018). However, the small chiral bias is unlikely to lead to a homochiral state and some prebiotic amplification mechanism is still required. This suggests enantio-selective bias in the evolution of the two living systems. Although the two forms of a chiral molecule have identical physical and chemical properties, the interaction with other chiral molecules may be different. Chiral molecules in living organisms exist almost exclusively as single enantiomers, a property that has a critical role in molecular recognition and replication processes. Therefore, it is a prerequisite for the origin of life. Left-handed and right-handed molecules of a compound will be formed in equal amounts (a racemic mixture) when synthesized in the laboratory in the absence of some directing template.

Assuming that amino acids, and some sugars, could be synthesized on dust particles in interstellar space, the observed optical activity may be a result of cosmic-ray bombardment. Two processes could be involved:

i. The first process might arise from asymmetric dust grains aligned with the magnetic field. Interstellar medium (ISM) polarization by aligned dust grains has been probed from the diffuse medium in the UV to heavily obscured sightlines in the near-IR and into dense clouds and heavily embedded sources using far-IR emission (for details, see Andersson et al., 2015). It should be kept in mind that not all environments produce the same amount of polarization. Although the dust mainly consists of silicates, amorphous carbon and small graphite particles in various simple shapes, in the dark, cold parts of molecular clouds, grain aggregates and mantles of volatile ice might form. Only a limited number of relatively large grain sizes (~0.01–1 μm) contribute to the polarization. This situation corresponds to a polarized target in the p–p elastic scattering process.
ii. Alternatively, polarized beam interaction could be considered. High-energy polarized protons in cosmic rays may be able to preferentially destroy one isomer on the dust particle surface because of significant asymmetry in proton (in cosmic rays)–proton (chirality center in amino acid or sugar) scattering. See Figure 4.14 for the schematic presentation of the process leading to preferential destruction of D-amino acids and L-sugars.

Single spin asymmetry in polarized p–p elastic scattering has been experimentally studied at high energies; see, for example, Adamczyk et al. (2013) for experiments done at Relativistic Heavy Ion Collider (RHIC). RHIC is one of the only two operating

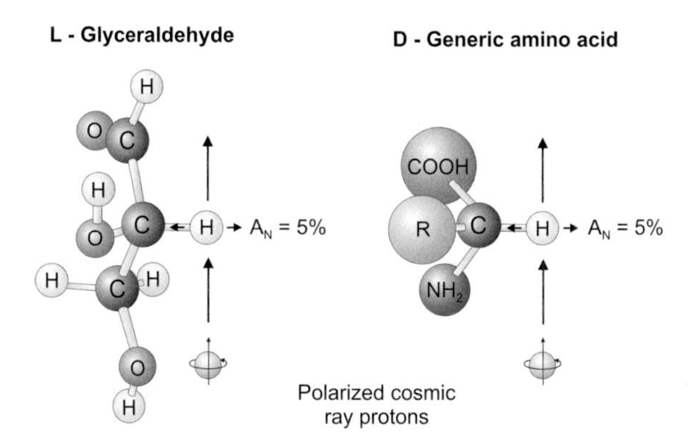

FIGURE 4.14 Schematic representation of the process of preferential destruction of D-amino acids and L-sugars.

heavy-ion colliders and the only spin-polarized proton collider ever built. It is located at Brookhaven National Laboratory (BNL) in Upton, New York. Polarization produced in p–p scattering has been observed even at low energies, 30 and 50 MeV (Batty et al., 1963).

If a sufficient enantiomeric excess existed in life's primordial molecular reservoir, it would almost certainly push life toward extreme bias of actual living beings (McGuire and Carroll, 2016). Meteoritic samples suggest that such an enantiomeric excess has been generated before Earth formation. The authors McGuire and Carroll (2016) presented a case of a meteorite that struck just outside of the city of Murchison, Victoria, Australia, on September 28, 1969. Its weight was ~100 kg, and it belongs to the class of carbonaceous chondrite and rich in organic compounds, including amino acids, many of them having an excess of left-handed enantiomers > 10%. It has been determined that meteorites, like the one that hit Murchison, date back to about the time of our solar system's formation. Also, the source of their molecular material can be traced to the cloud of gas and dust from which our solar system formed. There is a possibility of existence of a reasonable mechanism for generating an enantiomeric excess in that primordial cloud. In that case, the origins of life's enantiomeric bias could be linked to processes that occurred billions of years ago, before the solar system existed (McGuire and Carroll, 2016); see also Engel and Macko (1997) and McGuire et al. (2016).

Observation of CP formation in the light passage through the dust cloud would be direct evidence of the existence of homochiral molecules within the dust cloud (Gronstal, 2014). Since life on planet Earth is chiral, it is reasonable to assume that dust that formed our solar system contained already this information. At this point, we can put forward our fourth hypothesis: *Chirality is a sine qua non condition for the emergence of life.*

References

Abbas, S. H. and Schulze-Makuch, D. 2007. Synthesis of biologically important precursors on Titan. *Journal of Scientific Exploration* 21:673–687.

Abramoff, P. and Thomson, R. G. 1995, Jan. 1. *Kingdom Plantae: The Vascular Plants: Separate from Laboratory Outlines in Biology VI*, 6th Edition. W. H. Freeman, New York, NY, USA.

Adamczyk, L., Agakishiev, G., Aggarwal, M. M., et al. 2013. Single spin asymmetry AN in polarized proton-proton elastic scattering at $\sqrt{s}$=200 GeV. *Physics Letters B* 719:62–69.

Akchurin, N., Langland, J., Onel, Y., et al. 1993. Analyzing power measurement of pp elastic scattering in the Coulomb-nuclear interference region with the 200-GeV/c polarized-proton beam at Fermilab. *Physical Review D* 48:3026.

Ahmad, S. and Jensen, R. A. 1988. New prospects for deducing the evolutionary history of metabolic pathways in prokaryotes: Aromatic biosynthesis as a case-in-point. *Origins Life Evol Biosphere* 18:41–57.

Akter, K. F., Owens, G., Davey, D. E. and Naidu, R. 2005. Arsenic speciation and toxicity in biological systems. *Reviews of Environmental Contamination and Toxicology* 184:97–149.

Anbar, A. D. and Knoll, A. H. 2002. Proterozoic ocean chemistry and evolution: A bioinorganic bridge?. *Science* 297(5584):1137–1142.

Anbar, A. D. and Rouxel, O. 2007. Metal stable isotopes in paleoceanography. *Annual Review of Earth and Planetary Sciences* 35:717–746.

Andersson, B.-G., Lazarian, A. and Vaillancourt J. E. 2015. Interstellar dust grain alignment. *Annual Review of Astronomy and Astrophysics* 53:501–540.

Andreini, C., Banci, L., Bertini, I. and Rosato, A. 2008. Occurrence of copper proteins through the three domains of life: A bioinformatic approach. *Journal of Proteome Research* 7:209–216.

Andreini, C., Bertini, I. and Rosato, A. 2004. A hint to search for metalloproteins in gene banks. *Bioinformatics* 20:1373–1380.

Angelova, M. G., Petkova-Marinova, T. V., Pogorielov, M. V., Loboda, A. N., Nedkova-Kolarova, V. N. and Bozhinova, A. N. 2014. Trace element status (Iron, Zinc, Copper, Chromium, Cobalt, and Nickel) in iron-deficiency anaemia of children under 3 years. *Anemia*. Article ID 718089, (8 p). Doi: 10.1155/2014/718089.

Aral, H. and Vecchio-Sadus, A. 2008. Toxicity of lithium to humans and the environment - a literature review. *Ecotoxicology and Environmental Safety* 70:349–356.

Aravind, L. and Koonin, E. V. 1999. DNA-binding proteins and evolution of transcription regulation in the archaea. *Nucleic Acids Research* 27:4658–4670.

Archer, C. and Vance, D. 2006. Coupled Fe and S isotope evidence for Archean microbial Fe(III) and sulfate reduction. *Geology* 34:153–156.

Balamurugan, K. and Schaffner, W. 2006. Copper homeostasis in eukaryotes: teetering on a tightrope. *Biochimica et Biophysica Acta* 1763: 737–746.

Banerjee, R. and Ragsdale, S. W. 2003. The many faces of vitamin B12: Catalysis by cobalamin-dependent enzymes. *Annual Review of Biochemistry* 72:209–247.

Bánfalvi, G. 2011. Cellular effects of heavy metals. In *Heavy Metals, Trace Elements and Their Cellular Effects*, Springer, pp. 3–28. http://link.springer.com/chapter/10.1007%2F978-94-007-0428-2_1.

Bang, C. and Schmitz, R. A. 2015. Archaea associated with human surfaces: Not to be underestimated. *FEMS Microbiology Reviews* 39:631–648.

Barrinton, E. J. W. 1962. Hormones and vertebrate evolution. *Experientia* XVIII:201–209.

Batty, C. J., Gilmore, R. S. and Stafford, G. H. 1963. Measurement of the polarization in proton-proton scattering at 30 and 50 MeV. *Nuclear Physics* 45:481–491.

Bazilevsky, A., Alekseev, I., Aschenauer, E., et al. 2011. Measurements of the energy dependence of the analyzing power in pp elastic scattering in the CNI region. *Journal of Physics: Conference Series* 295:012096.

Beard, B. L., Johnson, C. M., Cox, L., Sun, H., Nealson, K. H. and Aguilar, C. 1999. Iron isotope biosignatures. *Science* 285:1889–1891.

Bell, S. D. 2005. Archaeal transcriptional regulation – variation on a bacterial theme? *Trends in Microbiology* 13:262–265.

Belyavskaya, N. A. 2004. Biological effects due to weak magnetic field on plants. *Advances in Space Research* 34:1566–1574.

Bennett, A., Rowe, R. I., Soch, N. and Eckhert, C. D. 1999. Boron stimulates yeast (*Saccharomyces cerevisiae*) growth. *Journal of Nutrition* 129:2236–2238.

Berg, J. M., Tymoczko, J. L. and Stryer, L. 2002. *Biochemistry*, 5th edition. W H Freeman, New York. ISBN-10: 0-7167-3051-0.

Blackmond, D. G. 2010. *The Origin of Biological Homochirality. Cold Spring Harbour Perspectives in Biology* Deamer, D and Szostak, J. W. (Eds.) Cold Spring Harbor Laboratory Press, Cold Spring Harbor, NY, USA, Vol. 2, pp. a002147.

Blaudez, D., Jacob, C., Turnau, K., et al. 2000. Differential responses of ectomycorrhizal fungi to heavy metals in vitro. *Mycological Research* 104(11):1366–1371.

Böck A., Forchhammer K., Heider J., Leinfelder W., Sawers G., Veprek B. and Zinoni F. 1991. Selenocysteine: the 21st amino acid. *Molecular Microbiology* 5:515–520.

Bolaños, L., Redondo-Nieto, M., Bonilla, I. and Wall, L. G. 2002. Boron requirement in the *Discaria trinervis* (*Rhamnaceae*) and *Frankia* symbiotic relationship. Its essentiality for Frankia BCU110501 growth and nitrogen fixation. *Physiologia Plantarum* 115:563–570.

Bonch-Osmolovskaya, E. A., Miroshnichenko, M. L., Kostrikina, N. A. Chernych, N. A. and Zavarzin, G. A. 1990, Nov. *Thermoproteus uzoniensis* sp. nov., a new extremely thermophilic archaebacterium from Kamchatka continental hot springs. *Archives of Microbiology* 154(6):556–559.

Bonilla, I., Blevins, D. and Bolaños, L. 2009. Boron functions in plants: Looking beyond the cell wall. Chapter 5, Essay 5.1 In Taiz, L., Zeiger, I., Møller, I. M. and Murphy, A. (Eds.) *Plant physiology and Development*, 6th Edition. Sinauer Assoc. Inc, 2015. An imprint of Oxford University Press, Oxford, UK.

Bonilla, I., Garcia-Gonzalez, M. and Mateo, P. 1990. Boron requirement in Cyanobacteria. Its possible role in the early evolution of photosynthetic organisms. *Plant Physiology* 94:1554–1560.

Bowen, H. J. M. 1966. *Trace Elements in Biochemistry.* Academic Press, New York.

Bowen, H. J. M. 1979. *Environmental Chemistry of the Elements.* Academic Press, London.

Boyd, R. N., Famiano, M., Onaka, T. and Kajino, T. 2018. Sites that can produce left-handed amino acids in the supernova neutrino amino acid processing model. *The Astrophysical Journal* 856(1): article id. 26, (5 p).

Bravar, A., Alekseev, I., Bunce, G., et al. 2005. *Spin Dependence in Elastic Scattering in the CNI Region.* Brookhaven National Laboratory. Report BNL-75128-2005-CP.

Brinkman, A. B., Bell, S. D., Lebbink, R. J., de Vos, W. M. and van der Oost, J. 2002. The Sulfolobus solfataricus Lrp-like protein LysM regulates lysine biosynthesis in response to lysine availability. *Journal of Biological Chemistry* 277:29537–29549.

Brock, T. D., Madigan, M. T., Martinko, I. M. and Parker, J. 1994. *Microbiology*. Prentice-Hall, Inc., Englewood Cliffs, NJ.

Bruland, K. W. 1989. Complexation of zinc by natural organic ligands in the central North Pacific. *Limnology & Oceanography* 34:269–285.

Budisa, N., Kubyshkin, V. and Schulze-Makuch, D. 2014. Fluorine-rich planetary environments as possible habitats for life. *Life (Basel)* 4(3):374–385.

Burne, R. A. and Y. Y. Chen. 2000. Bacterial ureases in infectious diseases. *Microbes and Infection*. 2:533–542.

Caetano-Anollés, G. and Caetano-Anollés, D. 2003. An evolutionarily structured Universe of protein architecture. *Genome Research* 13:1563–1571.

Cameron, V., House, C. H. and Brantley, S. L. 2012. A First Analysis of Metallome Biosignatures of *Hyperthermophilic Archaea*. *Archaea* Article 789278, (12 p). Doi: 10.1155/2012/789278.

Cammack, R. 1988. Nickel in metalloproteins. *Advanced Organic Chemistry* 32:297–333.

Canfield, D. E. 1998. A new model for Proterozoic ocean chemistry. *Nature* 396:450–453.

Cann, I. K. and Ishino, Y. 1999. Archaeal DNA replication: Identifying the pieces to solve a puzzle. *Genetics* 152(4):1249–1267.

Carney, G. E. and Bowen, N. J. 2004. p24 proteins, intracellular trafficking, and behavior: *Drosophila melanogaster* provides insights and opportunities. *Biology of the Cell* 96:271–278.

Chan, P.P., Holmes, A.D, Smith, A.M., Tran, D. and Lowe, T.M. 2012. The UCSC Archaeal genome browser: 2012 update. *Nucleic Acids Research* 40 (Database Issue):D646–D652.

Chen, T. S. S., Chang, C.-J. and Floss, H. G. 1981. On the biosynthesis of boromycin. *The Journal of Organic Chemistry* 46:2661–2665.

Chopra, A. and Lineweaver, C. H. 2009. The major elemental abundance differences between life, the oceans and the sun. *Australian Space Science Conference Series: 8th Conference Proceedings (DVD)*. National Space Society of Australia Ltd. ISBN 13:978-0-9775740-2-5. (www.tinyurl.com/ACCL08).

Chopra, A., Lineweaver, C. H., Brocks, J. J. and Ireland, T. R. 2010. Palaeoecophylostoichiometrics: Searching for the elemental composition of the last universal common ancestor. *Australian Space Science Conference Series: 9th Conference Proceedings*, Sidney, 28–30 September 2009 (DV). National Space Society of Australia Ltd, ISBN 13: 978-0-9775740. (www.tinyurl.com/ACetal10).

Claus, H. and Decker, H. 2006. Bacterial tyrosinases. *Systematic and Applied Microbiology* 29:3–14.

Cloud, P. E. 1972. A working model of the primitive Earth. *American Journal of Science* 272:537–548.

Cockell, C. 2015. *Astrobiology – Understanding Life in the Universe*. John Wiley & Sons Ltd. A companion website: www.wiley.com/go/cockell/astrology.

Crain, I. K. 1971. Possible direct causal relation between geomagnetic reversals and biological extinctions. *Bulletin of the Geological Society of America* 82:2603.

Crans, D. C., Smee, J. J., Gaidamauskas, E. and Yang, L. 2004. The chemistry and biochemistry of vanadium and the biological activities exerted by vanadium compounds. *Chemical Reviews* 104:849–902.

Crockford, S. J. 2009. Evolutionary roots of iodine and thyroid hormones in cell-cell signalling. *Integrative and Comparative Biology* 49(2):155–166.

Creanga, D. E., Poiata, A., Morariu, V. V. and Tupu, P. 2004. Zero-magnetic field effect in pathogen bacteria. *Journal of Magnetism and Magnetic Materials* 272–276:2442–2444.

Currie, H. A. and Perry, C. C. 2007. Silica in plants: Biological, biochemical and chemical studies. *Annals of Botany* 100:1383–1389.

Davies, R. E. and Koch, R. H. 1991. All the observed Universe has contributed to life. *Philosophical Transactions: Biological Sciences* 334(1271):391–403.

Deckert, G., Warren, P. V., Gaasterland, T., et al. 1998. The complete genome of the hyperthermophilic bacterium Aquifex aeolicus. *Nature* 392 (6674):353–358. Doi:10.1038/32831. PMID 9537320.

Delves, H. T. 1985. Assessment of trace element status. *The Journal of Clinical Endocrinology and Metabolism* 14(3):725–760.

de Marcellus, P., Meinert, C., Michel Nuevo, M., et al. 2011. Non-racemic amino acid production by ultraviolet irradiation of achiral interstellar ice analogs with circularly polarized light. *The Astrophysical Journal Letters* 727:L27 (6 p).

Dhawi, F., Al-Khayri, J. M. and Hassan, E. 2009. Static magnetic field influence on elements composition in date palm. *Research Journal of Agriculture and Biological Sciences* 5(2):161–166.

D'Hendecourt, L., Modica P., Meinert C., Nahon L. and Meierhenrich U. 2019. Interstellar ices: A possible scenario for symmetry breaking of extraterrestrial chiral organic molecules of prebiotic interest. arXiv:1902.04575v1 [astro-ph.EP].

Diekert, G., Konheiser, U., Piechulla, K. and Thauer, R. K. 1981. Nickel requirement and factor F430 content of methanogenic bacteria. *Journal of Bacteriology* 148:459–464.

Doukov, T. I., Iverson, T. M., Seravalli, J., Ragsdale, S. W. and Drennan, C. L. 2002. A Ni-Fe-Cu center in a bifunctional carbon monoxide dehydrogenase/acetyl-CoA synthase. *Science* 298:567–572.

Dupont, C. L., Butcher, A., Valas, R. E., Bourne, P. E. and Caetano-Anollés, G. 2010. History of biological metal utilization inferred through phylogenomic analysis of protein structures. *Proceedings of the National Academy of Sciences of the United States of America* 107:10567–10572.

Dupont, C. L., Yang, S., Palenik, B. and Bourne, P. E. 2006. Modern proteomes contain putative imprints of ancient shifts in trace metal geochemistry. *Proceedings of the National Academy of Sciences of the United States of America* 103(47):17822–17827.

Edgcomb, V. P., Molyneaux, S. J., Saito, M. A., et al. 2004. Sulfide ameliorates metal toxicity for deep-sea hydrothermal vent archaea. *Applied and Environmental Microbiology* 70:2551–2555.

EFSA-NDA. 2014. Scientific opinion on dietary reference values for chromium. *European Food Safety Authority, EFSA Journal* 12(10):3845.

Eitinger, T. 2004. In vivo production of active nickel superoxide dismutase from Prochlorococcus marinus MIT9313 is dependent on its cognate peptidase. *Journal of Bacteriology* 186:7821–7825.

Emsley, J. 1998. *The Elements*, 3rd edition, Clarendon Press, Oxford.

Engel, M. H. and Macko, S. A. 1997. Isotopic evidence for extraterrestrial non-racemic amino acids in the Murchison meteorite. *Nature* 389:265–268.

Epstein, E. 1994. The anomaly of silicon in plant biology. *Proceedings of the National Academy of Sciences of the United States of America* 91:11–17.

Ettema, T. J. G., Brinkman, A. B., Lamers, P. P., Kornet, N. G., de Vos, W. M. and van der Oost, J. Molecular characterization of a conserved archaeal copper resistance (cop) gene cluster and its copper-responsive regulator in Sulfolobus Solfataricus P2. *Microbiology* 152:1969–1979.

FAO. 2001. *Human Vitamin and Minerals Requirements, Report of a Joint FAO/WHO expert consultation. Bangkok, Thailand. Food and Agriculture Organization of the United Nations – World Health Organization.* Published by the Food and Nutrition Division, FAO Rome.

Fedonkin, M. A. 2008. Ancient biosphere: The origin, trends, and events. *Russian Journal of Earth Sciences* 10: ES1006. Doi:10.2205/2007ES000252.

Ferrer, M., Golyshina, O. V., Beloqui, A., Golyshin, P. N. and Timmis, K. N. 2007. The cellular machinery of *Ferroplasma acidiphilum* is iron-protein-dominated. *Nature* 445:91–94.

Fesenko, E. E., Mezhevikina, L. M., Osipenko, M. A., Gordon, R. Y. and Khutzian, S. S. 2010. Effect of the "zero" magnetic field on early embryogenesis in mice. *Electromagnetic Biology and Medicine* 29(1–2):1–8.

Fleischer, A., Titel, C. and Ehwald, R. 1998. The boron requirement and cell wall properties of growing and stationary suspension-cultured Chenopodium album L. cells. *Plant Physiology* 117(4):1401–1410.

Fontecave, M. 2006. Iron-sulfur clusters: ever-expanding roles. *Nature Chemical Biology* 2:171–174. Doi:10.1038/nchembio0406-171.

Frahm, J.-P. 2001. *Biologie der Moose.* Springer Spektrum.

Frausto da Silva, J. J. R. and Williams, R. J. P. 2001. *The Biological Chemistry of the Elements*, vol. 2nd, Oxford University Press, Oxford.

Frieden, E. 1984. A survey of the essential biochemical elements. In *Biochemistry of the Essential Ultratrace Elements* Friedmen, E. (Ed.) Plenum Press, New York. pp. 1–15.

Gadd, G. M. 1993. Interactions of fungi with toxic metals. *New Phytologist* 124(1): 25–60.

Gaidos, E. J., Nealson, K. H. and Kirschvink, J. L. 1999. Life in ice-covered oceans. *Science* 284:1631–1633. Doi: 10.1126/science.284.5420.1631.

Gaither, L. A. and Eide, D. J. 2001. Eukaryotic zinc transporters and their regulation. *Biometals* 14:251–270.

Gibson, R. S. 1989. Assessment of trace element status in humans. *Progress in Food & Nutrition Science* 13(2):67–111.

Globus, N. and Blandford, R. D. 2020. The chiral puzzle of life. *The Astronomical Journal Letters* 895:L11 (14 p).

Gold, B., Deng, H., Bryk, R., Vargas, D., Eliezer, D., Roberts, J., Jiang, X. and Nathan, C. 2008. Identification of a copper-binding metallothionein in pathogenic mycobacteria. *Nature Chemical Biology* 4:609–616.

Goodman, E. M., Greenebaum, B. and Marron, M. T. 1995. Effects of electromagnetic fields on molecules and cells. *International Review of Cytology* 158:279–338.

Greenwood, N. N. and Earnshaw, A. 1984. *Chemistry of the Elements*. Pergamon Press, Oxford, pp. 1167–1168.

Gregor, D. and Pfeifer, F. 2001. Use of a halobacterial bgaH reporter gene to analyze the regulation of gene expression in halophilic archaea. *Microbiology* 147:1745–1754.

Gribble, G. W. 2002. *Naturally Occurring Organofluorines. The Handbook of Environmental Chemistry*. Springer-Verlag, Berlin, Vol. 3N, pp. 121–136.

Grogan, D. W. 2004. Stability and Repair of DNA in Hyperthermophilic Archaea. *Current Issues in Molecular Biology* 6:137–144.

Gronstal, A. L. 2014. Light scattering on dust holds clues to habitability. *Astrobiology Magazine*, 25 Sep. 2014. https://www.astrobio.net/alien-life/light-scattering-dust-holds-clues-habitability/.

Gurpreet, B. 2014. *Elucidating the Biological Role of Silicon and Designing a Delivery System to Enhance Early Bone Mineralization*. PhD thesis, University of Birmingham, Birmingham.

Guss, J. M. and Freeman, H. C. 1983. Structure of oxidized popular plastocyanin at 1.6 Å resolution. *Journal of Molecular Biology* 169:521–563.

Hambidge, M. 1988. Assessing the trace element status of man. *Proceedings of the Nutrition Society* 47:37–44.

Hantke, K. 2005. Bacterial zinc uptake and regulators. *Current Opinion in Microbiology* 8:196–202.

Harris, E. D. 2000. Cellular copper transport and metabolism. *Annual Review of Nutrition* 20:291–310.

Hatfield, D. L., Carlson, B. A., Xu, X. M., Mix, H. and Gladyshev, V. N. 2006. Selenocysteine incorporation machinery and the role of selenoproteins in development and health. *Progress in Nucleic Acid Research and Molecular Biology* 81:97–142.

Hatfield, D. L. and Gladyshev, V. N. 2002. How selenium has altered our understanding of the genetic code. *Molecular and Cellular Biology* 22:3565–3576.

Hays, J. D. 1971. Faunal extinctions and reversals of the earth's magnetic field. *Bulletin of the Geological Society of America* 82:2433.

Hein, J. E. and Blackmond, D. G. 2012. On the origin of single chirality of amino acids and sugars in biogenesis. *Accounts of Chemical Research* 45(12):2045–2054.

Helmenstine, A. M. 2019, Mar. 21. Elemental composition of the human body by mass. ThoughtCo. thoughtco.com/elemental-composition-human-body-by-mass-608192.

HERAG. 2007. Health risk assessment guidance for metals. Fact Sheet, Essentiality. http://www.icmm.com/page/1213/health-risk-assessment-guidance-for-metals-herag.

Hochheimer, A., Hedderich, R. and Thauer, R. K. 1998. The formylmethanofuran dehydrogenase isoenzymes in *Methanobacterium wolfei* and *M. thermoautotrophicum*: Induction of the molybdenum isoenzyme by molybdate and constitutive synthesis of the tungsten isoenzyme. *Archives of Microbiology* 170:389–393.

Hochheimer, A., Hedderich, R. and Thauer, R. K. 1999. The DNA binding protein Tfx from Methanobacterium thermoautotrophicum: structure, DNA binding properties, and transcriptional regulation. *Molecular Microbiology* 31:641–650.

Hochheimer, A., Linder, D., Thauer, R. K. and Hedderich, R. 1996. The molybdcnum formylmethanofuran dehydrogenase operon and the tungsten formylmethanofuran dehydrogenase operon from Methanobacterium thermoautotrophicum. Structures and transcriptional regulation. *The FEBS Journal* 242:156–162.

Holden, J. F. and Adams, M. W. W. 2003. Microbe-metal interactions in marine hydrothermal environments. *Current Opinion in Chemical Biology* 7:160–165.

Huber, R., Wilharm, T., Huber, D., et al. 1992. *Aquifex pyrophilus gen. nov. sp. nov.*, represents a novel group of marine hyperthermophilic hydrogen-oxidizing bacteria. *Systematic and Applied Microbiology* 15(3):340–351.

Hugo Ramos-Vera, W., Labonte, V., Weiss, M., Pauly, J. and Fuchs, G. 2010. Regulation of autotrophic CO_2 fixation in the archaeon thermoproteus neutrophilus. *Journal of Bacteriology* 192(20):5329–5340.

Irving, H. M. N. H. and Williams, R. J. P. 1953. The stability of transition-metal complexes. *Journal of the Chemical Society*: 3192–3210. Doi:10.1039/JR9530003192.

Ishino, Y. and Cann, I. K. 1998. The euryarchaeotes, a subdomain of Archaea, survive on a single DNA polymerase: Fact or farce? *Genes & Genetic Systems* 73(6):323–336.

Jain, S., Caforio, A. and Driessen, A. J. M. 2014. Biosynthesis of archaeal membrane ether lipids. *Frontiers in Microbiology* 5:641. Doi: 10.3389/fmicb.2014.00641.

Johnson, F. M. 1972. Specroscopy of tetrabenzporphirin molecules and possible astrophysical implications. *Mémoires de la Société royale des sciences de Liège*, 6th Série, tome III, 391–407.

Johansson, L., Gafvelin, G. and Arnér, E. S. 2005. Selenocysteine in proteins-properties and biotechnological use. *Biochimica et Biophysica Acta* 1726:1–13.

Johnson, M. K., Rees, D. C. and Adams, M. W. W. 1996. Tungstoenzymes. *Chemical Reviews* 96(7):2817–2840.

Jones, W. J., D. P. Nagel, Jr. W. B. and Whitman. 1987. Methanogens and the diversity of archaebacteria. *Microbiological Reviews* 51:135–177.

Joyce, G. F., Visser, G. M., van Boeckel, C. A., van Boom, J. H., Orgel, L. E. and van Westrenen, J. 1984. Chiral selection in poly©-directed synthesis of oligo(G). *Nature* 310(5978):602–604.

Kabata-Pendias, A. and Mukherjee, A. B. 2007. *Trace Elements from Soil to Human*. Springer-Verlag, Berlin, pp. 87–93.

Karolchik, D., Hinrichs, A.S., Furey, T.S., Roskin, K.M., Sugnet, C.W., Haussler, D. and Kent, W.J. 2004. The UCSC table browser data retrieval tool. *Nucleic Acids Research* 32 (Database issue):D493–496.

Kelley, D. S., Baross, J. A. and Delaney, J. R. 2002. Volcanoes, fluids, and life at mid-ocean ridge spreading centers. *Annual Review of Earth and Planetary Sciences* 30:385–491.

Kerby, R. L., Ludden, P. W. and Roberts, G. P. 1997. In vivo nickel insertion into the carbon monoxide dehydrogenase of *Rhodospirillum rubrum*: Molecular and physiological characterization of *cooCTJ*. *Journal of Bacteriology* 179:2259–2266.

Kietzin, A. and Adams, M. W. W. 1996. Tungsten in biological systems. *FEMS Microbiology Reviews* 18:5–63.

Kim, H. Y. and Gladyshev, V. N. 2005. Different catalytic mechanisms in mammalian selenocysteine- and cysteine-containing methionine-R-sulfoxide reductases. *PLOS Biology* 3:e375.

Kletzin, A. and Adams, M. W. W. 1996. Tungsten in biological systems. *FEMS Microbiology Letters Reviews* 18:5–63.

Kobayashi, K. and Ponnamperuma, C. 1985. Trace elements in chemical evolution, I. *Origins of Life* 16:41–55.

Kobayashi, M. and Shimizu, S. 1999. Cobalt proteins. *European Journal of Biochemistry* 261:1–9.

Krassilov, V. A. and Schuster, R. M. 1984. Palaeozoic and mesozoic fossils. In Schuster, R. M. (Ed.) *New Manual of Bryology*. Hattori Botanical Laboratory, Nichinan, Japan, Vol. 2. pp. 1170–1193.

Kroneck, P. M. H. 2001. Binuclear copper A. In Messerschmidt, A., Huber, R., Poulos, T. and Wieghardt, K. (Eds.), *Handbook of Metalloproteins* John Wiley & Sons, Chichester, Vol. 2. pp. 1331–1341.

Kryukov, G. V., Castellano, S., Novoselov, S. V., Lobanov, A. V., Zehtab, O., Guigó, R. and Gladyshev, V. N. 2003. Characterization of mammalian selenoproteomes. *Science* 300:1439–1443.

Kryukov, G. V. and Gladyshev, V. N. 2004. The prokaryotic selenoproteome. *EMBO Reports* 5:538–543.

Kryukov, G. V., Kryukov, V. M. and Gladyshev, V. N. 1999. New mammalian selenocysteine-containing proteins identified with an algorithm that searches for selenocysteine insertion sequence elements. *Journal of Biological Chemistry* 274:33888–33897.

LibreTexts. 2020. www.bio.libretexts.org, Enantiomers. Last updated Aug. 15. 2020.

Lieberman, R. L., Arciero, D. M., Hooper, A. B. and Rosenzweig, A. C. 2001. Crystal structure of a novel red copper protein from *Nitrosomona europaea*. *Biochemistry* 40:5674–5681.

Liermann, L. J., Guynn, R. L., Anbar, A. and Brantley, S. L. 2005. Production of a molybdophore during metal-targeted dissolution of silicates by soil bacteria. *Chemical Geology* 220:285–302.

Lineweaver, C.H. and Chopra, A. 2012. What can life on earth tell us about life in the universe? Genesis – In The beginning: precursors of life, chemical models and early biological evolution In Seckbach, J. (Ed.) *Cellular Origin, Life in Extreme Habitats and Astrology*, Springer, New York, NY, Vol. 22, pp. 799–815.

Lineweaver, C. H. and Schwartzman, D. 2005. Cosmic thermobiology: Thermal constraints on the origin and evolution of life in the universe, In Seckbach, J. (Ed.) *Origins: Cellular Origins and Life in Extreme Habitats and Astrobiology*, Springer, New York, NY, Vol. 6, pp. 233–248. Springer.

Linnaeus, C. 1758. *Systema naturae per regna tria naturae :secundum classes, ordines, genera, species, cum characteribus, differentiis, synonymis, locis* (in Latin) 10th edition. Laurentius Salvius, Stockholm.

Lovatt, C. J. and Dugger, W. M. 1984. Boron. In Frieden, E. (Ed.) *Biochemistry of the Essential Ultratrace Elements*. Plenum Press, New York. pp. 389–421.

Lucantoni, M. and Luisi, P. L. 2012. On the universality of the living: A few epistemological notes. *Origins of Life and Evolution of Biospheres* 42:385–387.

L'vov, N. P., Nosikov, A. N. and Antipov, A. 2002. Tungsten containing enzymes. *Biochemistry (Moscow)* 67:196–200. Doi: 10.1023/A:1014461913945.

Mabberley, D. J. 2008. *Mabberley's Plant-Book: A Portable Dictionary of Plants, Their Classification and Usages*, 3rd edition Cambridge University Press, Cambridge.

MacPherson, I. S. and Murphy, M. E. 2007. Cell. *Cellular and Molecular Life Sciences* 64:2887–2899.

Madigan, M. and Martinko, J. (Eds.) 2005. *Brock Biology of Microorganisms*, 11th edition. Prentice-Hall. ISBN 0-13-144329-1.

Maguire, M. E. 2007. Magnesium, manganese, and divalent cation transport assays in intact cells. *Methods in Molecular Biology* 394:289–305.

Maria de Lourdes, T. M. and Rai, M. (Eds.) 2014. *Fungal Enzymes*. CRC Press, Taylor and Francis Group.

Mariscal, C. and Doolittle, W. F. 2015. Eukaryotes first: How could that be? *Philosophical Transactions of the Royal Society B* 370:20140322.

Maturana, H. R. and Varela, F. J. 1980. Autopoiesis and cognition. In Cohen, R. S. and Wartofsky, M. W (Eds.) *The Realization of the Living. Vol. 42 of Boston Studies in the Philosophy of Science*. D. Reidel Pub. Co., Boston, MA, USA.

McClendon, J. H. 1976. Elemental abundance as a factor in the origin of mineral nutrient requirements. *Journal of Molecular Evolution* 8:175–195.

McCleverty, J. A. 1994. *Encyclopedia of Inorganic Chemistry* King, R. B. (Ed.) John Wiley and Sons, New York. pp. 2304–2330.

McGuire, B. A. and Carroll, P. B. 2016. Mirror asymmetry in life and in space. *Physics Today* 69(11): 86. Doi: 10.1063/PT.3.3375.

McGuire, B. A., Carroll, P. B., Loomis, R. A., Finneran, I. A., Jewell, P. R., Remijan, A. J. and Blake, G. A. 2016. Discovery of the interstellar chiral molecule propylene oxide (CH_3CHCH_2O). *Science* 352:1449–1452.

Mendel, R. R. and Bittner, F. 2006. Cell biology of molybdenum. *Biochimica et Biophysica Acta* 1763:621–635.

Mertz, W. 1970. Some aspects of nutritional trace element research. *Federation Proceedings* 29:1482–1488.

Mertz, W. 1992. Chromium: History and nutritional importance. *Biological Trace Element Research* 32:3–8.

Mertz, W. 1998. Review of the scientific basis for establishing the essentiality of trace elements *Biological Trace Element Research* 66:185–191.

Messerschmidt A. (Ed.) 1997. *Multi-Copper Oxidases*. World Scientific, Singapore.

Messerschmidt, A., Huber, R., Poulos, T. and Wieghardt, K. (Eds.) 2001. Cupredoxins (type-1 copper proteins). In *Handbook of Metalloproteins*, John Wiley & Sons, Chichester, Vol. 2. pp. 1151–1241.

Morel, F. M. M., Milligan, A. J. and Saito, M. A. 2003. Marine bioinorganic chemistry: The role of trace metals in the oceanic cycles of major nutrients. In Holland, H.

D. and Turekian, K. K. (Eds.) *Treatise on Geochemistry.* Elsevier Academic Press, London, pp. 113–143.

Mukherjee, B., Patra, B., Mahapatra, S., Banerjee, P., Tiwari, A. and Chatterjee, M. 2004. Vanadium--an element of atypical biological significance. *Toxicology Letters* 150(2):135–43.

Mukund, S. and Adams, M. W. W. 1996. Molybdenum and vanadium do not replace tungsten in the catalytically active forms of the three tungstoenzymes in the Hyperthermophilic Archaeon *Pyrococcus furiosus. Journal of Bacteriology* 178:163–167.

Muller, H. J. 1927. Artificial transmutation of the gene. *Science* 66:84–87.

Mulkidjanian, A. Y. 2009. On the origin of life in the zinc world: 1. Photosynthesizing, porous edifices built of hydrothermally precipitated zinc sulfide as cradles of life on Earth. *Biology Direct* 4:26. Doi: 10.1186/1745-6150-4-26.

Murakami, M. and Hirano, T. 2008. Intracellular zinc homeostasis and zinc signaling. *Cancer Science* 99:1515–1522.

Murphy, C., Schaffrath, C. and O'Hagan, D. 2003. Fluorinated natural products: The biosynthesis of fluoroacetate and 4-fluorothreonine in *Streptomyces cattleya. Chemosphere* 52(2):455–461.

Musser, S. M. and Chan, S. I. 1998. Evolution of the cytochrome c oxidase proton pump. *Journal of Molecular Evolution* 46:508–520.

Mussiani, F., Dikiy, A., Semenov, A. Y. and Ciurli, S. 2005. Structure of the intermolecular complex between plastocyanin and cytochrome f from spinach. *Journal of Biological Chemistry* 280:18833–18841.

Nakamura, K. and Go, N. 2005. Function and molecular evolution of multicopper blue proteins. *Cellular and Molecular Life Sciences* 62:2050–2066.

Navas-Acien, A. and Guallar, E. 2008. Measuring arsenic exposure, metabolism, and biological effects: The role of urine proteomics. *Toxicological Sciences* 106(1):1–4.

Neilands, J. B. 1995. Siderophores: structure and function of microbial iron transport compounds. *Journal of Biological Chemistry* 270:26723–26726.

Neveu, M., Poret-Peterson, A. T., Lee, Z. M. P., Anbar, A. D. and Elser J. J. 2014. Prokaryotic cells separated from sediments are suitable for elemental composition analysis. *Limnology and Oceanography: Methods* 12:519–529.

Nielsen, F. H. 1990. New essential trace elements for the life science. *Biological Trace Element Research* 26(1):599–611.

Nielsen, F. H. 2009. Micronutrients in parenteral nutrition: Boron, silicon, and fluoride. *Gastroenterology* 137:S55–S60.

Nies, D. H. 1999. Microbial heavy-metal resistance. *Applied Microbiology and Biotechnology* 51(6):730–750.

Novoselov, A. A., Serrano, P., Forancelli-Pacheco, M. L. A., Chaffin, M. S., O'Malley-James, J. T., Moreno, S. C. and Batista-Ribeiro, F. 2013. From cytoplasm to environment: The inorganic ingredients for the origin of life. *Astrobiology* 13(3):294–302.

O'Hagan, D., Schaffrath, C., Cobb, S. L., Hamilton, J. T. and Murphy, C. D. 2002. Biochemistry: Biosynthesis of an organofluorine molecule. *Nature* 416(6878):279.

Okada, H., Alekseev, I., Bravar, A. et al. 2008, Jan. 9. Absolute polarimetry at RHIC. arXiv:0712.1389v2 [nucl-ex].

Ouhammouch, M., Dewhurst, R. E., Hausner, W., Thomm, M. and Geiduschek, E. P. 2003. Activation of archaeal transcription by recruitment of the TATA-binding protein. *Proceedings of the National Academy of Sciences of the United States of America* 100:5097–5102.

Papp, L. V., Lu, J., Holmgren, A. and Khanna, K. K. 2007. From selenium to selenoproteins: synthesis, identity, and their role in human health. *Antioxidants & Redox Signaling* 9:775–806.

Parr, A. J. and Loughman, B. C. 1983. Boron and membrane function in plants. In Robb, D. A. and Pierpoints, W. S. (Eds.) *Metals and Micronutrients Uptake and Utilization by Plants*. Academic Press, London, pp. 87–107.

Pechova, A. and Pavlata, L. 2007. Chromium as an essential nutrient: A review. *Veterinarni Medicina* 52(1):1–18.

Peres, T. V., Schettinger, M. R. C., Chen, P., Carvalho, F., Avila, D. S., Bowman, A. B. and Aschner, M. 2016. Manganese-induced neurotoxicity: A review of its behavioral consequences and neuroprotective strategies. *BMC Pharmacology and Toxicology* 17:57 (20 p).

Petersen, J. H. 2013, Apr. 28. *The Kingdom of Fungi*, 1st printing edition. Princeton University Press.

Pizzarello, S. 2006. The chemistry of life's origin: A carbonaceous meteorite perspective. *Accounts of Chemical Research* 39:231–237.

Pizzorno, L. 2015. Nothing boring about boron. *Integrative Medicine (Encinitas)* 14(4):35–48.

Pollard, A. S., Parr, A. D. and Loghman, B. C. 1977. Boron in relation to membrane function in higher plants. *Journal of Experimental Botany* 28:831–841.

Pope, M. T. 1987. *Comprehensive Coordination Chemistry* Wilkinson, G. (Ed.). Pergamon Press, Oxford-New York. pp. 1023–1060.

Porter, J. R. 1946. *Bacterial Chemistry and Physiology*. John Wiley and Sons, London.

Prasad, A. S. 1995. Zinc: An overview. *Nutrition* 11:93–99.

Proudfoot, A. T., Bradberry, S. M. and Vale, J. A. 2006. Sodium fluoroacetate poisoning. *Toxicological Reviews* 25(4):213–219.

Quattrocchi, U. 2000. *World Dictionary of Plant Names: Common Names, Scientific Names, Eponyms, Synonyms, and Etymology*. CRC Press, Boca Raton, FL.

Ragsdale, S. W. 1998. Nickel biochemistry, *Current Opinion in Chemical Biology* 2:208–215. Doi:10.1016/S1367-5931(98)80062-8.

Raven, P. H. 1992. *Biology of Plants*, 5th edition. Worth Publishers, New York.

Rayman, M. P. 2000. The importance of selenium to human health. *Lancet* 356:233–241.

Rehder, D. 1992. Structure and function of vanadium compounds in living organisms. *Biometals* 5(1):3–12.

Reysenbach, A.-L., Boone, D. R. and Castenholz, R. W. (Eds.) 2001. Aquificae phy. nov. In *Bergey's Manual of Systematic Bacteriology*, 2nd edition. Springer-Verlag, Berlin, pp. 359–367. ISBN 0-683-00603-7.

Ridge, P. G., Zhang, Y. and Gladyshev, V. N. 2008. Comparative genomic analyses of copper transporters and cuproproteomes reveal evolutionary dynamics of copper utilization and its link to oxygen. *PLOS One* 3:e1378. Doi: 10.1371/journal.pone.0001378.

Rodionov, D. A., Hebbeln, P., Gelfand, M. S. and Eitinger, T. 2006. Comparative and functional genomic analysis of prokaryotic nickel and cobalt uptake transporters: evidence for a novel group of ATP-binding cassette transporters. *Journal of Bacteriology* 188:317–327.

Rose, K. 2009, May 15. *The Kingdom Plantae*. CreateSpace Independent Publishing Platform, Gloucester, United Kingdom and Melbourne, Australia.

Rouault, T. A. and Klausner, R. D. 1996. Iron-sulfur clusters as biosensors of oxidants and iron. *Trends in Biochemical Sciences* 21:174–177. Doi: 10.1016/0968-0004(96)10024-4.

Sakurai, T. and Kataoka, K. 2007. Structure and function of type I copper in multicopper oxidases. *Cellular and Molecular Life Sciences* 64:2642–2656.

Sander, S. G., Koschinsky, A., Massoth, G., Stott, M. and Hunter, K. A. 2007. Organic complexation of copper in deep-sea hydrothermal vent systems. *Environmental Chemistry* 4:81–89.

Santamaria, A. B. and Sulsky, S. I. 2010. Risk assessment of an essential element: Manganese. *Journal of Toxicology and Environmental Health, Part A* 73(2):128–55. Doi: 10.1080/15287390903337118.

Schäfer, U. 2012. Evaluation of beneficial and adverse effects on plants and animals following lithium deficiency and supplementation, and on humans following lithium treatment of mood disorders. *Trace Elements and Electrolytes* 29(2):91–112.

Schallreuter, K. U., Kothari, S., Chavan, B. and Spencer, J. D. 2008. Regulation of melanogenesis--controversies and new concepts. *Experimental Dermatology* 17:395–404.

Scherer, P., Lippert, H. and Wolff, G. 1983. Composition of the major elements and trace elements of 10 methanogenic bacteria determined by inductively coupled plasma emission spectrometry. *Biological Trace Element Research* 5(3):149–163.

Schmidt, D. 2006. *Guide to Reference and Information Sources in Plant Biology*, 3rd edition. Libraries Unlimited, Westport, CT.

Schneider, K.L., Pollard, K.S., Baertsch, R., Pohl, A. and Lowe, T.M. 2006. The UCSC archaeal genome browser. *Nucleic Acids Research* 34(Database Issue):D407–D410.

Schrauzer, G. N. 2002. Lithium: occurrence, dietary intakes, nutritional essentiality. *Journal of the American College of Nutrition* 21:14–21.

Schulze-Makuch, D. and Irwin, L. N. 2008. *Life in the Universe: Expectations and Constraints*. Springer International Publishing, New York, NY, USA.

Scorei, R., 2012. Is boron a prebiotic element? A mini-review of the essentiality of boron for the appearance of life on earth. *Origins of Life and Evolution of Biospheres* 42:3–17.

Sczepanski, J. T. and Joyce, G. F. 2014. A cross-chiral RNA polymerase ribozyme. *Nature* 515:440–453.

Shi, W., Zhan, C., Ignatov, A., Manjasetty, B. A., Marinkovic, N., Sullivan, M., Huang, R. and Chance, M. R. 2005. Metalloproteomics: high-throughput structural and functional annotation of proteins in structural genomics. *Structure* 13(10):1473–1486.

Silver, S. and Phung, L. T. 2009. Heavy metals, bacterial resistance. In Schaechter, M. (Ed.) *Encyclopedia of Microbiology*, Elsevier, Oxford, pp. 220–227.

Silvera, S. A. N., and Rohan, T. E. 2007. Trace elements and cancer risk: A review of the epidemiologic evidence. *Cancer Causes Control* 18:7–27.

Smedley, P. L. and Kinniburgh, D. G. 2002. A review of the source, behavior and distribution of arsenic in natural waters. *Applied Geochemistry* 17:517–568.

Smyth, D. A. and Dugger, W. M. 1981. Cellular changes during boron deficient culture of the diatom *Cylindrotheca fusiformis*. *Physiologia Plantarum* 51:111–117.

Snyder, W. S., Cook, M. J., Tipton, I. H., Nasset, E. S., Karhausen, L. R. and Howells, G. P. 1975. *Reference Man: Anatomical, Physiological and Metabolic Characteristics*, 1st edition, ser. Report of Task Group on Reference Man - International Commission on Radiological Protection. New York: Pergamon Press Ltd., vol. 23, reprinted 1992.

Solomon, E. I., Sundaram, U. M. and Machonkin, T. E. 1996. Multicopper oxidases and oxygenases. *Chemical Reviews* 96:2563–2605.

Solomon, E. I., Szilagyi, R. K., George, S. D. and Basumallick, L. 2004. Electronic structures of metal sites in proteins and models: contributions to function in blue copper proteins. *Chemical Reviews* 104:419–458.

Srivastava, P. and Kowshik, M. 2013. Mechanisms of metal resistance and homeostasis in haloarchaea. *Archaea*: Article ID 732864, (16 p).

Stcherbic, V. V. and Buchatsky, L. P. 2015. *Living Matter: Algebra of Molecules*, CRC Press, Boca Raton, FL, USA. (160 p).

Steenbjerg, F. 1951. Yield curves and chemical plant analysis. *Plant Soil* 3:97–109.

Stephenson, S. L. 2010, Apr. 21. *The Kingdom Fungi: The Biology of Mushrooms, Molds, and Lichens*, 1st Edition. Timber Press, Portland, Oregon, USA.

Stetter, K. O. 1996. Hyperthermophilic prokaryotes. *FEMS Microbiology Reviews* 18:149–158.

Stetter, K. O., Fiala, G., Huber, G., Huber, R. and Segerer, A. 1990. Hyperthermophilic microorganisms. *FEMS Microbiology Reviews* 75:117–124.

Stiefel, E. L. 2002. The biogeochemistry of molybdenum and tungsten. In Dekker, M (Ed.) *Molybdenum and Tungsten: Their Roles in Biological Processes*. Marcel Dekker, New York, NY, pp. 1–29.

Stock, T. and Rother, M. 2009. Selenoproteins in Archaea and Gram-positive bacteria. *Biochimica et Biophysica Acta* 1790(11):1520–1532.

Stoeckenius, W. 1981. Walsby's square bacterium: Fine structure of an orthogonal procaryote. *Journal of Bacteriology* 148(1):352–360.

Sukdeo, N., S. L. Clugston, E. Daub and J. F. Honek. 2004. Distinct classes of glyoxalase I: metal specificity of the Yersinia pestis, Pseudomonas aeruginosa and Neisseria meningitidis enzymes. *Biochemical Journal* 384:111–117.

Todar, K. 2012. Nutrition and Growth of Bacteria. In *Todar's Online Textbook of Bacteriology*. www.textbookofbacteriology.net/nutgro.html.

Underwood, E. J., and Mertz, W. 1987. *Introduction in Trace Elements in Human and Animal Nutrition*, 5th Edition Mertz, E. (Ed.) Academic Press, Cambridge, Massachusetts, USA, pp. 1–19.

Valkovic, V. 1977. A possible mechanism for the influence of the geomagnetic field of the evolution of life. *Origins of Life* 8:7–11.

Valkovic, V. 1978. Elements essential for life. In Hemphill, D. D. (Ed.) *Trace Substances in Environmental Health* – XII. University of Missouri, Columbia.

Valković, V. 1990. Origin of trace element requirement by living matter. In Gruber, B. and Yopp, J. H. (Eds.) *Symmetries in Science IV*, Plenum Press, New York, pp. 213–242.

Valkovic, V. and Obhodas, J. 2020. The origin of chiral life. arXiv:2011.12145 [astro-ph. GA].

Van Gossum, A. and Neve, J. 1998. Trace element deficiency and toxicity. *Current Opinion in Clinical Nutrition & Metabolic Care* 1: 499–507.

Vila A. J. and Fernandez C. O. 2001. Copper in electron-transfer proteins. In Bertini I., Sigel A. and Sigel H. (Eds.) *Handbook on Metalloproteins*. Marcel Dekker, New York, pp. 813–856.

Von Damm, K. L., Edmond, J. M., Grant, B., Measures, C. I., Walden, B. and Weiss, R. F. 1985. Chemistry of submarine hydrothermal solutions at 21°N, East Pacific Rise,. *Geochimica et Cosmochimica Acta* 49:2197–2220.

Von Damm, K. L., Edmond, J. M., Measures, C. I. and Grant, B. 1985. Chemistry of submarine hydrothermal solutions at Guaymas Basin, Gulf of California. *Geochimica et Cosmochimica Acta* 49:2221–2237.

Wang, M. Yafremava, L. S., Caetano-Anollés, D., Mittenthal, J. E. and Caetano-Anollés, G. 2007. Reductive evolution of architectural repertoires in proteomes and the birth of the tripartite world. *Genome Research* 17:1572–1585.

WHO. 1996. *Trace Elements in Human Nutrition and Human Health*. World Health Organisation, Geneva.

WHO. 2002. *Principles and Methods for the Assessment of Risk from Essential Trace Elements*. Series Environmental Health Criteria No. 228. ISBN 92-4-157228-0. World Health Organisation, Geneva.

Wildhaber, I. and Baumeister, W. 1987. The cell envelope of Thermoproteus tenax: three-dimensional structure of the surface layer and its role in shape maintenance. *The EMBO Journal* 6(5):1475–1480.

Williams, R. J. P. 2001. Chemical selection of elements by cells. *Coordination Chemistry Reviews* 216–217:583–595.

Williams, R. J. P. and Fraústo da Silva, J. J. R. 2002. The involvement of molybdenum in life. *Biochemical and Biophysical Research Communications* 292(2):293–299.

Williams, R. J. P. and Fraústo Da Silva, J. J. R. 2003. Evolution was chemically constrained. *Journal of Theoretical Biology* 220(3):323–343.

Woese, C. R. 1981. Archaebacteria. *Scientific American* June 1.1981. See also Woese, C. R. Archaebacteria: The Third Domain of Life Missed by Biologists for Decades, Scientific American December 31, 2012.

Woese, C. R. 1987. Bacterial evolution. *Microbiological Reviews* 51:221–271.

Woese, C. R. and Fox, G. E. 1977. Phylogenetic structure of the prokaryotic domain: The primary kingdoms. *Proceedings of the National Academy of Sciences of the United States of America* 74:5088–5090.

Woese, C. R., Kandler, O. and Wheelis, M. L. 1990. Towards a natural system of organisms: proposal for the domains Archaea, Bacteria, and Eucarya. *Proceedings of the National Academy of Sciences of the United States of America* 87(12):4576–4579.

Wong, J. T., Chen, J., Mat, W. K., Ng, S. K. and Xue, H. 2007. Polyphasic evidence delineating the root of life and roots of biological domains. *Gene* 403:39–52.

Yin, P., Zhang, Z-M., Lv, H., et al. 2015. Chiral recognition and selection during the self-assembly process of protein-mimic macro-anions, *Nature Communications* 6, article number:6475. DOI: 10.1038/ncomms7475.

Young, A. H. 2011. More good news about the magic ion: lithium may prevent dementia. *The British Journal of Psychiatry* 198:336–337.

Zel'dovich, B., Saakyan, D. and Sobel'man, I. 1977. Energy difference between right-hand and left-hand molecules, due to parity nonconservation in weak interactions of electrons with nuclei. *JETP Letters* 25(2):94.

Zerkle, A. L., House, C. H. and Brantley, S. L. 2005. Biogeochemical signatures through time as inferred from whole microbial genomes. *American Journal of Science* 305:467–502.

Zhang, Y., Fomenko, D. E. and Gladyshev, V. N. 2005. The microbial selenoproteome of the Sargasso Sea. *Genome Biology* 6:R37.

Zhang, Y. and Gladyshev, V. N. 2005. An algorithm for identification of bacterial selenocysteine insertion sequence elements and selenoprotein genes. *Bioinformatics* 21:2580–2589.

Zhang, Y. and Gladyshev, V. N. 2008a. Molybdoproteomes and evolution of molybdenum utilization. *Journal of Molecular Biology* 379:881–899.

Zhang, Y. and Gladyshev, V. N. 2008b. Trends in selenium utilization in marine microbial world revealed through the analysis of the global ocean sampling (GOS) project. *PLOS Genetics* 4:e1000095.

Zhang, Y. and Gladyshev, V. N. 2009. Comparative genomics of trace element dependence in biology. *Chemical Reviews* 109:4828–4861.

Zhang, Y. and Gladyshev, V. N. 2010. General trends in trace element utilization revealed by comparative genomic analyses of Co, Cu, Mo, Ni, and Se. *Journal of Biological Chemistry* 285(5):3393–3405.

Zhang, Y. and Gladyshev, V. N. 2011. Comparative genomics of trace element dependence in biology. *Journal of Biological Chemistry*. Published online May 12. 2011, as Manuscript R110.172833. The latest version is at http://www.jbc.org/cgi/doi/10.1074/jbc.R110.172833.

Zhang, Y., Rodionov, D. A., Gelfand, M. S. and Gladyshev, V. N. 2009. Comparative genomic analyses of nickel, cobalt and vitamin B12 utilization. *BMC Genomics* 10:78. Doi: 10.1186/1471-2164-10-78.

Zheng, M. and Storz, G. 2000. Redox sensing by prokaryotic transcription factors, *Biochemical Pharmacology* 59(1):1–6. Doi: 10.1016/S0006-2952(99)00289-0.

Zhou, Z. and Zheng, S. 2003. The missing link in Gingko evolution. *Nature* 423:821–822.

Zlatanova, J. 1997. Archaeal chromatin: Virtual or real? *Proceedings of the National Academy of Sciences of the United States of America* 94:12251–12254.

Additional Readings

Balows, A., Truper, H., Dworkin, M., Harder, W. and Schleifer, K.-H. 1991. *The Prokaryotes: A Handbook on the Biology of Bacteria*, 2nd Edition, Vol. 1–4. Springer-Verlag, New York, NY.

Baum, D. A. 2015. Selection and the origin of cells. *BioScience* 65(7):678–684.

Breuker, A., Stadler, S. and Schippers, A. 2013. Microbial community analysis of deeply buried marine sediments of the New Jersey shallow shelf (IODP Expedition 313). *Federation of European Microbiological Societies* 85:578–592.

Cosic, I. 1994. Macromolecular bioactivity: Is it resonant interaction between macromolecules? – theory and applications. *IEEE Transactions on Biomedical Engineering* 41(12):1101–1114.

Cosic, I., Pirogova, E., Vojsavljevic, V. and Fang, Q. 2006. Electromagnetic properties of biomolecules. *FME Transactions* 34:71–80.

Deamer, D., Dworkin, J. P., Sandford, S. A., Bernstein, M. P. and Allamandola, L. J. 2002. The first cell membranes. *Astrobiology* 2(4):371–381.

Field, M. C. 2019. The kinetochore and the origin of eukaryotic chromosome segregation. *Proceedings of the National Academy of Sciences of the USA* 116(26):12596–12598.

Garcia-Gonzalez, M., Mateo, P. and Bonilla, I. 1988. Boron protection for O2 diffusion in heterocysts of *Anabaena sp.* PCC 7119. *Plant Physiology* 87:785–789.

Griffiths, G. 2007. Cell evolution and the problem of membrane topology nature reviews. *Molecular Cell Biology* 8:1018–1024.

Gelderblom, H. R. 1996. Structure and Classification of Viruses. In Baron, S. (Ed.) *Chapter 41 in Medical Microbiology*, 4th edition. The University of Texas, Galveston (7 p.).

Hotz, C., Lowe, N. M., Araya, M. and Brown, K.H. 2003. Assessment of the trace element status of individuals and populations: The example of zinc and copper. *The Journal of Nutrition* 133(5):15635–15685.

Koch, A. L. 2006. *The Bacteria: Their Origin, Structure, Function and Antibiosis*. Springer, Netherlands.

Koumandou, V. L., Wickstead, B., Ginger, M. L., van der Giezen, M., Dacks, J. B. and Field, M. C. 2013. Molecular palaeontology and complexity in the last eukaryotic common ancestor. *Critical Reviews in Biochemistry and Molecular Biology* 48(4):373–396.

Lane, N. and Martin, W. F. 2012. The origin of membrane bioenergetics. *Cell* 151:1406–1416.

Loomis, W. D. and Durst, R. W. 1992. Chemistry and biology of boron. *Biofactors* 3(4):229–239.

Matoh, T. 1997. Boron in plant cell walls. *Plant and Soil* 193(1–2):59–70.

Miller, E. P., Wu, Y. and Carrano, C. J. 2016. Boron uptake, localization, and speciation in marine brown algae. *Metallomics* 8:161–168.

Nielsen, F. H. 2002. Trace mineral deficiencies. In Berdanier C. D. (Ed.) *Handbook of Nutrition and Food*. CRC Press, Boca Raton, FL, pp. 1463–1487.

Nielsen, F. H. 2003. Trace elements. In Caballero, B., Trugo, L. and Finglas, P. (Eds.) *Encyclopedia of Food Sciences and Nutrition*, 2nd edition Academic Press, London, England, pp. 5820–5828.

Pandey, D. K., Singh, A. V. and Chaudhary, B. 2012. *Journal of Botany*, Article ID 375829, (9 p.) Doi: 10.1155/2012/375829.

Schrum, J. P., Zhu, T. F. and Szostak, J. W. 2010. The origins of cellular life. *Cold Spring Harbor Perspectives in Biology* 2:a002212.

Scorei, R. and Cimpoiaşu, V. M. 2006. Boron enhances the thermostability of carbohydrates. *Origin of Life and Evolution of Biospheres* 36:1–11.

Seshasayee, A. S. N. 2015, Mar. 5. *Bacterial Genomics: Genome Organization and Gene Expression Tools*, 1st edition, Kindle Edition. Cambridge University Press, Cambridge, England.

Spaargaren, D. H. 1991. The biological use of chemical elements: Selection on environmental availability and electron configuration. *Oceanologica Acta* 14(6):569–574.

Vosseberg, J., van Hooff, J. J. E., Marcet-Houben, M., et al. Timing the origin of eukaryotic cellular complexity with ancient duplications. 2020. *Nature Ecology & Evolution*, Articles. Doi: 10.1038/s41559-020-01320-z.

Zannoni, D. 2004. *Respiration in Archaea and Bacteria: Diversity of Prokaryotic Respiratory System*. Springer, Copyright 2004, Series: Advances in Photosynthesis and Respiration, Vol. 16.

5

Time and Place of the Origin of Life

5.1 Introduction

The present-day thinking about the phenomenon of life is summarized and presented in five points by Barrow et al. (2012), which could be modified as shown below.

1. Life is contained: All life we know is in cells, and all cells have a continuous, closed membrane that separates "inside" from "outside". The metabolism, growth and division of cells are essential to life on the Earth. Therefore, explaining the origin of cells is a central challenge in understanding the origin of life.
2. Life is out of equilibrium and requires a flow of energy: When the chemical and physical processes in living cells reach equilibrium, and there is no flux of energy through the cell, it is dead (i.e., not alive).
3. Life is self-replicating: The primary characteristic of the cell is that the division of a parent cell produced it, and, in most cases, it will also divide and produce other cells.
4. Life is adaptive: The cell can adapt its internal content so that it functions even when the outside environment changes. In some situations, it can also modify the external environment to make itself more comfortable.
5. Life appears to need water: All life we know involves molecules dissolved or organized in a medium that is mostly water. We do not know whether water is essential to all life or just to life as we know it. At present, we do not know exceptions to this.

According to this view, life is a spatially distinct, highly organized network of chemical reactions characterized by a set of remarkable properties that enable it to replicate itself and to adapt to changes in its environment. However, all of this describes how life on the Earth sustains rather than how it originated.

The chemical evolution of galaxies results in the changes of chemical composition (relative abundances of chemical elements) of stars, interstellar gas and dust. The increase in chemical element abundances with time (elements heavier than Li) provides a clock for galactic aging. The older the galaxy, the more heavy elements it contains. Looking at the living matter on the planet Earth, we see that life needs only some chemical elements

DOI: 10.1201/9781003181330-5

for its existence. All organisms use trace elements and provide proteins with unique coordination and catalytic and electron transfer properties.

Therefore, to shed some light on the problem of where and when life originated, we are putting forward ***the first hypothesis***. *The first hypothesis says:*

Life, as we know, is (H-C-N-O)-based and relies on the number of bulk (Na-Mg-P-S-Cl-K-Ca) and trace elements (Li-B-F-Si-V-Cr-Mn-Fe-Co-Ni-Cu-Zn-As-Se-Mo-I-W). It originated when the element abundance curves of the living matter and the Universe coincided. This coincidence occurring at a particular redshift could indicate the phase of the Universe when life originated, T_{origin}.

The remaining crucial question is which two-element abundance curves should be considered. The environmental abundance curve should be the abundance curve of either the part of the Universe or the individual galaxy where life supposedly originated at that time. The life abundance curve could be of the only life we know: the life on the planet Earth. Therefore, we should take that of the last universal common ancestor (LUCA) or some other primitive organism that can survive the planet-forming process, ignoring the fact that we do not know their element concentration factors. In other words, T_{origin} could be estimated from the minimum of χ^2 (best agreement) as defined by

$$\chi^2 = \frac{1}{N}\sum_{i=1}^{N}\left(C_{i,\,LUCA} - C_{i,\,environment}\right)^2 / C_{i,\,environment}$$

where C_i is the abundance of essential element i in the LUCA and in the environment where the LUCA originated. Analogously, T_{origin} could be estimated from calculating the maximum coefficient of determination (R^2), i.e., the percentage of the variance explained by the linear regression model between logarithmic values of abundances of essential elements in the LUCA and the environment of its origin.

Interstellar molecular clouds and circumstellar envelopes are places of complex molecular synthesis (Ehrenfreund et al., 2000; Kwok, 2004; van Dishoeck and Blake, 1998). In addition to gas, interstellar material also contains small micron-sized particles. Gas-phase and gas-grain interactions result in the formation of complex molecules. Surface catalysis on solid particles enables molecule formation and chemical pathways that cannot proceed in the gas phase because of reaction barriers.

A high number of molecules that are used in contemporary biochemistry on the Earth are present in the interstellar medium on surfaces of comets, asteroids, meteorites and interplanetary dust particles. Therefore, *our **second hypothesis** is that life originated in an interstellar molecular cloud with the critical role of dust particles and their environment.* The solar system we live in is the result of the gravitational collapse of a small part of a giant molecular cloud.

Assuming life originated long before the origin of the Earth, it could be hypothesized that life was brought to the Earth in the process of its formation! We must consider the role of cosmic dust mineral grains in this process. Boron played an essential role in this process since its primary purpose has been to provide thermal and chemical stability in hostile environments.

Rafelski et al. (2014) showed the rapid decline in the metallicity of damped Ly-α systems (DLAs) at z~5. They presented chemical abundance measurements for 47 DLAs, 30 at z>4, observed with the Echellette Spectrograph and Imager, and the High-Resolution Echelle Spectrometer on the Keck telescopes. DLA metallicity vs redshift shows a sharp decrease in metallicity at z>4.7. It looks that the properties of the dust in DLAs would not satisfy the requirements of the essential trace element signature of the LUCA (Cr, Mn, Fe, Co, Ni, Cu, Zn, Se, Mo, I and W) for z>4.7 (i.e., before $\approx-12.6\times10^9$years or $\approx1.1\times10^9$years after the Big Bang). At some point in time, the chemical evolution of the Universe resulted in an elemental abundance curve coinciding with the LUCA elemental signature. For any other time, $T\neq T_{origin}$, no additional such events could occur. The time when the coincidence of element abundance curves for the Universe and the LUCA occurred could not be older than $T\approx-12.6\times10^9$, a lower limit for T_{origin}, because the Universe was chemically too young to produce elements in the right elemental ratios as required by the LUCA. *Therefore, we put out the* ***third hypothesis*** *that, because of Universe aging, life originated only once.*

In addition to the text in previous chapters, this chapter provides supporting pieces of evidence for these hypotheses.

5.2 LUCA

The LUCA has been discussed in detail in Chapter 4. For example, Figure 4.5 shows a basic phylogenetic tree with assumed relationships between the three domains. Lines denote an evolutionary "distance" between the groups shown, indicating the time since they split from a common ancestor. They merely illustrate some examples. It can be seen that the tree is "rooted" in a single ancestor, LUCA. It should be noted that there are different types of phylogenetic trees. In some trees, the length of a branch reflects the amount of genetic change that has happened in a particular DNA sequence. For some other trees, branch length can represent chronological time and branching points can be determined from the fossil record. These trees are designed to convey a different type of information or combined information.

All organisms on the tree of life have a common ancestor. The most critical challenge is determining the nature of the LUCA. We shall repeat some of the discussion for the sake of completeness of the chapter. The first attempt to determine the nature of the LUCA was based on an analysis of the paralogue protein couple. These are proteins encoded by genes that have duplicated. If the gene duplication is found in all three domains, this implies that the duplication occurred before the three domains diverged. Thus, by examining the distribution of a variety of paralogous genes, one can infer when domains separated. This fact resulted in many scientists concluding that the tree of life is rooted in bacteria. The divergence between the archaea and eukaryotes occurred at a later time. A more recent examination using sequenced whole genomes to identify genes found in all three domains of life revealed approximately 100 common genes that were then used to construct a phylogenetic tree, with similar results.

In contrast to this view is the one which places the LUCA nearer the eukaryotes with a later divergence between bacteria and archaea. This thinking results from the fact that many complex processes involving RNA are found in eukaryotes, including the splicing of RNA, which has been hypothesized to be a relic of an ancient pre-biological RNA world. A concern with making these deep inferences is that biases can be introduced into tree reconstruction by varying rates of evolutionary change in different species. For example, the evolutionary rates of ribosomal RNA changes vary by up to 100 times in foraminifera.

When a universal tree of life is constructed, making special efforts to use only slowly evolving genes, then the root of life does not so clearly sit within the Bacteria and instead suggests that the rooting in the bacterial lineage could be an artifact. Part of the problem in defining the real evolutionary root of the tree is the low number of genes being considered. As a higher number of organism genomes are sequenced, the quantity of information that can be used to determine the root will increase and so the accuracy of the inferences.

A question of particular interest is what the functional capabilities of the LUCA were. Some biochemical pathways, such as the TCA cycle and components of electron transport chains, are very widely distributed and present in all three domains of life, suggesting that they are ancient and could have been present in the LUCA. An early analysis of the tree of life indicated that the first organisms were heat-loving. This situation led to numerous hypotheses to explain the observations.

It is a general belief that the LUCA had simple metabolism processes, with few essential amino acids and perhaps a few non-specific enzymes. Aromatic amino acids appear to have arisen sometime after the LUCA arose, as did more complex and specific enzymes (Ahmad and Jensen, 1988). It is suggested that the actual incorporation of iodine into essential biochemistry appears to have come after the rise of the LUCA (Barrinton, 1962; Crockford, 2009) and that its influence probably started with the LUCA itself.

Intense impact bombardment in the early history of the Earth was suggested to have been essential in the origin of life on the Earth; it created an evolutionary bottleneck for selecting the high-temperature-tolerant organisms. After bombardment, when impacts were more infrequent and the Earth cooled, life could have spread into a greater diversity of microorganisms adapted to growing at lower temperatures (Cockell, 2015).

However, the warm environment ancestry of life has been questioned. The heat-loving bacteria have acquired up to 24% of their genes from the archaea, suggesting considerable horizontal gene transfer. The DNA reverse gyrase gene, which encodes for an enzyme that introduces a supercoil to DNA, making it more difficult to unwind and conferring resistance to high temperatures, is an example of a gene that is likely to have been passed by horizontal gene transfer from archaea. It is found in bacteria such as *Thermotoga* and *Aquifex* that live in high-temperature environments. If the root of life is bacterial, then the existing data might argue against a hot origin of life if they acquired these genes from elsewhere (Cockell, 2015).

There are several complexities in unravelling the functional capabilities of the LUCA. First, the open question of where the tree of life is rooted must be resolved as only then can it be determined whether high-temperature adaptation is a primary inherited trait. Second, even if the tree of life does not have a root in organisms adapted to living at high temperatures, this does not rule out a hot origin of life. Namely, considerable time and extinct organisms might exist between the origin of life and the root of the tree, as defined by extant organisms.

All these considerations are done by assuming that the three existing domains of life define the LUCA's capabilities and characteristics. However, as pointed out by Cockell (2015), we cannot be sure that there were not domains that went extinct early in the history of life and took with them some crucial information.

The deepest roots of the 16S rRNA phylogenetic tree are hyperthermophilic (Lineweaver and Schwartzman, 2005; Wong et al., 2007). Presently existing organisms with the shortest branches are hyperthermophiles able to tolerate temperatures above 90°C. These organisms – *Aquifex*, *Thermotoga*, Nanoarchaeota, and Korarchaeota – could be the best choice of all terrestrial life for modelling the LUCA. Hyperthermophilic organisms closest to the root suggest that the LUCA was hyperthermophilic and that, by extension, the origin of life on the Earth was hyperthermophilic (Lineweaver and Chopra, 2012).

Although Fedonkin (2008) considers transitional metals as a key factor in life evolution on the Earth only, the philosophy used could be applied to the origin of life problem also. Different solubilities of some metal sulfides versus metal hydroxides (Di Toro et al., 2001) or metal sulfates in the modern ocean can give an idea of that dramatic change related to the oxygenation of the environments. The rise in free oxygen reduced the availability of some metals (such as W, V, Ni and Fe). In contrast, others (such as Mo, Cu and Zn) became more readily available (Frausto da Silva and Williams, 1997). The hypothetical sequence of the incorporation of metals into the enzymatic evolution in the early history of the biosphere should affect the metabolic development and domination of the particular physiology types. The replacement of the unavailable metals with those which were available seemed to be one of the significant ways in the early evolution of enzymes (Fedonkin, 2008).

The idea of a LUCA of all cells, or the progenitor, is the most important for the study of early evolution and life's origin, yet the information about how and where LUCA lived is still lacking. The physiology and possible habitat of the LUCA have recently been discussed (Weiss et al., 2016). The authors investigated all clusters and phylogenetic trees for 6.1 million protein-coding genes from sequenced prokaryotic genomes to reconstruct the microbial ecology of the LUCA. Among 286,514 protein clusters, they identified 355 protein families (~0.1%) that trace to the LUCA by using phylogenetic criteria. Since these proteins are not universally distributed, they can illuminate the LUCA's physiology. Their functions, properties and prosthetic groups depict the LUCA as anaerobic, CO_2-fixing, H_2-dependent with a Wood–Ljungdahl pathway, N_2-fixing and thermophilic. The Wood–Ljungdahl pathway (see Ragsdale, 2006) enables the use of hydrogen as an electron donor and carbon dioxide as an electron acceptor. It also allows them to be building blocks for biosynthesis. The LUCA's biochemistry was loaded with FeS clusters and radical reaction mechanisms. Its cofactors reveal dependence upon transition metals (Cr, Mn, Fe, Co, Ni, Zn, Mo and W), flavins, *S*-adenosyl methionine, coenzyme A, molybdopterins, corrins and selenium. The LUCA's genetic code required nucleoside modifications and *S*-adenosyl methionine-dependent methylations. The 355 phylogenies identify clostridia and methanogens, whose modern lifestyles resemble that of the LUCA, as basal among their respective domains. According to Weiss et al. (2016), the LUCA inhabited a geochemically active environment rich in H_2, CO_2 and iron.

5.3 Origin of Life on the Earth

Standard evolutionary biology assumes a time on the Earth between 3.5 and 4×10^9 years ago as life's starting point. A possible location is thought to be deep-ocean hydrothermal vents (Martin et al., 2008). It recognizes that the rapid emergence of life is a puzzle.

There are many scenarios of the origin of life on the Earth. For example, Mulkidjanian (2009) puts forward a scene proposing that life on the Earth emerged, powered by UV-rich solar radiation, on precipitated zinc sulfide (ZnS) similar to the one found in modern deep-sea hydrothermal vents. Under the high pressure of the ancient, carbon dioxide-dominated atmosphere, ZnS could precipitate at the surface of the continents, within the reach of solar light. The author suggested that the ZnS surfaces drove carbon dioxide reduction by solar radiation, yielding the building blocks for the first biopolymers. The same surfaces served as templates for the synthesis of longer biopolymers from simpler building blocks. They prevented the first biopolymers from photo-dissociation by absorbing from them the excess radiation. Also, the UV light may have favored the selective enrichment of photo-stable, RNA-like polymers.

Although the origins of life on the early Earth remain controversial, experimental evidence for the possible evolution of early life began to be accumulated with the work of Miller and Urey (Miller, 1953; Miller and Urey, 1959). The basic premise of their work, which is still central in many ongoing studies, is that pure organic compounds, including amino acids, were formed after a spark was applied to a flask with constituents. The process was thought to be present in the Earth's early atmosphere (e.g., methane, ammonia, carbon dioxide and water). It is thought that these pure organic compounds provided the "seeds" for the synthesis of more complex prebiotic organic compounds.

It should be mentioned that the Earth's magnetic field had existed since the planet's early days – long before life on the Earth began. Paleomagnetic studies of Australian red dacite and pillow basalt have estimated the magnetic field to be at least 3.5×10^9 years old (McElhinney and Senanayake, 1980; Buffett, 2000). This finding implies that life, if it originated on the planet Earth, has evolved in an existing magnetic field that protected the near-Earth environment from dangerous radiation and high-energy plasma from the Sun by partially blocking it out.

In addition to simple molecules formed in laboratory experiments performed in the 1950s, additional sources of life needed molecules started to be considered. These were the extraterrestrial sources, such as comets, meteorites and interstellar particles, postulated as possible sources for seeding the early Earth with simple molecules. Engel and Macko (1986), Galimov (2006) and some other claim that there is no firm evidence for the presence of prebiotic compounds in these extraterrestrial sources. The situation has been changed since; most compounds required for the origin of life have been detected in extraterrestrial bodies. The synthesis of adenosine triphosphate (ATP) is essential for life processes. It was assumed to be the most critical factor in the early stages of prebiotic evolution; see Figure 5.1 (Galimov 2001, 2004). The essential step in the assembly of complex molecules is the hydrolysis of ATP to adenosine diphosphate (ADP). As an example, we mention that the formation of peptides from amino acids and nucleic acids from nucleotides is linked to the ATP molecule. Another critical step in the prebiotic

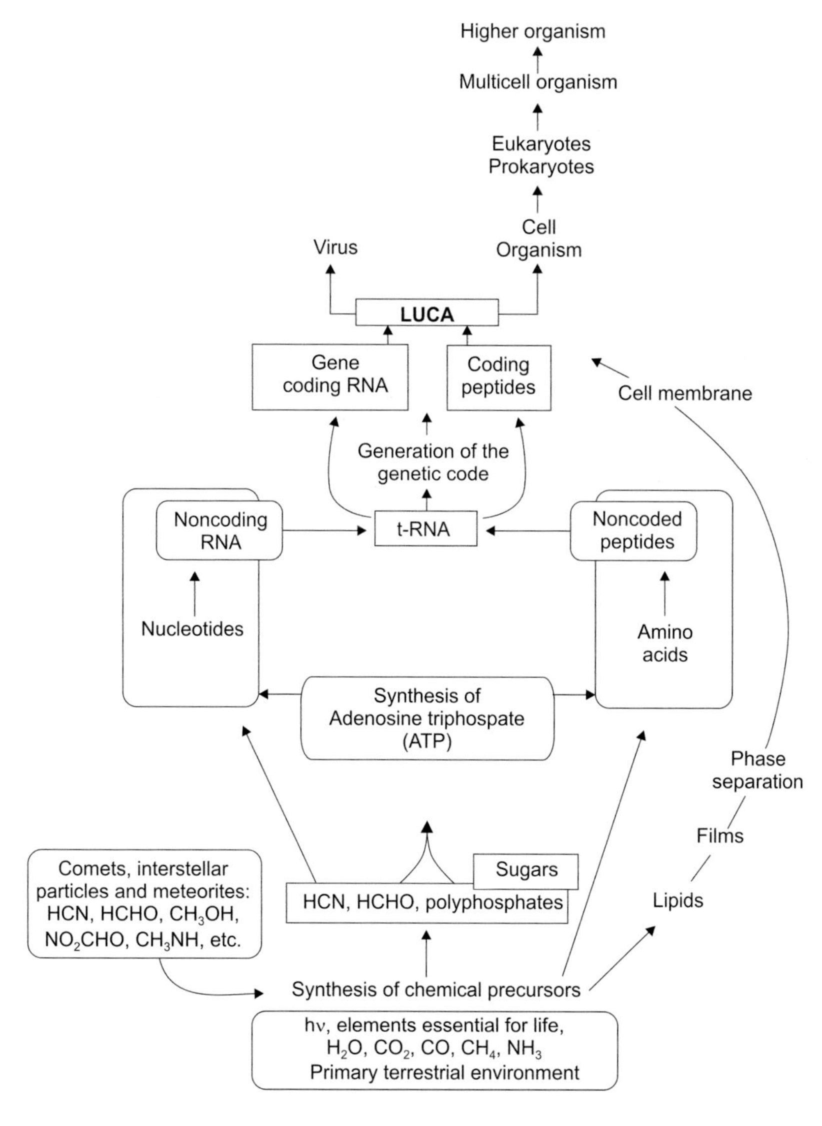

FIGURE 5.1 A scenario for the evolution of life focused on the notion that the synthesis of adenosine triphosphate (ATP) was most critical in the early stages of prebiotic evolution. (Following Galimov, 2001, 2004; see also http://press.princeton.edu/chapters/s9501.pdf.)

chemical evolution of life was the development of a molecule called transfer ribonucleic acid (tRNA) that allowed for genetic coding in primitive organisms. Some suspect that all life diverged from a common ancestor relatively soon after life began. The early eukaryotes split off from the archaebacteria some millions of years after their evolution. This belief has ground in DNA sequencing.

The Earth is at least 4.6×10^9 years old, but it was the microbial organisms that dominated the Earth for the first 70%–90% of its history (Woese, 1981; Woese et al., 1990). The composition of the atmosphere of the early Earth has extensively been debated; however, it is now accepted that a reducing environment prevailed (Sagan and Chyba, 1997; Galimov, 2005). The so-called banded iron formations (BIFs) first appeared in sediments deposited 3×10^9 years ago, during the early history of the Earth. BIFs include layers of iron oxides, magnetite or hematite, alternating with iron-poor layers of shale and chert. It is assumed that iron oxide formations resulted from the reaction between oxygen produced in cyanobacteria photosynthesis and dissolved reduced iron. The subsequent disappearance of BIFs in the geologic record approximately 1.8×10^9 years ago is believed to have resulted after a phase of oxygen level rising.

Levels of oxygen in the atmosphere began to rise at about 2.4 to 2.3×10^9 years ago (Farquhar et al., 2000; Bekker et al., 2004). Even during the mid-Proterozoic (ca. 1.8 to 0.8×10^9 years ago), the oceans were either anaerobic (no oxygen) or dysaerobic (low oxygen) compared to the oceans today (Arnold et al., 2004). The rise in oxygen in the oceans from this period to the present was due to increased production by photoautotrophic organisms, which transfer atmospheric CO_2 and dissolved inorganic C (e.g., HCO_3^-) into biomass by photosynthesis. Many of these early photoautotrophs were likely cyanobacteria (Brocks et al., 1999).

Here we should mention the approach by Fedonkin (2008), who considers transition metals as a critical factor of evolution. This factor is illustrated in Figure 5.2, modified from the mentioned work showing the universal tree of life with phylogenetic relationships between the living organisms within the three domains of Bacteria, Archaea and Eukarya.

Figure 5.2 introduces the concepts of the first eukaryotic common ancestor (FECA) and the last eukaryotic common ancestor (LECA) as two distinct forms. The FECA is commonly represented as the first descendant, on the eukaryotic branch, of the last common ancestral node of eukaryotes. The FECA is the ancestor of all eukaryotes that ever existed, whether extant or extinct. The LECA is, by definition, the ancestral state that gave rise to all extant eukaryotes. There are three possible forms of biological concepts of the LECA; they range from a single cell to a population to a consortium. O'Malley et al. (2019) elaborated on each biological concept and the major differences between them.

The succession of the branching and the branch length indicate the degree of difference among the gene sequences that code for RNA in the small subunit of the ribosomes (16S/18S rRNA). Thick lines on the base of the tree mark hyperthermophiles, organisms that successfully grow at high temperatures. This approach suggests the very early origin of the hyperthermophilic microorganisms in hot environments rich in iron, nickel and tungsten (indicated by their symbols), the metals that play a crucial role in the active sites of enzymes in these prokaryotic organisms.

Cyanobacteria occupy the final position in the domain Bacteria, which indicates the relatively late origin of this group and the oxygen photosynthesis. Oxygenation of the environment reduced the availability of iron, nickel and tungsten, while molybdenum, copper and zinc had become bioavailable.

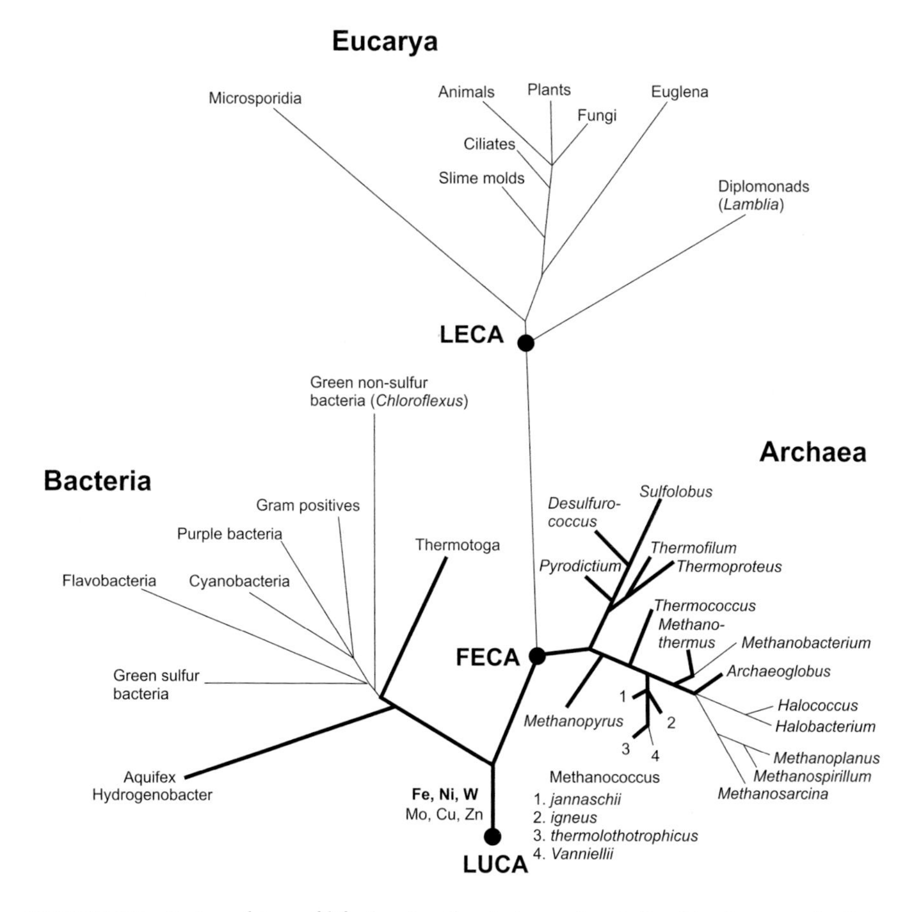

FIGURE 5.2 Universal tree of life showing the phylogenetic relationships between the living organisms within the three domains of Bacteria, Archaea and Eukarya. (Modified from Fedonkin, 2008.) FECA – first eukaryotic common ancestor; LECA – last eukaryotic common ancestor.

Recently, Chatterjee (2016) has discussed the possibility of the origin of life at hydrothermal impact crater-lakes. Submarine hydrothermal vents are considered as the likely habitats for the origin and evolution of early life on the Earth. The theory suffers from the "concentration problem" of cosmic and terrestrial biomolecules because of the large size of the Eoarchean global ocean. An alternative site would be small highly sequestered, hydrothermal crater-lakes that might have helped life on the early Earth. A model for the origin of life at hydrothermal crater-lakes is proposed by Chatterjee (2016). Impacts of meteors on the Eoarchean crust at the end of the Heavy Bombardment Period have probably played essential roles in the origin of life. Results that created hydrothermal crater-lakes on the Eoarchean crust represented the crucibles for chemistry with building blocks of life, which at the end led to the first organisms. In this scenario, life arose through hierarchical stages of increasing molecular complexity in multiple niches of crater basins. The building blocks of life originated in interstellar space during the explosion of a nearby star. According to this author, both comets and carbonaceous chondrites delivered building blocks of life and ice to the early Earth. All of this was accumulated in hydrothermal impact crater-lakes. Crater basins contained a spectrum of cosmic and terrestrial organic compounds powered by hydrothermal, solar, tidal and chemical energies, which helped the prebiotic synthesis. Self-assembled primitive lipid membranes floated at the water surface as a thick oil slick. The oldest fossils of thermophilic life ($\sim 3.5 \times 10^9$ years) have been detected in Archean greenstone belts in Greenland, Australia and South Africa. These locations represent most probably the relics of Archean craters. Monomers, nucleotides and amino acids were selected from random assemblies of the prebiotic soup. They were polymerized at pores of mineral surfaces with the coevolution of RNA and protein molecules to form the "RNA–protein world". Although the endosymbiotic model proposed by Chatterjee (2016) is speculative, it has intrinsic heuristic value.

Some more research groups have accepted the idea that some of the earliest habitable environments may have been submarine hydrothermal vents. For example, Dodd et al. (2017) have recently described putative fossilized microorganisms at least 3.77×10^9 and possibly 4.28×10^9 years old, which are found in sedimentary rocks containing iron oxide. The authors interpreted this finding as seafloor hydrothermal vent-related precipitates, from the Nuvvuagittuq Belt in Quebec, Canada, the earliest known rocks on the Earth (assumed to be 4.28×10^9 years old). The reported structures occur as micrometer-scale hematite tubes and filaments. Their morphologies and mineral assemblages are similar to filamentous microorganisms from modern hydrothermal vent precipitates. They are also similar to microfossils in younger rocks. The Nuvvuagittuq rocks material contains isotopically light carbon as carbonate and carbonaceous material. It occurs as graphitic inclusions in apatite blades intergrown among carbonate rosettes and magnetite granules. It is associated with carbonate in direct contact with putative microfossils. According to the authors, all of their observations of oxidized biomass are consistent with biological activity in submarine hydrothermal environments more than 3.77×10^9 years ago.

Organic chemistry on a planetary scale has probably transformed carbon dioxide and reduced carbon species delivered to the Earth. According to some models that propose the origin of life on the Earth, biological molecules that jump-started Darwinian evolution arose via this planetary chemistry. Most of these models assume that ribonucleic acid arose prebiotically, together with components for compartments and a primitive metabolism. Unfortunately, it has been very difficult to identify prebiotic chemistry that might have created RNA. Organic molecules have a well-known tendency to form multiple objects referred to as "tar" or "tholin". These mixtures are unsuited to support Darwinian processes and, indeed, have never been observed to yield a homochiral genetic polymer spontaneously. To date, the proposed solutions to this problem either involve too much direct human intervention to satisfy many sceptics or generate molecules that present evolutionary "dead ends" for standard conditions of temperature and pressure. Organic species, carbohydrates, have carbon, hydrogen and oxygen atoms in a ratio of 1:2:1 together with an aldehyde or ketone group. They are components of RNA, and their reactivity can both support interesting spontaneous chemistry as part of a "carbohydrate world" but also quickly form mixtures, polymers and tars (Benner et al., 2010).

According to Scorei and Cimpoias (2006), boron enhances the thermostability of carbohydrates. They have studied the effect of borate and pH upon the half-lives of ribose and glucose and found that under acidic conditions, the presence of boric acid increases the thermostability of ribose, while under primary conditions, glucose is favored.

The effect of boron on heterocystous and non-heterocystous dinitrogen-fixing cyanobacteria was examined by Bonilla et al. (1990). The absence of boron prevented growth and nitrogenase activity in cultures of *Nodularia* sp., *Chlorogloeopsis* sp. and *Nostoc* sp. Subsequent examinations of cultures showed changes in heterocyst morphology. However, cultures of non-heterocystous cyanobacteria, *Gloeothece* sp. and *Plectonema* sp., grown in the absence of boron did not show alteration in growth. These results imply the existence of boron requirement only in heterocystous cyanobacteria.

Because boric acid forms esters with cis-diols, Bonilla et al. (1990) suggested a possible role for B in the stabilization of the glycolipid inner layer of heterocysts by interacting with their -OH groups. The same has been proposed for higher plant cell membranes by Parr and Loughman (1983) and Pollard et al. (1977). Boron deficiency could lead to an alteration in the heterocyst envelope, which would facilitate O_2 diffusion and result in an inhibition of nitrogenase activity. This hypothesis is consistent with the inhibitory effect that B deficiency has only on heterocystous cyanobacteria. Furthermore, a drastic alteration in the protection from O_2 diffusion in the envelope of the heterocysts in Nodularia sp. cells has also been shown in this study.

The chemical element B is involved in the stabilization of the heterocyst structure. This result suggests that the essentiality of B is restricted to heterocystous cyanobacteria only. Given their biological antiquity, this might indicate that B was necessary for the early history of life. Finally, the findings show that cyanobacteria are adequate models for the study of mineral nutrient requirements concerning the origin of life (Bonilla et al., 1990).

Chopra and Lineweaver (2015) have described in their recent communications the average bulk elemental abundances in living life, which can yield an indirect estimate of the relative abundances of elements in the LUCA. Their results could give valuable hints about the stoichiometry of the environment where the LUCA existed. Perhaps clues of the processes involved in the origin and early evolution of life could also be obtained (Chopra et al., 2009). Chopra and Lineweaver (2015) conducted a meta-analysis of historical and recent studies that examined the elemental abundances in various taxa. The eukaryotic, bacterial and archaeal taxa across the existing tree of life were analyzed. Based on the elemental abundances therein and the phylogenetic relationship between the taxa, they derived their best estimate of the elemental composition of the LUCA. Their best rating for life-essential elements is shown in Figure 5.3.

In evaluating the bulk elemental composition of the LUCA, Chopra and Lineweaver (2015) used abundances for almost all of the biologically relevant elements, including the bulk elements (H, O, C and N), major elements (P, S, Na, K, Mg and Ca) and trace elements (Fe, Cu and Zn). In establishing an average elemental composition of life, they attempted to account for the differences in structure between species and other

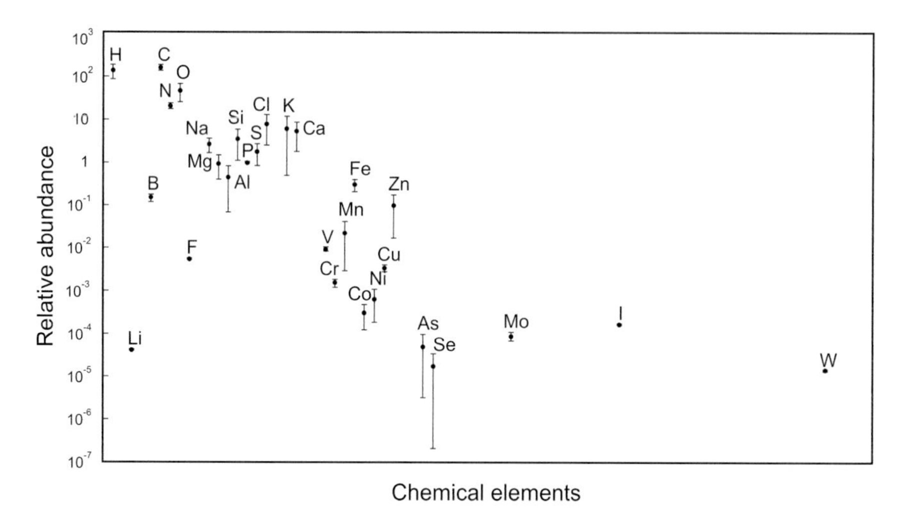

FIGURE 5.3 The bulk elemental composition of the LUCA as estimated by Chopra and Lineweaver (2015) from abundances in extant taxa. Abundances only for life-essential elements are shown.

phylogenetic taxa by weighting representative data sets so that the result represents the root of prokaryotic life. Variations in composition between data sets that can be attributed to different growth stages or environmental factors were used as estimates of the uncertainty associated with the average abundances for each taxon.

Quantitative estimates of the gene complement of the LUCA may be masked by ancient horizontal gene transfer events and polyphyletic gene losses. Biases in genome databases and methodological artifacts also represent drawbacks. Nevertheless, most reports agree that the LUCA resembled extant prokaryotes. A number of highly conserved genes are sequences involved in the synthesis, degradation and binding of RNA, including transcription and translation. Although the gene complement of the LUCA includes sequences that may have originated in different epochs, the outstanding conservation of RNA-related sequences supports the hypothesis that the LUCA was an evolutionary result of the so-called RNA–protein world. The available evidence suggests that the LUCA was not a hyperthermophile. However, it is currently not possible to assess its ecological niche or its mode of energy acquisition and carbon sources (Becerra, 2007).

The first successful laboratory synthesis of amino acids and related organic compounds was done by Miller (1953). Since then, several laboratory experiments have been performed, which revealed the importance of catalyzers, the role of nutrients, and physical and chemical environments on the primordial Earth to bear life (Dyson, 1982; Nisbet et al., 1995; Nisbet and Sleep, 2001; Nisbet, 2002; Russell and Arndt, 2005; Russell, 2007). Their primary assumption is that a habitable planet, like the Earth, must automatically yield life, which eventually leads to evolved life forms. This assumption has led to the speculations that life could be a common phenomenon in the Universe based on the number of discovered exoplanets (Maruyama et al., 2012).

Small exoplanets are common in the Milky Way galaxy (Fulton et al., 2017, and references therein). The recently launched Transiting Exoplanet Survey Satellite (TESS) is revolutionizing the field of exoplanet science by discovering planets around the nearest stars. The mass, atmosphere composition and other properties of small exoplanets will be measured by TESS (Dragomir et al., 2019). The planet–metallicity correlation is discussed by Fischer and Valenti (2005); they studied a subset of 850 stars and determined that fewer than 3% of stars with $-0.5 < [Fe/H] < 0.0$ have planets. Above solar metallicity, there is a smooth and rapid rise in the fraction of stars with planets. At $[Fe/H] > +3$ dex, 25% of the observed stars have detected gas giant planets. Higher stellar metallicity also appears to be correlated with the presence of multiple-planet systems and with the total detected planet mass. The authors suggested that stars with extrasolar planets do not have an accretion signature that distinguishes them from other stars; more likely, they are simply born in higher-metallicity molecular clouds (Fisher and Valenti, 2005).

McCabe and Lucas (2010) applied a simple stochastic model of evolution based on the requirement of passage through a sequence of "n" critical steps (Carter, 1983; Watson, 2008) to both terrestrial and extraterrestrial origins of life. The time at which humans have emerged in the habitable period of the Earth suggests a value of critical steps of $n=4$. By adding earlier evolutionary transitions, Maynard Smith and Szathmary (1995) gave an optimum fit with $n=5$, implying either that their initial developments are not critical or that habitability began around 6×10^9 years ago. According to them, the origin of life on Mars, or elsewhere within the solar system, is excluded by calculations. The simple anthropic argument argues that extraterrestrial life is scarce in the Universe because there was no time for it to evolve. However, the timescale can be extended if the migration of life-primary material to the Earth is possible. In the case, other transitions are included in the model allowing Earth migration; the start of habitability needs to be even earlier than 6×10^9 years ago. The present understanding of galactic habitability and dynamics does not exclude this possibility.

A simple stochastic model, proposed by Carter (1983), contains the assumption that evolution is controlled by a series of critical transitions or bottlenecks, which must be passed to transit from essential molecules to simple, sophisticated and afterward intelligent life. The anthropic model allowed the derivation of a probability density function and expectation time for the final step, based upon the habitable lifetime of the planet. Watson (2008) further analyzed the model to calculate probability functions and expectation times for every step. He identified seven critical transitions in biological evolution among the eight proposed by Maynard Smith and Szathmary (1995) and compared their probabilities with their estimated times of occurrence.

In 1961, Frank Drake (2003) proposed an equation to estimate the number of detectable extraterrestrial civilizations in our Milky Way galaxy, which has widely been referred to as "the Drake equation".

$$N = R^* f_p n_e f_l f_i f_c L$$

where

N = the number of detectable civilizations in the Milky Way galaxy.

R^* = the formation rate of stars having properties suitable for the development of intelligent life.
f_p = the fraction of those stars having planetary systems.
n_e = the number of planets having an environment suitable for life.
f_l = the fraction of planets on which life appears.
f_i = the fraction of life-bearing planets on which intelligent life has emerged.
f_c = the fraction of civilizations having a technology that releases detectable signals of their existence into space.
L = the length of time such civilizations is releasing detectable signals into space.

Although the Drake equation has long been used as a touchstone for astrobiology, the resulting large number for N has been questioned by many; see, for example, Maruyama et al. (2012), who discussed the specific capabilities of our planet to sustain life.

Its mix of astrophysical, biological and social aspects, with an enormous level of uncertainty, continues to provoke discussion. The critical transition model addresses an issue for any Drake-type equation, e.g., the Rare Earth equation (Ward and Brownlee, 2000). If any of the factors is zero, the result is then zero. The critical transition model breaks down the crucial biological steps, typically only represented by one or two factors, and introduces time dependence. In particular, it provides a reasonable fit to the timescales for the emergence of life on the Earth. Furthermore, if the possibility of life originating beyond the Earth is considered, it can provide useful clues as to the time that life might have emerged.

Recently, Frank and Sullivan III (2016) have addressed the cosmic frequency of technological species. Advances in the understanding of exoplanet properties provide strong constraints on all astrophysical terms in the Drake equation. Using these constraints and modifying the form and intent of the Drake equation, they set a firm lower value for the probability of the evolution of one or more technological species in the history of the observable Universe. They find that as long as the likelihood that a habitable-zone planet develops a technological species is more significant than $\sim 10^{-24}$, humanity is not the only time technical intelligence has evolved. This constraint has significant scientific and philosophical consequences.

The Drake equation can be rewritten as:

$$A = \left[N^* f_p n_p\right]\left[f_l f_i f_t\right] = N_{ast} f_{bt}$$

where A describes the total number of technological species that have ever evolved anywhere in the currently observable Universe. The equation is reduced to two factors, the first of which, N_{ast}, includes all factors involving astrophysics and represents the total number of habitable-zone planets. The second factor, f_{bt}, contains three factors involved with biology, evolution and "planetary sociology" and represents the total "biotechnical" probability that a given habitable-zone planet has ever evolved a technological species. The factor $f_{bt} = f_l f_i f_t$ is exceptionally predictable since we have no theory to guide any estimates. Additionally, we have only one known example of the occurrence and history of life, intelligence and technology.

According to Frank and Sullivan III (2016), only for values of f_{bt} lower than $2.5 \cdot 10^{-24}$ we are likely to be alone and singular in the history of the observable Universe. This limit on the value of f_{bt} should be considered to define a "pessimism line" in evaluations of the possibilities of the existence of technological civilizations on a cosmic scale. Cai et al. (2020) developed a model which identified a peak location for extraterrestrial intelligence at an annular region approximately 4 kpc from the Milky Way galactic center around 8×10^9 years, with complex life decreasing temporally and spatially from the peak point, with a high likelihood of intelligent life in the inner galactic disk. The simulated age distributions also suggest that most of the intelligent life in our galaxy are young, making their observation or detection a difficult job.

At this point, we should mention that after the historical discovery of the first planet outside our solar system, 51 Pegasi b orbiting star 51 Pegasi (see Mayor and Queloz, 1955), thousands of exoplanets are found orbiting nearby stars. The planning of the next generation of space-based and ground-based telescopes anticipates that some habitable planets can be identified in the coming decade. Even more foreseen is the chance to find signs of life on these habitable planets by way of bio-signature gases. The question is as follows: For which gases should one search? Answer to this question is offered by Seager et al. (2016). According to them, the first is a list of bio-signature gases present in the Earth's atmospheric spectrum O_2, CH_4 and N_2O. Also, one should consider these being produced at or able to accumulate to higher levels on exo-Earths, e.g., dimethyl sulfide and CH_3Cl. Life species on the Earth produces thousands of different gases (although most in very small quantities). Some might be produced and accumulate in an exo-Earth atmosphere to high levels, depending on the exo-Earth ecology and surface and atmospheric chemistry.

There is a possibility of their accumulation and possible false positives on exoplanets with atmospheres and surface environments different from the Earth's. An online community usage database to serve as a registry for volatile molecules, including biogenic compounds, is also provided by Seager et al. (2016). The authors have constructed a list of molecules being stable, maybe volatile, as pure compounds at standard temperature and pressure (STP). Materials made of small molecules are probably unstable, meaning more likely to be in gas form in a planetary atmosphere. They have constructed a list of molecules with up to $N=6$ non-hydrogen atoms stable in the presence of water and volatile at standard temperature and pressure. The file contains about 14,000 molecules: About 2,500 are composed of the six biogenic elements (C, N, O, P, S and H); about 900 are inorganics; and about 11,000 are halogenated compounds. This file forms a large set because of the large number of combinatorial possibilities of adding halogens to carbon skeletons. The list was constructed by a combinatorial approach and an intense database and literature search.

Marboeuf et al. (2014) presented works aiming at determining the chemical composition of planets formed in stellar systems of solar chemical composition. The main objective of their work was to provide valuable theoretical data for models of planet formation and evolution and future interpretation of the chemical composition of solar and extrasolar planets. The data include the chemical composition, ice–rock mass ratio, C/O molar ratio for planets in stellar systems and solar chemical composition. From an initial homogeneous structure of the nebula, they produce a wide variety of planets with chemical compositions varying with the mass of the disk and distance to the star.

Their volatile species are mainly H_2O, CO, CO_2, CH_3OH and NH_3. Planets with ice or ocean have systematically higher values of molecular abundances compared to big and rocky planets. Giant gas planets are depleted in highly volatile molecules (CH_4, CO and N_2) compared to planets with ice or oceans. The ice–rock mass ratio on icy or ocean and giant gas planets is equal, at maximum, to 1.01 ± 0.33 and 0.8 ± 0.5, respectively. This value is different from the usual assumptions made in planet formation models, which suggested this ratio to be 2–3. The C/O molar ratio in the atmosphere of gas giant planets is depleted by at least 30% compared to solar value.

Recently, the first discovery of a habitable-zone Earth-sized planet has been reported (Gilbert et al., 2020). They reported the discovery and validation of a three-planet system orbiting the nearby "" dwarf star TOI-700 positioned in the TESS (NASA's Transiting Exoplanet Survey Satellite) continuous viewing zone in the Southern Ecliptic Hemisphere. Three planets with radii in the range of $(1\text{–}2.6) \times R_{Earth}$ and orbital periods in the range of (9.98–37.43) days have been observed. The outermost planet, TOI-700 d, has a radius of $(1.19 \pm 0.11) \times R_{Earth}$ and resides within a conservative estimate of the host star's habitable zone. At that position, it receives a flux from its star that is ~86% of the Earth's insolation.

5.4 Origin of Life in the Universe

Life originating in physical or chemical environments very different from that of the Earth might select different groups of atoms and bonds to use in building biochemistry. For example, life on the Earth rarely uses the C–F bond, one of the strongest single bonds known. Life originating at much higher temperatures could use more C–F bonds to compensate for the greater instability of molecules at those temperatures and consequently would generate fluorocarbon bio-signature gases. Fluorocarbons are particularly interesting as signature molecules since they are anomalously volatile for their molecular weight (Seager et al., 2016).

On the other hand, carbon and molecules made from it have already been observed in the early Universe. During cosmic time, many galaxies undergo intense periods of star formation, during which elements carbon, oxygen, nitrogen, silicon and iron are produced. Many complex molecules, starting from carbon monoxide to polycyclic aromatic hydrocarbons, are detected in these systems, like in our galaxy. It is well established that interstellar molecular clouds and circumstellar envelopes are places where complex molecular synthesis happens (Ehrenfreund and Charnley, 2000; Kwok, 2004; van Dishoeck and Blake, 1998). In addition to interstellar gas, the material also contains small micron-sized particles. Gas-phase and gas-grain interactions lead to the formation of complex molecules on solid interstellar particles. Surface catalysis enables molecule formation since chemical pathways cannot proceed in the gas phase because of reaction barriers. A very high number of molecules that are used in contemporary biochemistry on the Earth have been found so far in the interstellar medium; planetary atmospheres; and surfaces of comets, asteroids, meteorites and interplanetary dust particles. Large quantities of this extraterrestrial material were delivered via comets and asteroids to young planetary surfaces during the heavy bombardment phase. Knowledge about the formation and evolution of organic matter in space is crucial to determine the prebiotic reservoirs available to the early Earth. It is equally important to reveal abiotic routes to prebiotic molecules in the Earth's environments (Ehrenfreund et al., 2011).

The so-called globular clusters are bound groups of about a million stars and stellar remnants. They are old, mostly isolated and very dense. Di Stefano and Ray (2016) considered how each of these unique features could influence the development of life, the evolution of intelligent life and the long-term survival of technological civilizations. They found that if they house planets, globular clusters provide ideal environments for advanced civilizations that can survive over long times. They, therefore, proposed methods to search for planets in globular clusters. If planets are found and if their assumptions are correct, searches for intelligent life are most likely to succeed when looking toward globular clusters. Di Stefano and Ray (2016) argue that globular clusters might be the best places in which distant life could be identified in our own or any other galaxy.

Our Milky Way galaxy contains about 150 globular clusters, most of them orbiting in the galactic periphery. They formed approximately 10×10^9 years ago on average. As a result, their stars composition has less of the heavy elements needed for making planets because elements such as iron and silicon must be created in earlier generations of stars. We could argue that this makes globular cluster stars less likely to host planets. It is known that only one planet has been found in a globular cluster so far. Another concern is that a crowded environment of the globular cluster would be a threat to any planets that formed in this area. A neighboring star could come too close and gravitationally disrupt a planetary system, flinging worlds into icy interstellar space.

5.4.1 Cosmic Dust

The cosmic background radiation, the remnant of the first event, the Big Bang, is substantially uniformly distributed in the Universe (COBE, 2015). The measured cosmic microwave background spectrum is almost perfect blackbody radiation with a temperature of 2.725 ± 0.002 K. This measurement matches the predictions of the hot Big Bang theory exceptionally well. It indicates that nearly all of the radiant energy of the Universe was released immediately after the Big Bang.

Contrary to radiation, the distribution of matter is highly non-uniform. Galaxies occupy 10^{-7} of the volume of the Universe, but contain most of the known matter. The sharp aggregation of matter means that the chemistry is occurring within galaxies. The small particles in the space between the stars have the surfaces convenient for the production of many simple and complex molecules. The existence of molecules in the low-density matter that exists between the stars has been first reported by Adams (1949), and the list of the observed molecules (see Table 5.1, after http://www.astro.uni-koeln.de/cdms/molecules) has been increasing since then. Currently, two molecules have been reported as detected whose detections have been questioned fairly convincingly in subsequent papers. These molecules are amino acetic acid, H_2NCH_2COOH, aka glycine, and 1,3-dihydroxypropanone, aka dihydroxyacetone. The authors consider these molecules as not yet detected. The detection of SH by vibrational spectroscopy of the S-type star R Andromedae is not included in this table as the molecule is present in the stellar atmosphere.

TABLE 5.1 Molecules in the Interstellar Medium or Circumstellar Shells, as of 01/2019

Number of Atoms	Molecules
2	H_2, AlF, AlCl, C_2[b], CH, CFH^+, CN, CO, CO[a], CP, SiC, HCl, KCl, NH, NO, NS, NaCl, OH, PN, SO, SO[a], SiN, SiO, SiS, CS, HF, HD, FeO(?), O_2, CF[a], SiH(?), PO, AlO, OH[a], CN[a], SH[a], SH, HCl[a], TiO, ArH^+, N_2, NO[a] (?), NS[a]
3	C_3[a], C_2H, C_2O, C_2S, CH_2, HCN, HCO, HCO[a], HCS[a], HOC[a], H_2O, H_2S, HNC, HNO, MgCN, MgNC, N_2H^+, N_2O, NaCN, OCS, SO_2, c-SiC_2, CO_2[a], NH_2, H_3^+[(a)], SiCN, AlNC, SiNC, HCP, AlOH, H_2O^+, H_2Cl^+, KCN, FeCN, HO_2, TiO_2, C_2N, Si_2C, HS_2, HCS, HSC, HCO
4	c-C_3H, I-C_3H, C_3N, C_3O, C_3S, C_2H_2[a], NH_3, HCCN, HCNH[a], HNCO, HNCS, $HOCO^+$, H_2CO, H_2CN, H_2CS, H_3O^+, c-SiC_3, CH_3[a], C_3N^-, PH_3, HCNO, HOCN, HSCN, H_2O_2, C_3H^+, NMgNC, HCCO, CNCN
5	C_5[a], C_4H, C_4Si, I-C_3H_2, c-C_3H_2, H_2CCN, CH_4[a], HC_3N, HC_2NC, HCOOH, H_2CNH, H_2C_2O, H_2NCN, HNC_3, SiH_4[a], H_2COH[a], C_4H^-, HC(O)CN, HNCNH, CH_3O, NH_4^+, H_2NCO^+, $NCCNH^+$, CH_3Cl
6	C_5H, I-H_2C_4, C_2H_4[a], CH_3CN, CH_3NC, CH_3OH, CH_3SH, HC_3NH[a], HC_2CHO, NH_2CHO, C_5N, I-HC_4H[a], I-HC_4N, c-H_2C_3O, H_2CCNH(?), C_5N^-, HNCHCN, SiH_3CN,C_5S(?)
7	C_6H, CH_2CHCN, CH_3C_2H, HC_5N, CH_3CHO, CH_3NH_2, c-C_2H_4O, H_2CCHOH, C_6H^-, CH_3NCO, HC_5O, $HOCH_2CN$
8	CH_3C_3N, HC(O)OCH_3, CH_3COOH, C_7H, C_6H_2, CH_2OHCHO, I-HC_6H[a], CH_2CHCHO(?), CH_2CCHCN, H_2NCH_2CN, CH_3CHNH, CH_3SiH_3
9	CH_3C_4H, CH_3CH_2CN, $(CH_3)_2O$, CH_3CH_2OH, HC_7N, C_8H, $CH_3C(O)NH_2$, C_8H^-, C_3H_6, CH_3CH_2SH(?), CH_3NHCHO(?), HC_7O
10	CH_3C_5N, $(CH_3)_2CO$, $(CH_2OH)_2$, CH_3CH_2CHO, CH_3CHCH_2O, CH_3OCH_2OH
11	HC_9N, CH_3C_6H, C_2H_5OCHO, $CH_3OC(O)CH_3$
12	c-C_6H_6[a], n-C_3H_7, i-C_3H_7CN, $C_2H_5OCH_3$(?)
>12	C_{60}[a], C_{70}[a], C_{60}^{+}[a], c-C_6H_5CN

Credit: The Cologne Database for Molecular Spectroscopy, http://www.astro.uni-koeln.de/cdms/molecules.

All molecules have been detected by rotational spectroscopy in the radiofrequency to far-infrared regions unless indicated otherwise.

[a] Molecules that have been recognized by their rotational–vibrational spectrum.

[b] Those detected by electronic spectroscopy only.

The list contains a new addition, HeH^+, as described in the paper by Güsten et al. (2019), which is found in the planetary nebula NGC 7027. This molecule is the first type ever formed in the Universe, around 10^5 years after the Big Bang. Its destruction created a path to the formation of molecular hydrogen through a series of reactions:

$$He^+ + H \rightarrow HeH^+ + h\upsilon$$

$$HeH^+ + e \rightarrow He + H$$

$$HeH^+ + H \rightarrow He + H_2^+$$

Molecular hydrogen is the molecule primarily responsible for the formation of the first stars.

TABLE 5.2 Extragalactic Molecules

Number of Atoms	Molecules
2	CH, CO, H_2[a], CH, CS, CH[a,b], CN, SO, SiO, CO[a], NO, NS, NH, OH[a], HF, SO[a], ArH[a]
3	H_2O, HCN, HCO[a], C_2H, HNC, N_2H[a], OCS, HCO, H_2S, SO_2, HOC[a], C_2S, H_2O[a], HCS[a], H_2Cl[a], NH_2
4	H_2CO, NH_3, HNCO, C_2H_2[a], H_2CS(?), HOCO[a], c-C_3H, H_3O[a], I-C_3H
5	c-C_3H_2, HC_3N, CH_2NH, NH_2CN, I-C_3H_2, H_2CCN, H_2CCO, C_4H
6	CH_3OH CH_3CN, HC_4H[a], HC(O)NH_2
7	CH_3CCH, CH_3NH_2, CH_3CHO
8	HC_6H
>8	c-C_6H_6[a]

Credit: The Cologne Database for Molecular Spectroscopy, http://www.astro.uni-koeln.de/cdms/molecules.

All molecules have been detected by rotational spectroscopy in the radiofrequency to far-infrared regions unless indicated otherwise.

[a] Molecules that have been identified by their rotational–vibrational spectrum.

[b] Those detected by electronic spectroscopy only.

In Table 5.2, extragalactic molecules are listed (modified on March 29, 2019). For more information on the Cologne Database for Molecular Spectroscopy (CDMS), Institute of Physics, Faculty of Mathematics and Natural Science, University of Köln, see publications by Müller et al. (2001, 2005).

To the lists shown in Tables 5.1 and 5.2, we should add glycine, NH_2CH_2COOH, the simplest amino acid and methylamine, NH_2CH_3, one of its possible precursors. They were detected in the coma of the comets by the Stardust mission (Elsila et al., 2009) and by Rosetta mission (Altwegg et al., 2016). Since the comets are the most primitive planetary bodies, the molecules found in our solar system have an interstellar origin. Ioppolo et al. (2020) investigated the "non-energetic" atom addition surface reaction as a formation route of these molecules. According to the authors, their extensive joint laboratory and modelling efforts strongly support the relevance of a non-energetic origin of glycine in interstellar ices.

Ioppolo et al. (2020) extended their finding by the suggestion "that surface substitution reactions of H atoms from the α carbon with larger side chains can potentially lead to the formation of other proteinogenic α-amino acids, the building blocks of the proteins relevant to life on Earth". This suggestion has been challenged by Valkovic and Obhodas (2020).

Interstellar dust is coupled to gas clouds and, as such, carried around the Milky Way. These clouds come in a wide variety of shapes, sizes, densities and temperatures. They can, however, be qualitatively classified into two basic categories: diffuse clouds and molecular clouds. The diffuse clouds are not distinguishable and are limited to a density less than about 300 hydrogen atoms per cm^3 with a temperature of 50–100 K. The molecular clouds can have temperatures of 20 K (even 10 K) and density above 300 hydrogen atoms per cm^3, including the densities at which clouds collapse to form stars. Diffuse clouds contain hydrogen in atomic form; molecular hydrogen is a dominant component of molecular clouds. They can be very dark, as illustrated by the Horsehead Nebula, and also by isolated blank regions in the sky, widely called Bok globules. These latter are probably concentrations of dust and gas, which are collapsing to form stars (Greenberg, 2002).

The energetic components of the environment affecting the dust (other than that associated with phenomena such as star formation and even more explosive supernovae) are the ultraviolet radiation from stars and the cosmic-ray particles. One of the significant effects of ultraviolet is that it heats the dust. The dust temperature in diffuse clouds could be about 15 K as a result of a dynamic balance between absorption and emission of radiation. However, in molecular clouds where it is shielded, the temperature may be as low as 5–10 K (Greenberg, 1971; Greenberg and Li, 1996).

Detailed comparisons of the B II (1,362 Å) and O I (1,355 Å) line shapes indicate that the B-to-O ratio is significantly higher in warm interstellar clouds than in cold clouds. These results support the incorporation of boron into dust grains in diffuse ISM (Howk et al., 2000). Ritchey et al. (2011) presented a survey of boron abundances in diffuse interstellar clouds. The data were obtained from observations made with the Space Telescope Imaging Spectrograph (STIS) of the Hubble Space Telescope.

The number of molecules already detected in extragalactic sources is large; see Tables 5.1 and 5.2. More may have to be added soon.

The main characteristics of interstellar solid particles are their size (sizes), shape, chemical composition and amount. These molecules must be determined by remote observations with telescopes in space and on the ground in a variety of optical wavelengths. Only by combining all these observations can one come to an accurate description. The most obvious consequence of the dust presence is the extraordinary blocking of the light of the stars. As seen from the average interstellar extinction curve, the light in red is reduced less than the light in the blue and the ultraviolet. From the shape of the extinction curve, one can deduce that the particles responsible for the visual extinction are about 0.1 μm in size (a mean radius) and those responsible for the hump and the far-ultraviolet disappearance are 10–100 times smaller (Li and Greenberg, 1997).

The principal molecular ingredients of interstellar dust are obtained by studying the way they absorb or emit infrared radiation. Luckily, the so-called fingerprint region of the infrared spectrum from 2.5 to 25 μm, where most vibrations of the molecular groups containing C, N, O and H occur, is accessible with modern telescopes.

The nuclei on which the solids can grow in space are mostly small silicate particles that are condensed in the atmospheres of cool, old (evolved) stars that are in their red giant expansion phase. While there may be other sources of such nuclei, they are not as identifiable nor as clearly seen as in the emission spectra of red giants. The spectra of red giants and supergiants have excess emissions at 10 and 20 μm produced by the heated dust, which is formed in and blown out with gas from the stars.

Dust grains in cold, dense interstellar clouds are considered to be mixtures of dust particles with molecular ices, water being the main constituent of these ices, accounting for more than 60% of the ice in most lines of sight (Whittet, 2003). Ices are believed to cover the surface of a dust core and/or to be physically mixed with dust. Potapov et al. (2020) provided evidence of the presence of solid-state water in the diffuse interstellar medium by combining the laboratory data and infrared observations. Their laboratory data can explain the observations, assuming reasonable mass-averaged temperatures for the protostellar envelopes and protoplanetary disks, demonstrating that a substantial fraction of water ice may be mixed with silicate grains.

The atomic composition of giant molecular clouds regions is determined by the history of nearby stars, which eject processed nuclear material via stellar winds and supernova explosions. The molecular structure reflects the balance between chemical evolution via reactions, destruction of molecules by light from stars or by cosmic rays, and condensation and subsequent reaction on dust grains.

Molecular clouds are characterized by low temperatures, on the order of 10 K. At these temperatures, there is insufficient energy for collisions to overcome any activation barrier to reaction. The only gas-phase chemical reactions that can proceed at such low temperatures are radical–radical reactions and ion–molecule reactions, both of which are barrierless. Interstellar gas clouds also have extremely low densities, which have drastic consequences for the collision frequency, and therefore the number of opportunities for chemistry to occur. The number of collisions is small, even in the densest regions; with densities of $10^6 cm^{-1}$, collision rates are around $5\times10^{-4}/s$, approximately one collision every half an hour. In less-dense regions, atoms and molecules may go for many weeks or even longer between collisions. Chemistry, therefore, occurs at a prolonged rate in interstellar space compared to the timescales on the Earth. The giant molecular clouds last for around 10–100 million years before they are dissipated by heat and stellar winds from stars forming within them. There is plenty of time for some quiet complex chemistry to occur, albeit at a rather slow rate (see, for example, Shaw, 2006; Vallence, 2014).

The very low collision frequency has significant consequences for the types of molecules that may form in interstellar space. Terrestrial concepts of molecular stability do not apply in this extremely non-reactive environment. Carbon does not need to have four bonds; in fact, there are many subvalent species, radicals, molecular ions and energetic isomers among the molecules observed in interstellar gas clouds. Carbon-containing compounds tend to be highly unsaturated, with many double and triple bonds and few branched chains. Polyynes (organic compounds with alternating single and triple bonds) are commonly observed, some with quite long chain lengths, for example,

$$H—C\equiv C—C\equiv C—C\equiv C—C\equiv C—C\equiv C—C\equiv N.$$

The collisions between atoms and molecules in interstellar space are extremely infrequent, and as such, conventional chemical principles such as "thermalization" are not relevant. It, therefore, makes more sense to consider chemistry in the interstellar medium in terms of individual collisions. Gas-phase molecular synthesis in interstellar clouds is believed to occur primarily via ion–molecule reactions, with some neutral reactions contributing. Since the molecular species identified from spectroscopic data are mostly neutral, the ionic species formed in these processes must become charge-neutral relatively quickly. The general scheme of molecular synthesis, therefore, looks something like the following (Vallence, 2014):

$$\text{Neutral gas} \xrightarrow{\text{ionization}} \text{Small Ions} \xrightarrow{\text{reaction}} \text{Large Ions} \xrightarrow{\text{neutralisation}} \text{Observed and ambient species}$$

Chemistry can also occur on the surface of dust grains. Surface-catalyzed reactions of this type turn out to be very important in the interstellar medium. Photoionization is a widespread process near stars. However, the high density of hydrogen and dust grains in molecular clouds prevents visible and UV light from penetrating very far. For this reason, molecular clouds often appear dark when viewed through a telescope. Infrared light can penetrate molecular clouds, and indeed, IR spectroscopy is a key method for identifying molecular species within these regions. However, infrared photons do not have sufficient energy to ionize neutral molecules. Instead, most ions within molecular clouds are formed through collisions with cosmic rays. Electrons fairly commonly attach to large carbon-based molecules (e.g., PAHs), yielding a negative ion. Sometimes electron attachment is dissociative, in which case the rate can be extremely high, with rate constants up to 10^{-7} cm^3/s. In non-dissociative attachment, the emission of a photon will generally be required to stabilize the ion. For example,

$$e^- + \text{PAH} \rightarrow \text{PAH}^- + h\upsilon$$

A wide variety of reaction mechanisms are present within the interstellar medium. These can be categorized into bond formation, bond breaking and rearrangement reactions. Apart from photo-dissociation, these processes are all bimolecular and usually diffusion-controlled, with rate constants of around 10^{-9} cm^3/s.

The discovery, starting in the 1970s, that the Universe is richly molecular rather than sparsely atomic is a striking triumph of radio astronomy. All the species listed in Tables 5.1 and 5.2 have been positively identified by high-resolution rotational spectroscopy, where the quality factor of the line (line frequency/line width) frequently exceeds 10^6. The polyatomic molecules attract the greatest interest since the majority of the observed species contained carbon. The rich chemistry primarily occurs in the dense molecular clouds. The standard method following the equilibrium arguments fails to reproduce the observed relative abundances of species. The observed relative abundances show clearly that equilibrium thermodynamic constraints are inappropriate because, in some instances, high-energy isomeric forms of species are more abundant (for example, HNC/HCN or $[CO+3H_2]/[CH_4+H_2O]$). In molecular clouds, temperature is ≈20°K, the typical value for $[H_2]$ is 10^5 cm^{-3}, and the predicted ratio of CO/CH_4 is 10^{-500}. Since CO is the second most abundant molecule, this rules out arguments of chemical equilibrium, which disagree with the observation by 500 orders of magnitude (Klemperer, 2008). According to this author, the carbon ion, C^+, is responsible for the rich chemistry observed. The efficient production of C^+ by cosmic-ray α-particles has as origin the lack of reactivity of He^+ with H_2. The rich organic chemistry is a direct consequence of helium (ion) chemistry.

Dust affects star formation through molecular gas formation; see Figure 5.4. Dust particles have an enormous contribution to the energy budget of a galaxy. In our galaxy, dust

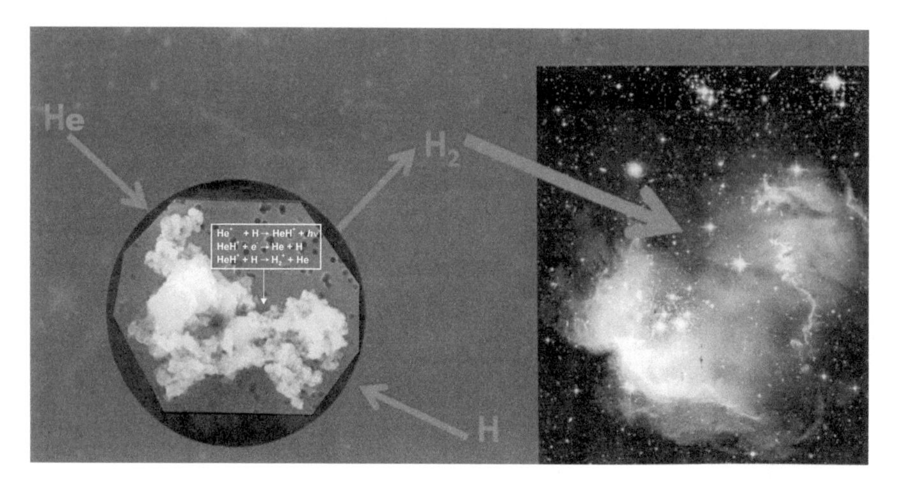

FIGURE 5.4 Schematics of molecules formation on dust particles.

presents less than 1% of mass, but contributes to about 30% of luminosity (Rémy-Ruyer et al., 2014). As galaxies evolve, their interstellar medium becomes continually enriched with metals, and this metal enrichment influences the subsequent star formation.

There are several interstellar dust models. Zubko et al. (2004) presented interstellar dust models that have been derived from fitting the far-ultraviolet to near-infrared extinction, the diffuse IR emission and the elemental abundance constraints. Different interstellar media imposed restrictions on the dust, including solar, F- and G-star, and B-star abundances; see Figure 5.5 showing elemental abundances.

The fitting problem is a typical inversion problem, in which the grain size distribution is the unknown, which the authors solve by using the method of regularization. The dust model of Zubko et al. (2004) contains various parameters such as values of PAHs concentration; the amounts of bare silicate, graphite and amorphous carbon particles; composite silicate particles; organic refractory material; water ice; and voids. The optical characteristics of these components were calculated using physical optical constants. As a particular case, they reproduced the results of Li and Draine (2001). However, their model required a large amount of silicon, magnesium and iron locked in dust particles, about 50 ppm, which is significantly higher than the upper limit imposed by solar abundances of these elements (34, 35 and 28 ppm, respectively). A significant conclusion of this paper is that a unique interstellar dust model that simultaneously explains the observed values of extinction, diffuse IR emission and abundance constraints does not exist. They reported several acceptable interstellar dust models that comply with these constraints. Their model is identical in composition with Li and Draine's (2001) model and contains PAHs, bare graphite and silicate grains, still with a size distribution optimized to comply with the abundance constraints. The second class of models also contains composite particles in addition to PAHs, graphite and silicate grains. Other categories of models contain amorphous carbon instead of graphite particles, or no carbon at all, except for that in PAHs. All classes are consistent with solar and F- and G-star abundances, but

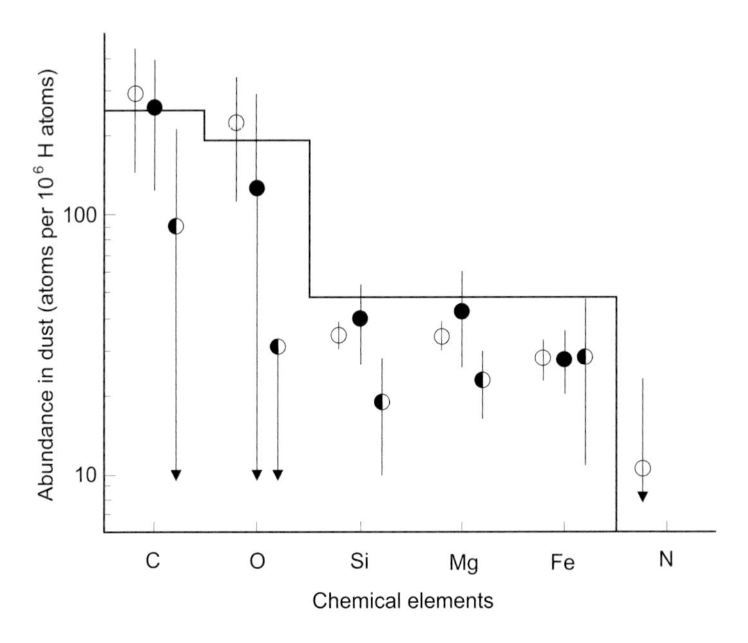

FIGURE 5.5 Elemental abundances for dust derived by assuming total interstellar abundances equal to those of the Sun (open circles), F- and G-stars (solid circles) or B-stars (half-solid–half-open circles). (After Zubko et al., 2004.) The solid line represents the calculations by Li and Draine (2001).

have greater difficulty fitting the B-star carbon abundance, which is better fitted with the later (no carbon) models. Additional observational constraints, such as the interstellar polarization or X-ray scattering, may be able to discriminate between the various interstellar dust models.

Krüger and Grün (2009) presented an overview of interstellar dust inside and outside the heliosphere. The Ulysses spacecraft measured interstellar dust in the solar system in the early 1990s after its flyby of Jupiter. Since then, the in situ dust detector onboard Ulysses continuously measured interstellar dust grains with masses up to 10^{-13} kg, which penetrated the solar system. The Ulysses measured the interstellar dust stream at high ecliptic latitudes between 3 and 5 AU. At the same time, the interstellar impactors measured interstellar dust with the *in situ* dust detectors onboard Cassini, Galileo and Helios crafts. They covered a heliocentric distance range of 0.3–3 AU in the ecliptic plane. The solar radiation pressure force alters the stream of interstellar dust in the inner solar system, gravitational focusing and interaction of charged grains with the time-varying interplanetary magnetic field. The dust grains are tracers of the physical conditions in the local interstellar cloud (LIC). Their *in situ* measurements imply the existence of a population of "big" interstellar grains (up to 10^{-13} kg), and a gas-to-dust mass ratio in the LIC that is a factor of 1.5–2 larger than the one derived from astronomical observations. These data are a measure of a concentration value for interstellar dust present in the very local interstellar medium. Until 2004, the interstellar dust flow direction measured by

Ulysses was close to the mean apex of the Sun's motion through the LIC, while in 2005, the data showed a 30° shift, the reason of which is presently unknown. Krüger and Grün (2009) reviewed the results from spacecraft-based in situ interstellar dust measurements in the solar system and discussed the implications for the physical and chemical states of the LIC (see also Draine, 2011).

Compiègne et al. (2011) described the Planck and Herschel missions while measuring the far-infrared to millimeter emission of dust. With combined IR data measured earlier, they provided for the first time the spectral energy distribution (SED) for the galactic interstellar medium dust emission. SED includes emissions from the mid-IR to the mm range, with unprecedented sensitivity and spatial scale down to ~30″. Such a global SED allows a systematic study of the dust evolution processes. Dust evolution directly affects the SED because evolution processes redistribute the dust mass among the grains of different sizes. Also, the dust SED is affected by variations of the radiation field intensity. Compiègne et al. (2011) published a numerical tool called DustEM that can predict the emission and extinction of dust grains based on their size distribution and their optical and thermal properties. To model dust evolution, the design of DustEM enables dealing with a variety of grain types, their structures and size distributions and being able to include new dust physics easily. The authors used DustEM tool to calculate the dust SED and dust extinction in the interstellar medium at high galactic latitude (DHGL), a natural reference SED that allows the study of dust evolution. They presented a coherent set of observations for the DHGL SED, which has been obtained by correlating the IR and H I 21-cm data. The dust components in their DHGL model are (i) PAHs, (ii) amorphous carbon and (iii) amorphous silicates. They used amorphous carbon dust instead of graphite because it better describes the observed high abundances of gas-phase carbon in some regions of the interstellar medium. Using the DustEM model, they illustrate how, in the optically thin limit, the IRAS/Planck HFI dust SED photometric band ratios can unravel the effect of the exciting radiation field intensity and constrain the abundance of small grains relative to larger grains. They also discuss the contributions of the different grain populations to the IRAS and Planck (and similarly to Herschel) results. Such information is needed to enable a study of the evolution of dust; to extract the thermal dust emission from CMB data systematically; and to analyze the radiation in the Planck polarized channels. The DustEM code described in their paper is publically available.

Meteorites provide an important record of the physical and chemical processes that occurred in the early solar system. To the scientists advocating the origin of life on the Earth, the delivery of organic matter by extraterrestrial material was an important source of organic carbon that has made a wide range of complex prebiotic molecules available for the emergence of life. The carbonaceous chondrites are a very primitive class of meteorite divided into eight different groups (CI, CM, CR, CH, CB, CO, CV and CK), which are further differentiated within these groups by a petrologic subtype number 1–6. Glavin et al. (2020) reported abundant extraterrestrial amino acids in the primitive CM carbonaceous chondrite Asuka 12236. They also observed large L-enantiomeric excesses of ~34%–64% for the protein amino acids, aspartic and glutamic acids and serine, while alanine was racemic. The obtained results are similar to previous amino acids analyses of the Murchison meteorite (Koga and Naraoka, 2017) and Tagish Lake meteorite (Kminek et al., 2002). The fact that only L-enantiomeric excesses have so far been observed in amino acids with a single asymmetric carbon in carbonaceous meteorites suggests that the origin of life on the Earth or elsewhere in our solar system may have been biased toward L-amino acid homochirality from the very beginning (Glavin et al., 2020).

As discussed earlier, biological polymers (e.g., nucleic acids and proteins) use D-sugars and L-amino acids exclusively, respectively. Therefore, it does not come as a surprise that Cooper and Rios (2016) showed that both rare and common sugar monoacids (aldonic acids) contain significant excesses of the D-enantiomer in multiple carbonaceous meteorites. Their findings also imply that meteoritic compounds and/or processes that operated on meteoritic precursors may have played an ancient role in the enantiomer composition of life's carbohydrate-related biopolymers.

In the end, let us mention that NASA's upcoming James Webb Space Telescope (to be launched sometime in 2021) will study dust-producing Wolf–Rayet binary stars. The telescope will detect the mid-infrared light that is exactly the wavelength of light needed to look at in order to study the dust and its chemical composition. Infrared wavelengths can slide between dust grains to reach the telescope, rather than being caught up bouncing around in the dust cloud. Webb will detect this light and allow astronomers to read the information it carries, including the signature of chemicals in the dusty environment, some of which may be the same chemicals that form the building blocks of life on the Earth.

This mission could give some information about the proposal put forward by Valkovic and Obhodas (2021) that chiral life possibly originated within the interstellar dust cloud. They considered only well-characterized cosmic environments, the ones for which the elemental abundance curves for elements could be constructed from the measured data or from existing models. They considered two cosmic environments: the solar neighborhood only and the Milky Way galaxy as a whole.

For the solar neighborhood, they considered the observational data from many authors as summarized by Kobayashi et al. (2020), as well as their GCE model calculations for individual chemical elements to Fe concentration ratios as a function of metallicity [Fe/H]. One can calculate concentration values ratios both for life-essential bulk elements and for essential trace elements as a function of metallicity [Fe/H]. This has been done for different [Fe/H] values corresponding to different times, T-values.

The calculated χ^2 for the agreement of the set of 19 life-essential elements (C, N, O, Na, Mg, Al, Si, P, S, K, Ca, V, Cr, Mn, Co, Ni, Cu, Zn and Mo) in solar neighborhood stars (taken as medium from the graphical representation of experimental values summarized by Kobayashi et al., 2020) and those in the LUCA (taken from the graphical representation in Chopra and Lineweaver, 2015) is presented in Figure 5.6 (top).

The same calculations for only seven elements (V, Cr, Mn, Co, Ni, Cu and Zn) are shown in Figure 5.6 (bottom). In both cases, $T_{origin} = (3–5) \times 10^9$ years is suggested by the calculated χ^2 distribution minima in Figure 5.6 (top and bottom). The same time period is also confirmed by the calculation of the regression coefficient, as shown in Figure 5.7.

In the evaluation by Valkovic and Obhodas (2021), data provided by the Galactic Archaeology with HERMES (GALAH) survey, a large-scale stellar spectroscopic survey of the Milky Way designed to deliver chemical information complementary to a large

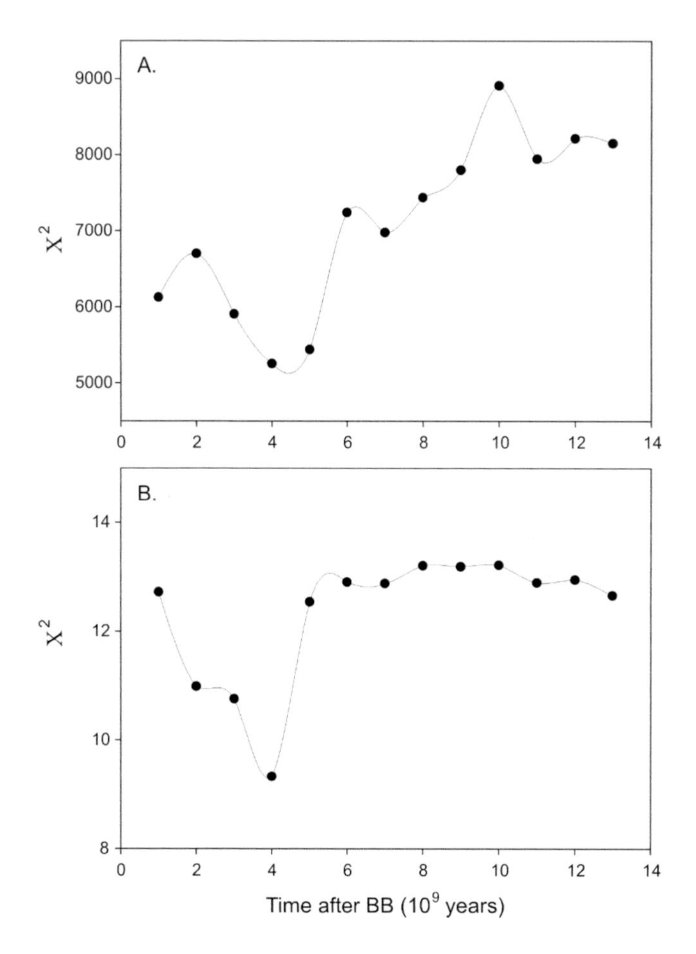

FIGURE 5.6 χ^2 minimum values for two sets of solar neighborhood data and the LUCA estimates, 19 elements in the top and 7 elements in the bottom, are presented.

number of stars covered by the Gaia mission, were used. Buder et al. (2018) presented the GALAH second public data release (GALAH DR2) containing 342,682 stars. For these stars, the GALAH collaboration provides stellar parameters and abundances for up to 23 elements. The calculated χ^2 for a set of 15 life-essential elements (C, O, Na, Mg, Al, Si, K, Ca, V, Cr, Mn, Co, Ni, Cu and Zn) taken from Buder et al. (2018) for different time periods and that for the same set of elements in the LUCA (taken from the graphical representation in Chopra and Lineweaver, 2015) is presented in Figure 5.8 (top). The same calculations for only seven elements (V, Cr, Mn, Co, Ni, Cu and Zn) are shown in Figure 5.8 (bottom). Again, in both cases, $T_{origin} = (3–5) \times 10^9$ years is suggested, which can be seen by the inspection of Figure 5.8.

The hypotheses define time T_{origin} as a time when conditions were right for life to originate in the primitive form and that happened only once in the history of the Universe.

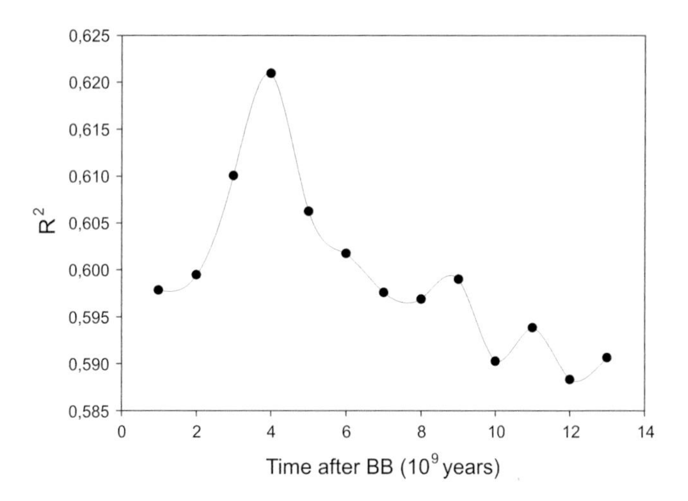

FIGURE 5.7 The regression coefficients obtained for the chemical composition of the solar neighborhood and the LUCA. Only 19 essential elements for which a substantial amount of data in the solar neighborhood are available are taken into consideration.

Exposure to cosmic rays and magnetic fields as well as nuclear physics laws made life chiral. Preliminary considerations indicate $T_{origin} = (4 \pm 1) \times 10^9$ years after the BB for the origin of life matter to form on the dust particles in the molecular clouds; let us call it the first universal stardust ancestor (FUSDA). The FUSDA could have survived planet formation processes and evolved in habitable zones of stars into the LUCA. During this process, it had to develop processes of element concentrations which should be seen in concentration factor dependence on environment properties, in particular, due to adaptation to different magnetic and gravitational fields and the availability of essential elements. This reasoning leads to the universality of the FUSDA but not the LUCA, who would probably have different characteristics in different habitable zones.

5.5 Extraterrestrial Intelligent Life

Huang (1959) first started studying the "habitability" of environments surrounding stars more than half a century ago. The idea of a circumstellar habitable zone (CHZ) is well defined by the demand for the presence of water as a necessary condition for life-as-we-know-it. The temperature must be in the range to allow water to be in a liquid state. This temperature range is a function of the luminosity of the star and the distance of the planet from it. A large amount of recent work, drawing on various disciplines (planetary dynamics, atmospheric physics, geology, biology, etc.), refined considerably our understanding of multiple factors that may affect the CHZ. Despite that progress, we should still consider the subject to be on its beginning (Chyba and Hand, 2005; Gaidos and Selsis, 2006; Prantzos, 2008).

Habitability, on a larger scale, was considered some time ago by Gonzalez et al. (2001). They argued that there exists a zone of enhanced habitability in the Milky Way, which

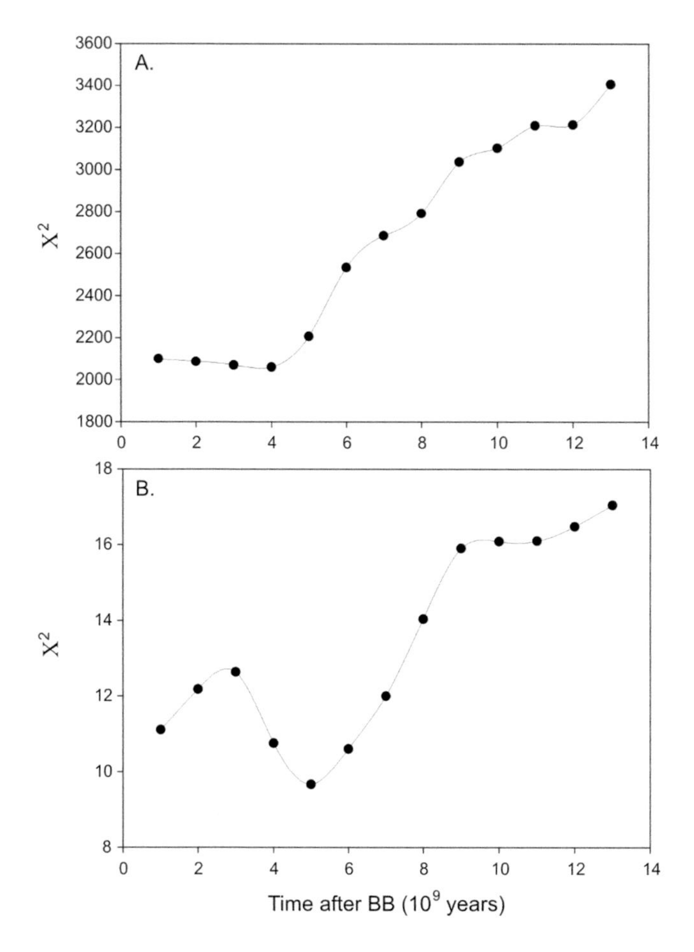

FIGURE 5.8 χ^2 minimum values for two sets of Milky Way galaxy data and the LUCA estimates, 15 elements in the top and 7 elements in the bottom, are presented.

they have termed the galactic habitable zone (GHZ). In these considerations, one of the more important factors is the metallicity of the interstellar matter, out of which a planetary system forms. This fact determines the masses of the terrestrial planets in the system (and probably also the gas giants). The assumption is that earthly planet mass scales with the surface density of solids in a disk; the lower metallicity leads to smaller planets. They estimated, very approximately, that a metallicity at least half that of the Sun is required to build a habitable terrestrial planet.

According to Gonzalez et al. (2001), the decrease in the interstellar medium abundances of the long-lived radioisotopes ^{40}K, ^{235}U, ^{238}U, and ^{232}Th relative to Fe is also meaningful. Heating by radioactive decays seems to be a requirement for the long-term maintenance of a planet's habitability. In this case, the carbon cycle provides climate stability. According to these authors, the ratios C/O, Si/Fe, Mg/Fe and S/Fe are of lesser

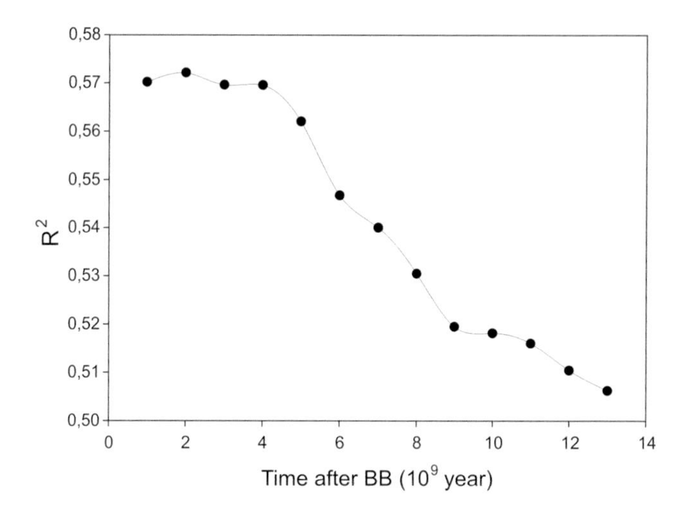

FIGURE 5.9 The R^2 estimates obtained for the chemical composition of the LUCA and the Milky Way throughout the duration of the galactic evolution. Only 15 essential elements for which a substantial amount of data in the Milky Way are available (Buder et al., 2018) are taken into consideration.

importance for habitability. They affect the water content, the core-to-mantle mass ratio and the state of the core of a terrestrial planet.

Based on galactic chemical evolution alone, the authors conclude that the thin disk near the star is the most probable place for Earth-like planets to be formed in the present time. As a rule, the inner disk should contain terrestrial planets more massive than the Earth and the outer disk is likely to include smaller terrestrial planets. Because of the metallicity dependence on giant planet formation, they should also be more common in the inner galactic disk. The bulge should contain many Earth-mass planets, but relatively few Earth-like planets, given the different mix of elements among its stars. The evolving concentration of the critical radioisotopes in the ISM establishes a window of time in the history of the Milky Way during which terrestrial habitable planets with long-lasting geological activity can exist. That window is slowly closing with a timescale of billions of years (Gonzales et al., 2001).

Some of these properties of the Milky Way galaxy are shown in Figure 5.10 (modified from Prantzos, 2008). The figure presents the chemical evolution of the Milky Way disk, obtained in the framework of a one-dimensional model with radial symmetry (Boissier and Prantzos, 1999; Hou et al., 2000). The left part of the figure, from top to bottom, shows radial profiles of the surface density of the gas, stars, star formation rate (SFR) and oxygen abundance. Profiles are shown for three epochs, namely at 1.5, 5 and 13×10^9 years; the last (marked by a thick curve in all figures) is compared to regions of the present-day Milky Way disk observations (shaded areas). The right part of the figure, from top to bottom, shows the same quantities plotted as a function of time for three different disk regions, located at galactocentric distances of 4, 8 and 16 kpc; the second one (thick curves in all panels of the right column) corresponds to the solar neighborhood.

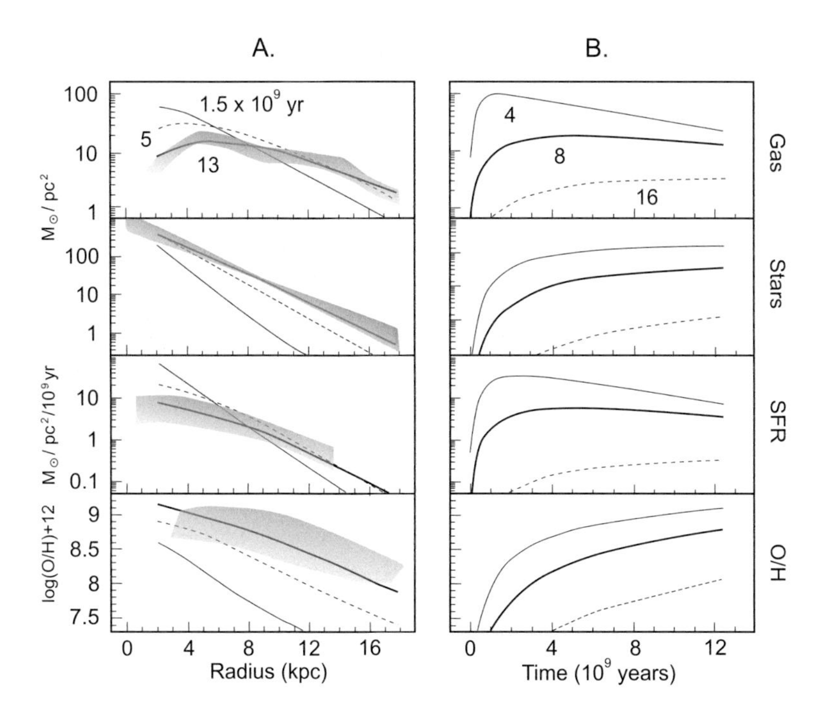

FIGURE 5.10 The chemical evolution of the Milky Way disk, obtained in the framework of a one-dimensional model with radial symmetry (Prantzos, 2008). For explanation, see text.

Spitoni and Matteucci (2015) discussed the galactic habitable zone of the Milky Way and M31. They define the galactic habitable zone as the region with sufficiently high metallicity to form planetary systems in which Earth-like planets could be found and that might be capable of sustaining life. They have assumed that the probability of developing habitable Earth-like planets depends on the [Fe/H] ratio, following the prescriptions of Prantzos (2008), the SFR and the supernova rate of the studied region. They define $P_{GHZ}(R, t)$ function of the galactic radius and time as the fraction of all stars having planets like the Earth (no gas giant planets) that survived nearby supernova explosions.

$$P(R,t) = \frac{\int_0^t SFR(R,t')P_E(R,t')P_{SN}(R,t')\,dt'}{\int_0^t SFR(R,t')\,dt'}$$

This number should be interpreted as the relative probability to have a complicated life around a star on the planet at a given position, as suggested by Prantzos (2008). In this equation, $P_{SN}(R, t')$ is the probability of surviving a SN explosion and $P_E(R, t')$ is the probability of having stars with Earth-like planets, but not gas giant planets that destroy the Earth-like planets. Finally, the total number of stars formed at a certain time t and distance R from the center of the galaxy hosting Earth-like planet with life is defined as

$$N_{\star life}(R,t) = P_{GHZ}(R,t) \times N_{\star tot}(R,t)$$

where $N_{\star tot}(R, t)$ represents the total number of stars created up to time t at the distance R.

In conclusion, Spitoni and Matteucci (2015) reported that the principal consequence of the gas radial inflows is an enhancement of the number of stars hosting a habitable planet concerning the "classical" model results in the region with maximal probability for this occurrence. This fact is due to the increase in gas toward the inner part because of radial inflows, which leads to larger SFR values. Models with radial gas inflows have no threshold in the star formation. All results are obtained by taking into account the supernova destruction processes. In particular, the authors find that in the Milky Way, the maximum number of stars hosting habitable planets is at 8 kpc from the galactic center. The model with radial flows predicts a number that is 38% larger than what is foreseen by the classical model.

The determination of whether Earth-like planets are common or rare acts as a gauge in the question of life in the Universe. Petigura et al. (2013) performed a search of Kepler photometry for transiting planets to measure the occurrence of planets as a function of the orbital period, P, and planet radius, R_P. They restricted the survey to a set of Sun-like stars (GK type) that are the most amenable to the detection of Earth-size planets. They defined GK-type stars as those with surface temperatures T_{eff}=4,100–6,100 K and gravities log g=4.0–4.9 (log g is the base ten logarithms of a star's surface gravity measured in cm/s^2). Their search for planets was additionally confined to the brightest stars observed by Kepler (K_p=10–15 mag). Altogether, 42,557 stars (Best42k) have the lowest photometric noise, making them candidates for the detection of Earth-like planets. When such a planet crosses in front of its star, it causes a partial dimming proportional to the fraction of the stellar disk being blocked, $\delta F=(R_p/R_\odot)^2$, where $R_\odot$ is the radius of the star. Viewed by a distant observer, the Earth would dim the Sun by ≈100 parts per million (ppm), lasting 12 hours every 365 days (Petigura et al., 2013).

The quality of the survey for small planets is a complicated function of P and R_P. It decreases with increasing P and reducing R_P due to fewer transits and less dimming. It is hazardous to replace this injection and recovery assessment with noise models to determine C. Such models are not sensitive to the normalization of C and only provide relative completeness. Models also may not capture the complexities of a multistage transit-finding pipeline that is challenged by noise. Measuring the occurrence of small planets with long periods requires injection and recovery of synthetic transits to determine the absolute detectability of the small signals buried in noise (Petigura et al., 2013).

In conclusion, Earth-size planets are common in the Kepler field. Assuming that the stars in the Kepler field are representative of stars in the solar neighborhood, Earth-size planets should be common around nearby Sun-like stars. If one were to adopt a 22% occurrence rate of Earth-size planets in the habitable zones of Sun-like stars, then the nearest such planet would be expected to be in an orbit of a star less than 12 ly from the Earth which can be seen by the bare eye. Future instrumentation to take spectra need only to observe a few dozen nearby stars to detect a sample of Earth-size planets residing in the HZs of their host stars (Petigura et al., 2013). Estimates of the occurrence of

Earth-like planets also appear in several other works, including Catanzarite and Shao (2011), Traub (2012) and Dong and Zhu (2013).

Solutions have profoundly influenced the search for extraterrestrial intelligence in agreement with the Drake equation, which gives an integer value for the number of communicating civilizations resident in the Milky Way, and by the Fermi paradox, stated as: "If they are there, where are they?" Both rely on using average values of critical parameters, such as the mean signal lifetime of a communicating civilization. A more precise answer must take into account the distribution of stellar, planetary and biological characteristics of the galaxy, as well as the stochastic nature of evolution itself. Forgan (2009) outlines a method of Monte Carlo realization that does this and hence allows an estimation of the distribution of critical parameters in the search for extraterrestrial intelligence, as well as allowing quantification of their errors. Also, it provides a means for competing theories of life and intelligence to be compared quantitatively.

According to Chopra and Lineweaver (2016), there is no evidence that an advanced technological civilization has colonized our galaxy. Archaeological excavations have not found any remains of the alien spaceships, and the radio searches for extraterrestrial intelligence have not been successful so far (Tarter, 2001). Assuming that once life emerges, it evolves toward intelligence and technological civilizations, we would be faced with Fermi's paradox. The so-called Fermi paradox claims that if technical life existed anywhere else, we would see evidence of its visits to the Earth. Since we do know this evidence, such life does not exist, or some different explanation is required. Enrico Fermi, however, never published any text on this topic. On one occasion, he is known to have mentioned it by asking: "Where is everybody?" – suggesting that we do not see any extraterrestrials on the Earth because interstellar travel may not be feasible. Still, he does not indicate that intelligent extraterrestrial life does not exist or indicate its absence is paradoxical. The sentence "they are not here; therefore, they do not exist" was first published by Hart and Zuckerman (1982). They claimed that interstellar travel and colonization of the galaxy would be inevitable if intelligent extraterrestrial life existed, and its absence here is proof that it does not exist anywhere. It looks like the Fermi paradox originates in Hart's argument, not in Fermi's question (Gray, 2015).

Hanson (1998) introduced the idea of a Great Filter, describing the possible bottlenecks in the assumed progression: molecular chemistry to life, life to intelligence, and intelligence to galactic colonization. If the emergence of life is an infrequent and challenging process, then an emergence bottleneck could resolve Fermi's paradox. However, if technological civilizations inescapably destroy themselves, this self-destruction bottleneck could also solve the so-called Fermi's paradox.

The prerequisites and components needed for life seem to be abundantly available in the Universe. See the official Exoplanet Archive (exoplanetarchive.ipac.caltech.edu). However, the Universe does not look to be full of life. According to Chopra and Lineweaver (2016), the most common explanation for this is a low probability for the origin of life (an emergence bottleneck). They presented an alternative Gaian bottleneck explanation: If life emerges on a planet, it only rarely evolves quickly enough to be able to regulate greenhouse gases and albedo, thereby maintaining surface temperatures compatible with liquid water and habitability. Such a bottleneck suggests that (i) extinction is the cosmic default for most life that has ever originated on the surfaces of wet rocky planets in the Universe and (ii) rocky planets need to be inhabited to remain habitable (bootstrapping?). In the Gaian bottleneck model, the sustenance of planetary habitability is connected with the biological regulation of surface volatiles, not with the distance to the host star and its luminosity (Chopra and Lineweaver, 2016). In conclusion, life may be rare in the Universe, not because it is challenging to get started, but because habitable environments are extremely difficult to maintain during the first 10^9 years.

Morrison and Gowanlock (2015) suggested that the inner Milky Way galaxy should logically be a prime target for searches of extraterrestrial intelligence and that any civilizations that may have emerged there are potentially much older than our own. In their work, they extend the assessment of habitability to consider the potential for life to further evolve to the point of intelligence – termed the propensity for the emergence of intelligent life, φ_I. They assumed φ_I was strongly influenced by the time durations available for evolutionary processes to proceed undisturbed by the nearby supernovae. The time interval between supernova events provide windows of opportunity for the evolution of intelligence. They developed a model that allowed analyses of these window times to generate a metric φ_I and examination of the spatial and temporal variations of this metric. Even under the assumption that long-time durations are required between sterilizations to allow for the emergence of intelligence, their model suggests the inner galaxy provides the highest number of opportunities for intelligence to arise. This fact is due to the substantially higher number density of habitable planets in this region, which outweighs the effects of a higher supernova rate in the area. Their model also shows φ_I increasing with time. Intelligent life appeared at approximately the present time at the Earth's galactocentric radius; however, a similar level of opportunity for evolution was available in the inner Galaxy more than 2×10^9 years ago, according to Morrison and Gowanlock (2015).

By using a simplified Bayesian model, Snyder-Beattie et al. (2021) demonstrated that expected evolutionary transition time likely exceeds the lifetime of the Earth, perhaps by many orders of magnitude. This would suggest that intelligent life in the Universe is exceptionally rare, assuming that intelligent life elsewhere requires analogous evolutionary transitions. An interesting remark made by the authors says, "if we were to find life on Mars, it would have emerged extremely early and had a common ancestor with life on Earth".

While the inner Galaxy has a higher overall propensity for intelligent life, as defined in the study by Morrison and Gowanlock (2015), it should be noted that this does not imply any degree of actual inhabitancy. The emergence of life and intelligence may be infrequent events, and their occurrence on the Earth may be a statistical outlier. No other form of intelligence (or a life of any kind) may have arisen elsewhere in our galaxy. However, the alternative – that life and intelligence do exist elsewhere in our galaxy – is also possible, and the results of this study suggest this may be the more probable scenario.

More discussion of the habitability of planets can be found in the chapter of the book by Cockell (2015).

5.6 Life in the Universe

An exciting approach to this subject is presented by McCabe and Lucas (2010). Table 5.3 summarizes some of the timings for critical events and transitions through the history of the Universe and the Earth; see also Frebel (2007), Valley et al. (2002), Maynard-Smith and Szathmary (1995) and Watson (2008). Although some may be regarded as controversial, they are used as best estimates in subsequent considerations. The origin of life in the galaxy, more specifically the start of galactic habitability, is taken to have occurred $(8.5 \pm 5) \times 10^9$ years ago, where the central value of 8.5×10^9 years is supported by the work of Lineweaver et al. (2004), Prantzos (2008) and Mattson (2009). The indicated range of uncertainty allows for the most extreme conceivable values from the first stars in the Universe to life on the Earth.

Also, an extrapolation of the genetic complexity of organisms to earlier times, as described by Sharov (2012) and Sharov and Gordon (2013), suggests that life began before the Earth was formed. Life may have originated with systems having single heritable elements that are functionally equivalent to a nucleotide. The genetic complexity, measured by the number of non-redundant functional nucleotides, is expected to have grown exponentially

TABLE 5.3 Astrobiological Timings

Time in the Past 10^9 Years	Time Since Habitability on the Earth (10^9 Years)	Critical Event/Transition
13.7		Big Bang
13.7 -		Nucleosynthesis of H/He after ~100 s
13.7 -		Atoms formed after 0.3×10^6 years (CMBR)
13.6		First stars expected
13.2[1]		First (halo) stars in galaxy observed
(8.5 ± 5)[2]		Existence of galactic habitable zone
$(8.3^{+2.8}–1.8)$[3]		Galactic thin disk formed
4.57		Formation of the Sun
4.55		Formation of the Earth
$(4.3^{+0.1})$[4]		Liquid water on the Earth
$(4.0^{+0.1} – 0.2)$[5]	0.0	Late Heavy Bombardment (Earth habitability)
	0.2	Proto-cells (molecules in a compartment)
	0.4	Chromosomes (RNA)
$(3.5 - 3.0) \pm 0.4$	0.5	Emergence of prokaryotes
$2.8^{+0.4} - 2.6$		Oxygen photosynthesis
2.2 – 2.0		Atmosphere becomes oxidizing
$1.9^{+0.8} - 1.2^{-0.4}$	2.1	Prokaryotes to eukaryotes
	2.5	Asexual to sexual populations
$1.2^{+0.7} – 0.6$	2.8	Cell differentiation: protists to animals, plants
0.55		Microscopic to macroscopic life
0.0001–0.0002	4.0	Primates to man
In future (?)	5.0	Predicted CO_2 loss (inhabitability)

Source: Modified from McCabe and Lucas (2010).

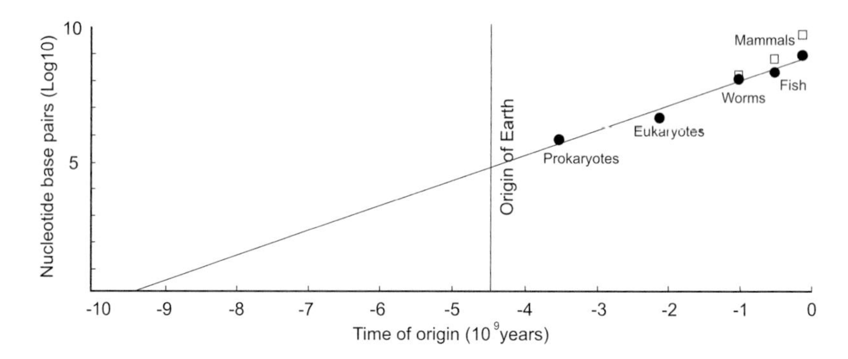

FIGURE 5.11 The complexity of organisms, as measured by the length of functional non-redundant DNA per genome counted by nucleotide base pairs (bp), increases linearly with time (Sharov, 2012). Time is counted backward in 10^9 years before the present (time 0).

TABLE 5.4 History of Universe Milestones

Time/10^9 Years	Event	Reference
–13.75	Big Bang	Jarosik et al. (2011)
–13.55	Dark Ages ending	Zheng et al. (2012)
–13.2	EGSY8p7 galaxy	Zitrin et al. (2015)
–13.2	HE 1523-0901, halo star	Frebel (2007)
-9.75 ± 1.0	Origin of life	Valkovic and Obhodas (2021)
-9.7 ± 2.5	Origin of DNA	Sharov (2006, 2012)
-8.5 ± 5	Start of galactic habitability	McCabe and Lucas (2010)
–4.6	Formation of the Sun	
–4.56	Formation of the Earth	Houdek and Gough (2011)
0	Now	

because of several positive feedback factors: (i) gene cooperation, (ii) duplication of genes with their subsequent specialization (via expanding differentiation trees in multicellular organisms) and (iii) emergence of novel functional niches associated with the existing genes. Genetic complexity is measured by the length of functional non-redundant DNA per genome and nuclide base pairs. Linear regression of genetic complexity (on a semi-log scale) extrapolated back to the time of only one base pair gives the time of the origin of life to be $(9.7\pm2.5)\times10^9$ years ago; see Figure 5.11 (Sharov, 2006, 2012; Sharov and Gordon, 2013).

Table 5.4 shows a schematic view of the development of the Universe since the Big Bang. It shows the proposed estimate for the origin of life to be $(9.7\pm2.5)\times10^9$ years ago. The "Dark Ages" probably have ended at -13.55×10^9 years (Zheng et al., 2012), with the Big Bang at -13.75×10^9 years (Jarosik et al., 2011). The most distant galaxy yet observed is at $z=8.68$ (redshift), or more than 13.2×10^9 years old, dating just 550×10^6 years from the start of our Universe (see Zitrin et al., 2015). The authors used the W. M. Keck telescope in Mauna Kea, Hawaii, to discover the EGSY8p7 galaxy via its Lyman-alpha emissions, with a significance of about 7.5 standard deviations.

The scenario for the time evolution of life is proposed based on the following assumptions:

1. Life originated and evolved to the prokaryote, probably in an environment different from the one on the Earth.
2. Life probably needed an extended period (approx. 5×10^9years) to reach the complexity of bacteria.
3. Intelligent life in our Universe did not exist before us. As a consequence, the proposal of the intelligent aliens seeding life on the Earth does not hold.
4. The Earth was seeded by panspermia during its formation.
5. The experimental replication of the origin of life from scratch may have to emulate many cumulative rare events.
6. The Drake equation estimating the number of civilizations in the Universe is not correct; see also Sharov and Gordon (2013).

We should mention the work of Panov (2005), who discusses the scaling law for the biological evolution and the hypothesis of the galaxy origin of life. The order of the life-related major events on the Earth obeys the scaling law. An estimation of the duration of the pre-biological chemical evolution is obtained by an extrapolation of the scaling law of the biospheric development. The proposed period of interstellar pre-biological panspermia ($\sim0.2\times10^9$years) is much shorter than the estimated duration of the pre-biological chemical evolution. According to Panov (2005), the hypothesis of the lengthy pre-biological chemical evolution implies that:

a. the pre-biological evolution and the origin of life may be a self-consistent galaxy process and not a process localized on single planets.
b. life needs the same chemical elements and the same chirality everywhere in the galaxy.

If life originated before the origin of the Earth, then the Earth was contaminated with bacterial spores in a process called panspermia (see also Wallis and Wickramasinghe, 2004; Schroeder, 2015).

We should consider the role of cosmic dust mineral grains in this process. Some important facts should be taken into account. Complex molecules of organic matter, which are the building blocks for life, have been documented in intergalactic dust clouds, comets and meteorites. In addition, Richter et al. (1999) reported the detection of molecular hydrogen. The formation of molecular hydrogen requires the presence of dust (and therefore gas, enriched in heavy elements). The absorption of molecular hydrogen in a high-velocity cloud along the line of sight to the Large Magellanic Cloud was discovered. They also derived for the same cloud an iron abundance, which is half of the solar value. From these data, they concluded that gas in this cloud originated in the disk of the Milky Way.

With these estimates of the production ratio of U^{235}/U^{238} in the r-process, Burbidge et al. (1957) calculated the age of these uranium isotopes (and by inference of the other r-process isotopes) using their observed terrestrial abundance ratio. Their best value for the time of a single-event synthesis of U^{235} and U^{238} is thus 6.6×10^9years ago. Assuming these isotopes were produced in a single supernova, this is its date. For the change of a factor of two in the production ratio of the uranium isotopes, this figure changes by 0.85×10^9 years.

Houdek and Gough (2011) attempted a seismic calibration of standard solar models intending to improve earlier estimates of the main-sequence age $t_\odot$ and the initial heavy element abundance Z_0. The long-term goal has been an achievement of the precision of distinguishing between planet formation and the creation of the Sun. The best estimate is around $(4.60 \pm 0.04) \times 10^9$ years. This age is close to the previous preferred value, in particular, the age adopted for Christensen-Dalsgaard's Model S. The present-day surface heavy element abundance lies between the values quoted by Asplund et al. (2009) and Caffau et al. (2009).

A hypothesis on the origin of life by considering the elemental requirements of living matter was put forward by Valkovic (1990). Life, as known to us, is (H-C-N-O)-based and relies on the number of bulk and trace elements originated when the essential trace element abundance curve of the living matter, namely the LUCA, and the galactic abundance curve in the Fe region coincided.

This coincidence occurring in the DLA system at a particular redshift could indicate the phase of the Universe when life originated. It is proposed to look into the redshift region $z = 0.5$–4.7 (approximately $t = -5.2 \times 10^9$ to -12.6×10^9 years). DLA systems in this region have been studied, and one should use these data to study the evolution of chemical element abundances in galaxies to determine the time when the Universal element abundance curve and the element abundance curve of the LUCA coincided. The genetic code has transmitted the characteristic properties of the latter, while the Universe element abundance curve changed as the galaxies aged.

It is, in principle, possible to determine the chemical element composition of a galaxy as a function of position and time by measuring elemental abundances of stars with different birthplaces and ages, assuming that their atmospheres represent the composition of the gas from which they were formed. This type of study may give valuable information about the chemical evolution of galaxies and even about the structure of the matter in the very early phases of the Universe.

H and He were present very early on in the Universe, while all metals (except for a tiny fraction of Li) were produced through nucleosynthesis in stars. The fraction by mass of heavy elements is denoted by Z.

- The Sun's abundance is $Z_{Sun} \sim 0.02$,
- Most of the metal-poor stars in the Milky Way have $Z \sim 10^{-5}$–$10^{-4}\ Z_{Sun}$.

Stars use H and He in their nuclei to produce heavy elements. These elements are partially returned to the interstellar gas at the end of the star's life through winds and supernovae explosions. Some fraction of the metals are locked into the remnant of the star. This fact implies that the chemical abundance of the gas in a star-forming galaxy should evolve with time. The metal abundance of the gas and subsequent generations of stars should increase with time. The changes in the chemical element abundances in a galaxy can serve as a clock for galactic aging. On average, very old stars contain less iron than younger stars.

Of our interest are, in particular, Fe-peak elements. The elements in the Fe-peak are created in various late burning stages (see Woosley and Weaver, 1995), as well as in supernovae. The Fe-peak abundance trends reported by Hollek et al. (2011) follow those of other halo star samples and generally indicate a successive increase of these elements over time (e.g., McWilliam, 1997).

Gas dynamics and star formation are processes that regulate the gas-phase oxygen abundance (metallicity) of star-forming galaxies. The oxygen in the Universe is formed in the late-stage evolution of massive stars. Oxygen scattered into the interstellar medium by supernovae and stellar winds enhances the metallicity of galaxies as they expand their stellar mass. In most star-forming galaxies at $z \lesssim 2$, the augment in stellar mass is dominated by inflows of gas from the intergalactic medium; see Noeske et al. (2007) and Whitaker et al. (2012). At the same time, outflows of gas are ubiquitously observed in star-forming galaxies at $z \approx 3$ or lower (see, for example, Weiner et al., 2009; Steidel et al., 2010). Because metallicity is established by the interplay of gas flows and star formation, contemplation of the chemical evolution of galaxies provide essential constraints for these physical processes in models of galaxy evolution (see, for example, Zahid et al., 2012; Møller et al., 2013).

According to Yuan et al. (2012), there is a clear evolution in the mean and median metallicities of star-forming galaxies as a function of redshift. The mean metallicity falls by ≈ 0.18 dex for the redshift from 0 to 1 and falls further by ≈ 0.16 dex for the redshift from 1 to 2. [The symbol dex represents a conventional notation for decimal exponent, i.e., dex $(2.35) = 10^{2.35}$.] More rapid evolution is seen in the range of $z \approx 1 \rightarrow 3$ than in $z \sim 0 \rightarrow 1$ for the high-mass galaxies ($10^{9.5}\ M_{\odot} < M^{*} < 10^{11}\ M_{\odot}$), with almost twice as much enrichment in the range of $z \approx 1 \rightarrow 3$ than in $z \approx 1 \rightarrow 0$. The metallicity of stars in our galaxy ranges from Fe/H = −4 to +0.5 dex, while the solar iron abundance is $\varepsilon(Fe) = 7.51 \pm 0.01$ dex. The average values of the Fe/H ratio in the solar neighborhood, the halo and the galactic bulge are −0.2, −1.6 and −0.2 dex, respectively. Detailed abundance analysis shows that the galactic disk, halo and bulge have peculiar abundance patterns for O, Mg, Si, Ca and Ti and some neutron-capture elements. These signatures show that the galactic environment has an important role in chemical element evolution and that supernovae come in a variety of flavors with a range of element yields (Zahid et al., 2013).

The metallicity may be a general indicator for finding in which part of the Universe one should look for the evolution pattern of essential trace elements (e.g., Cr, Mn, Fe, Co, Ni, Cu, Zn, Se, Mo, I and W) that resembles the signature of the LUCA. Attention should be paid to the regions with iron abundance similar to the solar value. However, more appropriate would be the measurement of individual element yields as a function of redshift.

Let us mention some of the reports which might contain relevant information. For example, Hollek et al. (2011) presented the abundances or upper limits of 20 elements (Li, C, Mg, Al, Si, Ca, Sc, Ti, Cr, Mn, Fe, Co, Ni, Zn, Sr, Y, Zr, Ba, La and Eu) for 16 new stars and four standard stars derived from high-resolution, high-S/N MIKE spectra via traditional manual analysis methods using the MOOG code. They found that, except for Mg, abundances match well with those of other halo stars reported in the literature, e.g., Cayrel et al. (2004). Hollek et al. (2001) report, among other ratios for elements of our interest: Cr/Fe, Mn/Fe, Ni/Fe and Zn/Fe as a function of star's metallicity measured by

Fe/H ratio. It should be mentioned that new stars have [Fe/H] < −3.6, and there the metallicity distribution function severely drops.

Roediger et al. (2013) presented an extensive literature compilation of age, metallicity and chemical abundance information for the 41 galactic globular clusters (GGCs) studied by Schiavon et al. (2005). They estimate that their compilation incorporates all relevant analyses from the literature up to mid-2012.

Despite many questions still awaiting an answer, we can conclude that the distribution of life-essential trace elements holds the best possible clue to the origin of life issues.

The fact that the Earth contains heavy elements such as gold, lead or uranium (all heavier than iron) shows that our Sun is a second- or third-generation star, 6.6×10^9 years old, preceded by at least one supernova explosion of another nearby star. Molecular cloud of gas and dust particles containing prebiotic molecules, maybe even the LUCA, underwent the process of gravitational collapse during which the planets and the Earth were formed some 4.56×10^9 years ago.

Interstellar molecular clouds and circumstellar envelopes are locations for the synthesis of different molecules; see, for example, Ehrenfreund et al. (2000), Kwok (2004) and van Dishoeck and Blake (1998). In addition to gas, interstellar material also contains small micron-sized particles. Gas-phase and gas-grain interactions result in the formation of complex molecules. It is surface catalysis on solid interstellar particles that enables molecule formation and chemical pathways that cannot proceed in the gas phase owing to reaction barriers. A high number of molecules that are used in contemporary biochemistry on the Earth are discovered in the interstellar medium, on the surfaces of comets, asteroids, meteorites and interplanetary dust particles. Large quantities of extraterrestrial material were delivered via comets and asteroids to young planetary surfaces during the heavy bombardment phase. The study of the formation and evolution of organic matter in space is a critical factor in the determination of the prebiotic reservoirs available to the early Earth. It is equally important to reveal abiotic routes to prebiotic molecules in the planet Earth environment.

Previously, it was concluded that the properties of the dust in DLA systems (considering the metallicity only) would not satisfy the requirements of the essential trace element signature of the LUCA (Cr, Mn, Fe, Co, Ni, Cu, Zn, Se, Mo, I and W) for $z > 4.7$ (i.e., before $\approx -12.6 \times 10^9$ years or $\approx 1.1 \times 10^9$ years after the Big Bang). Therefore, $T \approx -12.6 \times 10^9$ years is a lower limit for T_{origin}, the time when the coincidence of abundance curves occurred. For any other time, $T \neq T_{origin}$, no additional such events could occur. Therefore, we can conclude that, because of Universe aging, life originated only once.

Also, an extrapolation of the genetic complexity of organisms to earlier times suggests that life began before the Earth was formed (Sharov, 2012; Sharov and Gordon, 2013). Life may have started from systems with only single heritable elements. Extrapolation back to only a single DNA base pair results in the time of the origin of life being $(9.7 \pm 2.5) \times 10^9$ years ago. This assumption could be tested by the proposed hypothesis by comparing elemental abundances of distant galaxies observed at z (redshift) values corresponding to this age (or some other) with essential trace element requirements of the LUCA.

There are other scenarios (McCabe and Lucas, 2010) for the origin of life in the galaxy, or more specifically, the start of galactic habitability is taken to have occurred $(8.5 \pm 5) \times 10^9$ years ago. The central value of 8.5×10^9 years is deduced from the work of Lineweaver et al. (2004), Prantzos (2008) and Mattson (2009). The range of uncertainty covers the most extreme conceivable values for period from the first stars formed in the Universe to the origin of life on the Earth.

Assuming the life originated long before the origin of the Earth, we have to hypothesize life being brought to the Earth in the process of its formation! The participation of cosmic dust mineral grains in this process must be considered. Boron played an essential role in this process since its primary purpose has been to provide thermal and chemical stability in hostile environments. The comparison of galactic cosmic-ray abundances with solar system nuclei abundances shows 10^5 discrepancies in boron abundances, indicating that most of the boron is of secondary origin. The low boron abundance in the solar system is because boron cannot be created from the spallation of carbon. Boron is primarily synthesized in the ISM. The boron atoms produced by spallation reactions are stably locked within interstellar graphite grains (Ramadurai and Wickramasinghe, 1975). Such a significant difference in boron abundance indicates that the decision on boron essentiality has probably been made outside the solar system.

The process that turns enormous clouds of cosmic dust into newborn planets over millions of years has recently been observed directly. A protoplanet in the making around a young star (2×10^6 years old), LkCa15, in the neighborhood of Taurus, 450 ly from the Earth, has been spotted (Sallum et al., 2015). So far, more than 2,000 exoplanets have been discovered and confirmed. These could be locations where life might sustain and evolve, providing the appropriate duration of habitability conditions.

Although many exoplanets might have a magnetic field similar to the planets in our solar system (except Venus), the direct detection of an exoplanetary magnetic field has failed so far (Turner et al., 2020). Measuring the magnetic field of an exoplanet will give valuable information to constrain its interior structure, both composition and thermal state, its atmospheric escape and the nature of star–planet interaction. Additionally, the magnetic field might contribute to the sustained habitability of exoplanet by deflecting energetic stellar wind particles and cosmic rays.

Turner et al. (2020) presented observations of the exoplanetary systems 55 Cancri, υ Andromedae and τ Boötis made by low-frequency array of low-band antenna (1,090 MHz). They tentatively detected a circularly polarized emission from the τ Boötis system in the range of 14–21 MHz with a flux density of ~890 mJy and with a statistical significance of $\sim 3\sigma$. They also detected a slowly variable circularly polarized emission from τ Boötis in the range of 21–30 MHz with a flux density of 400 mJy and with a statistical significance $>8\sigma$. The source of the detected emission is probably the τ Boötis planetary system, radio emission from the exoplanet τ Boötis b via the cyclotron maser mechanism. Assuming a planetary origin, the authors derived a maximum surface polar magnetic field for τ Boötis b to be between ~5 and 11 G. The signals for τ Boötis b range from 190 to 890 mJy with an emitted power of 6.3×10^{14}–2.0×10^{16} W and a brightness temperature of $(0.42\text{–}2.0) \times 10^{18}$K (Turner et al., 2020).

References

Adams, W. S. 1949. Observations of interstellar H and K, molecular lines, and radial velocities in the spectra of 300 O and B stars. *Astrophysical Journal* 109:354.

Ahmad, S. and Jensen, R. A. 1988. New prospects for deducing the evolutionary history of metabolic pathways in prokaryotes: Aromatic biosynthesis as a case-in-point. *Origins of Life and Evolution of Biospheres* 18:41–57.

Altwegg, K., Balsiger, H., Bar-Nun, A., et al. 2016. Prebiotic chemicals – amino acids and phosphorus – in the coma of comet 67P/Churyumov-Gerasimenko. *Science Advances* 2(5):e1600285.

Arnold, G. L, Anbar, A. D, Barling, J. and Lyons, T. W. 2004. Molybdenum isotope evidence for widespread anoxia in mid-Proterozoic oceans. *Science* 304:87–90.

Asplund, M., Grevesse, N., Sauval, A. J. and Scott, P. 2009. The chemical composition of the sun. *Annual Review of Astronomy & Astrophysics* 47(1):481–522.

Barrinton, E. J. W. 1962. Hormones and vertebrate evolution. *Experientia* XVIII:201–209.

Barrow, J. D., Morris, S. C., Freeland, S. J. and Harper, C. L. (Eds.). 2012. *Fitness of the Cosmos for Life, Biochemistry and Fine-Tuning.* Cambridge University Press, Cambridge, 526 p.

Becerra, A., Delaye, L., Islas, S. and Lazcano, A. 2007. The very early stages of biological evolution and the nature of the last common ancestor of the three major cell domains. *Annual Review of Ecology, Evolution, and Systematics* 38:361–379.

Benner, S. A., Kim, H.-J., Kim, M.-J., and Ricardo, A. 2010. Planetary Organic Chemistry and the Origins of Biomolecules. *Cold Spring Harbor Perspectives in Biology* 2(7):a003467. Doi: 10.1101/cshperspect.a003467.

Boissier, S. and Prantzos, N. 1999. Chemo-spectrophotometric evolution of spiral galaxies - I. The model and the Milky Way. *Monthly Notices of the Royal Astronomical Society* 307:857–876.

Bonilla, I., Garcia-Gonzalez, M. and Mateo, P. 1990. Boron requirement in cyanobacteria. Its possible role in the early evolution of photosynthetic organisms. *Plant Physiology* 94:1554–1560.

Brocks, J. J., Logan, G. A., Buick, R. and Summons, R. E. 1999. Archean molecular fossils and the early rise of eukaryotes. *Science* 285:1033–1036.

Buder, S., Martin Asplund, M., Duong, L., et al. 2018, Apr. 17. The GALAH survey: Second data release. *Monthly Notices of the Royal Astronomical Society*: 1–38. arXiv:1804.06041v1 [astro-ph.SR].

Buffett, B. A. 2000. Earth's core and geodynamo. *Science* 288(5473):2007–2012.

Burbidge, E. M., Burbidge, G. R., Fowler, W. A. and Hoyle, F. 1957. Synthesis of the elements in stars. *Reviews of Modern Physics* 29:547–649.

Caffau, E., Ludwig, H.-G. and Steffen, M. 2009. Solar abundances and granulation effects. *Memorie della Società Astronomica Italiana* 80:643–646.

Cai, X., Jiang, J. H., Fahy, K. A., Yuk L. and Yung, Y. L. 2020. A statistical estimation of the occurrence of extraterrestrial intelligence in the Milky Way galaxy. arXiv:2012.07902 [astro-ph].

Carter, B. 1983. The anthropic principle and its implications for biological evolution. *Philosophical Transactions of the Royal Society A* 310:347–363.

Catanzarite, J. and Shao, M. 2011. The occurrence rate of earth analog planets orbiting sun-like stars. *Astrophysical Journal* 738(2):151–160.

Cayrel, R., Depagne, E., Spite, M., Hill, V., Spite, F., François, P., Plez, B., Beers, T., Primas, F., Andersen, J., Barbuy, B., Bonifacio, P., Molaro, P. and Nordström, B. 2004. First stars V - Abundance patterns from C to Zn and supernova yields in the early Galaxy. *Astronomy and Astrophysics* 416:1117–1138.

Chatterjee, S. 2016. A symbiotic view of the origin of life at hydrothermal impact crater-lakes. *Physical Chemistry Chemical Physics* 18(30):20033–20046. Doi: 10.1039/c6cp00550k.

Chopra, A., Lineweaver, C. H., Brocks, J. J., Ireland, T. R. Palaeoecophylostoichiometrics: Searching for the Elemental Composition of the Last Universal Common Ancestor. Australian Space Science Conference Series: 9th Conference Proceedings, Sydney, 28–30 September 2009 (DV). National Space Society of Australia Ltd, ISBN 13: 978-0-9775740 (2009).

Chopra, A. and Lineweaver, C. H. 2015. An estimate of the elemental composition of LUCA. *Astrobiology Science Conference 2015*, Chicago, IL, June 15–19, 2015. Paper 7328.

Chopra, A., and Lineweaver, C. 2016. The case for a Gaian bottleneck: The biology of habitability. *Astrobiology* 16(1):7–22.

Chyba, C. F. and Kevin P. Hand, K. P. 2005. ASTROBIOLOGY: The study of the living universe. *Annual Review of Astronomy and Astrophysics* 43:31–74.

COBE, Cosmic Background Explorer. 2015. http://lambda.gsfc.nasa.gov/product/cobe/.

Cockell, C. S. (Ed.). 2015. *Dissent, Revolution and Liberty Beyond Earth*. Springer, part of Springer Nature Switzerland AG.

Compiègne, M., Verstraete, L., Jones, A., Bernard, J.-P., Boulanger, F., Flagey, N., Le Bourlot, J., Paradis, D. and Ysard, N. 2010. The global dust SED: tracing the nature and evolution of dust with DustEM. *Astronomy and Astrophysics* 525:A103 (14 pp).

Cooper, G. and Rios, A. C. 2016. Enantiomer excesses of rare and common sugar derivatives in carbonaceous meteorites. *PNAS* 113(24):E3322–E3331.

Di Stefano, R. and A. Ray, A. 2016. Globular clusters as cradles of life and advanced civilizations. *The Astrophysical Journal* 827:54 (12 pp).

Di Toro, D. M., Allen, H. E., Bergman, H. L., Meyer, J. S., Paquin, P. R. and Santore, R. C. 2001. Biotic ligand model of the acute toxicity of metals. 1. Technical basis. *Environmental Toxicology and Chemistry* 20(10):2383–2396.

Dodd, M., Papineau, D., Greene, T., Slack, J. F., Rittner, M., Pirajno, F., O'Neil, J. and Little, C. T. S. 2017. Evidence for early life in Earth's oldest hydrothermal vent precipitates. *Nature* 543(7643):60–64.

Dong, S., and Zhu, Z. 2013. Fast rise of "Neptune-size" planets (4-8 R_Earth) from P ~ 10 to ~250 days—Statistics of Kepler planet candidates up to ~0.75 AU. *Astrophysical Journal* 778(1):53–63.

Dragomir, D., Teske, J., Günther, M. N. et al. 2019. The longest period TESS planet yet: A sub-Neptune transiting a bright, nearby K dwarf star. arXiv:1901.00051v1 [astro-ph. EP 31 Dec 2018, draft version January 3, 2019.].

Draine, B. T. 2011. Large dust grains in the ISM. Leiden 2011.03.01. http://www.strw.leidenuniv.nl/HERSCHELDUST/draine_lorentz_110301.pdf.

Drake, F. D. 2003, Sep. 29. The drake equation revisited: Part I. *Astrobiology Magazine*.
Dyson, F. J. 1982. A model for the origin of life. *Journal of Molecular Evolution* 18:344–350.
Edvardsson, B., Andersen, J., Gustafsson, B., Lambert, D. L., Nissen, P. E. and Tomkin, J. 1993. The chemical evolution of the galactic disk - part one - analysis and results. *Astronomy and Astrophysics* 275:101–152.
Ehrenfreund, P. and Charnley, S. B. 2000. Organic molecules in the interstellar medium, comets, and meteorites: A voyage from dark clouds to the early earth. *Annual Review of Astronomy and Astrophysics* 38:427–483.
Ehrenfreund, P., Spaans, M. and Holm, N. G. 2011. The evolution of organic matter in space. *Philosophical Transactions of the Royal Society A* 369:538–554.
Elsila, J. E., Glavin, D. P. and Dworkin, J. P. 2009. Cometary glycine in samples returned by Stardast. *Meteoritics & Planetary Science* 44:1323–1330.
Engel, M. and Macko, S. 1986. Stable isotope evaluation of the origins of amino acids in fossils. *Nature* 323:531–533.
Farquhar, J., Bao, H. and Thiemens, M. 2000. Atmospheric influence of earth's earliest sulfur cycle. *Science* 289(548):756–758.
Fedonkin, M. A. 2008. Ancient biosphere: The origin, trends, and events. *Russian Journal of Earth Sciences* 10 ES1006, pp. 1–9. Doi: 10.2205/2007ES000252.
Forgan, D. H. 2009. A numerical testbed for hypotheses of extraterrestrial life and intelligence. *International Journal of Astrobiology* 8(02):121–131.
Frank, A. and Sullivan, III, W. T. 2016. A new empirical constraint on the prevalence of technological species in the universe. *Astrobiology* 16(5):359–362.
Frausto da Silva, J. J. R. and Williams, R. J. P. 1997. *The Biological Chemistry of the Elements – The Inorganic Chemistry of life*. Oxford University, Oxford, UK, 672 p.
Frebel, A. 2007. Discovery of HE 1523-0901, a strongly r-process-enhanced metal-poor star with detected uranium. *The Astrophysical Journal Letters* 660:L117.
Fulton, B. J., Petigura, E. A., Howard, A. W., et al. 2017. The California-Kepler survey. III. A gap in the radius of distribution of small planets. Preprint: https://arxiv.org/abs/1703.10375.
Gaidos, E. and Selsis, F. 2006. From protoplanets to protolife: The emergence and maintenance of life. http://arXiv.org/abs/astro-ph/0602008.
Galimov, E. M. 2001. *Phenomenon of Life: Between Equilibrium and Nonlinearity - Origin and Principles of Evolution (in Russian)*. Editorial URSS, Moscow, 256 p.
Galimov, E. M. 2004. Phenomenon of life: Between equilibrium and nonlinearity. *Origins of Life and Evolution of the Biosphere* 34:599–613.
Galimov, E. M. 2005. Redox evolution of the Earth caused by a multi-stage formation of its core. *Earth and Planetary Science Letters* 233(3–4):263–276.
Galimov, E. M. 2006. Phenomenon of life. Origin and principles of evolution. URSS Publ. Hall, Moscow (in Russian). Engl. Translation. *Geochemistry International* 44(1):S1–S95.
Gilbert, E. A., Barclay, T., Schlieder, J. E., et al. 2020. The first habitable-zone earth-sized planet from TESS. I. Validation of the TOI-700 system. *The Astronomical Journal* 160:116 (21 p).

Glavin, D. P., McLain, H. L., Dworkin, J. P., et al. 2020. Abundant extraterrestrial amino acids in the primitive CM carbonaceous chondrite Asuka 12236. *Meteoritics and Planetary Science* 1–28 Doi: 10.1111/maps. 13560.

Gonzalez, G., Brownlee, D. and Ward, P. 2001. The galactic habitable zone: Galactic chemical evolution. *Icarus* 152:185–200.

Gratton, R.G., Carretta, E., Claudi, R., Lucatello, S. and Barbieri, M. 2003. Abundances for metal-poor stars with accurate parallaxes. I. Basic data. *Astronomy and Astrophysics* 404:187–210.

Gray, R. H. 2015. The Fermi paradox is neither Fermi's nor a paradox. *Astrobiology* 15(-3):195–199. Doi: 10.1089/ast.2014.1247.

Greenberg, J. M. 1971. Interstellar grain temperatures effects of grain materials and radiation fields. *Astronomy & Astrophysics* 12:240–249.

Greenberg, J. M. 2002. Cosmic dust and our origins. *Surface Science* 500:793–822.

Greenberg, J. M. and Li, A. 1996. Evolution and emission of cold, warm and hot dust populations in diffuse and molecular clouds, In Block, D. L. and Greenberg, J. M. (Eds.) *New Extragalactic Perspectives in the New South Africa*. Kluwer, Dodrecht. pp. 118–134.

Güsten, R., Wiesemeyer, H., Neufeld, D., et al. 2019. Astrophysical detection of the helium hydride ion HeH_+. *Nature* 568:357–359.

Hart, M. H. and Zuckerman, B. 1982. *Extra-Terrestrials, Where Are They? Second Edition, 1995.* Cambridge University Press, Cambridge.

Hanson, R. 1998. The great filter—are we almost past it? http://mason.gmu.edu/~rhanson/greatfilter.html or https://archive.is/J02C9 (accessed January 21, 2016).

Hollek, J. K., Frebel, A., Roederer, I. U., Sneden, C., Shetrone, M., Beers, T. C., Kang, S.-J. and Thom, C. 2011. The chemical abundances of stars in the halo (cash) project. II. A sample of 14 extremely metal-poor stars. *The Astrophysical Journal* 742(1):54 (19 p).

Hou, J., Prantzos, N. and Boissier, S. 2000. Abundance gradients and their evolution in the Milky Way disk. *Astronomy & Astrophysics* 362:921–936.

Houdek, G. and Gough, D. O. 2011. On the seismic age and heavy element abundance of the Sun. *Monthly Notices of the Royal Astronomical Society (MNRAS)* 418(2):1217–1230.

Howk, J. C., Sembach, K. R. and Savage, B. D. 2000. The abundance of interstellar boron. *The Astrophysical Journal* 543:278–283.

Huang, S.-S. 1959. Life outside the solar system. *American Scientist* 47:393–402.

Ioppolo, S., Fedoseev, G., Chuang, K.-J., et al. 2020. A non-energetic mechanism for glycine formation in the interstellar medium. *Nature Astronomy Articles*, Doi: 10.1038/s41550-020-01249-0.

Iwamoto, K., Brachwitz, F., Nomoto, K., Kishimoto, N., Umeda, H., Hix, W. R. and Thielemann, F.-K. 1999. Nucleosynthesis in Chandrasekhar mass models for Type IA supernovae and constraints on progenitor systems and burning-front propagation. *The Astrophysical Journal Supplement Series* 125:439–462.

Jarosik, N., Bennett, J. Dunkley, B., et al. 2011. Seven-year Wilkinson Microwave Anisotropy Probe (WMAP) observations: Sky maps, systematic errors, and basic results. *Astrophysical Journal Supplement Series* 192(2):1–15.

Klemperer, W. 2008. Interstellar chemistry. *Proceedings of the National Academy of Sciences of the United States of America* 103(33):12232–12234.

Kminek, G., Botta, O., Glavin, D. P. and Bada, J. L. 2002. Amino acids in the Tagish Lake meteorite. *Meteoritics and Planetary Science* 37:697–701.

Kobayashi, C., Karakas, A. I. and Lugaro, M. 2020. The origin of elements from carbon to uranium. *The Astrophysical Journal* 900:179 (33p).

Koga, T. and Naraoka, H. 2017. A new family of extraterrestrial amino acids in the Murchison meteorite. *Scientific Reports* 7:636. Doi: 10.1038/s41598-017-00693-9.

Publications of the Astronomical Society of Japan (PASJ) 1. arXiv:1205.6005v1 [astro-ph. GA 27 May 2012.]

Krüger, H. and Grün, E. 2009. Interstellar dust inside and outside the heliosphere. *Space Science Reviews* 143:347–356.

Kwok, S. 2004. The synthesis of organic and inorganic compounds in evolved stars. *Nature* 430:985–991.

Li, A. and Draine, B. T. 2001. Infrared emission from interstellar dust. II. The diffuse interstellar medium. *The Astrophysical Journal* 554:778–802.

Li, A. and Greenberg, J. M. 1997. A unified model of interstellar dust, *Astronomy & Astrophysics* 323:566–584.

Lineweaver, C.H. and Chopra, A. 2012. What can Life on earth tell us about life in the universe? Genesis – In Seckbach, J. (Ed.) The Beginning: Precursors of Life, Chemical Models and Early Biological Evolution *Cellular Origin, Life in Extreme Habitats and Astrology.* Springer-Verlag GmbH, Heidelberg, Germany, Vol. 22, pp. 799–815..

Lineweaver, C. H., Fenner, Y. and Gibson, Y. K. 2004, Jan. 2. The galactic habitable zone and the age distribution of complex life in the Milky Way. *Science* 303:59–62.

Lineweaver, C. H. and Schwartzman, D. 2005. Cosmic thermobiology: Thermal constraints on the origin and evolution of life in the universe, In Seckbach, J. (Ed.) *Origins: Cellular Origins and Life in Extreme Habitats and Astrobiology.* Springer-Verlag GmbH, Heidelberg, Germany, Vol. 6, pp. 233–248.

Marboeuf, U., Thiabaud, A., Alibert, Y., Nahuel Cabral, N. and Benz, W. 2014. From planetesimals to planets: Volatile molecules. *Astronomy and Astrophysics* 570:A36 (16 p).

Martin, W., Baross, J., Kelley, D. and Russell, M. J. 2008. Hydrothermal vents and the origin of life. *Nature Reviews Microbiology* 6:805–814.

Maruyama, S., Ikoma, M., Genda, H., Hirose, K., Yokoyama, T. and Santosh, M. 2012. The naked planet Earth: Most essential prerequisite for the origin and evolution of life. *Open Access funded by China University of Geosciences* (Beijing). Doi: 10.1016/j.gsf.2012.11.001.3.

Mattson, L. 2009. *On the Existence of a Galactic Habitable Zone and the Origin of Carbon.* Swedish Astrobiology Meeting, Lund. http://videos.nordita.org/conference/SwAN2009/Mattsson.pdf.

Maynard Smith, J. and Szathmary, E. 1995. *The Major Transitions in Evolution.* Oxford University Press, ISBN 019850294X.

Mayor, M. and Queloz, D. 1995. A Jupiter-mass companion to a solar-type star. *Nature* 378(6555):355–359.

McCabe, M. and Lucas, H. 2010. On the origin and evolution of life in the galaxy. *International Journal of Astrobiology* 9(04):217–226. http://arxiv.org/ftp/arxiv/papers/1104/1104.4322.pdf.
McElhinney, T. N. W. and Senanayake, W. E. 1980. Paleomagnetic evidence for the existence of the geomagnetic field 3.5 Ga ago. *Journal of Geophysical Research* 85:3523–3528.
McWilliams, A. 1997. Abundance ratios and galactic chemical evolution. *Annual Review of Astronomy and Astrophysics* 35:503–556.
Miller, S. J. 1953. A production of amino acids under possible primitive earth conditions. *Science* 117:528–529.
Miller, S. L. and Urey, H. 1959. Organic compound synthesis on the primitive earth. *Science* 130:245–251.
Møller, P., Fynbo, J. P. U., Ledoux, C. and Nilsson, K. K. 2013. Mass–metallicity relation from z=5 to the present: Evidence for a transition in the mode of galaxy growth at z=2.6 due to the end of sustained primordial gas infall. *Monthly Notices of the Royal Astronomical Society* 430:2680–2687.
Morrison, I. S. and Gowanlock, M. G. 2015. Extending galactic habitable zone modeling to include the emergence of intelligent life. *Astrobiology* 15(8):683–696.
Müller, H. S. P., Schlöder, F., Stutzki, J. and Winnewisser, G. 2005. The cologne database for molecular spectroscopy, CDMS: A useful tool for astronomers and spectroscopists. *Journal of Molecular Structure* 742:215–227.
Müller, H. S. P., Thorwirth, S., Roth, D. A. and Winnewisser, G. 2001. The cologne database for molecular spectroscopy, CDMS. *Astronomy & Astrophysics* 370:L49–L52.
Nisbet, E. G. 2002. Fermor lecture: The influence of life on the face of Earth – garnets and moving continents. In Fowler, C. M. R., Ebinger, C. J., Hawkesworth, C. J. (Eds.) *The Early Earth: Physical, Chemical and Biological Development*. Geological Society London Special Publication, London, UK, pp. 275–307.
Nisbet, E. G., Cann, J. R. and Vandover, C. L. 1995. Origins of photosynthesis. *Nature* 373:479–480.
Nisbet, E. G. and Sleep, N. H. 2001. The habitat and nature of early life. *Nature* 409:1083–1091.
Noeske, K. G., Weiner, B. J., Faber, S. M., et al. 2007. Star formation in AEGIS field galaxies since z=1.1: The dominance of gradually declining star formation, and the main sequence of star-forming galaxies. *The Astrophysical Journal* 660(1):L43–L46.
O'Malley, M. A., Leger, M. M., Wideman, J. G. and Ruiz-Trillo, I. 2019. Concepts of the last eukaryotic common ancestor. *Nature Ecology & Evolution* 3:338–344.
Panov, A. D. 2005. Scaling law of the biological evolution and the hypothesis of the self-consistent Galaxy origin of life. *Advances in Space Research* 36:220–225.
Parr, A. J. and Loughman, B. C. 1983. Boron and membrane function in plants. In Robb, D. A. and Pierpoints, W. S. (Eds.) *Metals and Micronutrients Uptake and Utilization by Plants* Academic Press, London, pp. 87–107.
Petigura, E. A., Howard, A. W. and Marcy, G. W. 2013. Prevalence of earth-size planets orbiting Sun-like stars. *Proceedings of the National Academy of Sciences* 110(48):19273–19278.

Potapov, A., Bouwman, J., Jäger, C., et al. 2020. Dust/ice mixing in cold regions and solid-state water in the diffuse interstellar medium. *Nature Astronomy Articles*. Doi: 10.1038/s41550-020-01214-x.

Prantzos, N. 2008. On the “galactic habitable zone”. *Space Science Reviews* 135:313–322.

Prochaska, J. X., Howk, J. C. and Wolfe, A. M. 2003a. The elemental abundance pattern in a galaxy at z=2.626. *Nature* 423:57–59.

Prochaska, J. X., Gawiser, E., Wolfe, A. M., Castro, S. and Djorgovski, S. G. 2003b. The age-metallicity relation of the Universe in neutral gas: The first 100 damped Lyα systems. *Astrophysical Journal* 595:L9–L12.

Rafelski, M., Neeleman, M., Fumagalli, M., Wolfe, A. M. and Prochaska, J. X. 2014. The rapid decline in metallicity of damped Lyα systems at z~5. *The Astrophysical Journal Letters* 782:L29 (6 p).

Ragsdale, S. W. 2006. Metals and their scaffolds to promote difficult enzymatic reactions. *Chemical Reviews* 106: 3317–3337.

Ramadurai, S. and Wickramasinghe, N. C. 1975. The mystery of the cosmic boron abundance. *Astrophysics and Space Science* 33:L41–L44.

Rémy-Ruyer, A., Madden, S. C., Galliano, F., Lebouteiller, V. and Jones, A. 2014. Probing the impact of metallicity on the dust properties in galaxies. *Contribution to Physique Chimie du Milieu Interstellaire (PCMI) Metting AstroRennes 2014*, Rennes, Brittany, France, Oct. 27–30. 2014.

Richter, P., de Boer, K. S., Widmann, H., Kappelmann, N., Gringel, W., Grewing, M. and Barnstedt, J. 1999. Discovery of molecular hydrogen in a high-velocity cloud of the Galactic halo. *Nature* 402:386–387.

Ritchey, A. M. Federman, S. R., Sheffer, Y. and Lambert, D. L. 2011. The abundance of boron in diffuse interstellar clouds. *The Astrophysical Journal* 728: 70 (37 p).

Roediger, J. C., Courteau, S., Graves, G. and Schiavon, R. P. 2013, Oct. 11. Constraining stellar population models - I. Age, metallicity and abundance pattern compilation for galactic globular clusters. arXiv:1310.3275v1 [astro-ph.GA].

Russell, M. J. 2007. The alkaline solution to the emergence of life: Energy, entropy and early evolution. *Acta Biotheoretica* 55:133–179.

Russell, M. J., and Arndt, N. T. 2005. Geodynamics and metabolic cycles in the Hadean. *Biogeosciences* 2:97–111.

Sagan, C. and Chyba, C. 1997. The early faint sun paradox: Organic shielding of ultraviolet-labile greenhouse gases. *Science* 276(5316):1217–1221.

Sallum, S., Follette, K. B., Eisner, J. A., Close, L. M., Hinz, P., Kratter, K., Males, J., Skemer, A., Macintosh, B., Tuthill, P., Bailey, V., Defrère, D., Morzinski, K., Rodigas, T., Spalding, E., Vaz A. and Weinberger, A. J. 2015. Accreting protoplanets in the LkCa 15 transition disk. *Nature* 527:342–344.

Schiavon, R. P., Rose, J. A., Courteau, S. and MacArthur, L. A. 2005. A library of integrated spectra of galactic globular clusters. *The Astrophysical Journal Supplement Series* 160:163–175.

Schroeder, C. 2015, Nov. 13. Explainer: What is interplanetary dust and can it spread the ingredients of life? University of Stirling. Published on http://the conversation.com.

Scorei, R. and Cimpoias, V. M. 2006. Boron enhances the thermostability of carbohydrates. *Origins of Life and Evolution of Biospheres* 36:1–11.

Seager, S., Bains, W. and Petkowski, J. J. 2016. Toward a list of molecules as potential biosignature gases for the search for life on exoplanets and applications to terrestrial biochemistry. *Astrobiology* 16(6):1–21.

Sharov, A. A. 2006. Genome increase as a clock for the origin and evolution of life. *Biology Direct* 1:17.

Sharov, A. A. 2012. A short course on biosemiotics: 2. Evolution of natural agents: Preservation, development, and emergence of functional information. Second Life®: Embryo Physics Course: http://embryogenesisexplained.com/2012/04/a-short-course-on-biosemiotics-2.html.

Sharov, A. A. and Gordon, R. 2013. Life before Earth. arXiv:1304.3381 [physics.gen-ph], Cornell University Library.

Shaw, A. M. 2006. *Astrochemistry – From Astronomy to Astrobiology.* John Wiley & Sons, Inc., New Jersey, USA.

Sneden, C., Cowan, J.J. and Gallino, R. 2008. Neutron-capture elements in the early galaxy. *Annual Review of Astronomy and Astrophysics* 46:241–288.

Sneden, C., Cowan, J. J., Lawler, J. E., et al. 2003. The extremely metal-poor, neutron capture–rich star CS 22892-052: A comprehensive abundance analysis. *The Astrophysical Journal* 591:936–953.

Snyder-Beattie, A., Sandberg, A., Drexler, K. E. and Bonsall, M. B. 2021. The timing of evolutionary transitions suggests intelligent life is rare. *Astrobiology* 21(3) (14 p). Doi: 10.1089/ast.2019.2149.

Spitoni, E. and Matteucci, F. 2015. The effect of radial gas flows on the chemical evolution of the Milky Way and M31. arXiv:1502.01836v1 [astro-ph.GA].

Steidel, C. C., Erb, D. K., Shapley, A. E., et al. 2010. The structure and kinematics of the circumgalactic medium from far-ultraviolet spectra of $\tilde{z}$= 2–3 galaxies. *The Astrophysical Journal* 717:289–322.

Tarter, J.C. 2001. The search for extra-terrestrial intelligence. *Annual Review of Astronomy and Astrophysics* 39:511–548.

Traub, W. 2012. Terrestrial, habitable-zone exoplanet frequency from Kepler. *The Astrophysical Journal* 745(1):20–29.

Turner, J. D., Zarka, P., Griessmeier, J.-M., et al. 2020. The search for radio emission from the exoplanetary systems 55 Cancri, υ Andromedae, and τ Boötis using LOFAR beam-formed observations. *Astronomy and Astrophysics* (29 p). Doi: 10.1051/0004-6361/201937201.

Valkovic, V. 1990. Origin of trace element requirements by living matter. In Gruber, B. and Yopp, J. H. (Eds.) *Symmetries in Science IV.* Plenum Press, New York, pp. 213–242.

Valkovic, V. and Obhodas, J. 2020. The origin of chiral life. arXiv:2011.12145 [astro-ph.GA], Submitted on 23 Nov 2020.

Valkovic, V. and Obhodas, J. 2021. Origins of chiral life in interstellar molecular clouds. Doi: 10.21203/RS.3.RS-376137/V1Corpus ID: 233531530.

Vallence, C. 2014. Fundamentals of atmospheric chemistry and astrochemistry: An introduction to astrochemistry. http://vallance.chem.ox.ac.uk/pdfs/AstrochemistryLectureNotes2014.pdf.

Valley, J. W., Peck, W. H., King, E. M. and Wilde, S. A. 2002. A cool early earth. *Geology* 30:351–354.

van Dishoeck, E. F. and Blake, G. 1998 Chemical evolution of star-forming regions. *Annual Review of Astronomy and Astrophysics* 36:317–368.

Wallis, M. K. and Wickramasinghe, N. C. 2004. Interstellar transfer of planetary microbiota. *Monthly Notices of the Royal Astronomical Society* 348:52–61.

Ward, P. D. and Brownlee, D., 2000. *Rare Earth: Why Complex Life is Uncommon in the Universe.* Copernicus Books. Springer Verlag, New York. ISBN 0-387-98701-0.

Watson, A. J. 2008. Implications of an anthropic model for the evolution of complex life and intelligence. *Journal of Astrobiology* 8:1–11.

Weiner, B. J., Coil, A. L., Prochaska, J. X., et al. 2009. Ubiquitous outflows in DEEP2 spectra of star-forming galaxies at z=1.4. *The Astrophysical Journal* 692:187–211.

Weiss, M. C., Sousa, F. L., Mrnjavac, N., Neukirchen, S., Roettger, M., Nelson-Sathi, S. and Martin, W. F. 2016. The physiology and habitat of the last universal common ancestor. *Nature Microbiology* 1, Article number: 16116.

Whitaker, K. E., van Dokkum, P. G., Brammer, G. and Franx, M. 2012. The star-formation mass sequence out to z=2.5. *The Astrophysical Journal Letters* 754: L29 (6 p).

Whittet, D. C. B. 2003. *Dust in Galactic Environment.* IOP Publishing, Bristol.

Woese, C. R. 1981, Jun. 1. Archaebacteria. *Scientific American* . 244:98–122.

Woese, C. R., Kandler, O. and Wheelis, M. L. 1990. Towards a natural system of organisms: Proposal for the domains Archaea, Bacteria, and Eucarya. *Proceedings of the National Academy of Sciences of the United States of America* 87(12):4576–4579.

Wong, J. T., Chen, J., Mat, W. K., Ng, S. K. and Xue, H. 2007. Polyphasic evidence delineating the root of life and roots of biological domains. *Gene* 403:39–52.

Woosley, S. E. and Weaver, T. A. 1995. The Evolution and Explosion of Massive Stars. II. Explosive Hydrodynamics and Nucleosynthesis. Report UCRL-ID-12211006. Lawrence Livermore National Laboratory. Livermore, CA, USA.

Yuan, T. T., Kewley, L. J. and Richard, J. 2012, Nov. 27. The metallicity evolution of star-forming galaxies from redshift 0 to 3: Combining magnitude limited survey with gravitational lensing. arXiv:1211.6423v1 [astro-ph.CO].

Zahid, H. J., Dima, G. I., Kewley, L. J., Erb, D. K. and Davé, R. 2012. A census of oxygen in star-forming galaxies: An empirical model linking metallicities, star formation rates, and outflows. *The Astrophysical Journal* 757: 54 (22 p).

Zahid, H. J., Geller, M. J., Kewley, L. J., Hwang, H. S., Fabricant, D. G. and Kurtz, M. J. 2013. The chemical evolution of star-forming galaxies over the last 11 billion years. *The Astrophysical Journal Letters* 771:L19 (6 p).

Zheng, W., Postman, M., Zitrin, A., et al. 2012. A magnified young galaxy from about 500 million years after the Big Bang. *Nature* 489(7416):406–408.

Zitrin, A., Labbe, I., Belli, S., Bouwens, R. Ellis, R. S., Roberts-Borsani, G., Stark, D. P., Oesch, P. A. and Smit, R. 2015. Lyα emission from a luminous z=8.68 galaxy:

implications for galaxies as tracers of cosmic reionization. *The Astrophysical Journal Letters*. 810L12 (6 pp).

Zubko, V., Dwek, E. and Arendt, R. G. 2004. Interstellar dust models consistent with extinction, emission, and abundance constraints. *The Astrophysical Journal Supplement Series* 152:211–249.

Additional reading

Abazajian, K. N., Adelman-McCarthy, J. K., Agüeros, M. A., et al. 2009. The seventh data release of the sloan digital sky survey. *The Astrophysical Journal Supplement Series* 182:543–558.

Araki, M., Takano, S., Kuze, N., et al. 2020. Observations and analysis of absorption lines including J=K rotational levels of CH_3CN: The envelope of Sagittarius B2(M). *Monthly Notices of the Royal Astronomical Society* 497:1521–1535.

Bach-Møller, N. and Jørgensen, U. G. 2021. Orbital eccentricity-multiplicity correlation for planetary systems and comparison to the Solar system. *Monthly Notices of the Royal Astronomical Society* 500(1):1313–1322.

Berkowitz, R. 2019. Iron-rich object closely orbits a white dwarf. *Physics Today* 72(6):14–16.

Bhattacharyya, A. 1943. On a measure of divergence between two statistical populations defined by their probability distributions. *Bulletin of Calcutta Mathematical Society* 35:99–109. MR 0010358.

Booth, A. S., Walsh, C., Ilee, J. D., Notsu, S., Qi, C., Nomura, H. and Akiyama, E. 2019. The first detection of ${}_{13}C_{17}O$ in a protoplanetary disk: A robust tracer of disk gas mass. *Astrophysical Journal Letters* 882:L31 (7 p).

Brayson, S., Kunimoto, M., Kopparapc, R. K., et al. 2020, Nov. 3. The occurrence of rocky habitable zone planets around solar-like stars from kepler data. arXiv:2010.14812v2 [astro-ph.EP].

Edmunds, M. G. and Phillipps, S. 1997. Global chemical evolution - II. The mean metal abundance of the Universe. *Monthly Notices of the Royal Astronomical Society* 292:733–747.

Fischer, D. A. and Valenti, J. 2005. The planet-metallicity correlation. *The Astrophysical Journal* 622(2):1102–1117.

Forbes, D. A. and Bridges, T. 2010. Accreted versus in situ Milky Way globular clusters. *Monthly Notices of the Royal Astronomical Society* 404:1203–1214.

Furukawa, Y., Chikaraishi, Y., Ohkouchi, N., et al. 2019. Extraterrestrial ribose and other sugars in primitive meteorites. PNAS Latest Articles, 6 p. www.pnas.org/cgi/doi/10.1073/pnas.1907169116.

Gilbert, E. A., Barclay, T., Schlieder, J. E., et al. 2020. The first habitable-zone earth-sized planet from TESS. I. Validation of the TOI-700 system. *The Astronomical Journal* 160:116 (21 p).

Gillon, M., Jehin, E., Lederer, S. M., et al. 2016. Temperature earth-sized planets transiting a nearby ultracool dwarf star. *Nature* 533(7602): 221–224.

Grevesse, N. and Sauval, A. J. 1998. Standard solar composition. *Space Science Reviews* 85(1/2):161–174.

Heger, A. and Woosley, S. E. 2010. Nucleosynthesis and evolution of massive metal-free stars. *The Astrophysical Journal* 724(1):341–373.

Horneck, G. 1993. Responses of *Bacillus subtilis* spores to space environment: Results from experiment in space. *Origins of Life and Evolution of Biospheres* 23(1):37–52.

Horneck, G., Bucker, H. and Reitz, G. 1994. Long-term survival of bacterial spores in space. *Advances in Space Research* 14(10):41–45.

Horneck, G., Rettberg, P., Reitz, G., Wehner, J., Eschweiler, U., Strauch, K., Panitz, C., Starke, V. and Baumstark-Khan, C. 2001. Protection of bacterial spores in space, a contribution to the discussion on panspermia. *Origins of Life and Evolution of the Biosphere* 31:527–547.

Hoyle, F. and Wickramasinghe, C. 1981. *Evolution from Space*. Published by Simon & Schuster, Inc., New York, NY.

Jordan, S. F., Rammu, H., Zheludev, I. N., et al. 2019. Promotion of protocell self-assembly from mixed amphiphiles at the origin of life. *Nature Ecology & Evolution*, articles. Doi: 10.1038/s41559-019-1015-y.

Joseph, R. and Schild, R. 2010. Origins, evolution, and distribution of life in the cosmos: Panspermia, genetics, microbes, and viral visitors from the stars. *Journal of Cosmology* 7: 1616–1670.

Kobayashi, K. and Ponnamperuma, C. 1985. Trace elements in chemical evolution. *Origins of Life and Evolution of the Biosphere* 16:41–55.

Krot, A. N., Nagashima, K., Lyons, J. R., Lee, J.-E. and Bizzarro, M. 2020. Oxygen isotopic heterogeneity in the early solar system inherited from the protosolar molecular cloud. *Science Advances* 6:eaay2724.

Lane, N., Allen, J. F. and Martin, W. 2010. How did LUCA make a living? Chemiosmosis in the origin of life. *BioEssays* 32:271–280.

Lu, L., Sargent, W. L. W., Barlow, T. A., Churchill, C. W. and Vogt, S. S. 1996. Abundances at high redshifts: The chemical enrichment history of damped Lyalpha galaxies. *Astrophysical Journal Supplement* 107:475–519.

Maiolino, R., Nagao, T., Grazian, A., et al. 2009. LSD: Lyman-break galaxies Stellar populations and dynamics – I. Mass, metallicity and gas at z~ 3.1. *Monthly Notices of the Royal Astronomical Society* 398:1915–1931.

Mannucci, F., Cresci, G., Maiolino, R., Marconi, A. and Gnerucci, A. 2010. A fundamental relation between mass, star formation rate and metallicity in local and high-redshift galaxies. *Monthly Notices of the Royal Astronomical Society* 408:2115–2127.

Mannucci, F., Cresci, G., Maiolino, R., et al. 2009, LSD: Lyman-break galaxies Stellar population and Dynamics - I. Mass, metalicity and gas at z~3.1. *Monthly Notices of the Royal Astronomical Society* 398:1915–1931.

Manser, C. J., Gänsicke, B. T. and Siegfried Eggl, S., et al. 2019. A planetesimal orbiting within the debris disc around a white dwarf star. *Science* 364:66–69.

Navarro Silvera, S. A. and Rohan, T. E. 2007. Trace elements and cancer risk: A review of the epidemiologic evidence. *Cancer Causes Control* 18:7–27.

Oba, Y., Takano, Y., Naraoka, H., Watanabe, N. and Kouchi, A. 2019. Nucleobase synthesis in interstellar ices. *Nature Communications* 10:4413. Doi: 10.1038/s41467-019-12404-1.

Oba, Y., Takano, Y., Naraoka, H. et al. 2020. Extraterrestrial hexamethylenetetramine in meteorites – a precursor of prebiotic chemistry in the inner solar system. *Nature Communications* 11:6243. Doi: 10.1038/s41467-020-20038-x.

Pettini, M. 2003. Element abundances through the cosmic ages. Lectures given at the XIII Canary Islands Winter School of Astrophysics "Cosmochemistry: The Melting Pot of Elements". http://www.ast.cam.ac.uk/pettini/canaries13. arXiv:astro-ph/0303272.

Phillipps S and Edmunds M. G. 1996. Global chemical evolution - I. QSO absorbers and the chemical evolution of galaxy discs. *Monthly Notices of the Royal Astronomical Society* 281:362–368.

Prochaska, J. X. and Wolfe, A. M. 1999. Chemical abundances of the damped Lyα systems at z>1.5. *Astrophysical Journal Supplement* 121:369–415.

Prochaska, J. X., Wolfe, A. M., Tytler, D., et al. 2001. The UCSD HIRES/Keck I damped Lyα abundance database. I. The data. *Astrophysical Journal Supplement* 137:21–73.

Rana, N. C. 1991. Chemical evolution of the galaxy. *Annual Review of Astronomy and Astrophysics* 29:129–162

Rafelski, M., Wolfe, A. M., Prochaska, J. X., Neeleman, M. and Mendez, A. J. 2012. Metallicity evolution of damped Ly-α systems out to z≈5. *The Astrophysical Journal* 755:89 (21 p).

Sahai, N., Kaddour, H., Dalai, P., Wang, Z., Bass, G. and Gao, M. 2017. Mineral surface chemistry and nanoparticle-aggregation control membrane self-assembly. *Scientific Reports* 7:43418. Doi: 10.1038/srep43418. www.nature.com/scientificreports.

Schulze-Makuch, D., Heller, R. and Guinan, E. 2020. In search for a planet better than earth: Top contenders for a superhabitable world. *Rapid Communication, Astrobiology* 20(12):11 p. Doi: 10.1089/ast.2019.2161.

Smith, K. E., House, C. H., Arevalo Jr., R. D., Dworkin, J. P. and Callaha, M. P. 2019. Organometallic compounds as carriers of extraterrestrial cyanide in primitive meteorites. *Nature Communications* 10:2777 Doi: 10.1038/s41467-019-10866-x.

Tinsley, B. M. 1979. Stellar lifetimes and abundance ratios in chemical evolution. *The Astrophysical Journal* 229:1046–1056.

Totani, T. 2020. Emergence of life in an inflationary Universe. *Nature Scientific Reports* 10:1671. Doi: 10.1038/s41598-020-58060-0.

Twarog, B. A. 1980. The chemical evolution of the solar neighborhood. II - The age-metallicity relation and the history of star formation in the galactic disk. *Astrophysical Journal*, Part 1, 242:242–259.

Vladilo, G. 2002a. Chemical abundances of damped Lyα systems: A new method for estimating dust depletion effects. *Astronomy and Astrophysics* 391:407–415.

Vladilo, G. 2002b. A scaling law for interstellar depletions. *Astrophysical Journal* 569:295–303.

Wagner, K., Boehle, A., Pathak, P., et al. 2021. Imaging low-mass planets within the habitable zone of α Centauri. *Nature Communications* 12:922 (7 p). Doi: 10.1038/s41467-021-21176-6.

Wanajo, S. and Janka, H. T. 2011, Dec. 7. The r-process in the neutrino-driven wind from a black-hole torus. arXiv:1106.6142v2 [astro-ph.SR].

Westby, T. and Conselice, C. J. 2020. The astrobiological copernican weak and strong limits for intelligent life. *The Astrophysical Journal* 896:58 (18 p).

Wickramasinghe, J., Wickramasinghe, C., Napler, W. 2010. *Comets and the Origin of Life*. World Scientific Publishing Co., Singapore.

Wickramasinghe, N. C., Tokoro, G. and Syroeshkin, A.V. 2018. Confirmation of microbial ingress from space. *Advances in Astrophysics* 8(4):266–270.

Wlodarczyk-Sroka, B. S., Garrett, M. A., and Siemion, A. V. P. 2020, Aug. 28. Extending the Breakthrough Listen nearby star survey to other stellar objects in the field. *Monthly Notices of the Royal Astronomical Society* arXiv:2006.09756v2 [astro-ph. IM]. 10 p. preprint.

Yung, Y. L., Chen, P., Nealson, K., et al. 2018. Methane on mars and habitability: Challenges and responses. *Astrobiology* 18(10), Doi: 10.1089/ast.2018.1917.

6

Multiverse Cosmological Models and Anthropic Principle

6.1 Introduction

The Big Bang model is the widely accepted theory of the evolution of the Universe today. Its essential feature is the emergence of the Universe from a state of extremely high temperature and density – the so-called Big Bang occurred some 13.8×10^9 years ago. The idea of this type of Universe was first discussed by Friedmann (1922) and Lemaître (1931a, b). Lemaitre (1931a, b) anticipated most of the contents of the present-day standard cosmological model. He included the acceleration of the expansion due to repulsive dark energy, the interpretation of the cosmological constant as vacuum energy or the possible non-trivial topology of space.

All Big Bang models are based on the observations and experiments whose results have been extrapolated as far as possible into the past constructed by the process of hypothesis and calculations. The Big Bang model is now almost universally accepted by the scientific community. It could explain the large number of observations made by the grand telescopes and the results of experiments carried out with particle accelerators and retrace the principal stages in the creation of the Universe, a process which took 13.81×10^9 years. It appears that in the standard cosmological model, the Universe is dominated by the so-called dark energy and dark matter, forms of matter and energy that we are yet to understand fully but whose existence is supported by observational evidence (see Figure 6.1). The ordinary matter present now has an almost negligible cosmic relevance compared to dark matter and dark energy.

This way to this situation was paved by many significant discoveries recognized by Nobel Prizes given to the main actors. For example, the Nobel Prize in Physics 1967 was given to Hans Bethe for his contributions to the theory of nuclear reactions – in the first place, his discoveries concerning the energy production in stars; see Bethe (1968) and references therein.

The Nobel Prize in Physics **1974** was divided between Martin Ryle for his observations and inventions in the field of radio astrophysics, in particular of the aperture synthesis technique (see Longair, 2016), and Antony Hewish for his decisive role in the discovery of pulsars (Hewish 1952, 1975).

DOI: 10.1201/9781003181330-6

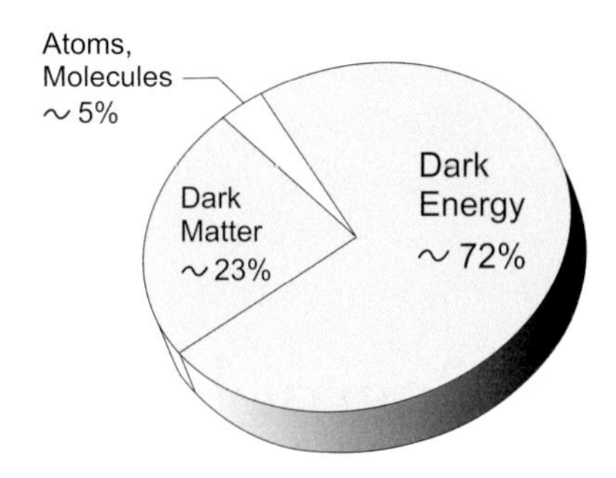

FIGURE 6.1 The composition of the universe today.

The Nobel Prize in Physics **1978** was divided again with one half going to P. L. Kapitsa "for his basic inventions and discoveries in the area of low-temperature physics" and the other half jointly to A. A. Penzias and R. W. Wilson "for their discovery of cosmic microwave background radiation". Penzias and Wilson (1965a, b) were investigating the use of high-altitude echo-balloons in communication when they detected an unusual persistent background noise in their experiments. After some time, it became clear that they had discovered the relic background radiation, called cosmic microwave background (CMB), because it peaks in the microwave part of the electromagnetic spectrum.

S. Chandrasekhar and W. A. Fowler shared the Nobel Prize in Physics **1983**. Chandrasekhar received the prize for his theoretical studies of the physical processes of importance to the structure and evolution of stars (see Chandrasekhar, 1958, 2005). Fowler obtained half of the prize for his theoretical and experimental studies of the nuclear reactions essential for the formation of chemical elements in the Universe (see Burbidge, 1957; Fowler, 1984).

It passed ten years before in **1993** Russell Alan Hulse and Joseph Hooton Taylor Jr. shared the Nobel Prize in Physics. They obtained it for a new type of pulsar, a discovery that has opened up new possibilities for the study of gravitation (see Hulse and Taylor, 1975).

It was **2002** when the Nobel Prize Committee decided that the prize for physics should be divided between three persons: Raymond Davis Jr., Masatoshi Koshiba and Riccardo Giacconi. The first two obtained it for pioneering contributions to astrophysics, in particular for the detection of cosmic neutrinos (see Davis 1953, 1964; Fukuda et al., 1998, 1999). Giacconi obtained his share of the Nobel Prize for pioneering contributions to astrophysics, which have led to the discovery of cosmic X-ray sources. He shaped the tools and culture of modern astronomy, which centers on large facilities that serve the community (see Giacconi and Rossi, 1960; Giacconi, 2008).

J. C. Mather and G. F. Smoot obtained the Nobel Prize in Physics **2006** for the discovery of the blackbody nature and anisotropy of the CMB radiation. Data from the COBE satellite proved that the CMB has a blackbody spectrum (Mather et al., 1990), with a temperature of 2.725 K., and in 1992, the existence of anisotropies was also revealed; see Smoot et al. (1992).

The Nobel Prize in Physics **2011** was shared by Brian P. Schmidt and Adam G. Riess. They obtained the prize for the discovery of the accelerating expansion of the Universe through observations of distant supernovae. Their work, together with the data collected by WMAP satellite (2003–2006) and European Planck Telescope (2013–2015), produced the so-called concordance cosmological model from which the age of the Universe was suggested to be 13.81×10^9 years.

The Nobel Prize in Physics **2019** was shared by James Peebles, Michel Mayor and Didier Queloz. The contributions of Peebles to modern cosmology are summarized in his books: Physical Cosmology (published in 1971), The Large-Scale Structure of the Universe (published in 1980) and Principles of Physical Cosmology (published in 1993). His various contributions are the study of primordial helium abundance and the primordial fireball (Peebles, 1966a, b), structure formation and galaxy evolution, CMB, dark matter, dark energy and inflation (see, for example, Ratra and Peebles, 2003). Michel Mayor and Didier Queloz discovered the first extrasolar planet (51 Pegasi b) orbiting a Sun-like star, 51 Pegasi (Mayor and Queloz, 1995). Their discovery of a planet outside our solar system, the so-called exoplanet, started a revolution in astronomy; over 4,000 exoplanets have since been found in the Milky Way.

Finally, the Nobel Prize in Physics **2020** has recognized achievements in the theoretical and experimental understanding of the black holes. A half of the prize went to Roger Penrose (University of Oxford) "for the discovery that black hole formation is a robust prediction of the general theory of relativity".

Penrose won half the prize for proving that under very general conditions, the collapsing matter would trigger the formation of a black hole described in his ground-breaking paper (Penrose, 1965). This rigorous result opened up the possibility that the astrophysical process of gravitational collapse, which occurs when a star runs out of its nuclear fuel, would lead to the formation of black holes in nature. He was also able to show that at the heart of a black hole must lie a physical singularity – an object with infinite density, where the laws of physics simply break down. At the singularity, our concepts of space, time and matter fall apart. Penrose invented new mathematical concepts and techniques while developing this proof. Equations that Penrose derived in 1965 have been used by physicists studying black holes ever since. In fact, just a few years later, Stephen Hawking, together with Penrose, used the same mathematical tools to prove that the Big Bang cosmological model had a singularity at the initial moment. These are results from the celebrated Penrose–Hawking singularity theorem. The fact that mathematics demonstrated that astrophysical black holes might exist in nature energized the quest to search for them using astronomical techniques. Indeed, since Penrose's work in the 1960s, numerous black holes have been identified.

His books motivated many young scientists to follow his steps. Penrose became interested in the definition of consciousness and wrote two books in which he argued that quantum mechanics is required to explain the conscious mind (Penrose, 1989, 1994). He also wrote an extensive overview of mathematics and physics (Penrose, 2004). In yet another book (Penrose, 2010), he posited his theory of conformal cyclic cosmology, formulating the Big Bang as an endlessly recurring event.

The other half was jointly awarded to Andrea Ghez (University of California, Los Angeles) and Reinhard Genzel (Max Planck Institute for Extraterrestrial Physics) "for the discovery of a supermassive compact object at the center of our galaxy". Both of them led separate research teams, which in the 1990s identified a black hole at the center of the Milky Way.

Since the 1990s, Ghez and Genzel have each led groups that have mapped the orbits of stars close to the galactic center. These studies led them to conclude that an extremely massive, invisible object must be dictating the stars' frantic movements. The object known as Sagittarius A* is the most convincing evidence yet of a supermassive black hole at the center of the Milky Way (see Ghez et al., 2003, Genzel et al., 1994, 2010).

Here we have introduced two entities: dark matter and dark energy. It is believed that ~23% of the mass–energy of the Universe consists of dark matter; see Figure 6.1. Dark matter is quite different from ordinary matter, atoms and molecules that make the world around us. Namely, dark matter only interacts with gravity; in other words, it neither reflects, emits nor obstructs light or any other electromagnetic radiation. This fact is the reason it cannot be observed directly. However, the Hubble telescope studies of how the cluster of galaxies bends the light that passes through them indicate the position of the hidden mass. In 2007, it was possible to construct the first three-dimensional map of the large-scale distribution of dark matter in the Universe (https://spacetelescope.org/news/heic0701/). This map showed that ordinary matter (mainly in the form of galaxies) also accumulates the densest concentrations of dark matter.

Elaboration of the rate of expansion of the Universe suggests that dark energy is by far the most significant part of the Universe's mass–energy content; it makes 73% of the known Universe; see Figure 6.1. Hubble telescope studies of the expansion rate of the Universe have found that the expansion is speeding up; this is explained by dark energy working against gravity to push the Universe apart ever faster. Today, the true nature of dark energy is still a mystery.

We should note that in a recent study, Chen et al. (2019) presented the data for H_0, Hubble constant, which is the measure of the expansion rate of the Universe, from three time-delay gravitational lens systems with adaptive optics imaging. Their result shows a problem with the Standard Model (SM) of cosmology, which shows the Universe was expanding very fast early in its history. The expansion rate slowed down due to the gravitational attraction of dark matter, and now the expansion is speeding up again because of dark energy! The crisis in cosmology could be explained by the fact that Hubble constant is constant everywhere in space at a given time; however, it is not constant in time.

All the laws of the Universe have particular constants associated with them. Examples of these constants are the gravitational constant, the speed of light, the electric charge, the mass of the electron and Planck's constant. Some are derived from physical laws; the speed of light, for example, comes from Maxwell's equations. However, for most, their values are arbitrary. The rules would still operate if the constants had different values, although the resulting interactions would be radically different.

An aspect of anthropic reasoning that has attracted plenty of attention is the use in cosmology in explaining the apparent fine-tuning of our Universe. "Fine-tuning" means that a set of cosmological parameters or fundamental physical constants have such values that if they had been only slightly different, the Universe would not have the presence of intelligent life. In the classical Big Bang model, for example, the early expansion velocity seems fine-tuned. Had it been very slightly higher, the Universe would have expanded too rapidly and no galaxies would have formed. There would only have been a very low-density hydrogen gas getting more and more dispersed as time went by. In such a Universe, life could not evolve. Had the initial expansion velocity been very slightly less, then the Universe would have collapsed soon after the Big Bang and there would have been no life now. The fact that our Universe has just the right conditions for life appears to be "balancing on a knife's edge" (Leslie, 1989). Several other parameters seem fine-tuned in the same meaning, including the ratio of the electron to proton mass, the neutron–proton mass difference, the magnitudes of force strengths, the smoothness of the early Universe and perhaps even the metric signature of spacetime (Tegmark, 1996).

Therefore, we can say that some completely arbitrary properties of the Universe appear to have values which, if changed only very slightly, would preclude the existence of beings like ourselves or, more speculatively, any intelligent life form whatsoever. The term "completely arbitrary" means not being determined by or calculable from any known law of physics.

Here are some examples: the ratio of masses of the proton and electron and the relative strength of the four fundamental forces.

- If the gravitational constant, which determines the strength of gravity, were lower, stars would have insufficient pressure to overcome the Coulomb barrier to start thermonuclear fusion (i.e., stars would not shine). If higher, stars will burn faster, using up all the fuel before life had a chance to evolve.
- The strong force coupling constant holds particles together in the nucleus of the atom. If weaker, multi-proton nuclei would not hold together and hydrogen would be the only element in the Universe. If stronger, all elements with smaller mass and charge than iron would be scarce.
- The electromagnetic coupling constant determines the strength of the electromagnetic force that couples electrons to the nucleus. If smaller, no electrons would be held in orbits. If stronger, electrons will not make bonds with other atoms. Either way, there will be no molecules.

These constants are all critical for the formation of the basic building blocks of life. The range of values for these constants is very narrow (1%–5%) for the combination of constants. Outside this range, life (in particular, intelligent life) would be impossible.

This apparent fine-tuning of the fundamental constants of nature is disturbing to many scientists since it can be interpreted as evidence the Universe was designed just for us – that the Universe has a purpose and that we are it. This sentence seems to step back to a prescientific world view, which all the evidence of hundreds of years of science has refuted point by point.

The situation became even worse with the cosmological discoveries of the 1980s. The two critical cosmological parameters are the cosmic expansion rate (Hubble's constant, determining the age of the Universe) and the cosmic density parameter (Ω, determining the acceleration of the Universe and its geometry). The cosmic density parameter specifies the three possible shapes the Universe might have a flat Universe (Euclidean with zero curvature), a spherical or closed Universe (or positive curvature) and a hyperbolic or open Universe (or negative curvature).

6.2 Standard ΛCDM Cosmological Model

In the standard ΛCDM (Lambda cold dark matter) cosmological model, the cosmological expansion is dominated by dark energy (a cosmological constant) and cold dark matter. The model has had multiple triumphs in explaining cosmological observations. One can say that the ΛCDM cosmological model and SM of particles and fields are the two pillars of our understanding of the Universe. In spite of this, several discrepancies have persisted. The most notorious is the difference between the values of the Hubble constant H_0, which parameterizes the expansion rate of the Universe, as measured locally and those inferred from CMB measurements. Cosmology can be strongly affected by the production and dynamics of particles beyond those of the SM. Of particular interest are sterile neutrinos. Sterile or right-handed neutrinos are an essential part of the seesaw mechanism (Minkowski, 1977; Yanagida, 1980) that generates small masses for active neutrinos. While the original seesaw mechanism invoked a large-scale mass, there are models in which a low mass scale is generated naturally (Kusenko et al., 2010). Sterile neutrinos provide an excellent probe of early Universe cosmology (Gelmini et al., 2020a, 2000b).

Previous galaxy surveys in the context of the cold dark matter paradigm have shown that the mass of the dark matter halo and the total stellar mass are coupled through a function that varies smoothly with mass. Their average ratio M_{halo}/M_{stars} has a minimum of about 30 for galaxies with stellar masses near that of the Milky Way (~5×10^{10} solar masses) and increases both toward lower masses and toward higher masses (Behroozi et al., 2013; Moster et al., 2010). Van Dokkum et al. (2018) reported an observation of a galaxy, NGC 1052-DF2, lacking dark matter. They estimated that M_{halo}/M_{stars} for this galaxy is approximately unity (and consistent with zero), a factor of a least 400 lower than expected. This galaxy demonstrates that dark matter is not always coupled with baryonic matter on a galactic scale. This finding demonstrates that dark matter does not always trace the bulk of the baryonic mass, which in clusters is in the form of gas. Van Dokkum et al. (2019) reported a discovery of a second galaxy in this class, residing in the same group: NGC 1052-DF4, which closely resembles NGC 1052-DF2 in terms of its size, surface brightness and morphology. The origin of these objects with an apparent lack of dark matter is not presently understood.

The local Universe observations find a value of the Hubble constant H_0 that is larger than the value inferred from the CMB and other early Universe measurements, assuming known physics and the ΛCDM cosmological model. Gelmini et al. (2020) showed that additional radiation in active neutrinos produced just before BBN by an unstable sterile neutrino with a mass of 10 MeV can alleviate this discrepancy. The necessary masses and couplings of the sterile neutrino, assuming it mixes primarily with ν_τ and/or ν_μ neutrinos, are within the reach of Super-Kamiokande detector (see http://www-sk.icrr.u-tokyo.ac.jp/sk/index-e.html) as well as upcoming laboratory experiment at CERN (see NA62 experiment, at https://home.cern/science/experiments/na62).

A precise measurement of the Hubble constant provides essential information about the cosmological energy content and could further reveal significant new features related to fundamental physics. Multiple independent measurements of the Hubble constant in the local Universe, from Cepheids and type Ia supernovae, which are nearly independent of the cosmological expansion history, lead to a value $H_0=74.03\pm1.42$ km/s/Mpc (Riess et al., 2019). In contrast, the value inferred from Planck satellite CMB data and baryon acoustic oscillations (BAO) data, assuming that the standard ΛCDM cosmological model describes the expansion history since recombination, is $H_0=67.66\pm0.42$ km/s/Mpc (Planck Collaboration, 2018). The discrepancy is significant at the $\sim 5\sigma$ level. The underestimation of systematic effects does not seem responsible for this discrepancy. This substantiates a possible cosmological origin of the tension in the H_0 measurements (Gelmini et al., 2020).

Although the ΛCDM cosmological model has been successful in explaining many cosmological observations on the largest scale of the Universe, it is not so successful on the small scales. Among the problems the model does not handle are the following:

- Phase-space correlation of satellite galaxies.
- Coupling between the visible, baryonic matter in the galaxy and the observed dynamics dominated by dark matter halo at large radii.

This situation has resulted in several drastically different ideas, resulting in different theories of gravity. For example, Milgrom (1983) proposed the possibility that there is not much dark matter in galaxies and galaxy systems. If a certain modified version of the Newtonian dynamics is used to describe the motion of bodies in a gravitational field of a galaxy, the observational results could be reproduced with no need to assume hidden mass in appreciable quantities. Various characteristics of galaxies would result in no further assumptions. This is called the Milgromian dynamics (MOND) paradigm, which modifies the standard laws of dynamics at low accelerations corresponding to weak gravitational fields rather than assuming the existence of dark matter.

It is the strong equivalence principle (SEP) that distinguishes general relativity from other viable theories of gravity (Chae et al., 2020). The SEP demands that the internal dynamics of a self-gravitating system under free fall in an external gravitational field is not dependent on the external field strength. The authors claim that they detected a systematic downward trend in the weak gravity part of the radial acceleration relation at the right acceleration predicted by the external field effect of the MOND modified gravity. Tidal effects from neighboring galaxies in the ΛCDM context are not strong enough to explain these phenomena. They are not predicted by the existing ΛCDM models of galaxy formation and evolution. The results published by Chae et al. (2020) point to the breakdown of the SEP by supporting modified gravity theories beyond general relativity.

6.2.1 Dark Matter

The substance called dark matter interacts gravitationally with itself and with ordinary matter. Its presence is inferred indirectly from the properties of the CMB, from the distribution and velocities of stars, galaxies and galaxy clusters, and from gravitational lensing of distant light sources.

Gravitational lensing and dispersion measurements of galaxy clusters, the largest bound systems that have been observed, show that dark matter is the dominating mass component. Detailed studies of half a dozen or so merging galaxy clusters have clearly ruled out possible alternative explanations involving modifications of the gravitational law and are now starting to probe the properties of dark matter itself. We also know that dark matter exists in our own galaxy, the Milky Way, which shows rotational velocities that are independent of radius at high radii, just as in any other spiral galaxy we observe. This flat rotation curve is clearly inconsistent with that expected from Kepler's laws, but is naturally explained by the fact that galaxies are immersed in a halo of dark matter that dominates their mass.

For several decades, the favorite dark matter candidate was the weakly interacting massive particle (WIMP). It was assumed that this particle interacts with ordinary matter particles by the weak force, and it was assumed that its mass is in the interval 1–100 mass of the proton. WIMPs could have been produced in collisions of known particles shortly after the Big Bang in an amount consistent with cosmological and astrophysical observations.

Dark matter can be expected to have weak couplings to standard matter, so that it can be searched for with laboratory experiments. This direct search for dark matter is pursued with a variety of complementary technologies and experiments. The XENON project, in particular, is one of the most sensitive direct searches for dark matter. The XENON dark matter experiment is installed underground at the Laboratori Nazionali del Gran Sasso of INFN, Italy. The XENON experiment is a 3,500 kg liquid xenon detector to search for the elusive dark matter.

XENON's main goal has always been to detect a particular type of dark matter particle called weakly interacting massive particle (WIMP). If a WIMP flying through the tank knocked a xenon atom hard enough, it could produce detectable light. The published report (XENON Collaboration, 2020), presenting data taken over one year, reported 285 events in an energy range where the expected were only 232 background events. The excess was within 3.5 sigma statistical significance. However, the reported events had a large electron-to-photon ratio, indicating that they originated in interactions with atom electrons (see also Lin, 2020). Theorists already reacted to this potential signal with explanations involving new physics (Rini, 2020). A possibility that is receiving increased attention is that of a particle that has a mass between 1 keV/c^2 and the mass of the proton (Essig, 2020). This particle is supposed to interact with regular matter by some new force rather than the weak force.

Recently, Calmet and Kuipers (2021) have shown that quantum gravity leads to lower and upper bounds on the masses of dark matter candidates. They presented theoretical bounds on dark matter masses. These bounds depend on the spins of the dark matter candidates and the nature of interactions in the dark matter sector. For the singlet scalar dark matter, they report the mass range to be 10^{-3} eV $\leq m \leq 10^{7}$ eV. The lower bound comes from limits on fifth force type interactions and the upper bound from the lifetime of the dark matter candidate (Calmet and Kuipers, 2021).

Atomic responses to general dark matter–electron interactions in the detector materials has been discussed by Catena et al. (2020). This has been done with the aim to interpret the results of direct detection experiments searching for signals of dark mater–electron interactions. Materials' responses to dark matter–electron interactions can be quantified in terms of the overlap between initial-state and final-state electron wave functions. The authors found that the rate at which atoms can be ionized by dark matter–electron scattering can, in general, be expressed in terms of four independent atomic responses. They applied their theoretical considerations to dark matter particles masses in the 1–1,000 MeV/c^2 range and on their interactions with electrons in argon and xenon materials.

Axion is a spin-zero chargeless massive particle suggested as a solution to the strong charge–parity problem in quantum chromodynamics. It turns out that the axions are also a promising cold dark matter candidate and offer an explanation for matter–anti-matter asymmetry in the early Universe (see Darling, 2020a). Searches for either ordinary matter or dark matter axions have not yielded positive results so far. Referring to the existence of axionic dark matter, only different limits have been reported (Darling, 2020a, b; Jeong et al., 2020; Foster et al., 2020, and references therein).

Free neutrons with a lifetime of $\tau \approx 14$ minutes decay by weak force to

$$n \rightarrow p + e^- + \bar{\upsilon}_e.$$

There has been much experimental work to determine the precise neutron lifetime by two sets of techniques. The so-called beam approach is based on the measurement of the neutron flux of a cold neutron beam after passing through a region where the emitted protons are detected. In this approach, the β-decay rate is measured directly since the possible other neutron-disappearing processes are on the level of 10^{-3} or smaller. The other approach is the so-called bottle experiment in which ultra-cold neutrons are stored (confined) by gravitational force in a material or magnetic trap. By measuring the neutron loss rate in the trap, one can obtain a neutron lifetime. It turns out that these two types of measurements result in different values for the neutron lifetime. The 1% difference between the results of the two approaches is serious with the precise measurements:

a. In "beam" measurement (Yue et al., 2013), $\tau = 887.7 \pm 1.2$ s.
b. In "bottle" measurement (Pattie et al., 2018), $\tau = 877.7 \pm 0.7$ s.

Many theoretical attempts to resolve this 1% neutron lifetime difference have been published. The two, searching for the physics beyond the standard model, have attracted our attention:

i. Neutron decay to black matter particle, X, via the decay $n \rightarrow X + \gamma$ with constraints 937.900 MeV $< m_X <$ 938.783 MeV for the mass of the dark particle and 0.782 MeV $< E_\gamma <$ 1.664 MeV for the photon energy.
ii. $n - \tilde{n}$ oscillations, where $\tilde{n}$ is "dark" neutron.

Both have, so far, not been experimentally confirmed; contrarily, both have been shown to be very unlikely solutions to the problem. There is a recent effort considering n–n′

(neutron–mirror neutron) oscillations without harming the established physics (Tan, 2019).

Mirror matter was originally proposed by Lee and Yang (1956) in order to restore the parity symmetry of the Lagrangian of the SM. The most natural way to do so is to add to the existing Lagrangian its parity-symmetric counterpart so that the whole Lagrangian is invariant under the parity transformation, each part transforming into the other. This corresponds to reintroducing all the known fields with the same coupling constants, but with opposite parities. The result is a new group of particles, called mirror particles, which is an exact duplicate of the ordinary group, but where ordinary particles have left-handed interactions, mirror particles have right-handed interactions (Foot et al., 1991). As a consequence, the three-gauge interactions act separately in each group, the only link between them being gravity. Because mirror baryons, just like their ordinary counterparts, are stable and can be felt only through their gravitational effects, the mirror matter scenario provides an ideal interpretation of the dark matter.

In his paper, Tan (2019) proposed a new mechanism of n–n′ oscillations. His model can explain the observed difference of neutron lifetime measurements without harming the other established physics laws with n–n′ mixing strength of 2×10^{-5}. In his paper, the author showed that the mass difference of n–n′ doublet is about 2×10^{-6} eV/c^2 in the framework of the mirror matter theory with slightly broken mirror symmetry. He also demonstrated that both groups contributed to the observed dark matter-to-baryon matter ratio of $\Omega_{dark}/\Omega_B=5.4$. Actually, he has extended the SM with mirror matter and used this formulation for understanding the dark matter and neutron lifetime puzzle. Extensions of this model to other open problems are feasible.

There is a significant number of theoretical speculations where the dark matter is made entirely or partly of mirror matter. For example, Ciarcelluti and Wallemacq (2014) presented numerical simulations of CMB and large-scale structure in the case in which the cosmological dark matter is made entirely or partly made of mirror matter. They demonstrated that cosmological models with pure mirror matter, mirror matter mixed with cold dark matter (CDM) and pure CDM are equivalent concerning the CMB and large scale structure power spectra, as a consequence of the fact that mirror matter and collisional WIMPs have the same behavior at linear scales.

To others like Kusenko et al. (2020), primordial black holes are viable for dark matter, provided their masses are in the currently unconstrained "sublunary" mass range. The issue is not yet closed!

The existence of dark matter has been postulated to resolve discrepancies between astrophysical observations and accepted theories of gravity. In particular, the measured rotation curve of galaxies provided much experimental support to the dark matter concept. However, most theories used to explain the rotation curve have been restricted to the Newtonian potential framework, disregarding the general relativistic corrections associated with mass currents. Ludwig (2021) showed that the gravitomagnetic field produced by the currents modifies the galactic rotation curve, notably at large distances. The coupling between the Newtonian potential and the gravitomagnetic flux function results in a nonlinear differential equation that relates the rotation velocity to the mass density. The solution of this equation reproduces the galactic rotation curve without recourse to obscure dark matter components, as exemplified by three characteristic cases: (i) dwarf galaxy NGC 1560, (ii) spiral galaxy NGC 3198 and (iii) lenticular galaxy NGC 3115. The effects attributed to dark matter can simply be explained by the gravitomagnetic field produced by the mass currents.

6.3 Multiverse Cosmological Models

The fine-tuning of parameters in the early Universe is required to reproduce our present-day Universe. If so, this might suggest that our Universe might be a region within an eternally inflating super-region. Many other areas beyond our observable Universe could exist, with each such region governed by a different group of physical parameters than the ones we have measured in our Universe. A collision between these regions should leave signatures in anisotropy in the CMB. Chary (2015) discussed the possibility of detection of such an anomaly in the CMB.

The Universe extends only as far as light can travel in the 13.8×10^9years since the Big Bang (that would be 13.8×10^9 ly). The spacetime beyond that distance is considered to be its separate Universe. In such a way, a multitude of Universes exists next to each other in a giant patchwork of worlds.

Multiverse ideas have been discussed more by philosophers and less by scientists for a long time. Two theoretical developments in string/M theory and eternal inflation have put scientific interest to them again. String theory speculates a vast number of alternative low-energy physical worlds, and the speculation alone does not mean that all such worlds are physically possible. The existence of these other worlds, or "pocket Universes", as Susskind (2003) has called them, is rendered plausible when accounted for inflationary Universe cosmology. According to this nowadays standard model, the Universe at or near its origin possessed tremendous vacuum energy (or Λ). A Λ term in Einstein's gravitational field equations acts like a repulsive force, causing the Universe to expand exponentially. This so-called inflationary episode may have lasted no longer than about 10^{-35} seconds in our "pocket Universe". After that period, the primordial vacuum energy decayed to the presently observed value, releasing the energy difference as heat. After this inflationary episode, the hot Big Bang model remains much the same, involving the early synthesis of helium, galaxy and star formation, etc.

In a more popular variant known as eternal inflation (Vilenkin, 1983; Linde, 1983), the Universe we live in is just one particular vacuum bubble within a vast, most likely infinite assemblage of bubbles or pocket Universes. This multiverse of Universes is characterized by frenetically continuing inflation in the overall super-structure, driven by exceedingly high vacuum energies. Here and there, "bubbles" of low-energy vacuum would nucleate quantum mechanically from the eternally inflating region and evolve into pocket Universes. When putting together eternal inflation with the complex landscape of string theory, there is a mechanism for generating Universes with different local laws, i.e., different low-energy physics. Each bubble nucleation proceeding from tremendous vacuum energy represents a symbolic "ball" rolling down randomly by the landscape from some dizzy height and ending up in one of the valleys or vacuum states. So the ensemble of physical laws available from string theory becomes actualized as an ensemble of pocket Universes, each with its low-energy physics. The total number of this type of Universes may be infinite, and the complete variety of possible low-energy physics finite but significant. According to Linde (1987), the size of a typical inflationary region is enormous, even by Hubble radius standards. The expansion rate is exponential with an e-folding time of $2\pi M_P^2/m^2$, with M_P being the Planck mass and m being the mass of the scalar inflation field, the energy density of which drives the inflationary expansion (Davies, 2004).

According to Davies (2004), the existence of a multiverse does not rest on the validity of string theory or even inflationary cosmology. Instead, it is a generic property of an attempt to explain low-energy physics as the product of particular quantum states in combination with a model of the Universe originating in a Big Bang.

Several other multiverse theories exist in the literature. There are many logically and physically assumed models in which an ensemble of Universes can be described. Of interest are the brane theories, in which "our Universe" is regarded as a three-dimensional sheet or brane embedded in a higher-dimensional space (Randall and Sundrum, 1999). In string theory and supergravity theories, a brane is a physical object that generalizes the notion of a point particle into higher dimensions. A point particle can be considered as a brane of dimension zero, while a string can be considered as a brane of dimension one. It is also possible to consider higher-dimensional branes. In dimension *p*, these are called *p*-branes. The word "brane" has its origin in the term "membrane", which refers to a two-dimensional brane.

This theory further suggests these brane Universes aren't always parallel and out of reach. Sometimes, they might collide, causing repeated Big Bangs that reset the Universes over and over again.

Tegmark (2003, 2004) proposed an extreme version of the multiverse. In addition to Universes with all possible values of the fundamental "constants" of physics, Tegmark speculates about Universes with entirely different laws of physics, including those describable by unconventional mathematics such as fractals. He suggests that all logically possible Universes exist; however, the vast majority of such Universes would not support life and so go unobserved.

Ideas of the multiverse, together with the anthropic principle attempting to explain fine-tuning, are still regarded with high suspicion, or even hostility, among the majority of physicists, although it has some notable apologists. There is a consensus that such explanations should not impede searches for more satisfying descriptions of the nature of the existing physical laws and parameters.

The list of publications expressing skepticism about the concept of parallel Universes is long, and we shall mention only some. Maybe the best known skeptic is White (2000), whose thoughts can be summarized with his paragraph in which he questioned the test of the existence of the other Universes. All cosmologists accept that some regions of the Universe lie beyond the reach of our telescopes. However, credibility reaches a limit with the idea that there is an infinite number of Universes. If one follows this line of thinking, more must be accepted on faith, and less is open to scientific verification. Extreme multiverse explanations are, therefore, similar to theological discussions. Invoking the infinity of unseen Universes to explain the unusual features of the one we do see is similar to invoking an invisible Creator. The multiverse theory is dressed up in scientific language, but in essence, it requires the same jump of faith.

The simple fact that our Universe is fine-tuned gives us no reason to assume that there are Universes other than ours. Assuming there is just one Universe, it is surprising that it is life-permitting. Compared to this extremely improbable outcome of the Big Bang, it is more probable that there is a cosmic designer who might adjust the physical parameters to allow for the evolution of life. So the fine-tuning facts challenge us to question whether the Big Bang was merely an accident. White (2000) finishes his arguments with the following statement: "So our good fortune to exist in a life-permitting Universe gives us no reason to suppose that there are many Universes".

Carr and Ellis (2008) presented their differing views on whether speculations about other Universes are part of legitimate science!

Ellis (2011) kept his skepticism about the concept of parallel Universes contemplating the multiverse as being an opportunity to reflect on the nature of science and the ultimate nature of existence. Parallel Universes may exist or may not exist; the case is not proved yet. This uncertainty will stay with us for some time. However, nothing is wrong with scientifically based philosophical speculation, which is what multiverse proposals are, but they should be named as such.

Multiverse theories raise severe philosophical problems about the nature of reality and the nature of consciousness and observation. Attempts to improve the discussion and provide a more rigorous definition of concepts such as the number of Universes, and objective descriptions of infinite sets of Universes, have not progressed far. Anyhow, the multiverse idea has probably earned a place in physical science, and new physical theories are considered in the future; their consequences for biophilic and multiple cosmic regions will likely be eagerly assessed (Davies, 2004).

The existence of parallel Universes looks like the idea used by science fiction writers, with only some relevance to modern theoretical physics. However, the idea of a "multiverse" made up of an infinite number of parallel Universes has been considered as a possibility – although a matter of debate. The accent is now on the effort to find a way to test this hypothesis, including the search for signs of collisions among Universes.

On the other hand, Vilenkin (2011) argued that the case for parallel Universes is a solid scientific idea. For a century, cosmologists have been studying only the aftermath of the Big Bang, i.e., the process of how the Universe expanded and cooled down and how galaxies were gradually pulled together by gravity. However, the Big Bang event itself has come into focus only relatively recently. It is studied by the theory of inflation, which was first developed by Guth (1981), Linde (1983), and others, and has resulted in a radically new global view of the Universe.

Its central postulate is inflation, a period of super-fast, accelerated expansion in earliest cosmic history. It is so incredibly fast that in a fraction of a second, a tiny subatomic speck of space is blown to dimensions much higher than the entire currently observable region. At the end of the inflation period, the energy that drove the expansion ignites a hot fireball of particles and radiation. This process is what is called the Big Bang.

The end of inflation started by quantum, probabilistic processes and does not occur everywhere at once. In our cosmic neighborhood, inflation ended 13.8×10^{9} years ago, but it is probably going on in faraway parts of the Universe, and other "normal" regions like ours are continually being formed. The new areas appear as tiny, microscopic bubbles and immediately start growing. The bubbles keep growing without bound simultaneously being pushed apart by the inflationary expansion, making in such a way space for the formation of more bubbles. This never-ending process is known as eternal inflation. We are contained in one of the bubbles and can observe only a small part of it. We cannot reach the expanding boundaries of our bubble; therefore, we live in a self-contained bubble Universe (Vilenkin, 2011).

This picture of the Universe, or multiverse, explains the long-standing mystery of why the constants on nature laws appear to be fine-tuned for the emergence of life. The intelligent observers exist only in those bubbles in which, by pure chance, the constants happen to be just right for life evolution. The rest of the multiverse remains without life, but no one is there to complain about that. Observational tests of the multiverse picture may be possible. Aguirre and Johnson (2009), Kleban (2011) and others have speculated that a collision of our Universe with another bubble in the multiverse could produce an imprint in the cosmic background radiation. The result is a round spot of higher or lower radiation intensity. Detection of such an area with the predicted intensity profile would be direct evidence for the existence of other bubble Universes.

According to Lim (2015), the theory of parallel Universes is not just mathematics; it is a science that can be tested. All the Universes predicted by string theory and inflation are contained in the same physical space, contrary to many theoretical speculations, which are only mathematical constructions; therefore, they can overlap or collide. They inevitably must collide, leaving possible signatures behind, which we can try to find. The details of the signatures depend intimately on the models used. They could be cold or hot spots in the CMB or anomalous voids in the distribution of galaxies. Since collisions with another Universe must occur in a particular direction, it is expected that any signature will break the uniformity of the observable Universe. Scientists are actively pursuing these signatures. Some are searching for it by trying to find imprints in the CMB, the afterglow of the Big Bang. However, no such signatures have been seen so far. Others are looking for indirect evidence such as gravitational waves, which are ripples in spacetime as massive objects pass by. Such waves would directly prove the existence of inflation, which ultimately strengthens the support for the multiverse theory.

The first observational tests of eternal inflation are reported by Feeney et al. (2011). According to the infinite inflation scenario, our observable Universe resides inside a single bubble embedded in a vast inflating multiverse. In their paper, the first observational tests of eternal inflation are presented by performing a search for cosmological signatures of collisions with other Universes in CMB data from the Wilkinson Microwave Anisotropy Probe (WMAP) satellite. The WMAP, a NASA Explorer mission, was launched in June 2001 with the mission to perform fundamental measurements of cosmology – the study of the properties of our Universe as a whole. They concluded that the 7-year WMAP data do not warrant augmenting ΛCDM model with bubble collisions, constraining the average number of detectable bubble collisions on the full sky to be less than 1.6 with 68% probability. They suggested that data from the Planck satellite can be used to test the bubble collision hypothesis more definitively.

We should conclude with the statement that the multiverse theory is still in its infancy, and some conceptual problems remain to be resolved.

6.4 Anthropic Principle

The multiverse may help explain one of the more paradoxes about our world, called the anthropic principle; this reflects the fact that we are here to observe it. To some, our Universe looks surprisingly fine-tuned for life. Without its perfect alignment of the physical constants, everything from the power of the force of binding electrons to atoms to the relative weakness of gravity, planets and stars, biochemistry and life itself would be impossible. Atoms would not stick together in a Universe with more than four dimensions. If ours were the only Universe made by a Big Bang, these life-friendly properties would seem impossibly unlikely. But in a multiverse containing many Universes, a small number of life-friendly ones would arise by chance, and we could just happen to reside in one of them. The Universe we observe is bio-friendly, or we would not be watching it. This statement is a tautology. It appears since an account is taken of the sensitivity of biology to the laws of physics and the initial cosmological conditions. This phenomenon is the so-called fine-tuning problem, and it has been discussed for some decades. If the laws of physics differed only slightly from their observed form, then life as we know it, and possibly any kind of life, would be impossible (Hall et al., 2014).

Some authors argue that the anthropic principle is true regardless of whether there is a multiverse or not. According to them, the anthropic principle is valid irrespective of what fundamentally is the correct explanation for the values of parameters in the theories. The reason it is often mentioned in combination with the multiverse is that proponents of the multiverse argue it is the only explanation. No further clarification is needed or necessary to look for; see Bostrom, (2002a, 2002b).

The term "anthropic principle" was coined by Carter (1974). Carter's concept of the anthropic principle, as demonstrated by the various uses, is appropriate. However, his definitions and explanations of it are often unclear. Although Carter himself was never in doubt about how to apply the principle, he is missing a philosophically transparent explanation done in such a manner to enable all his readers to do the same (Bostrom, 2002). The trouble starts with the name. Anthropic reasoning has nothing to do with *Homo sapiens*. Calling the principle "anthropic" is misleading and has indeed misled some authors (e.g., Gale, 1981; Gould, 1985; Worrall, 1996).

Carter (1983) has expressed regrets about not using a different name, suggesting that maybe names such as "the psycho-centric principle", "the cognizability principle" or "the observer self-selection principle" would be better. The time for terminological reform has probably gone, but emphasizing these concerns should help prevent misunderstandings.

Carter defined two versions of the anthropic principle: the strong anthropic principle (SAP) and the weak anthropic principle (WAP).

The WAP states that our location in the Universe is privileged to the extent of being compatible with our existence as observers. While the SAP says that our Universe, together with fundamental parameters on which it depends, must admit the creation of inside observers at some stage.

Carter's original formulations have been attacked for being mere duplications, therefore incapable of doing any explanatory work whatever, and for being speculative (and lacking any empirical support). Often the WAP is accused of being speculative and the SAP lacking any empirical support.

According to Barrow and Tipler (1988), there are three grades of the anthropic principle:

- Following the WAP, the observed values of all physical and cosmological quantities are not equally probable. Their values are restricted by the requirement that there must exist sites where life can evolve, provided that the Universe be old enough for it to have already done so.
- Following the SAP, the Universe must have such properties which allow life to develop within it at some stage in its history.
- Following the final anthropic principle (FAP), an intelligence capable of information processing will come into existence in the Universe, and, once in existence, it will probably never die out.

The FAP is pure speculation, and it has no claim on any special methodological status. The appearance, to the contrary, is what induced Gardner's mockery. Namely, Gardner (1986) suggested that the FAP should be more accurately named CRAP, the completely ridiculous anthropic principle. It may be possible to interpret the FAP only as a scientific hypothesis, and that is indeed what Barrow and Tipler (1988) were set out to do.

Leslie (1989) proposed that the AP, WAP and SAP could all be understood as redundancies and that the difference between them was often purely verbal. In his explication, the AP says that "Any intelligent living beings that there are can find themselves only where creative life is possible" (Leslie, 1989). The WAP then postulates that, within a Universe, observers find themselves only at locations where observers are likely. The SAP assumes that observers find themselves only in worlds that allow observers to exist. "Universes" mean roughly the vast spacetime regions that might be more or less causally disconnected from other spacetime regions. The definition of a Universe is not precise; neither is the distinction between the WAP and the SAP. The WAP talks about where within a life-permitting Universe, we should expect to find ourselves. In contrast, the SAP talks about in what kind of Universe in an ensemble of Universes we should expect to find ourselves. In this interpretation, the two principles are fundamentally similar, differing only in scope (Bostrom, 2002).

One way to explain this fine-tuning is to invoke the anthropic principle, which, in its weakest form, states that "We observe the Universe with properties which permit us to exist because otherwise, we wouldn't be here to do the observing". The anthropic principle cannot be refuted – it is a logical tautology (or circular reasoning) that one cannot observe conditions in the Universe that preclude one's existence. But invoking the anthropic principle is, in a way, almost as unsettling as arguing that the values of the physical constants were preset by a benign Creator so that we could eventually inhabit the Universe. Either explanation permanently removes the subject matter, in this case, the values of the physical constants, from the domain of science, for in neither case would we expect ever to be able to calculate the values of the constants from first principles.

The book by Barrow and Tipler (1986) introduced anthropic reasoning to a broader audience. Still, it also introduces the terminological disorder by minting several new "anthropic principles", some of which have no connection to observation selection effects as defined by Bostrom (2002). A total of over 30 anthropic principles have been formulated in the book. Some of them have been defined several times over by different authors, and some even by the same authors on different occasions. The result has been some pretty big confusion concerning what the whole thing is. Some reject anthropic reasoning as being an obsolete and irrational form of anthropocentrism. Some hold that anthropic inferences are based on elementary mistakes in probability calculus. Some argue that at least some of the anthropic principles are tautological and therefore indisputable. Some have dismissed tautological principles as empty and thus of no interest or ability to do analytical work. Others have insisted that, though analytically true, anthropic principles can also be exciting and illuminating. Others still try to derive empirical predictions from these same principles and regard them as testable hypotheses (Bostrom, 2002).

Let us come back to an interesting example from the history of nuclear physics and cosmic nucleosynthesis. In 1953, Fred Hoyle (Hoyle et al., 1953) realized that to make enough carbon inside the stars, a resonance state of the carbon-12 nucleus at 7.68 MeV had to exist, which at the time was not known experimentally. Hoyle said, "since we exist, then carbon must have an energy level at 7.6 MeV" – anthropic prediction! According to the SAP, the Universe must have those properties which allow life to develop within it at some stage in its history (Carter, 1974).

Some authors (see Kragh, 2010) argue that the excited levels in ^{12}C and other atomic nuclei cannot be used as an argument for the predictive power of the anthropic principle. Kragh (2010) claims that following the historical circumstances of the prediction and its subsequent experimental confirmation shows that Hoyle did not associate the energy level in the carbon with life at all. Only after the emergence of the anthropic principle did it become common to see Hoyle's prediction as anthropically significant. The anthropic myth has no basis in historical fact. It is also doubtful if the excited energy levels in carbon-12 or other atomic nuclei could be used as evidence of the predictive power of the anthropic principle, as has been done by several physicists and philosophers.

It is worthwhile mentioning that in the book by Barrow and Tipler (1986), an entire chapter argues that man – *Homo sapiens* – is most likely the only intelligent species in the Milky Way galaxy.

Hoyle realized that a remarkable chain of coincidences – the stability of beryllium, the existence of an advantageous resonance level in ^{12}C and the non-existence of an appropriate level in ^{16}O – were necessary, and remarkably fine-tuned, conditions for our survival and indeed the existence of any carbon-based life in the Universe. For the ratio, Hoyle deduced ^{12}C:^{16}O = 1:3.

Yet another example relates to hydrogen and water. If the nuclear force were only about 4% stronger, the two protons could form a bound state but unstable against decay via the weak force to deuterium. This situation would have tremendous implications for primordial nucleosynthesis and the chemical composition of the Universe. As pointed out by Dyson (1971), the primordial soup of protons would rapidly transform into deuterium, which would then synthesize 100% helium, leaving no hydrogen in the Universe. With hydrogen absent, there would be no water, thought to be an essential ingredient of life. There would be no stable hydrogen-burning stars like our Sun to sustain a biosphere.

Also, let us mention the strength of gravity. Gravitation is much weaker than the other forces of nature. The power of gravity sets the time and distance scale of the Universe. The Universe is big because the force of gravity is so weak. A big Universe is necessary for an old Universe, and an old Universe (several billions of years) is a prerequisite for the emergence of life. If gravitation were, for example, 100 times stronger, the Universe would collapse before observers had time to evolve (Davies, 2004).

6.5 Origin of Life

The question of the origin of life has interested people for centuries. All current views on this problem can be classified into different areas of human knowledge, including natural sciences, philosophy and theology (religion). It is a great interest to look at them closer and to classify all the theories about the origins of life. In this way, one could then see links existing between them and relationships that indicate their nature. Nowadays, driving forces of pre-biological chemical evolution and the mode of explanation of the transition "non-life into life" give a great variety of solutions. The differences between the theories, and the current controversies in the scientific community are of secondary importance for the systematization in comparison with several much more profound philosophical assumptions underlying the origin of life studies. The proposal by Świeżyński (2016) to organize and classify different types of the theories of the genesis of life allows for extracting conceptions of various kinds: metaphysical and scientific, and also for comparing them with each other. Some of them provide an answer to the question of the emergence of life in general. Most discuss the issue of the origin of life on Earth only. From the perspective of contemporary scientific research on the origin of life, it seems unusual that two main ideas, spontaneous generation and panspermia, are still present as assumptions of specific theories, but have been modified. Thus, a "philosophical key" seems to be the most appropriate to systematize all kinds of opinions on the origin of life. His paper is an attempt to justify the position adopted. The most important conclusion is that the philosophical implications, which are found within the scientific theories of the origin of life, indicate that this problem is not just the strictly scientific one. It is a philosophical problem, too; thus, it cannot be fully solved merely by referring to the practical aspect of biogenesis (Świeżyński, 2016).

By accepting the SAP, which claims that the Universe must have such properties that allow life development within it at some stage in its history, we can put forward a hypothesis:

- Life, as we know, which is (H-C-N-O)-based, relying on the number of bulk (Na-Mg-P-S-Cl-K-Ca) and trace elements (Cr-Mn-Fe-Co-Ni-Cu-Zn-Se-Mo-I-W), and possibly (Li-B-F-Si-V-As), originated when two-element abundance curves coincided.

TABLE 6.1 Abundance Ratios

Abundance Ratios	Values	Ratio
$(C/O)_{univ.}/(C/O)_{life}$	0.5/0.37	1.35
$(C/O)_{solar}/(C/O)_{life}$	0.54/ 0.37	1.46
$(C/O)_{crust}/(C/O)_{life}$	0.001/0.37	0.0027
$(C/O)_{seawater}/(C/O)_{life}$	$16\times10^{-8}/0.37$	43×10^{-8}
$\Sigma(H, C, N, O)_{univ.}/\Sigma(H, C, N, O)_{life}$	76.6/96.857	0.791
$\Sigma(H, C, N, O)_{crust}/\Sigma(H, C, N, O)_{life}$	47.6025/96.857	0.491
$\Sigma(H, C, N, O)_{seawater}/\Sigma(H, C, N, O)_{life}$	99.305/96.857	1.025
$\Sigma(Na, Mg, P, S, Cl, K, Ca)_{univ.}/\Sigma(Na, Mg, P, S, Cl, K, Ca)_{life}$	0.1201/3.5214	0.034
$\Sigma(Na, Mg, P, S, Cl, K, Ca)_{seawater.}/\Sigma(Na, Mg, P, S, Cl, K, Ca)_{life}$	3.3197/3.5214	0.943
$\Sigma(Cr, Mn, Fe, Co, Ni, Cu, Zn, Se, Mo, I, W)_{univ.}/\Sigma(Cr, Mn, Fe, Co, Ni, Cu, Zn, Se, Mo, I, W)_{life}$	0.1186396/0.009	13.18
$\Sigma(Cr, Mn, Fe, Co, Ni, Cu, Zn, Se, Mo, I, W)_{seawater}/\Sigma(Cr, Mn, Fe, Co, Ni, Cu, Zn, Se, Mo, I, W)_{life}$	0.09179/0.009	10.20
$\Sigma(Li, B, F, Si, V, As)_{univ.}/\Sigma(Li, B, F, Si, V, As)_{life}$	0.0701414/0.0079	7.793
$\Sigma(Li, B, F, Si, V, As)_{seawater}/\Sigma(Li, B, F, Si, V, As)_{life}$	0.00205245/0.0079	0.2598

The remaining question is, which two-element abundance curves? The universal abundance curve should be the abundance curve for the galaxy where life originated at that time. The life abundance curve should be that of the LUCA or some other primitive organism that can survive the intergalactic transfer (panspermia).

Table 6.1 presents an effort made to estimate the closeness of different abundance curves (data from Emsley (1998, 2001), http://www.seafriends.org.nz/oceano/seawater.htm, and http://periodictable.com/Properties/A/UniverseAbundance.html).

References

Aguirre, A. and Johnson, M. C. 2009. A status report on the observability of cosmic bubble collisions. arXiv: 0908.4105v2 [hep-th].

Barrow, J. D. and Tipler, F. J. 1988. *The Anthropic Cosmological Principle.* Oxford University Press, Oxford. ISBN 0-19-282147-4.

Behroozi, P. S., Wechsler, R. H. and Conroy, C. 2013. The average star formation histories of galaxies in dark matter halos from z=0–8. *The Astrophysical Journal* 770:57 (36 p).

Belot, G., Earman, J. and Ruetsche, L. (1999). The hawking information loss paradox: The anatomy of a controversy. *British Journal for the Philosophy of Science* 50(2):189–229.

Bethe, H. A. 1968. Energy production in stars. *Science New Series* 161(3841): 541–547:

Bostrom, N. 1997. Investigations into the Doomsday argument. Preprint, http://www.anthropic-principles.com/preprints/inv/investigations.html.

Bostrom, N. 2002a. *Anthropic Bias – Observation Selection Effects in Science and Philosophy.* Routledge, New York.

Bostrom, N. 2002b. Existential risks: Analyzing human extinction scenarios and related hazards. *Journal of Evolution and Technology* 9(1):1–30.

Burbidge, E. M., Burbidge, G. R., Fowler, W. A. and Hoyle, F. 1957. Synthesis of the elements in stars. *Reviews of Modern Physics* 29(4):547–650.

Calmet, X. and Kuipers, F. 2021. Theoretical bounds on dark matter masses. *Physics Letters B* 814:136068.

Carr, B. and Ellis, G. 2008. Universe or multiverse. *Astronomy & Geophysics* 49:2.29–2.33.

Carter, B. 1974. Large number coincidences and the anthropic principle in cosmology. *Confrontation of Cosmological Theories with Data*. M. S. Longair, Reidel, Dordrecht, pp. 291–298.

Carter, B. 1983. The anthropic principle and its implications for biological evolution. *Philosophical Transactions of the Royal Society A* 310:347–363.

Catena, R., Emken, T., Spaldin, N. A. and Tarantino, W. 2020. Atomic responses to general dark matter-electron interactions. *Physical Review Research* 2:033195 (27 p).

Chae, K.-H., Lelli, F., Desmond, H., et al. 2020. Testing the strong equivalence principle: Detection of the external field effect in rotationally supported galaxies. *The Astrophysical Journal* 904:51 (20 p).

Chary, B. 2015. Spectral variations of the sky: Constraints on alternative universes. arXiv:1510.00126v2 [astro-ph.CO].

Chandrasekhar, S. 1958. *An Introduction to the Study of Stellar Structure*. Dover, New York. ISBN 978-0-486-60413-8.

Chandrasekhar, S. 2005. *Principles of Stellar Dynamics*. Dover, New York. ISBN 978-0-486-44273-0.

Chen, G. C.-F, Fassnacht, C.D., Suyu, S.H., et al. 2019. A SHARP view of H0LiCOW: H_0 from three time-delay gravitational lens systems with adaptive optics imaging. *Monthly Notices of Royal Astronomical Society* 490:1743–1773.

Ciarcelluti, P. and Wallemacq, Q. 2014. Is dark matter made of mirror matter? Evidence from cosmological data. *Physics Letters B* 729:62–66..

Davis, R. 1953. Attempt to detect the antineutrinos from a nuclear reactor by the 37Cl (ν, e−) 37Ar reaction. *Physical Review* 97(3):766–769.

Davis, R. 1964. Solar neutrinos II, experimental. *Physical Review Letters* 12(11): 303–305.

Davies, P. C. W. 2004. Multiverse cosmological models. *Modern Physics Letters A* 19(10):727–743.

Darling, J. 2020a. New limits on axionic dark matter from the magnetar PSR J1745-2900. *The Astronomical Journal Letters* 900: L28 (10 p).

Darling, J. 2020b. Search for axionic dark matter using the magnetar PSR J1745-2900. *Physical Review Letters* 125:121103.

Dyson, F. J. 1971. Energy in the universe. *Scientific American*, September 225:51–59.

Ellis, G. 2011. Does the multiverse really exist? *Scientific American* 305:38–43.

Emsley, J. 1998. *The Elements*, 3rd edition. Clarendon Press, Oxford, UK.

Emsley, J. 2001. *Nature's Building Blocks*. Oxford University Press, Oxford, UK.

Essig, R. 2020. The low-mass dark matter frontier. *Physics* 13:172. Doi:10.1103/Physics.13.172.

Feeney, S. M., Johnson, M. C., Mortlock, D. J. and Peiris, H. V. 2011. First observational tests of eternal inflation. *Physical Review Letters* 107: 071301.

Foster, J. W., Kahn, Y., Macias, O., et al. 2020. Green Bank and Effelsberg radio telescope searches for axion dark matter conversion in neutron star magnetospheres. *Physical Review Letters* 125(17):171301.

Fowler, W. A. 1984. Experimental and theoretical nuclear astrophysics: The quest for the origin of the elements. *Reviews of Modern Physics* 56:149–179.

Friedman, A. 1922. Über die Krümmung des Raumes (English translation: On the curvature of space). *Zeitschrift für Physik* 10 (1):377–386.

Foot, R., Lew, H. and Volkas, R. R. 1991. A model with fundamental improper spacetime symmetries. *Physics Letters B* 272(1–2):67–70.

Fukuda Y, et al. (Super-Kamiokande Collaboration). 1998. Evidence for oscillation of atmospheric neutrinos. *Physical Review Letters* 81(8):1562–1568.

Gale, G. 1981. The anthropic principle. *Scientific American* 245(June):154–171.

Gardner, M. 1986. WAP, SAP, FAP & PAP. *New York Review of Books* 33(8), May 8:22–25.

Gelmini, G. B., Kusenko, A. and Takhistov, V. 2020a. Possible hints of sterile neutrinos in recent measurements of the hubble parameter. arXiv:1906.10136v2 [astro-ph.CO] 14 Sep 2020.

Gelmini, G. B., Lu, P. and Takhistov, V. 2020a. Visible sterile neutrinos as the earliest relic probes of cosmology. *Physics Letters B* 800:135113. arXiv:1909.04168 [hep-ph].

Genzel, R., Eisenhauer, F. and Gillessen, S. 2010. The galactic center massive black hole and nuclear star cluster. arXiv:1006.0064 [astro-ph.GA].

Genzel, R.., Hollenbach, D. and Townes, C.H. 1994. The nucleus of our galaxy. *Reports on Progress in Physics* 57(5): 417–479.

Ghez, A. M., Duchêne, G. and Matthews, K. 2003. The first measurement of spectral lines in a short-period star bound to the galaxy's central black hole: A paradox of youth. *The Astrophysical Journal* 586(2):L127–L131. arXiv:astro-ph/0302299.

Giacconi, R. 2008. *Secrets of the Hoary Deep*. The Johns Hopkins University Press, Baltimore.

Giacconi, R. and Rossi, B. 1960. A telescope for soft X-ray astronomy. *Journal of Geophysical Research* 65: 773–775.

Gould, S. J. 1985. *The Flamingo's Smile: Reflections in Natural History*. Penguin Books, London.

Guth, A. H. 1981. Inflationary universe: A possible solution to the horizon and flatness problems. *Physical Review D* (Particles and Fields) 23(2):347–356.

Hall, M. J. W., Deckert, D.-A. and Wiseman, H. M. 2014. Quantum phenomena modeled by interactions between many classical worlds. *Physical Review X* 4:041013.

Hawking, S. 1974. The anisotropy of the universe at large times. In Longair M. S. (Ed.) *Confrontation of Cosmological Theories with Observational Data*. Reidel, Dordrecht.

Hewish, A. 1975. Pulsars and high density physics. *Science* 188(4193):1079–1083.

Hewish, A. 1952. *The Fluctuations of Galactic Radio Waves* (Ph.D. thesis). The University of Cambridge, Cambridge, England.

Hoyle, F., Noel, D., Dunbar, F., Wenzel, W. A. and Whaling, W. 1953. A state in C12 predicted from astrophysical evidence. *Physical Review* 92:1095–1098.

Hulse, R. A. and Taylor, J. H. 1975. Discovery of a pulsar in a binary system. *Astrophysical Journal* 1958(2):L51–L53.

Jeong, J., Youn, S. W. and Bae, S. 2020. Search for invisible axion dark matter with a multiple-cell haloscope. *Physical Review Letters* 125(22):221302 (6 p).

Kleban, M. 2011. Cosmic bubble collisions. *Classical and Quantum Gravity* arXiv: 1107.2593.

Koshiba, M., et al. 1999. Constraints on neutrino oscillation parameters from the measurement of day-night solar neutrino fluxes at super-Kamiokande. *Physical Review Letters* 82(9):1810.

Kragh, H. 2010. When is a prediction anthropic? Fred Hoyle and the 7.65 MeV carbon resonance. Preprint, http://philsci-archive.pitt.edu/id/eprint/5332, Date deposited: 04 May 2010.

Kusenko, A., Sasaki, M., Sugiyama, S., Takada, M., Takhistov, V. and Vitagliano, E. 2020. Exploring primordial black holes from multiverse with optical telescopes. *Physical Review Letters* 125: 181304.

Kusenko, A., Takahashi, F. and Yanagida, T. T. 2010. Dark matter from split seesaw. *Physics Letters B* 693(2010):144–148.

Lee, T. D. and Yang, C. N. 1956. Question of parity conservation in weak interactions. *Physical Review* 104(1):254–258.

Lemaître, G. 1931a. The beginning of the world from the point of view of quantum theory. *Nature* 127(3210):706–706.

Lemaître, G. 1931b. The evolution of the universe: Discussion. *Nature* 128(3234):699–701.

Leslie, J. 1989. *Universes*. Routledge, London and New York.

Leslie, J. 1996. The anthropic principle today. In R. Hassing (Ed.) *Final Causality in Nature and Human Affairs*. Catholic University Press, Washington, DC.

Lim, E. 2015, Sep. 2. The theory of parallel universes is not just maths – it is science that can be tested. *The Conversation*.

Lin, T. 2020. Dark matter detector delivers enigmatic signal. *Physics* 13: 135. Doi: 10.1103/Physics.13.135.

Linde, A. D. 1983. Chaotic inflation. *Physics Letters* 129B(3,4): 177–181.

Longair, M. 2016. *Maxwell's Enduring Legacy: A Scientific History of the Cavendish Laboratory*. 1st edition. Cambridge University Press, Cambridge, UK.

Ludwig, G. O. 2021. Galactic rotation curve and dark matter according to gravitomagnetism. *European Physical Journal C* 81:186 (25 p).

Mather, J., Cheng, E. S., Eplee Jr, R. E. et al. 1990. A preliminary measurement of the cosmic microwave background spectrum by the Cosmic Background Explorer (COBE) satellite. *Astrophysical Journal* 354:L37–L40.

Mayor, M. and Queloz, D. 1995. A Jupiter-mass companion to a solar-type star. *Nature* 378(6555):355–59.

Milgrom, M. 1983. A modification of the Newtonian dynamics as a possible alternative to the hidden mass hypothesis. *Astrophysics Journal* 270:365–370.

Minkowski, P. 1977. $\mu \rightarrow e\gamma$ at a rate of one out of 10_9 muon decays? *Physics Letters B* 67(4):421–428.

Moster, B. P., Somerville, R. S., Maulbetschet, C., et al. 2010. Constraints on the relationship between stellar mass and halo mass at low and high redshift. *Astrophysical Journal* 710:903–923.

Pattie, R. W., Callahan, N. B., Cude-Woods, C., et al. 2018. Measurement of neutron lifetime using a magneto-gravitational trap and in situ detection. *Science* 360(6389):627–632.

Peebles, P. J. E. 1966a. Primordial helium abundance and the primordial fireball. I. *Physical Review Letters* 16(10):410–413.

Peebles, P. J. E. 1966b. Primordial helium abundance and the primordial fireball. II. *Astrophysical Journal* 146:542–552.

Penrose, R. 1989. *The Emperor's New Mind: Concerning Computers, Minds and The Laws of Physics*. Oxford University Press, (480 p). Oxford, England.

Penrose, R. 1965. Gravitational collapse and spacetime singularities. *Physical Review Letters* 14(3):57–59.

Penrose, R. 1994. *Shadows of the Mind*, 1st edition. Oxford University Press, (457 p).

Penrose, R. 2004. *The Road to Reality*. Jonathan Cape, Penguin Random House, London.

Penrose, R. 2010. *Cycles of Time: An Extraordinary New View of the Universe*. The Bodley Head (UK), Alfred A. Knopf (US), (288 p).

Penzias, A.A. and Wilson, R. W. 1965a. A measurement of excess antenna temperature at 4080 Mc/s. *Astrophysical Journal Letters* 142: 419–421.

Penzias, A.A. and Wilson, R. W. 1965b. A measurement of the flux density of CAS A at 4080 Mc/s. *Astrophysical Journal Letters* 142:1149–1154.

Planck Collaboration. 2018. Planck 2018 results. VI. Cosmological parameters. arXiv:1807.06209 [astro-ph.CO].

Randall, L. and Raman Sundrum, R. 1999. Large mass hierarchy from a small extra dimension. *Physical Review Letters* 83:3370.

Ratra, B. and Peebles, P. J. E. 2003. The cosmological constant and dark energy. *Reviews of Modern Physics* 75 (2):559–606.

Riess, A. G., Casertano, S., Yuan, W., et al. 2019. Large magellanic cloud cepheid standards provide a 1% foundation for the determination of the hubble constant and stronger evidence for physics beyond ΛCDM. *Astrophysical Journal* 876:85 (13 p), arXiv:1903.07603 [astro-ph.CO].

Rini, M. Theorists react to potential signal in dark matter detector. *Physics* 13:s132. Doi: 10.1103/Physics.13.s132.

Smoot, G., Bennett, C. L., Kogut, A., et al. 1992. Structure in the COBE differential microwave radiometer first-year maps. *Astrophysical Journal* 396:L1–L4.

Susskind, L. 2003. The anthropic landscape of string theory. arXiv:hep-th/0302219.

Świeżyński, A. 2016, Jul. 25. Where/when/how did life begin? A philosophical key for systematizing theories on the origin of life. *International Journal of Astrobiology*, FirstView Articles, pp. 1–9. Doi: 10.1017/S1473550416000100, Pub. online Cambridge University Press, Cambridge, UK.

Tan, W. 2019, Sep. 11. Neutron oscillations for solving neutron lifetime and dark matter puzzles. *Physics Letters* B 797:134921. arXiv:1902.018337v9 [physics.gen-ph].

Tegmark, M. 1996. Does the universe in fact contain almost no information? *Foundations of Physics Letters* 9(1):25–42.

Tegmark, M. 2003, May. Parallel Universes. *Scientific American*:41–51.

Tegmark, M. 2004. *Science and Ultimate Reality: From Quantum to Cosmos*. Barrow, J. D., Davies, P. C. W. and Harper, C. L. (Eds). Cambridge University Press, Cambridge, UK.

Tipler, F. J. 2003. Intelligent life in cosmology. *International Journal of Astrobiology* 2(2):141–148.

van Dokkum, P., Danieli, S., Abraham, R., Conroy, C., and Romanowsky, A. J. 2019, Mar. 20. A second galaxy missing dark matter in the NGC 1052 group. *The Astrophysical Journal Letters* 874:L5 (8 p), Doi: 10.3847/2041-8213/ab0d92.

van Dokkum, P., Danieli, S., Cohen, Y., et al. 2018. A galaxy lacking dark matter. *Nature* 555:629–632.

Vilenkin, A. 1983. Birth of inflationary universes. *Physical Review* D27: 2848–2855.

Vilenkin, A. 2011, July. The case for parallel universes. *Scientific American*:1–12.

Whitaker, M. A. B. 1988. On hacking's criticism of the wheeler anthropic principle. *Mind* 97(386):259–264.

White, R. 2000. Fine-tuning and multiple universes. *Noûs* 34(2):260–276.

Wilson, P. A. 1991. What is the explanandum of the anthropic principle? *American Philosophical Quarterly* 28(2):167–73.

Wilson, P. A. 1994. Carter on anthropic principle predictions. *British Journal for the Philosophy of Science* 45:241–253.

XENON Collaboration. 2020. Excess electronic recoil events in XENON1T. *Physical Review D* 102:072004.

Yanagida, T. 1980. Horizontal symmetry and masses of neutrinos. *Progress of Theoretical Physics* 64(3):1103–1105.

Yue, A. T., Dewey, M. S., Gilliam, D. M., et al. 2013. Improved determination of neutron lifetime. *Physical Review Letters* 111(22): 222501.

Additional Reading

Abazajian, K. N., Horiuchi, S., Kaplinghat, M., Keeley, R. E. and Macias, O. 2020. Strong constraints on thermal relic dark matter from Fermi-LAT observations of the galactic center. *Physical Review D* 102:043012 (24 p).

Aguirre, A. 2001. The cold big-bang cosmology as a counter-example to several anthropic arguments. Physics preprint archive astro-ph/0106143.

Barrow, J. D. 1983. Anthropic definitions. *Quarterly Journal of the Royal Astronomical Society* 24:146–153.

Carmona, A., Ruiz, J. C. and Neubert, M. 2021. A warped scalar portal to fermionic dark matter. *The European Physical Journal C* 81:58–78.

Carter, B. 1989. *The Anthropic Selection Principle and the Ultra-Darwinian Synthesis*. The Anthropic Principle. Bertola, F. and Curi, U. Cambridge University Press, Cambridge, pp. 33–63.

Catena, R., Emken, T., Spaldin, N. A. and Tarantino, W. 2020. Atomic responses to general dark matter-electron interactions. *Physical Review Research* 2:033195. Doi: 10.1103/Phys. Rev. Research.2:033195 (27 p).

Chen, G. C.-F. 2019. A SHARP view of H_0LiCOW: H_0 from three time-delay gravitational lens systems with adaptive optics imaging. *Monthly Notices of the Royal Astronomical Society* 490(2):1743–1773.

Chen, X., Loeb, A., and Xianyu, Z.-Z. 2019, Feb. 17. Unique fingerprints of alternatives to inflation in the primordial power spectrum. arXiv:1809.02603v3 [astro-ph.CO].

Chen, X., Namjoo, M. H. and Wang, Z. 2019. arXiv:1509.03930v2 [astro-ph.CO] 22 Jan 2016.

Chiang, Y.-K., Makiya, R., Ménard, B. and Komatsu, E. 2020. The cosmic thermal history probed by Sunyaev–Zeldovich effect tomography. *The Astrophysical Journal* 902:56 (17 p).

Coc, A., Pospelov, M., Uzan, J.-P., et al. 2014, May 7. Modified big bang nucleosynthesis with non-standard neutron sources. arXiv:1405:1718v1 [hep-ph].

Croker, K. S. and Weiner, J. L. 2019. Implications of symmetry and pressure in Friedmann cosmology. I. Formalism. *The Astrophysical Journal* 882:19 (15 p) Doi: 10.3847/1538-4357/ab32da.

Danieli, S., van Dokkum, P., Conroy, C., Abraham, R. and Romanowsky, A. J. 2019, Apr. 1. Still missing dark matter: KCWI high-resolution stellar kinematics of NGC1052-DF2. *The Astrophysical Journal Letters* 874:L12 (8 p) Doi: 10.3847/2041-8213/ab0e8c.

De Duve, C. 2008. C. 2008. *How Biofriendly is the Universe? Chapter 10 in Fitness of the Cosmos for Life* Barrow, J. D., Morris, S. C., Freeland, S. J. and Harper, Jr., C. L. (Eds.) Cambridge University Press, pp. 169–196.

Earman, J. 1987. The SAP also rises: A critical examination of the anthropic principle. *Philosophical Quarterly* 24(4):307–17.

Ezquiaga, J. M. and Zumalacárregui, M. 2020. Gravitational wave lensing beyond general relativity: Birefringence, echoes, and shadows. *Physical Review D* 102:124048.

Freedman, W. L. 2000. W. L. 2000. The Hubble constant and the expansion age of the universe. *Physics Letters* 333(1–6):13–31.

Gale, G. 1996. G. 1996. *Anthropic-Principle Cosmology: Physics or Metaphysics? Final Causality in Nature and Human Affairs*. Hassing, R. (Ed.) Catholic University Press, Washington, DC.

Greene, B. 2011. *The Hidden Reality: Parallel Universes and the Hidden Laws of the Cosmos*. Knopf, New York, NY, USA.

Ivanov, M. M., Kovalev, Y. Y., Lister, M. L., Panin, A. G., Pushkarev, A. B., Savolaineni, T., and Troitskya, S. V. 2019. Constraining the photon coupling of ultra-light dark-matter axion-like particles by polarization variations of parsec-scale jets in active galaxies. *Journal of Cosmology and Astroparticle Physics*, 059. ArXiv ePrint: 1811.10997.

Jedamzik, K. and Pospelov, M. 2009. Big Bang nucleosynthesis and particle dark matter. *New Journal of Physics* 11:105028 (20 p).

Jordan, S. F., Rammu, H., Zheludev, I. N., Hartley, A. M., Maréchal, A. and Lane, N. 2019. Promotion of protocell self-assembly from mixed amphiphiles at the origin of life. *Nature Ecology & Evolution*. Doi: 10.1038/s41559-019-1015-y.

Kanitscheider, B. 1993. Anthropic arguments—are they really explanations? In Bertola, F. and Curi, U. (Eds.) *The Anthropic Principle: Proceedings of the Venice Conference on Cosmology and Philosophy.* Cambridge University Press, Cambridge.

Macquart, J.-P., Prochaska, J. X., McQuinn, M., et al. 2020. A census of baryons in the universe from localized fast radio bursts. *Nature* 581:391–395.

Maturana, H. R. and Varela, F. J. 1972. *Autopoiesis and Cognition.* D. Reidel Publishing Company, Dordrecht, Holland.

McMullin, E. 1993. Indifference principle and anthropic principle in cosmology. *Studies in the History of the Philosophy of Science* 24(3):359–389.

Meissner, U.-G. 2015. Anthropic considerations in nuclear physics. *Science Bulletin* 60(1):43–54.

Meneghetti, M., Davoli, G, Bergamini P., et al. 2020. An excess of small-scale gravitational lenses observed in galaxy clusters. *Science* 369(6509):1347–1351.

Nadler, E. O. et al. 2020. Milky way satellite census. III. Constraints on dark matter properties from observations of milky way satellite galaxies. DES-2020–546, FERMILAB-PUB-20–277-AE, SLAC-PUB-117554. arXiv:2008.00022v1 [astro-ph.CO]31 Jul 2020.

Peebles, P. J. E. 1971. *Physical Cosmology.* Princeton University Press, Princeton, NJ.

Peebles, P. J. E. 1980. *The Large-Scale Structure of the Universe.* Princeton University Press, Princeton, NJ.

Pesce, D. W., Braatz, J. A., Reid, M. J., et al. 2020. The megamaser cosmology project. XIII. Combined hubble constant constraints. *The Astrophysical Journal Letters* 891:L1 (9 p).

Sampaio-Santos, H., Zhang, Y., Ogando, R. L. C., et al. 2020. Is diffuse intracluster light a good tracer of the galaxy cluster matter distribution? *Monthly Notices of the Royal Astronomical Society* 501(1):1300–1325.

Sicilian, D., Cappelluti, N., Bulbul, E., et al. 2020. Probing the Milky Way's dark matter halo for the 3.5 keV line. *The Astrophysical Journal*, 905:146 (30 p).

Susskind, L. 2006. *The Cosmic Landscape: String Theory and the Illusion of Intelligent Design.* Back Bay Books, New York, NY, USA.

Swinburne, R. 1990. Argument from the fine-tuning of the universe. In Leslie, J. (Ed.) *Physical Cosmology and Philosophy.* Collier Macmillan, New York, pp. 154–73.

Tagmark, M. 2003. Parallel universes. In Barrow, J. D., Davies, P. C. W., and Harper, C. L. (Eds.), *Science and Ultimate Reality: From Quantum to Cosmos, honoring John Wheeler's 90th Birthday.* Cambridge University Press, Cambridge, UK.

Tanimura, H., Aghanim, N., Kolodzig, A. et al. 2020. First detection of stacked X-ray emission from cosmic web filaments. *Astronomy and Astrophysics* 648:L2 (p. 7).

Tin, W. 2019. Neutron oscillations for solving neutron lifetime and dark matter puzzles. arXiv:1902.01837v9 [physics.gen-ph] 11 Sep 2019.

Tipler, F. J. 1982. Anthropic-principle arguments against steady-state cosmological theories. *Observatory* 102:36–39.

Vilenkin, A. 2006. *Many Worlds in One: The Search for Other Universes.* Hill and Wang, New York, NY, USA.

Wang, J., Bose, S., Frenk, C. S., Gao, L., Jenkins, A., Springel, V. and White, S. D. M. 2020. Universal structure of dark matter haloes over a mass range of 20 orders of magnitude. *Nature* 585:39–42.

Weaver, D. and Villard, R. 2019, Apr. 27. Important threshold crossed in mystery of the universe's expansion rate. SciTechDaily; https://scitechdaily.com/important-threshold-crossed-in-mystery-of-the-Universes-expansion-rate/.

Weinberg, S. 1987. Anthropic bound on the cosmological constant. *Physical Review Letters* 59(22):2607–2610.

Yoshikawa, K., Tanaka, S., Yoshida, N., and Saito, S. 2020. Cosmological Vlasov-poisson simulations of structure formation with relic neutrinos: Nonlinear clustering and the neutrino mass. *The Astronomical Journal* 904:159 (16 p).

Zhang, Y., Yanny, B., Palmese, A., et al. 2019. Dark energy survey year 1 results: Detection of intracluster light a redshift ~ m0.25. *The Astronomical Journal* 874:165 (19 p).

7

Open Problems – Laboratory Experiments

7.1 Nuclear Physics Experiments

Nuclear physics experiments can shed light on many of the problems discussed in previous chapters. This approach is possible because of the existence of a variety of atomic particle accelerators and radiation counting devices, and associated electronic modules.

The non-existence of nuclei of atomic weight A=5 and A=8 complicates the nucleosynthesis models. Therefore, we must study the details of the structure of these unstable nuclei.

7.1.1 (n, 2n) Reactions on Light Elements: $^{10}B(n, 2n)^9B$

Because of the charge symmetry of nuclear force, the mirror nuclei, which have an exchanged number of protons and neutrons, are expected to have nearly identical structures with analogue states at almost the same excitation energy. Many mirror pairs such as ^7Li–^7Be, ^{13}C–^{13}N, ^{15}N–^{15}O, ^{17}O–^{17}F and ^{19}F–^{19}Ne are found to have nearly identical energy levels (Kroepfl and Browne, 1967). However, the comparison of the energy levels in the mirror pair of mass-9 nuclei does not always provide strong evidence for the existence of analogue levels in the ^9B nucleus to that of the ^9Be nucleus at the similar excitation energy.

The latest compilation of the energy levels for the mass-8 nuclei is depicted in Figure 7.1.

However, the first excited ½$^+$ state of ^9B remains elusive (Fortune and Sherr, 2013). As shown in Figure 7.2, the first excited ½$^+$ state of ^9Be lies at 1.685 MeV, which is just above the threshold for the breakup into ^8Be+n. A tentative broad level at 1.6 MeV, with the width of about 0.7 MeV, is expected for the analogue state in ^9B, but experimentally, it has not been confirmed with certainty.

The search for the first excited state of ^9B is crucial because it improves our understanding of the role of mirror nuclei and other factors such as the Coulomb displacement energy.

Also, this might contribute to a better understanding of BBN element synthesis. It is well known that BBN gives high ^7Li cosmic abundance compared to observation. There are

DOI: 10.1201/9781003181330-7

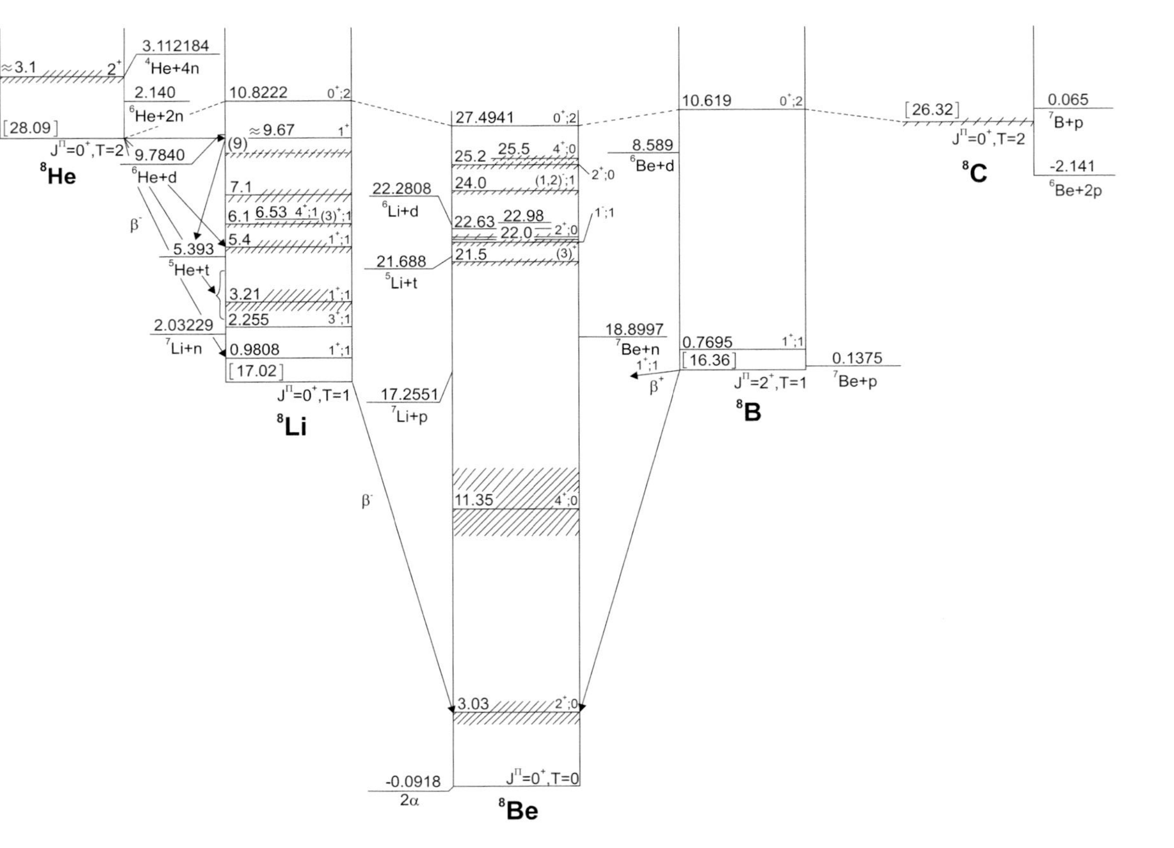

FIGURE 7.1 Isobar diagram, A = 8. The energy diagrams for individual isobars have been shifted vertically to compensate for the neutron–proton mass difference and the Coulomb energy, $E_C = 0.60Z(Z-1)/A^{1/3}$. Dashed lines connect the levels which are presumed to be isospin multiplets. (After Tilley et al., 2004; and Triangular Universities Nuclear Laboratory (TUNL) Nuclear Data Evaluation Project "A = 8 Energy Levels Diagrams", available at http://www.tunl.duke.edu/nucldata/figures/08figs/menu08.shtml.)

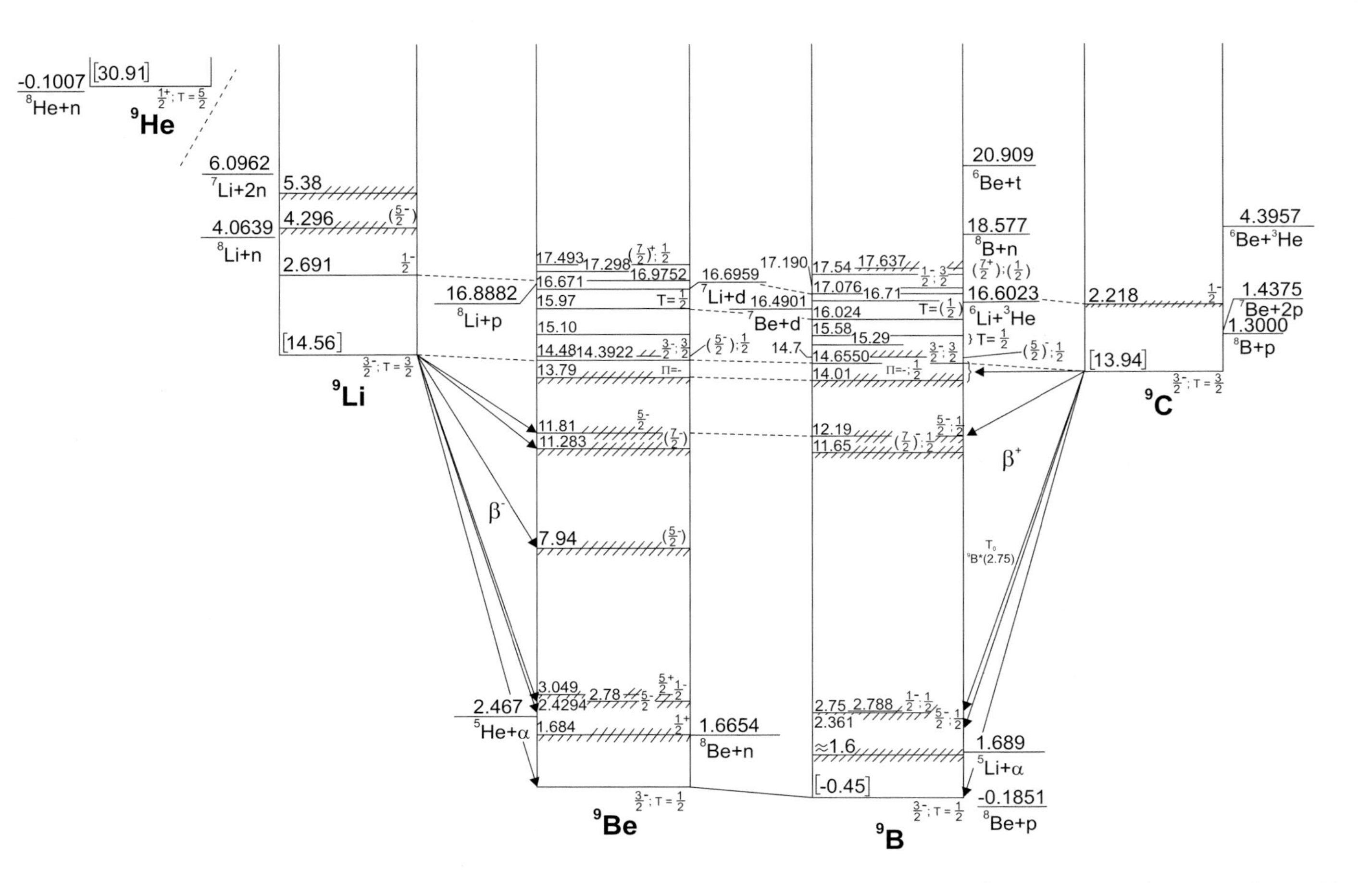

FIGURE 7.2 A = 9 energy level diagrams. (After Tilley et al., 2004, and TUNL Nuclear Data Evaluation Project “A = 9 Energy Levels Diagrams”, available at http://www.tunl.duke.edu/nucldata/figures/09figs/menu09.shtml.)

TABLE 7.1 Experimental Data and Theoretical Calculations of the First Excited State in ^{9}B

Reference	Reaction	First Excited State in ^{9}B	
		E_x (MeV)	Γ (MeV)
Marion and Levin (1959)	$^{9}Be(p,n)^{9}B$	1.4	1
Saji (1960)	$^{9}Be(p,n)^{9}B$	1.4	1
Spencer et al. (1960)	$^{10}B(^{3}He,\alpha)^{9}B$	Did not find	Did not find
Symons and Treacy (1962)	$^{12}C(p,\alpha)^{9}B$	1.7±0.2	1
Bauer et al. (1964)	$^{9}Be(p,n)^{9}B$	Did not find	Did not find
Teranishi and Furubayashi (1964)	$^{9}Be(p,n)^{9}B$	1.7	–
Farrow and Hay (1964)	$^{10}Be(p,d)^{9}B$	Did not find	Did not find
Islam and Tracy (1965)	$^{12}C(p,\alpha)^{9}B$	1.50±0.05	Broad
Slobodrian et al. (1967)	$^{9}Be(p,n)^{9}B$	1.4	–
Kroepfl and Browne (1967)	$^{9}Be(^{3}He,t)^{9}B$	1.5	0.7
Anderson et al. (1970)	$^{9}Be(p,n)^{9}B$	1.4	0.0
Gul et al. (1970)	$^{7}Li(^{3}He,n)^{9}B$	Did not find	Did not find
Chou et al. (1978)	$^{9}Be(p,n)^{9}B$	Did not find	Did not find
Byrd et al. (1983)	$^{9}Be(p,n)^{9}B$	Did not find	Did not find
Sherr and Bertsch (1985)	Calculation	0.9	1.4
Kadija et al. (1987)	$^{9}Be(^{3}He,t)^{9}B$	1.16±0.5	1.0±0.2
Burlein et al. (1988)	$^{9}Be(^{6}Li,^{6}He)^{9}B$	1.32±0.08	0.86±0.26
Arena et al. (1988)	$^{10}B(^{3}He,\alpha)^{9}B$	1.8±0.2	0.9±0.3
Catford et al. (1992)	$^{9}Be(^{6}Li,^{6}He)^{9}B$	Did not find	Did not find
Efros and Bang (1999)	Calculation	1.1	1.5
Akimune et al. (2001)	$^{10}B(^{3}He,\alpha)^{9}B$	1.8±0.22	0.6±0.3
Scholl et al. (2011)	$^{9}Be(^{3}He,t)^{9}B$	1.85±0.13	0.7±0.27
Baldwin et al. (2012)	$^{6}Li(^{6}Li,t)^{9}B$	0.8–1.0	1.5
Karki (2013)	$^{9}Be(p,n)^{9}B$	Did not find	Did not find
Fortune and Sherr (2013)	$^{9}Be(^{6}Li,^{6}He)^{9}B$	1.27(1.31)	1.38

many possible improvements in the model, including the chain of reactions with ^{9}B participation as ^{9}B compound nuclear system in $d+^{7}Be \rightarrow ^{9}B$ (Paris et al., 2013; Broggini et al., 2012).

The experimental and theoretical work performed so far is summarized in Table 7.1. The properties of the mass-9 system, in which the ^{9}B partner is particle unbound, even in the ground state, have been difficult to determine; the ground state is unbound to break up into $p+^{8}Be$ by 186 keV. There has been an extensive theoretical and experimental effort directed toward predicting and observing the low-lying states of ^{9}B, especially the first excited $\frac{1}{2}^{+}$ state. The unbound $\frac{1}{2}^{+}$ state at 1.68 MeV in the mirror ^{9}Be has been known for many years, and yet the existence and properties of the state in ^{9}B are not apparent. The state is hard to define because it is difficult to excite and very broad. Energy levels of the ^{9}B nucleus are shown in Figure 7.3 (after Tilley et al., 2004).

Using a 14 MeV neutron beam, one can hope to determine the location of the first excited state in ^{9}B. Differential cross sections could be measured by using the experimental setup, which is shown in Figure 7.4 for the reaction $^{10}B(n, 2n)^{9}B$, $Q=-8.4371$ MeV.

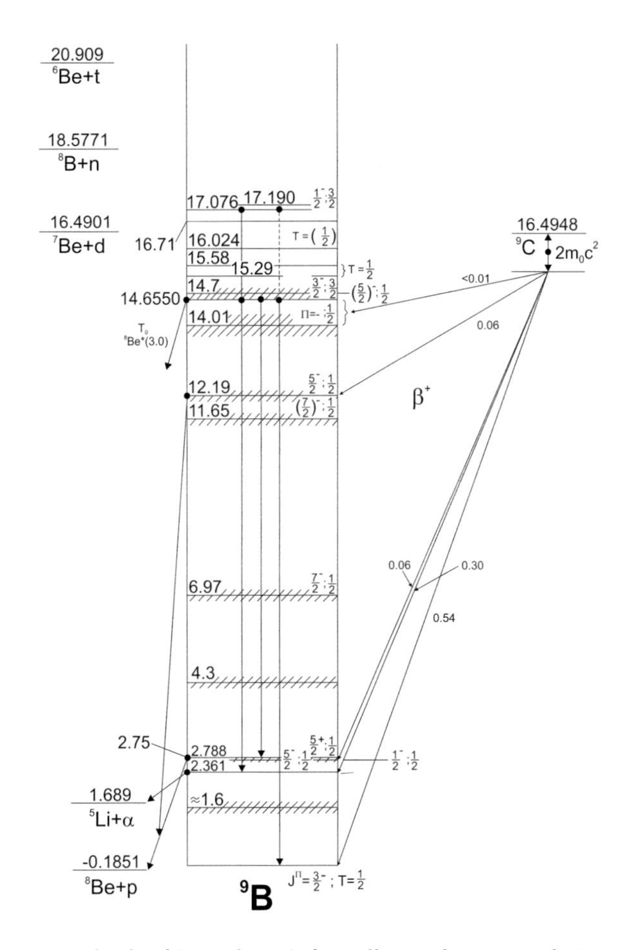

FIGURE 7.3 Energy levels of ^{9}B nucleus. (After Tilley et al., 2004, and TUNL Nuclear Data Evaluation Project "A=9 Energy Levels Diagrams", available at http://www.tunl.duke.edu/nucldata/figures/09figs/menu09.shtml.)

Using the experimental setup shown in Figure 7.4, one could measure n–n coincidences with two neutron detectors placed outside the cone of tagged neutron beam and, from time-of-flight measurements, deduce the neutron energies and calculate the ^{9}B missing mass spectrum. This approach could lead to much more precise information about the low-lying states. The simple setup prepared for the measurements of n–n coincidences is shown in Figure 7.5.

The study of the $^{10}B(n, 2n)^{9}B$ reaction by measurements of n–n coincidences induced by 14 MeV tagged neutrons, as shown in Figure 7.4, could be performed by an experimental setup shown in Figure 7.5 The calculation could be made to estimate the coincidence counting rate of the two neutron detectors and alpha detectors. The following parameters for the experiment are assumed: the distance between neutron detectors and boron target d=100 cm, 10 cm boron target length, 3″×3″ Ne213 neutron detectors and boron

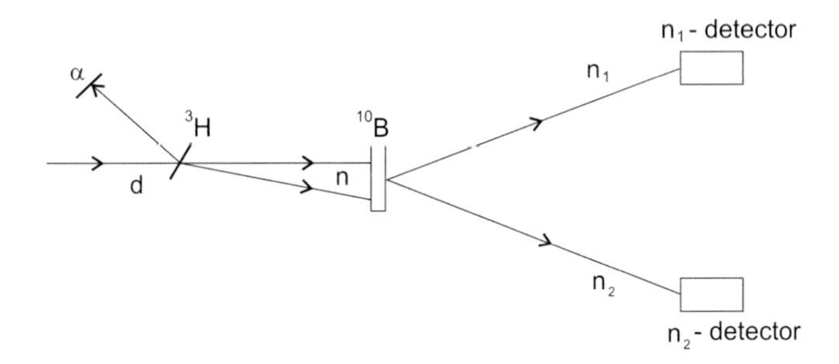

FIGURE 7.4 A sketch of the experimental setup for the measurement of $^{10}B(n, 2n)^9B$ reaction.

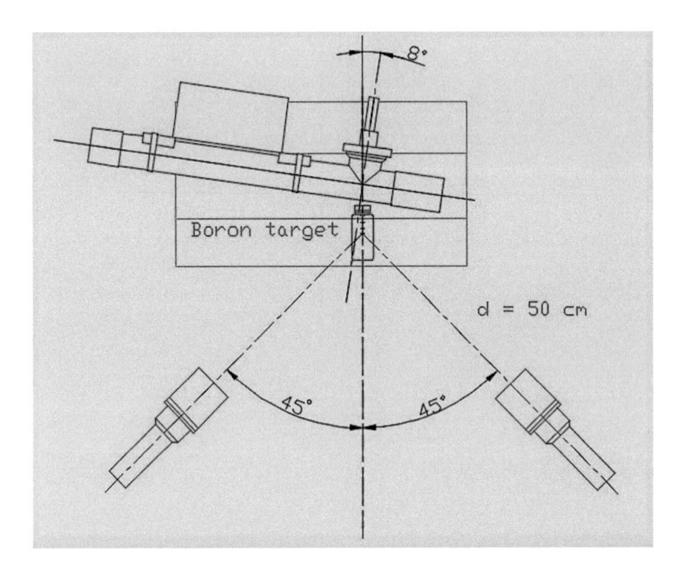

FIGURE 7.5 Experimental setup for the measurement of neutron–neutron coincidences induced by 14 MeV tagged neutrons from $^{10}B(n, 2n)^9B$ reaction.

carbide $^{10}B_4C$ as a boron target. Time-of-flight measurements could measure the velocities of neutrons by using the associated alpha-particle method. The result of calculations is $I = 2 \times 10^{-5}$ c/s per alpha segment.

Preliminary measurements with d=50 cm and $10 \times 10 \times 10\,cm^3$ plastic scintillators were performed at Ruđer Bošković Institute in Zagreb (Valković, et al., 2018). The proper functioning of the experimental and electronic setup was ensured by the measurement of events from the bombardment of a piece of graphite ($5 \times 5 \times 5\,cm^3$) corresponding to the first excited state of ^{12}C at 4.44 MeV excitation energy. In that case, one neutron detector detected the inelastically scattered neutrons from $^{12}C(n, n'\gamma)^{12}C$ reaction, while the other the 4.44 gamma rays. The peak at 4.44 MeV of excitation ^{12}C energy was seen. Unfortunately, the peak position strongly depends on the assumed location of the reactions occurring in neutron detectors as well as in the target, requiring $d \geq 100$ cm. The measurements were repeated for $d=100$ cm, but in that case, the number of counts was reduced to the factor of 16 and the first excitation level was failed to be detected.

Another problem was found to be the random coincidence counting rate, which causes a typical triple coincidence rate at around 2 Hz, much higher than that calculated for $d=50$ cm. It is believed that the use of a new ING27 instead of an old API120 neutron generator would partially solve the problem. Also, the Monte Carlo calculation shows that the number of neutrons scattered on the laboratory walls and hitting the detectors again was around 16% of those hitting the detector directly from the neutron generator.

Based on the experience gained so far, the following improvements to the experimental setup are proposed:

- Increasing distance to $d=100$ cm;
- Using an ING27 neutron generator with a 3×3 segmented alpha detector;
- $C_{10}B_4$ (enriched boron carbide) as the target pressed as much as possible, covering the alpha detector solid angle;
- $10\times10\times10$ cm^3 plastic neutron detectors (32 pieces);
- Associated electronics;
- Protecting shield.

7.1.2 (n,^{2}n) Reaction on d Followed by ^{2}n-Induced Reactions on Li Isotopes

Although the bound state of two neutrons does not exist, the known effects of final state interaction (fsi) in reactions with three particles in the final state may result in reactions of type (2n, x) on some light nuclei.

The reaction n+d→p+n+n for $E_n=14$ MeV shows a strong n–n final state interaction. Therefore, two neutrons can, in principle, interact with the target nucleus as a single projectile. The two neutrons (^{2}n) induced a reaction on Li isotopes (^{6}Li and ^{7}Li) could result in the following chain of reactions:

i. ^{6}Li+^{2}n→^{8}Li→^{8}Be+e+ν followed by ^{8}Be →2α.
ii. ^{7}Li+^{2}n→^{9}Li→^{9}Be+e+ν followed by ^{9}Be → n+2α, as indicated in Figure 7.6.

Here the symbol ^{2}n represents the neutron–neutron final state interaction (fsi) as observed in the n+d→p+n+n reaction for $E_n=14$ MeV bombarding energy.

As discussed in Chapter 1, the most important unresolved problem in nuclear astrophysics is the so-called cosmological lithium problem. It refers to the large discrepancy between the abundance of primordial ^{7}Li predicted by the standard BBN theory and the value inferred from the so-called Spite plateau in halo stars. The predictions of the BBN theory successfully reproduce the observations of all primordial abundances except for ^{7}Li. The abundance of ^{7}Li is overestimated by more than a factor of 3. Nuclear reactions shown in Figure 7.6 can contribute to the destruction of Li isotopes.

Another candidate to consider is an experiment in which one brings ^{2}H and Li targets/nuclei as close as possible. One possibility is using lithium hydrides ^{6}Li^{2}H and ^{7}Li^{2}H.

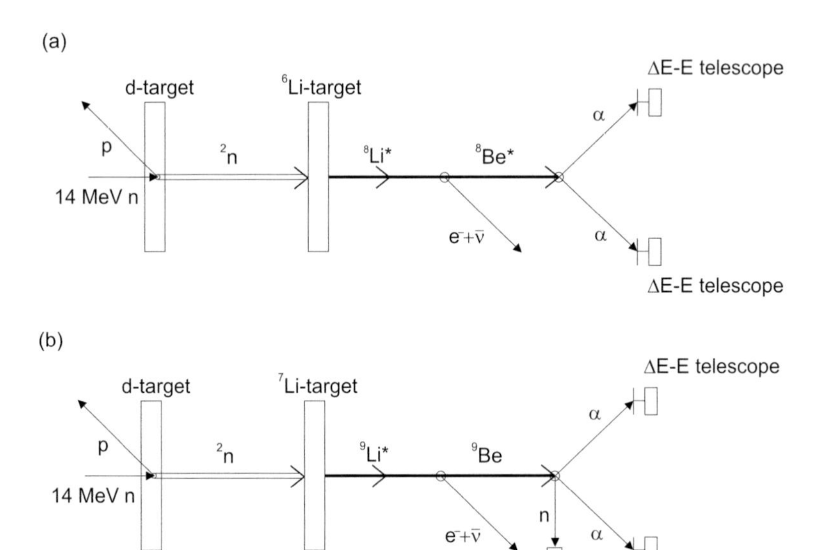

FIGURE 7.6 (a) The $^6Li+^2n \rightarrow ^8Li \rightarrow ^8Be+e+\nu$ process followed by $^8Be \rightarrow 2\alpha$ reaction and (b) $^7Li+^2n \rightarrow ^9Li \rightarrow ^9Be+e+\nu$ reaction followed by $^9Be \rightarrow n+2\alpha$ decay.

Lithium hydride has the chemical formula LiH and the crystal structure fcc (NaCl type), as shown in Figure 7.7.

Another possible target is ^9Be, as indicated in Figure 7.8:

The estimated number of counts in $^7Li+^2n \rightarrow ^9Li \rightarrow ^9Be+e+\bar{\upsilon}$ followed by $^9Be \rightarrow n+2\alpha$ nuclear reactions is 0.1 c/s for ING27 equipped with a 3×3 segmented alpha detector, beam intensity 10^8 n/s and ring gamma detector. During BBN, in the period of transition from quark–gluon plasma to baryonic matter, the probability of ^2n formation [n–n fsi or n–n bound state (?)] is probably increased. This consideration would have led to Li distraction via reactions (i) and (ii). Investigation of these problems fits missions of heavy-ion colliders. One should elaborate a proposal for the detection of neutrons in heavy-ion collisions by measurements not only multiplicity but also correlations and fsi phenomena.

Since the primordial ^7Li is mainly produced by β-decay of ^7Be ($t_{1/2}$ = 53.2 days), the abundance of ^7Li is mostly determined by the production and destruction of ^7Be. The neutron-induced reactions can also play a role in the destruction of ^7Be, in particular the $^7Be(n, \alpha)^4He$ reaction in the energy range of interest for BBN, in particular between 20 and 100 keV. The recent activity in solving the "lithium problem" in Big Bang nucleosynthesis is focused on the role that putative resonances play in the resonance-enhanced destruction of ^7Li.

Energy level diagrams of ^8Li and ^9Li are shown in Figures 7.9 and 7.10, respectively. In these figures, energy values are plotted vertically in MeV, based on the ground state as zero (after Tilley et al., 2004). For the ^8Li and ^9Li diagrams, all levels are represented by

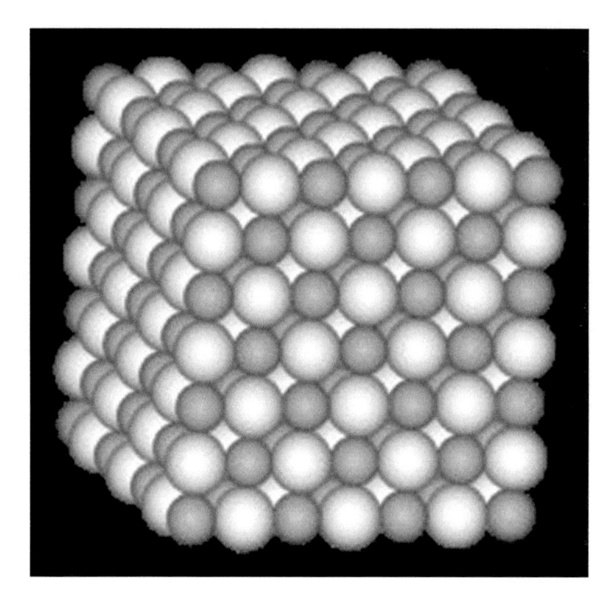

FIGURE 7.7 Crystal structure of lithium hydride with lattice constant equal to a = 0.40834 nm.

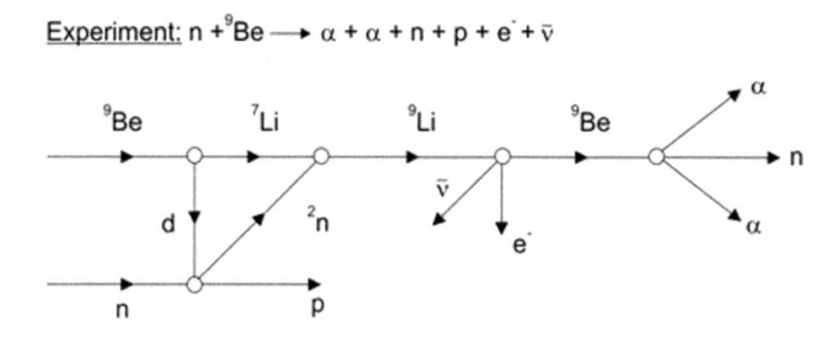

FIGURE 7.8 Incoming neutron interacting with d in ^{9}Be nucleus leading to ^{7}Li+2n interaction and a sequence of decays as shown.

discrete horizontal lines. The values for total angular momentum J, parity π and isobaric spin T, which are reasonably well established, are indicated on the levels; less specific assignments are enclosed in parentheses. In the original paper (Tilley et al., 2004), only some thin-target excitation functions are shown, with yield plotted in the horizontal direction and the bombarding energy in vertical, for reactions in which ^{8}Li and ^{9}Li are the compound nuclei. Bombarding energies are indicated in the laboratory reference frame, while the excitation function is scaled into the CM reference frame so that resonances are aligned with levels. Excited energy levels of the residual nuclei involved in these reactions have generally not been shown. For nuclear reactions in which the present nucleus occurs as a remaining product, excitation functions have not been shown. Q-values and threshold energies are based on atomic masses from Audi et al. (2003).

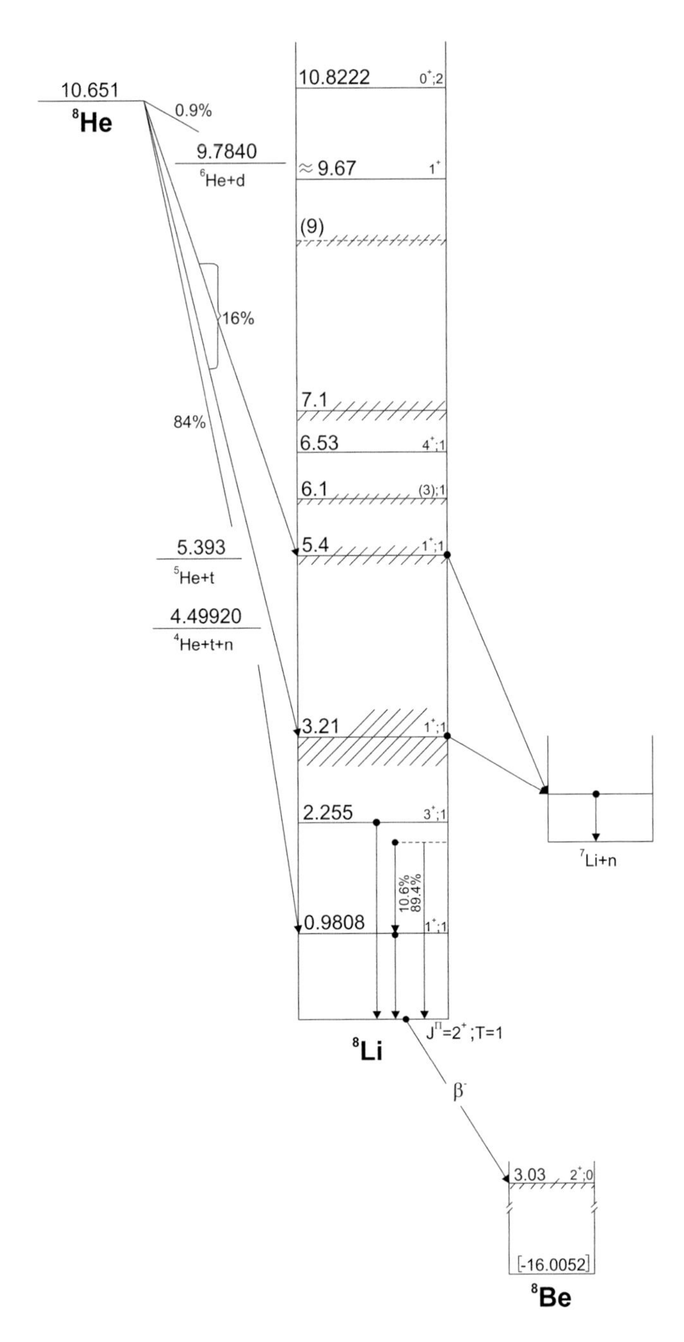

FIGURE 7.9 Energy levels of ^{8}Li. (After Tilley et al., 2004, and TUNL Nuclear Data Evaluation Project "A=9 Energy Levels Diagrams", available at http://www.tunl.duke.edu/nucldata/figures/08figs/menu08.shtml.)

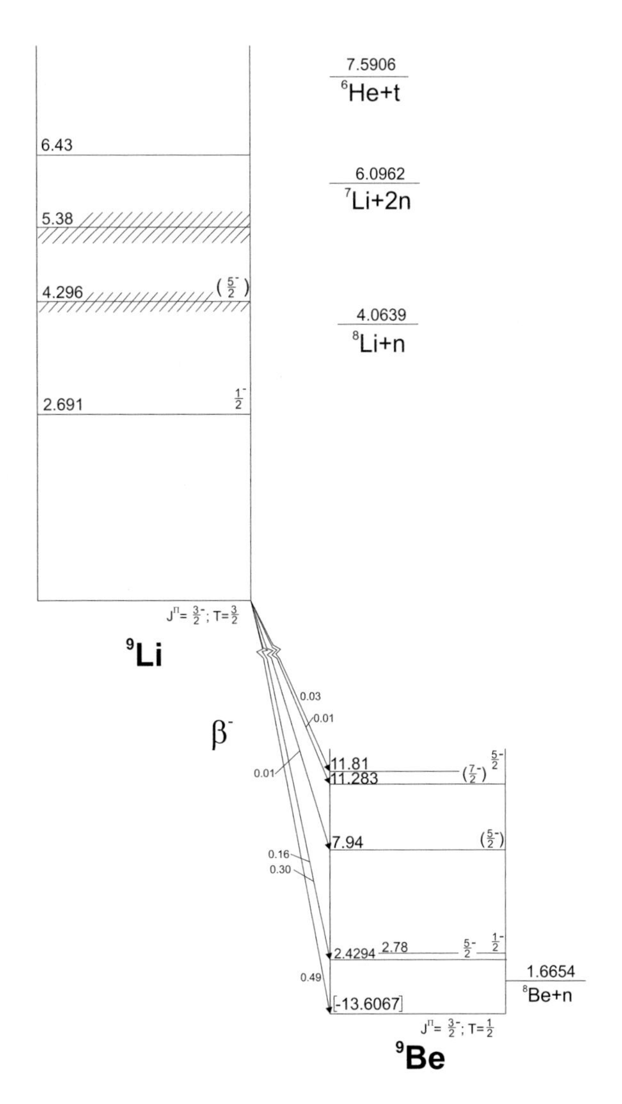

FIGURE 7.10 Energy levels of ^{9}Li. (After Tilley et al., 2004, and TUNL Nuclear Data Evaluation Project "A=9 Energy Levels Diagrams", available at http://www.tunl.duke.edu/nucldata/figures/09figs/menu09.shtml.)

7.2 Experiments with Boron

Cosmic or interplanetary dust particles (IDPs) have been collected in the stratosphere by NASA U-2, ER-2 and WB-57F aircrafts. Particles were collected on oil-coated Lexan surfaces since 1981 and are curated at the Johnson Space Center (JSC). Many thousands of 2–100 μm particles have been collected, and it is believed that both comet and asteroid samples are present and available (NASA, 2012).

The cosmic dust particles are preserved as fine particles dispersed on the silicone oil [n($(CH_3)_2SiO$)]-coated surfaces of the cosmic dust samplers. Particles from these collectors are rinsed with the organic solvent hexane in order to remove the remaining silicon oil. However, some residual oil may remain on the particles. Cosmic dust has also recently been collected using oil-free polyurethane foam substrates, and other oil-free substrates may be used in the future.

Because of the presence of terrestrial contaminants, the JSC curation staff is rarely sure of the origin of any particular extraterrestrial particle. A subset of the collected particles is examined by optical and scanning electron microscopies and energy-dispersive X-ray spectrometry to determine the structure and chemical composition of individual dust grains. Extraterrestrial particles are often identified by their characteristic chemical compositions and are termed "cosmic". Cosmic-type particles are identified as those having one of the three following sets of properties:

i. Irregular to spherical, opaque, dark-colored and composed mostly of Fe with minor S and Ni.
ii. Irregular to spherical, translucent to opaque, dark-colored and containing various proportions of Mg, Si and Fe with traces of S and Ni.
iii. Irregular to faceted blocks, transparent to translucent and containing mostly Mg, Si and Fe with traces of S and Ni.

Some collected extraterrestrial particles may have chemical compositions that differ from these classes. The types of samples available from JSC are individual whole particles, and very rarely, entire small collection surfaces. Complete small cosmic dust collections are available to qualified investigators.

7.2.1 Boron Concentration Measurements

The high measured abundance of boron in type I carbonaceous (C1) chondrites may be a result of the presence of graphite grains in the primitive solar nebula being irradiated by high-energy nucleons at some stage of their history. The boron atoms produced by spallation reactions are stably locked inside interstellar graphite grains and make an essential contribution to the boron abundance of C1 chondrites. Positive detection of boron in meteoritic material encompasses a wide range of values of B/H. The values range from $\approx 10^{-10}$ in ordinary chondrites to $38–59 \times 10^{-10}$ in C1 chondrites (Cameron et al., 1973). The amount of boron abundance in cosmic dust particles could be related to the time of particle exposure to the cosmic-ray flux.

7.2.2 Preferential Destruction of Enantiomers

The two forms of a chiral molecule, enantiomers, have identical physical and chemical properties. Still, the interaction with other chiral molecules may be different, just as a left hand interacts differently with left- and right-hand gloves. Chiral molecules in living organisms exist almost exclusively as single enantiomers, a property that has a critical role in molecular recognition and replication processes. Therefore, it is a prerequisite for the origin of life. Left- and right-handed molecules of a compound will be formed in equal amounts (a racemic mixture) when synthesized in the laboratory in the absence of some directing template.

The existence of the single chirality of biological molecules, exclusively left-handed amino acids and right-handed sugars, presents us with two questions (Blackmond, 2010):

i. What served as the original template for the biased production of one enantiomer over the other in the chemically austere and racemic environment of the prebiotic world?
ii. How was this sustained and propagated to result in the biological world of single chirality that surrounds us?

The review by Blackmond (2010) focuses primarily on the second question: the plausible mechanisms for the evolution of molecular chirality as exemplified by the D-sugars and L-amino acids found in living organisms today.

Laboratory synthesis of these molecules starting from optically inactive materials yields (1:1) mixtures of L- and D-isomers. However, only L-amino acids are found in proteins in living organisms. Here is the obvious question: Could there have been some asymmetric factors in the environment that influenced the choice when life began? There have been several suggestions, including (i) polarized light, (ii) optically active quartz and (iii) natural radioactivity.

The last scheme assumed the origin of life in a radioactive environment. The result of β-radioactivity is longitudinally polarized β-rays, which produce circularly polarized bremsstrahlung. This radiation was believed to have preferentially destroyed D-isomers. Some recent experiments have not supported this assumption.

"Symmetry breaking" is the term used for describing the occurrence of a difference between left and right enantiomeric molecules. This imbalance is measured in terms of the enantiomeric excess, or ee, where $ee=(R-S)/(R+S)$. R and S stand for concentrations of the right- and left-handed molecules, respectively. Speculations for how an imbalance might have come about could be grouped as either terrestrial or extraterrestrial and further subdivided into either random or deterministic. The evidence of enantiomeric excess in amino acids found in chondritic meteor deposits (Pizzarello, 2006) allows the hypothesis that the initial imbalance is older than our world.

Assuming that amino acids could be synthesized on dust particles in interstellar space, the observed optical activity may be a result of cosmic-ray bombardment. High-energy polarized protons in cosmic rays may be able to preferentially destroy one isomer because of significant asymmetry in proton (in cosmic rays)–proton (in amino acid) scattering. Single spin asymmetry in polarized proton–proton elastic scattering has experimentally been studied at high energies; see, for example, Adamczyk et al. (2013) for experiments done at Relativistic Heavy Ion Collider (RHIC). RHIC is one of only two operating heavy-ion colliders and the only spin-polarized proton collider ever built. It is located at Brookhaven National Laboratory (BNL) in Upton, New York. Polarization produced in proton–proton scattering has been observed even at low energies, 30 and 50 MeV (Batty et al., 1963).

Interstellar dust appears to play a critical role in the formation of interstellar molecules. Molecules may be formed on or in grain surfaces. Many molecules have been observed in interstellar space. Johnson (1972) reported the observation of interstellar porphyries (molecule $MgC_{46}H_{30}N_6$). The existence of interstellar molecules suggests the following: (i) Such molecules support or are the metabolic products of an interstellar biota. (ii) Such molecules participating in planetary condensation from the interstellar medium can make a significant contribution to the origin of terrestrial life.

The scientific consensus is that if a sufficient enantiomeric excess existed in life's primordial molecular reservoir, it would almost certainly push life toward the extreme bias of actual living beings (see McGuire and Carroll, 2016). Meteoritic samples suggest that such an enantiomeric excess has been generated before the Earth's formation. The authors present a case of a meteorite that struck just outside of the city of Murchison, Victoria, Australia, on September 28, 1969. Its weight was ~100 kg, and it belongs to the class of carbonaceous chondrite and was rich in organic compounds, including amino acids, many of them having an excess of left-handed enantiomers >10%. It has been determined that meteorites like the one that hit Murchison date back to about the time of our solar system's formation. Also, the source of their molecular material can be traced to the cloud of gas and dust from which our solar system formed. There is a possibility of existence of a reasonable mechanism for generating an enantiomeric excess in that primordial cloud. In that case, the enantiomeric bias of the origins of life could be linked to processes that occurred billions of years ago, before the solar system existed (McGuire and Carroll, 2016); see also Engel and Macko (1997) and McGuire et al. (2016).

7.3 Effects of Magnetic Field on Living Matter

7.3.1 Introduction

Roughly one-third of all living species became extinct at the close of the Cretaceous, a period marked by the resumption of polarity reversals of the Earth's magnetic field following a very long period of regular magnetic activity. There is mounting evidence that a correlation exists between major faunal extinctions and geomagnetic polarity reversals. The validity of this correlation in recent geological time seems to have been well established by studies of fossil species of single-celled marine microorganisms (Hays, 1971).

Several mechanisms linking changes in the geomagnetic field (GMF) with the effects on living organisms have been proposed. Most are based on the assumption that during polarity reversals, the dipole component of the GMF weakens or disappears for periods of a few thousand years, allowing a much higher flux of both solar protons and galactic cosmic rays to bombard the Earth's surface. Other mechanisms include climate change and the effects of a considerable reduction in the content of ozone in the atmosphere, which would increase exposure to ultraviolet radiation. Direct magnetic field effects on growth have been discussed (Hays, 1971; Crain, 1971).

A mechanism through which the GMF might have directly influenced the extinction of the species has also been proposed (Valković, 1977). The concentration factor dependence on magnetic field intensity was assumed to explain geomagnetic effects on the functioning of living organisms; see Figure 7.11.

During polarity reversals, GMF intensity decreases by an order of magnitude. During this time, species are affected by magnetic field intensity of the order of magnitude of H_1.

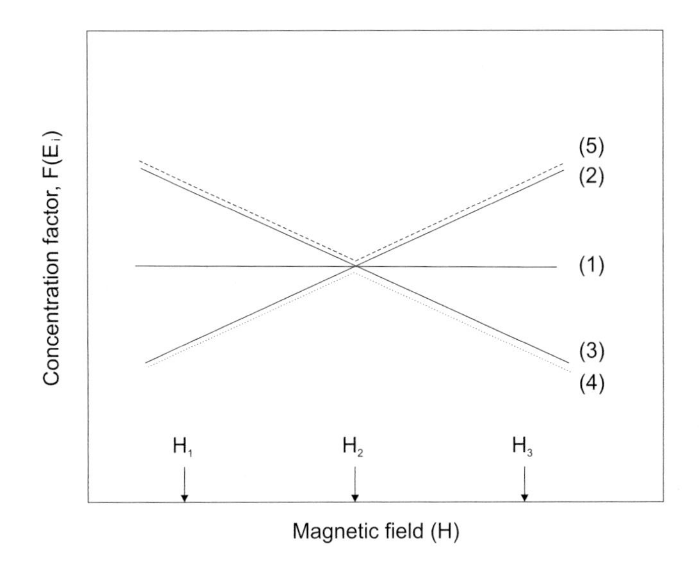

FIGURE 7.11 Some possibilities for functional relationship $F_i = F_i$ (H). Curve (1) represents F_i=const.; curves (2) and (3) represent $F_i = C \cdot H$, while curves (4) and (5) have maximum (minimum) for H_2 value.

If they were living in an environment that had provided them with elements, they would have been in the range of adequacy when the Earth's magnetic field was H_0. Because of the assumed dependence of concentration factor on magnetic field intensity, they will be in the range of deficiency or toxicity when the magnetic field is H_1 (depending on the slope of the assumed functional dependence). A prolonged deficiency or toxicity for several generations may lead to disastrous effects, including the species' disappearance. This effect can happen for one or more essential elements. Magnetic field dependence of concentration factors can bring living organisms into the range of deficiency or toxicity without changing trace element availability in the environment. One can imagine a species living in a situation such that a supply of a given essential trace element is near the edge of the range of adequacy. In this case, only small disturbances are needed to have an inadequate supply of the essential trace element and consequent development of abnormalities.

In experiments reported previously (Valković, 1980), the organism used was a respiratory-deficient mutant of *M. bacilliformis*, which also has lost the ability to grow as mycelium. Instead, it exists as spherical cells that reproduce only by budding. The mutant is sensitive to temperature (dies at 30°C) and is also vulnerable to vigorous mechanical shaking. It grows relatively slowly in a defined medium at the expense of alcoholic fermentation. The four components of the medium were sterilized separately (at 21°C for 15 minutes) and then mixed aseptically. The culture was prepared, and after 3–4 days, inoculation was done by the transfer of starter culture and medium into specially designed flasks.

In the experiments, the growth of the culture took place at room temperature in round-bottom flasks containing 14 mL of medium. Each flask was inoculated to have 10^3

cells/mL; six flasks were placed in solenoids whose magnetic fields were extremely uniform over the active region of growth, and six flasks were used as controls. Growth (increase in cell number) was monitored daily by measurements of the turbidity of the cell suspensions. After 7–10 days, when the cell number reached saturation, the cells were harvested.

Trace elements were analyzed by proton-induced X-ray emission spectroscopy. Targets were exposed to a 3 MeV proton beam and characteristic X-rays; a Si(Li) detector detected X-rays. The concentrations of most essential elements were measured simultaneously.

The measured data for Mn/Zn and Cu/Zn ratios indicate that the concentration ratios for these elements within the microorganisms might be dependent on magnetic field intensity. Both Mn/Zn and Cu/Zn ratios decreased with the increase in the magnetic field (Mn/Zn for 5%; Cu/Zn for 13%). However, this decrease was still within the experimental error. Another measurement was performed to verify this result. The result is shown in Figure 7.12. Although the absolute values for Cu/Zn concentration ratios are slightly higher than in previous experiments, and the Cu/Zn ratio seems to follow the trend observed in the first experiment, it decreases with the increase in magnetic field intensity. Similar, but not certain, conclusions are also valid for Mn/Zn and Fe/Zn concentration ratios. The Ni/Zn concentration ratio does not show a tendency toward a decrease while the magnetic field intensity increases.

In conclusion, the reported data strongly indicated the necessity for the continuation of such research. Other groups should repeat the experiments with the same microorganisms and the same medium, but with better control of all other parameters (temperature, target preparation and data analysis). The magnetic field should also be varied in more delicate steps. In addition to the measurement of element concentration ratios, the absolute concentrations of essential trace elements should be measured. Experiments with other microorganisms and differently prepared media should also be done.

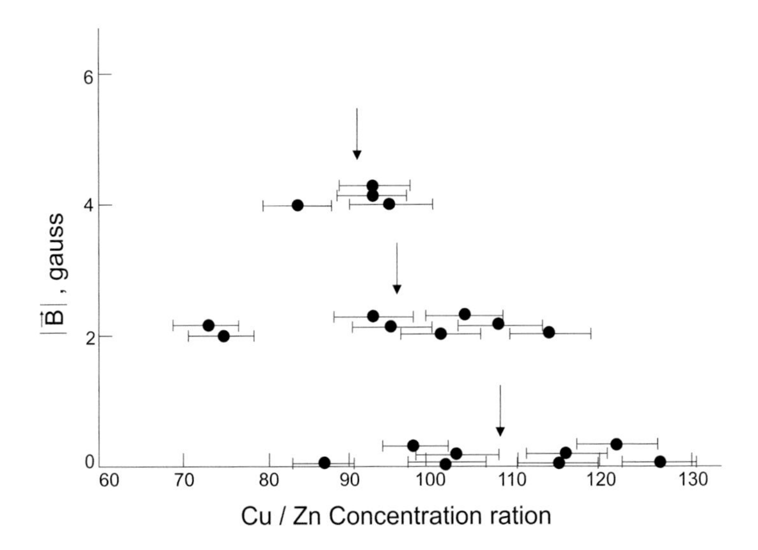

FIGURE 7.12 Magnetic field dependence of Cu/Zn concentration ratio for *M. Bacilliformis*.

In a recent publication by Fu et al. (2014), magnetic field in another field is proposed to have played an essential role in some of the most enigmatic processes of planetary formation by mediating the rapid accretion of disk material onto a central star and the creation of the first solids. However, there were no experimental constraints on the intensity of these fields until Fu et al. (2014) show that dusty olivine-bearing chondrules from the Semarkona meteorite were magnetized in a nebular field of $54 \pm 21\,\mu T$. This intensity supports chondrule formation by nebular shocks or planetesimal collisions rather than by electric current, the x-wind, or another mechanism near the Sun. This fact implies that the background magnetic fields in the terrestrial planet-forming region are likely to be 5–54 μT (Fu et al., 2014).

7.3.2 Biological Effects of Very Low Magnetic Fields

The GMF is an essential component of the environment for all living organisms on the planet Earth. GMF is continuously acting on living systems and influencing many biological processes. There are many local differences in the strength and direction of the GMF. On the surface of the Earth, the vertical component is maximal at the magnetic pole (about 67 μT) and is zero at the magnetic equator. The horizontal part is maximal at the magnetic equator, about 33 μT, while it is zero at the magnetic poles (Kobayashi et al., 2004).

Mass extinction events profoundly changed the Earth's biota during the early and late Mesozoic, and terrestrial plants were among the most severely affected group. Several plant families were wiped out, while some new families emerged and eventually became dominant. The properties of the GMF during the Mesozoic and late Paleozoic, or more precisely between 86 and 276.5 million years before, are of particular interest. Its virtual dipole moment (VDM) has been significantly reduced compared to today's values (Shcherbakov et al., 2002).

The time of origin of the GMF has important implications for the thermal evolution of the planetary interior and the habitability of the early Earth. It has been proposed that detrital zircon grains from Jack Hills, Western Australia, provide evidence for a GMF as early as 4.2×10^9 years. However, combined paleomagnetic, geochemical and mineralogical studies performed by Borlina et al. (2020) on Jack Hills zircons indicated that most have poor magnetic recording properties and secondary magnetization carriers that postdate the formation of the zircons. Therefore, the authors concluded that the existence of the geodynamo before 3.5×10^9 years ago remains unknown.

During the Earth's history, the GMF had several changes of polarity, with the so-called geomagnetic reversals characterized by persistent times with the same polarity. These events occurred some hundred times since the Earth's formation, and the mean time between a reversal and the next one has been estimated at around 300,000 years. Because the present normal polarity started approximately 780,000 years ago and a significant field decline has been ongoing during the last 1,000 years, an imminent geomagnetic reversal should not be so unexpected (De Santis et al., 2004). The time distributions of geomagnetic reversals in the last 5 million years are shown in Figure 7.13.

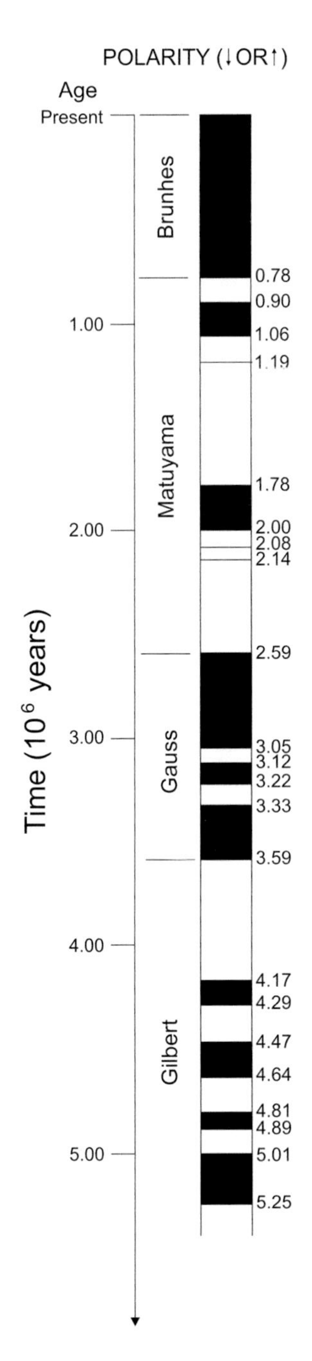

FIGURE 7.13 Geomagnetic polarity during the last 5 million years. Dark areas denote the periods when the polarity matches today's normal polarity; light areas denote the periods when the polarity is reversed.

The vast range of times estimated for various reversals at different times and places indicates the inadequacy of the current data set to resolve such issues as how long an individual reversal takes, whether the time taken varies with location, and with different inversions, and also whether it changes along with reversal rate. A detailed analysis of the geomagnetic reversals found in the Glatzmaier–Roberts dynamo (Coe et al., 2000; Coe and Glen, 2004) indicates that reversal times vary widely. These variations depend on criteria used to determine the onset of the reversal, but also with location, for successive reversals, and may exhibit different behaviors depending on the thermal boundary conditions.

Sedimentary and volcanic paleomagnetic records are complementary, but both are always incomplete and give only lower bounds on how rapidly changing and complex the behavior of the reversing field may have been. Geodynamo simulations provide a theoretical approach to the question. Still, limitations in computer power have prevented the researchers from operating near the parameter regime appropriate for the core, with concomitant loss of spatial and temporal resolution. Nonetheless, the combined evidence from all three approaches appears sufficient to conclude that at least some reversals are much more complicated than typically portrayed, with episodes of oscillatory and very rapid field change.

The scientific literature is loaded with studies of the influence of magnetic fields on biological systems. These studies are often motivated by suggested health hazards of electromagnetic fields that accompany the distribution and use of electrical power. The majority of them report definite effects. However, in the few cases in which independent replication has been attempted, the initial results have usually proved irreproducible (after Hore, 2012).

With no theoretical mechanism to guide experimental design, the majority of these investigations have been unsatisfactory. An exception is a series of articles by Buchachenko and Kouznetsov, and their associates (Buchachenko et al., 2005, 2008, 2010; Buchachenko and Kuznetsov, 2008). In the first paper, the phosphoglycerate kinase (PGK) is found to be controlled by a $^{25}Mg^{2+}$-related magnetic isotope effect. The PGK is present in all living organisms as one of the two ATP-generating enzymes. In the absence of Mg, no enzyme activity occurs (Varga et al., 2012). The Mg^{2+} nuclear spin selectivity manifests itself in PGK-directed ADP phosphorylation. This activity has been proven by a comparison of ATP synthesis rates estimated in reaction mixtures with different Mg isotopes. Both pure $^{25}Mg^{2+}$ and $^{24}Mg^{2+}$ species, and their combination were used in experiments. In the presence of $^{25}Mg^{2+}$, ATP production was found to be 2.6 times higher compared with the yield of ATP obtained by $^{24}Mg^{2+}$-containing PGK-based catalytic systems. A vital element of the chemical mechanism proposed is a non-radical pair formation in which $^{25}Mg^{+}$ radical cation and phosphate oxyradical are involved.

Although the membrane bioenergetics are universal, the phospholipid membranes of archaea and bacteria (the deepest branches in the tree of life) are fundamentally different. This divergence in membrane chemistry is reflected in other stark differences between the two domains, including ion pumping and DNA replication (Sojo et al., 2014). Contrary to this paradoxical difference in membrane composition is the universality of membrane bioenergetics (Lane and Martin, 2012). Mostly all cells power ATP synthesis through the chemiosmotic coupling, in which the ATP synthase (ATPS) has power due to electrochemical differences in H^+ (or Na^+) concentration across membranes (Mitchell, 1961). The ATPS is universally conserved (Stock et al., 1999) and shares the same deep phylogenetic split as the ribosome, suggesting that both were present in the LUCA (Lane et al., 2010).

Lee and Heroux (2014) proposed that the magnetic field acts on ATPS, an enzyme that catalyzes the production of ATP, the energy source for all living cells. The chemical formula of ATP is $C_{10}H_{16}N_5O_{13}P_3$. The ATP molecule is composed of three essential components. A sugar molecule, ribose, is positioned at the center. Adenine (a group consisting of linked rings of carbon and nitrogen atoms) is attached to the side. A string of phosphate groups (see May, 1997) is attached to the other side of the sugar. These phosphate groups are the key to the activity of ATP; see Figure 7.14.

Note that the Nobel Prize in Chemistry 1997 was shared by Walker, Boyer, and Skou for the determination of the detailed mechanism by which ATP shuttles energy. ATP functions by losing the endmost phosphate group when instructed to do so by an enzyme. In this reaction, a lot of energy is released, which the organism can then use for building proteins, contract muscles, etc. The products of this reaction are adenosine diphosphate (ADP) and the phosphate group, such as orthophosphate, HPO_4, which is attached to another molecule (e.g., alcohol):

$$\mathrm{ATP} + \mathrm{H_2O} \rightarrow \mathrm{ADP} + \mathrm{HPO_4}$$

Lee and Heroux (2014) concluded that the biological effects of magnetic fields are connected to a change in the structure of water that impedes the flux of protons in ATPS channels. It should be noted that Semikhina and Kiselev (1988) reported some 30 years ago the evidence that magnetic fields can influence the structure of water at levels as low as 25 nT. The influence of mT ELF magnetic fields on the water in conductometric sensors was also investigated (Sojo et al., 2014). The studied frequencies were in the band of 1–50 Hz, alternative magnetic field inductance was 0–2 mT, and constant magnetic field

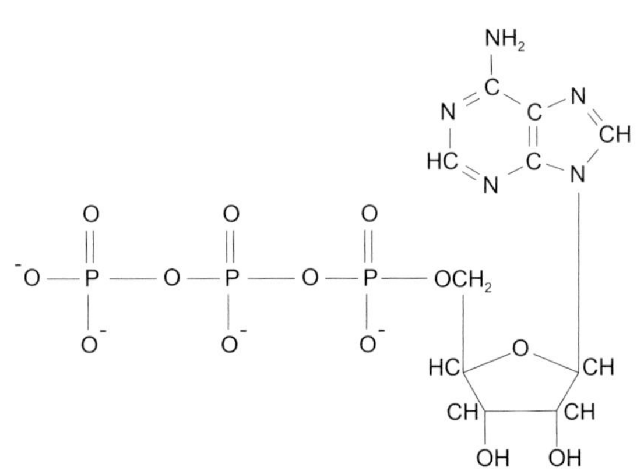

FIGURE 7.14 ATP consists of a base; in this case, adenine, ribose and a phosphate chain.

inductance was 0.07–3.5 mT. Substantial changes in water parameters at the frequencies 6.5 Hz and 7.5 Hz, and fewer variations at 12–14 Hz were observed.

In the work reported by Lee and Heroux (2014), five cancer cell lines were exposed to ELF MFs, in the range of 0.025–5 μT, and the cells were examined for karyotype changes after 6 days. All cancer cells lost chromosomes from MF exposure, with a mostly flat dose–response. Constant MF exposures for 3 weeks allowed a rising return to the baseline, unperturbed karyotypes. From this point, small MF increases or decreases are again capable of inducing karyotype contractions (KCs). Their data show that the KCs are caused by MF interference with mitochondria's ATPS. The process is compensated for by the action of adenosine monophosphate-activated protein kinase (AMPK). The effects of MFs are similar to those of the ATPS inhibitor, oligomycin. They are amplified by metformin, an AMPK stimulator, and attenuated by resistin, an AMPK inhibitor. Over environmental MFs, KCs of various cancer cell lines show exceptionally wide and flat dose–response, except for those of erythroleukemia cells displaying a progressive rise from 0.025 to 0.4 μT. They conclude that the biological effects of MFs are related to an alteration in the structure of water that impedes the flux of protons in ATPS channels. These results may be environmentally significant because of the central roles ATPS and AMPK play in human physiology, particularly in their connections with diabetes, cancer and longevity.

Maffei (2014), in his review article on the effects of magnetic field on plant growth, development and evolution, has concluded that laboratory studies done so far have shown that the fields can produce or modify a wide range of phenomena. Understanding the diversity of the reported effects is the main problem. In recent years, the following types of physical processes or models underlying hypothetically primary mechanisms of the interaction of MF responses in biological systems have been proposed:

a. classical and quantum oscillator models;
b. cyclotron resonance model;
c. interference of quantum states of bound ions and electrons;
d. coherent quantum excitations;
e. biological effects of torsion fields accompanying MF;
f. biologically active metastable states of liquid water;
g. free radical reactions and other "spin" mechanisms;
h. parametric resonance model;
i. stochastic resonance as an amplifier mechanism in magnetobiology and other random processes;
j. phase transitions in biophysical systems displaying liquid crystal ordering;
k. bifurcation behavior of solutions of nonlinear chemical kinetics equations;
l. radio-technical models, in which biological structures and tissues are portrayed as equivalent electric circuits; and
m. macroscopic charged vortices in the cytoplasm.

Furthermore, mechanisms combining these concepts and models cannot be excluded (Belyavskaya, 2004). It is suggested that a prolonged exposure of plants to a weak magnetic field may result in different effects at the cellular, tissue and organ levels.

7.3.3 Experiments with the Varying Magnetic Field Intensities

In the laboratory, low MF can be created by different methods, including shielding (surrounding the experiment by ferromagnetic material having high magnetic permeability, which deviates MF and concentrates it in the metal) and compensation by using Helmholtz coils.

In any case, during the coming magnetic field reversal, the living species will be exposed to much smaller (in the nT–μT range) magnetic field intensity. Also, daily variations of magnetic field intensity are observed (see, for example, Price, 1969; Okeke and Hamano, 2000). The main GMF, which is slowly varying, originates within the Earth. More rapid variations, with periods from seconds to days, are produced by processes above the Earth. A geomagnetic observatory is a place where the measurements of the Earth's magnetic field are recorded accurately and continuously, with a time interval of one minute or less. The site of the observatory must be free of magnetic noise and remain so for the foreseeable future. The earliest magnetic observatories where continuous vector observations were made began operation in the 1840s. There are many locations of currently operating magnetic observatories. In this work, the data reported by the observatory located in Lonjsko Polje, Croatia (N=44.5919°, E=16.6592°), operated by the Department of Geophysics, Faculty of Science, University of Zagreb, were used. For an illustration of daily variations of intensity of the Earth's magnetic field, see Figure 7.15, which shows the variations as observed on November 8, 2017, in Lonjsko Polje (LON).

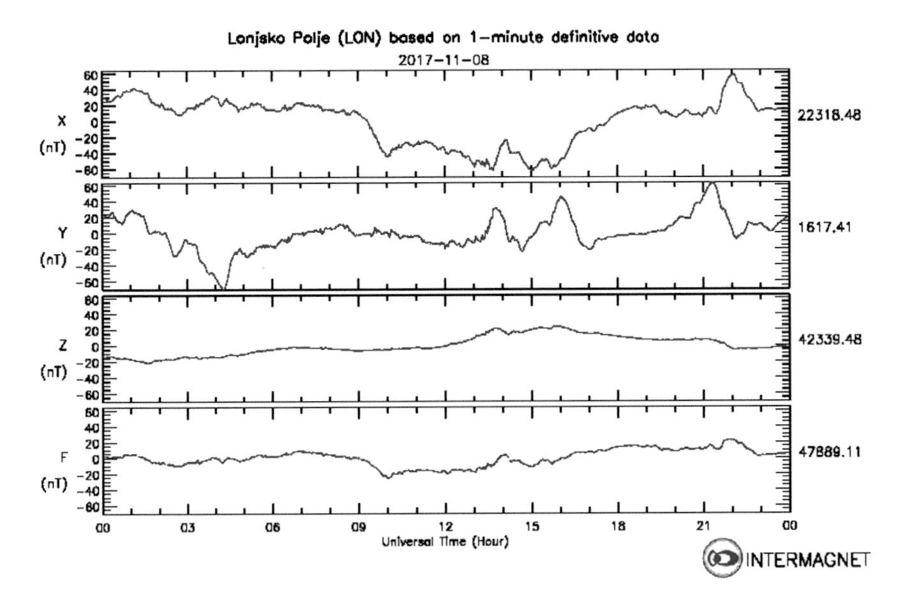

FIGURE 7.15 Daily variations of the Earth's magnetic field as measured at the observatory in Lonjsko Polje on November 8, 2017. We thank the Faculty of Science, Department of Geophysics, University of Zagreb, for supporting its operation and INTERMAGNET for promoting high standards in magnetic observatory practice (www.intermagnet.org).

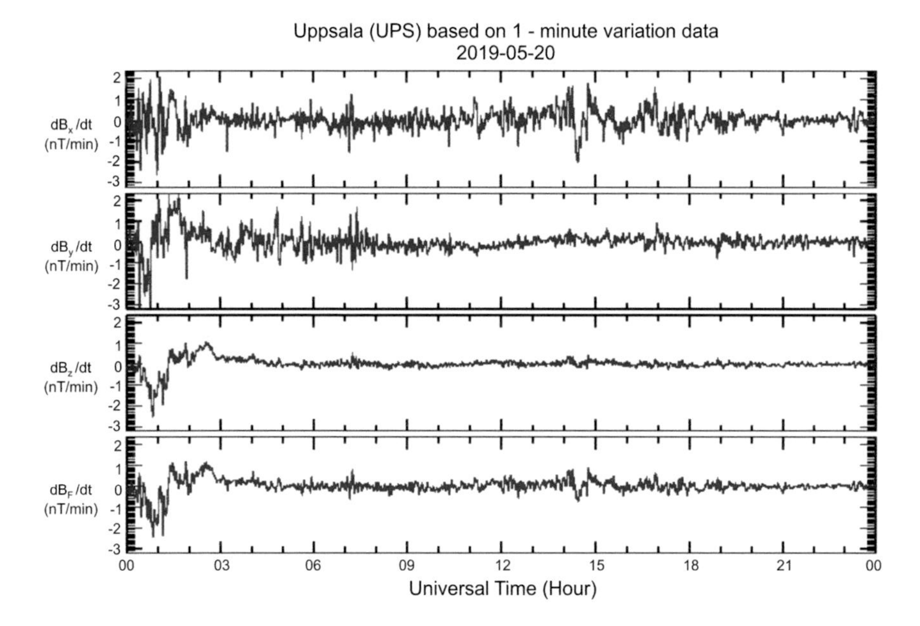

FIGURE 7.16 Magnetic field data measured in the observatory Uppsala (Fiby), Sweden, on May 20, 2019, representing a "calm day". We thank Sveriges geologiska undersökning (SGU, Geological Survey of Sweden) for supporting its operation and INTERMAGNET for promoting high standards in a magnetic observatory practice (www.intermagnet.org).

As another example, we show the short-term variations in the Earth's magnetic field for a calm day in Figure 7.16 and for a disturbed day on the same location in Figure 7.17. The data shown are ground observations, as reported by Richter (2010).

Magnitudes of variations in the x, y and z components of the Earth's magnetic fields during the "calm day" are in the range <10 nT, while in the "disturbed day", all the parts can frequently vary for more than 150 nT in a very short time.

7.3.3.1 Magnetic Shields – Zero Gauss Chamber

The commercial MuMETAL Zero Gauss Chambers are available from the company Magnetic Shield Corp (www.magnetic-shield.com). Ambient magnetic fields and the Earth's GMF (0.25–0.65 G) are attenuated to milligauss levels inside a multilayer chamber ($F_a \approx 1{,}500$). To achieve the lowest magnetic field levels within the chamber and to provide long-term stability and uniformity of the internal magnetic field, the periodic use of a degaussing coil is recommended. The field intensity could be changed with the use of three-axis Helmholtz coil. The required experiments should cover the field intensity downward from 50 μT to target 0.1 nT. The dimensions of the commercially available MuMETAL chamber are shown in Figure 7.18.

A three-axis Helmholtz coil assembly, a bipolar power supply (BOP) and a three-axis magnetometer can be used to dynamically cancel the Earth's magnetic field in a control volume at the center of the coil assembly. The magnetometer sensor located in the

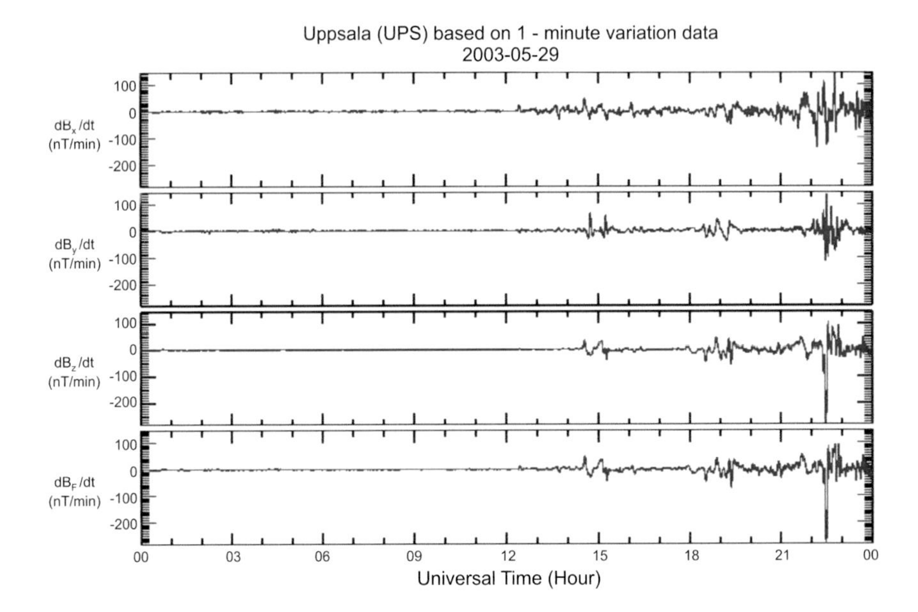

FIGURE 7.17 Magnetic field data as measured in observatory Uppsala (Fiby), Sweden, on May 20, 2019, representing a "disturbed day". We thank Sveriges geologiska undersökning (SGU, Geological Survey of Sweden) for supporting its operation and INTERMAGNET for promoting high standards in magnetic observatory practice (www.intermagnet.org).

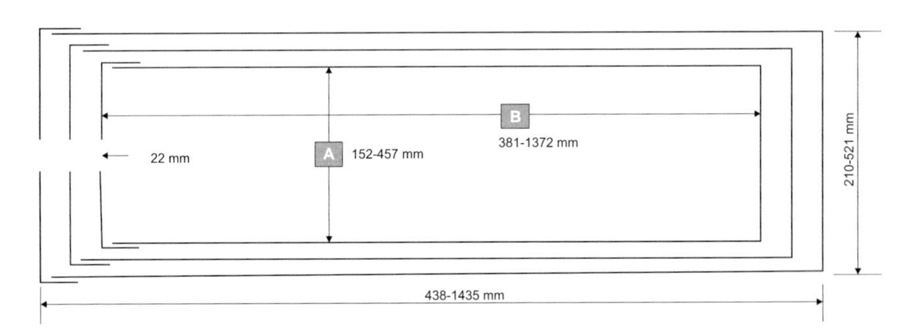

FIGURE 7.18 A schematic of the commercially available MuMETAL Zero Gauss Chamber (Magnetic Shield Corp – www.magnetic-shield.com).

Helmholtz coil assembly near the control volume could be used as a null detector in a negative feedback loop that drives the Helmholtz coils.

The sensor measures the difference between the Earth's field and the opposing Helmholtz coil field. This difference signal is applied to the input of the BOP that is driving the Helmholtz coil. If the difference signal indicates that the ambient field is greater than the Helmholtz coil field, the BOP increases the magnitude of the Helmholtz coil field. Likewise, if the difference signal indicates that the ambient field is less than the Helmholtz coil field, the BOP reduces the magnitude of the Helmholtz coil field. Eventually, a steady-state condition is reached where the difference between the two fields is tiny, in fact, near zero.

How close to zero the difference field becomes depends on the forward gain of the feedback loop. Such a system can usually attenuate the Earth's magnetic field from 50–60 μT (500–600 mG) to a few nT (1 nT = 0.01 mG = 1 gamma). This system automatically attenuates any changes in the ambient Earth's field as well as local disturbances that produce a uniform change in the control volume (MEDA, 1999).

Usually, the control volume is in the middle of the Helmholtz coil. The size of the control volume depends on the size of the Helmholtz coil and the desired zero-field error. A typical control volume for a 4 m (side length) square Helmholtz coil might be a 40 cm diameter sphere about its center.

The magnetometer electronics unit excites the sensor and processes its output. The magnetometer output is an analogue signal that is proportional to the field detected by the sensor. The output of the magnetometer drives the BOP, which, in turn, drives the Helmholtz coil. Such a system would attenuate the ambient field by a ratio of 1:30,000. In the presence of an ambient field of 60 μT, the field in the control volume would be on the order of 2 nT. Even though the sensor is offset from the control volume, the attenuation factor described above will be achieved. The Helmholtz coil field is relatively uniform over a large volume, so the difference between the field at the center of the coil and the sensor will be small. A bias signal can be added to the magnetometer sensor field to trim the control volume field to the desired zero-field level (MEDA, 1999).

The company Bartington Instruments Ltd. (https://www.bartington.com) has several interesting items on the market. Mu-metal cylindrical shield distorts the magnetic field around its layers, shielding the experimental setup from external magnetic fields. Open- and capped-end versions of this type of shields are produced, with the capped-end available in two diameters. They also have additional shielding at both ends of the sample chamber and are therefore less dependent on orientation. The open-end version of the shield is most efficient when the axis of the shield is aligned in a magnetic east–west direction. Both types will typically attenuate the Earth's field of ~50,000 nT to a level of ~1 nT, enabling the user to measure remanent magnetism in small components or errors in magnetic sensors.

The Bartington Instruments' HC1 Helmholtz Coil System is designed to generate stable and homogeneous magnetic fields in up to three axes. The system consists of HC1 Helmholtz Coils with a diameter of 500 mm, PA1 Power Amplifier and CU1 Control Unit, which interfaces with a PXI control system (or other compatible National Instruments acquisition card) that controls the operation of the system. Each pair of coils produces a magnetic field in one specific axis – X, Y or Z. The two-axis version is also available with any combination: X+Y, X+Z or Y+Z.

The two coils generate a homogeneous magnetic field, up to ±500 μT at DC and up to ±100 μT at 3 kHz. The coils produce a homogeneous field to 0.1% in a volume of 260 cm^3. A supporting table is provided to hold the experimental setup inside the homogeneity volume. The HC1 Helmholtz coils can also be purchased separately from the rest of the system.

The CU1 Control Unit that interfaces with the PXI control system allows the control of the system from a LabVIEW™-based software. At the same time, compensation potentiometers on the PA1 are used to cancel the local DC magnetic field.

Also, HC2 1m Helmholtz Coils, HC9 350mm Helmholtz Coils and HC16 350mm Helmholtz Coils are available. HC2 coil pairs generate a homogeneous magnetic field, up to ±250μT at DC and up to ±100μT at 300kHz. The 1m diameter HC2 allows for testing of big items, and it provides a field homogeneity to 1% in a volume of 14,000 cm^3 and uniformity to 0.1% in a volume of 2,500 cm^3. A table is provided to hold the test item inside the central homogeneous volume. HC9 coil pairs generate a homogeneous magnetic field, up to ±1μT at DC and at 440kHz and >±100μT at 5kHz. The HC9 allows for the testing of small sensors at higher fields and provides a field homogeneity to 1% in a volume of 1,150 cm^3 and uniformity to 0.1% in a volume of 175 cm^3. HC16 coil pairs generate a homogeneous magnetic field, up to ±1 mT at DC and at 440 Hz and >±100μT at 5kHz. The HC16 allows for the testing of small sensors at higher fields and provides a field homogeneity to 1% in a volume of 1,150 cm^3 and uniformity to 0.1% in a volume of 175 cm^3.

The price of Bartington system including three-layer magnetic shields with closed ends – 200mm, three-axis Helmholtz coil with a diameter of 500mm, PA1 Power Amplifier for three-axis Helmholtz coils, the control unit for Helmholtz coil system, CU2, PXI system for cui/hcs including DAQ card, Mag-13MS, three-axis magnetic field sensor, 13ms&mc cable 5m, i3ms 3-axis magnetic field sensor mounting, hcl raiser adaptor and spacers – is of the order of 55,000 Euros.

The CU2 is a closed-loop compensation module that will allow the system to automatically compensate for external variations of field sensed by the reference sensor included in the module (see http://www.bartington.com/presentation/helmholtz-coil-control-system), which should be positioned close to the center of the coils. Additionally, the calibration of the CU2 would also compensate for the background DC field as well as any orthogonality errors, which would usually be done on the PA1. The CU2 option dramatically simplifies the setup and the use of the equipment and improve the field stability (also attenuate 50 Hz main field variations). If one were to use the equipment to calibrate other sensors, the only requirement would be to ensure that the axes of the sensor and that of the reference magnetometer are as perfectly aligned as possible.

Here, we shall describe the initial testing of the experimental configuration with a homemade pilot setup (early model), as shown in Figure 7.19.

7.3.4 Possible Effects of Magnetic Field on Concentration Factor

Living cells possess electric charges produced by ions or free radicals, which act as internal magnets. These internal magnets can be affected by outside magnetic fields, which can orient unpaired electrons. It has been established that external magnetic fields influence both activation of ions and polarization of dipoles in living cells (Goodman et al., 1995; Belyavskaya, 2004). It is assumed that treatments with magnetic field enhance seed potency, influencing the biochemical processes with free radicals and stimulating the

FIGURE 7.19 Pilot Helmholtz coil assembly on the top of one of the three power supplies.

activity of proteins and enzymes (Dhawi et al., 2009). Fesenko et al. (2010) studied the effect of "zero" magnetic field (0.2 μT) on early embryogenesis in mice and found several abnormalities. The effects of "zero" magnetic field have also been studied in pathogen bacteria by Creanga et al. (2004).

The Earth's dipole has decreased in strength by about 10% since C. F. Gauss measured it in 1840. According to the British Geological Survey (2018), the present trend in the strength of the magnetic field would suggest zero dipole moment in about 1,500–1,600 years. Contrary opinions are stating that although decreasing rapidly, the Earth's magnetic field is probably not now reversing but only oscillating (Cox, 1969).

Possible exogenetic (cosmic) factors associated with the reversal of axial geomagnetic dipole have been discussed by Nagi and Tiwari (1985). Their results question the widely accepted theory of randomness of geomagnetic reversal for a long period sequence. Their results postulate long-term cyclicity with reversal periods of 285, 114, 64(71), 47 and 34(32) million years with distinct resolution. The maximum spectral power is found for the cosmic year (285 million years) term: the period of a complete revolution of the solar system around the Milky Way galactic center. The other reversal periods correspond to different properties of the solar system galactocentric radial motion.

On the cosmic scale, the average equipartition strength of the total fields for a sample of 74 spiral galaxies is B=9±2 μG. The average field strength of 21 bright galaxies observed since 2000 is B=17±3 μG. Dwarf galaxies host fields of similar strength as spirals if their star formation rate per volume is similarly high. Blue compact dwarf galaxies are radio-bright with equipartition field strengths of 10–20 μG. Spirals with moderate star-forming activity and moderate radio surface brightness such as M31 and M33, our Milky Way's neighbors, have B≈6 μG. In "grand-design" galaxies with massive star formation, such as M51, M83 and NGC 6946, B≈15 μG is a typical average strength of the total field (Beck and Wielebinski, 2013).

7.3.5 Boron Determination – A Review of Analytical Methods

A comprehensive review of the published methods of sample preparation, determinant purification and determination of boron concentration and isotope composition in a sample is published by Sah and Brown (1997). The most usual methods for the determination of B concentration are spectrophotometric and plasma source spectrometric methods. Most of the used spectrophotometric methods are based on colorimetric reactions of B with azomethine-H, curcumin or carmine. These methods suffer from many interferences and have low sensitivity and precision. The application of nuclear reaction and atomic emission/absorption spectrometric (AES/AAS) methods has remained limited because these methods have reduced sensitivity and suffer from severe memory effects and interferences. There is a large number of published nuclear reaction methods; however, only prompt γ-ray spectrometry has been of practical use. The prompt γ-ray process can determine B concentration in original samples, which makes this method especially useful for medical applications, including boron neutron-capture therapy. However, this is a time-consuming method and not suitable for the detection of low levels of B. The inductively coupled plasma optical emission spectrometry (ICP-OES) represents a new dimension in B determination because of its simplicity, sensitivity and multielement capability. On the other hand, it suffers interferences and is not adequately sensitive for some nutritional and medical applications involving animal tissues that are naturally low in B. All methods involving the measurement of B isotopic composition require a mass spectrometer. Thermal ionization mass spectrometry (TIMS) and secondary ion mass spectrometry (SIMS) have been used to measure the isotopic composition of B; however, these methods are time-consuming and require extensive sample preparation and purification. With the development of inductively coupled plasma mass spectrometry (ICP-MS), most of the drawbacks of earlier methods were overcome, and the capability of measuring B isotopes made the following possible: (i) B concentration determination by isotope dilution, (ii) verification of B concentration by isotope fingerprinting in routine analysis and (iii) determination of total B concentration and B isotope ratio for biological tracer studies in the same run. Therefore, the plasma source MS appears to be the method of choice among the present-day technologies.

Both the ICP-MS and ICP-OES methods can be used for the determination of the boron concentration in water (Kmiecik et al., 2016). The best analytical methods for determining boron in biological materials are inductively coupled plasma atomic emission spectroscopy (ICP-AES), ICP-MS and colorimetric carmine method (see ATSDR, 2009). The most common analytical procedure to analyze boron in biological materials involves digestion of the sample in hot acid or base, followed by the analysis of the resulting solution by ICP-AES.

Several nuclear reaction analysis (NRA) methods have been reported for boron measurement. Some of these methods may be of merely academic significance, having little practical value for routine B determination. All these methods involve the bombardment of B nuclei and the measurement of the reaction product(s). For convenience, the reported NRA methods are divided into two classes: (i) neutron activation analysis (NAA) and (ii) other NRA methods (Sah and Brown, 1997).

In NAA, the sample is bombarded with neutrons, the elements of interest are made radioactive, and the quantity of the element is determined by measuring the radioactivity or radioactive decay products. NAA is a nondestructive method capable of handling solid samples with multielement detection capability and generally low detection limits. However, it is not suitable for sample mass or liquid volumes that pose a threat of radioactive leaks after activation.

The nuclear methods for B determination were reviewed in a paper by Pillay and Peisach (1992). The measurements by the NAA methods require access to a nuclear reactor for the production of thermal neutrons for the bombardment to convert isotope(s) of interest in a sample to radioisotopes. Boron is an exception to this strategy. The activation of ^{10}B by an incident beam of thermal neutrons does not make it radioactive, but causes the following neutron-capture reaction:

$$^{10}\mathrm{B} + \text{neutron} \rightarrow {}^{7}\mathrm{Li} + \alpha\ (2.31\ \mathrm{MeV}) + \gamma\ (478\ \mathrm{KeV}).$$

This reaction involves only the ^{10}B isotope, which has approximately 20% abundance in naturally occurring B. All NAA methods for B determination are based on the measurement of one or more products (α-particles and γ-photons) of this reaction. Perhaps the essential method based on the measurement of α-particles is neutron activation mass spectrometry (NA-MS), while that based on the measurement of gamma rays is prompt γ-ray spectrometry.

NA-MS applied for the simultaneous determination of Li and B in biological materials is described by Iyengar et al. (1990) and Clarke et al. (1987). The sample was placed in an ultrapure polyethylene "liner" and freeze-dried. The liners containing the freeze-dried samples were placed in lead containers and evacuated to about 10^{-5} Pa. The lead vessels were pinched-sealed following evacuation for neutron irradiation. A static mass spectrometer was used to measure ^{4}He (from ^{10}B) and ^{3}He (from ^{6}Li) generated by the NA reaction. The error rate at $\geq$1 ppm B concentration was 1%–5%, but at 6 ppb B, the error rate increased to 75%. This sensitivity is not adequate for BNCT and some nutritional and environmental applications.

The α-track etching, also called neutron-capture radiography, is generally used to determine the microscopic distribution of the ^{10}B isotope in tissues. The sample containing ^{10}B is placed in contact with a detector film and is irradiated with neutrons. Following irradiation, the film is stained, reversed and etched with KOH or NaOH. This technique has been used for mapping the distribution of natural B in histological sections of mouse tissue (Laurent-Pettersson et al., 1992) and in parenchyma cells of clover leaves (Martini and Thellier, 1993). The quantification of B is possible using an image analyzer.

Neutron depth profiling (NDP) is the method used for the near-surface analysis of isotopes that undergo neutron-induced positive Q-value charged particle reactions such as $^{10}B(n, \alpha)^{7}Li$ for B determination, where ^{10}B is the target isotope, n is the neutron as an irradiation source and α-particles and ^{7}Li are the products of the reaction. Lamaze et al. (1993) used the NDP method to measure B in CVD diamond surfaces. The samples were irradiated with cold neutrons, and the resulting particles escaping from the surface of the sample were detected with a silicon surface barrier detector.

Methods based on the measurement of prompt γ-rays are extensively used for the measurement of ^{10}B (Pillay and Peisach, 1992; Matsumoto and Aizawa, 1990; Matsumoto et al., 1991). The gamma ray emitted from the disintegrating ^{10}B nuclei due to the action of the neutron is detected. This method is also nondestructive; however, it is not sensitive for the detection of low B levels (generally, <5 μg/g) in the sample. As the B concentration in the sample decreases, the counting time to achieve desired precision increases logarithmically. For example, the counting time necessary to achieve 1% precision was 10 hours for a sample containing 1 ppm B and 50 hours for a sample containing 0.5 ppm B.

In the eventual future experiments, the counter telescope (or some parts of it) could be used for B concentration and concentration distribution measurements. All the proportional counters would be filled with CO_2 gas at the pressure of 10 cm Hg. The role of dE/dx gas proportional counters is to measure the energy loss of the detected particles and to define their direction. In the series of possible future experiments with cyanobacteria, it is proposed to follow the experimental procedure used by Bonilla et al. (1990) in the study of boron requirement in cyanobacteria. Here we present their methods.

The sources and properties of the cyanobacteria strains used in their research are listed in Table 7.2. The *Chlorogloeopsis* PCC 6912 and *Gloeothece* PCC 6501 were received from the Department of Biochemistry (University of Seville, Spain). The *Plectonema calothricoides*, strain No. 1463-4, was from the Göttingen University Algal Culture Collection. The cyanobacteria *Nostoc* UAM 205 and Nodularia Ml have recently been isolated. *Nostoc* sp. was isolated in soil samples from a rice field in Valencia, Spain, and *Nodularia* sp. from a mountain stream in the north of Madrid, Spain.

In the work presented by Bellack (2011), cultures of aerobic dinitrogen-fixing cyanobacteria were grown in a medium without combined nitrogen, placed in 1 L polyethylene bottles at a temperature of 26°C and in the following conditions:

TABLE 7.2 Some Properties of the Cyanobacteria Used in Experiments by Bonilla et al. (1990)

Cyanobacteria	Collection	Organism Structure	Nitrogen Fixation	Heterocysts Formed
Gloeothece sp.	PCC 6501	Unicellular	Aerobic	–
P. calothricoides	Göttingen No. 1463-4	Filamentous	Anaerobic	–
Nodularia sp.	M1	Filamentous	Aerobic	+
Nostoc sp.	UAM 205	Filamentous	Aerobic	+
Chlorogloeopsis sp.	PCC 6912	Filamentous	Aerobic	+

The *Chlorogloeopsis* sp. cultures were grown with continuous air bubbling under a constant light intensity of 90 μE/m²/s.

The *Nostoc* sp. and *Nodularia* sp. cultures were grown with continuous air bubbling under a constant light intensity of 30 μE/m²/s.

The *Gloeothece* sp. cultures were grown without air bubbling under a constant light intensity of 3 μE/m²/s.

For the normal growth of *P. calothricoides*, stock cultures were kept in a nitrate medium (Kratz and Myers, 1955) and bubbled with air under a constant light intensity of 90 μE/m²/s. For dinitrogen fixation experiments, cultures were transferred to a medium lacking combined nitrogen, as described above. Media without a combined nitrogen source and boron-free cells should be prepared and processed as described by Mateo et al. (1986).

It is proposed to perform experiments on the effects of magnetic field on concentration factors in some Archaea species. The procedures described in the dissertation of Bellack (2011) should be followed. We shall describe some of the elements involved. Of our interest should be the composition of media used in her research.

In the work by Bonilla et al. (1990), chemicals were purchased from Merck. Culture density was determined at 600 nm by a Hitachi 150-20 spectrophotometer. For dry weight determination, cells were collected on 0.45 μm filters, washed and dried at 70°C for 24 hours.

Nitrogenase activity was determined by acetylene reduction and carried out as described in Mateo et al. (1986). Ethylene production was estimated by injection into a Shimadzu GC-8A gas chromatograph.

The anaerobic induction of nitrogenase activity in *P. calothricoides* was performed as follows. Cells produced during the exponential growth phase were harvested, washed with sterile material and resuspended in a fresh nitrate-free sterile medium. Cultures were grown in polyethylene bottles sealed with rubber stoppers and flushed for 10 minutes with argon before injecting 5% CO_2 into the flask. The assay was started by injecting acetylene to a 10% concentration. The bottles were maintained at 26°C under cycles of 16 hours light and 8 hours dark. Nitrogenase activity was determined by using 0.5 mL gas samples removed with a syringe at the indicated time, and their ethylene content was measured.

The examination of living filaments by light microscopy was conducted with preparations mounted in water. Micrographs were taken with Olympus BH-2. For transmission electron microscopy, cells were fixed with glutaraldehyde (2%) in 0.1 M phosphate buffer, pH 7.2, and post-fixed with OsO_4 (2%). Dehydration was carried out with water–acetone solutions. Samples embedded in Vestopal were sectioned with a diamond knife, stained with uranyl acetate and observed in a Philips EM 300 transmission electron microscope (Buchachenko et al., 2008).

7.3.6 Modern Magnetic Field Measurement Devices

Articles by Richter (2017) and Tumanski (2013) give a compact overview of a representative choice of modern magnetometers and describe their specific characteristics. An overview of the noise of different kinds of magnetometers is presented by Richter

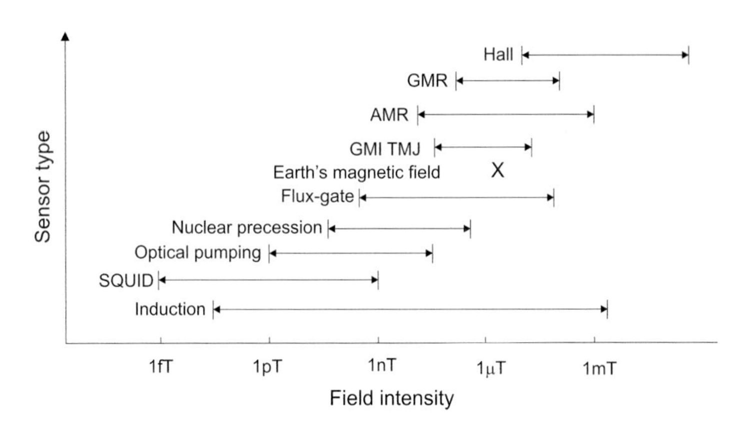

FIGURE 7.20 Field intensity regions of applicability of different sensors. (AMR, anisotropic magnetoresistive sensor; GMI, giant magnetoimpedance sensor; Hall, Hall sensor; TMJ, tunnel magnetoresistance.)

(2017). Typical areas of application of the primary magnetic field sensors are shown in Figure 7.20 (after Tumanski, 2013).

7.3.7 Experimental Setup

Here we shall describe a preliminary experiment designed to measure the effects of varying small magnetic fields on *Bacillus subtilis* subsp. *subtilis* DSMZ No. 10 obtained from Deutsche Sammlung von Mikroorganismen und Zellkulturen GmbH. The bacterium was pregrown overnight (16 hours/35°C) at nutrient agar plates (Biolife, Italy). The biomass was suspended in autoclaved nutrient broth (Biolife, Italy) prepared from the same bottle of the dehydrated medium. The suspension was divided into sterilized glass flasks. One flask was exposed to the magnetic field, while the other flask served as a control.

The number of total bacteria and the number of spores were determined in triplicate at the beginning and end of each experiment. One milliliter of bacterial suspension was serially diluted in sterile saline, and 0.1 mL from dilutions was inoculated onto nutrient agar plates. Plates were incubated (24 hours/35°C) following the counting of bacterial colonies (CFUs – colony-forming units). Several total viable bacteria were expressed as CFU/mL of the sample. The rest of the sample was pasteurized (15 minutes/80°C), and the above-mentioned procedure was repeated to calculate the number of spores in a sample. The initial number of *B. subtilis* in all experiments averaged $1.5 \pm 0.7 \times 10^6$ CFU/mL with 0.5‰ of spores.

A prepared flask with *B. subtilis* in nutrient solutions was placed in the center of three orthogonal pairs of Helmholtz coils, as shown in Figure 7.21.

The whole setup was placed within a wooden box covered inside with sheets of mu-metal, nickel permalloy, magnetic shielding foil, mu-metal 0.1 T. This arrangement allowed more comfortable control of the Helmholtz coils-produced compensations of the Earth's magnetic field components. The three components (x, y and z) of the resulting

FIGURE 7.21 Details of the experimental setup. (a) Helmholtz coils – front view. (b) Helmholtz coils – side view. (c) Power supplies for coils and voltmeters for magnetometers. (d) Flask with *B. subtilis* – side view. (e) Flask with *B. subtilis* – top view. (f) Box containing experimental arrangement, outside the small box for control flask.

magnetic field in the center of the setup were measured by three miniature fluxgate sensors powered by batteries placed outside of the hood containing the experimental setup in the wooden box. Magnetic field components were measured by recording the voltages induced in fluxgate sensors (with 1 mV corresponding to 50 nT). Each of the eight experiments was made for a total of 24 hours. During this period, the magnetic field in the central position has varied, and these variations have defined the region of the magnetic field values for the particular experiment. At the beginning and the end of the experiments, the following parameters were determined: the total number of viable bacteria in control and probe, number of spores, rate of bacterial multiplication and concentrations of K using EDXRF as an analytical tool.

The aim of the experiments performed by the author of this book and his collaborators (Mr. Robert Kollar, Prof. Jasna Hrenović and Dr. Jasmina Obhođaš) was to establish the existence of direct magnetic field effects on the growth of microorganisms. As a test organism, the bacterium *Bacillus subtilis* was chosen. *B. subtilis* is one of the best characterized bacteria and is used as a model organism for gram-positive bacteria. *B. subtilis* is a gram-positive rod-shaped facultatively aerobic bacterium, which is ubiquitous within the biosphere (Earl et al., 2008). *B. subtilis* produces endospores that allow its survival on extreme environmental conditions, including heat and desiccation. In the soil, the natural environment of *B. subtilis*, the bacterium continuously encounters various changing environmental conditions, including drastic differences in oxygen tension.

Bacillus subtilis was grown in a glass flask placed in the center of three Helmholtz coils for 24 hours. The control flask was placed in the same time interval outside the controlled area in an identical flask in the same medium. Experiments performed until now were all done in the mT region. We have performed eight experiments in the nT region. For each experiment, the values of average magnetic field intensity were $H_{average}/nT = 97.5$, 174.7, 228, 286, 309.2, 366.4, 687 and 1,115. This is shown in Figure 7.22 and Table 7.3.

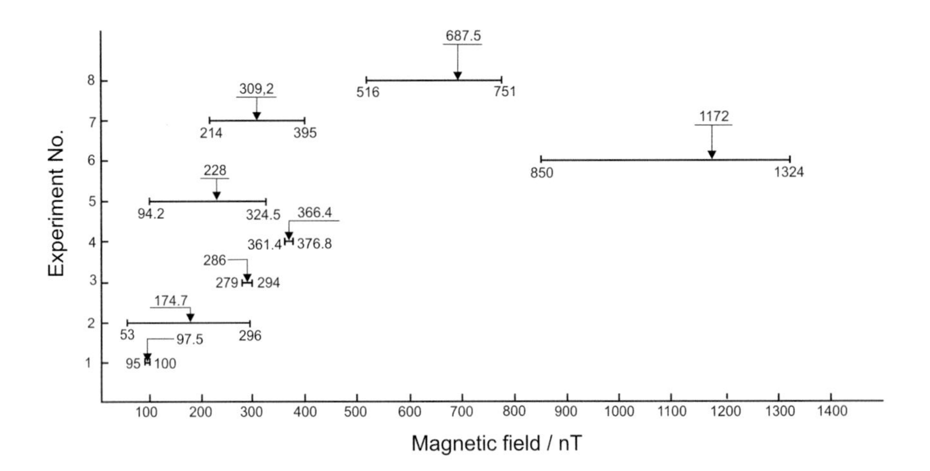

FIGURE 7.22 Eight experiments performed: For each experiment, the range and medium magnetic field intensity are indicated.

TABLE 7.3 Experiments on the Effect of Magnetic Field on *Bacillus subtilis* Arranged by Field Intensity

Experiment	Time Interval	$H_{initial}$/nT	$H_{control}$/nT	H_{final}/nT	H_{range}/nT	$H_{average}$/nT	H_{Earth}/nT Station
#1	06/07.11.2017. 10:30-10:35(+1)	95	-	100	95–100	97.5	47 901±37
#2	07/08.11.2017. 10:55-10:55(+1)	53	175	296	53–296	174.7	47 889±158
#5	24/25.01.2018. 13:30-13:30(+1)	94	303	324.5	94–325	228	47 913±65
#3	20/21.11.2017. 10:30-9:44(+1)	285	279	294	279–294	286	47 900±92
#7	20/21.02.2018. 10:24-10:20(+1)	214	395 358	270	214–394	309.2	47 912±38
#4	21/22.11.2017. 10:53-10:53(+1)	361	377	361	361–377	366.4	47 900±62
#8	21/22.02.2018 10:25-10:25(+1)	707,3	718.4; 744.5; 750.9	516,5	516.5–750.9	687.5	47 915±35
#6	30/31.01.2018 9:30-9:06(+1)	850	1.324	1.172	850–1.324	1.115	47 912±60

The information on the values of the Earth's magnetic field intensities during the period of measurements was obtained through INTERMAGNET – the global network of observatories monitoring the Earth's magnetic field. At their web site, one can achieve close to real-time data and information from several geomagnetic observatories around the world. We used the data provided by the station located in Lonjsko Polje, Croatia, N=44.5919° E=16.6592° (LON), operated by Faculty of Science, Department of Geophysics, University of Zagreb.

7.3.7.1 Growth of *Bacillus subtilis* in Reduced Magnetic Field Intensity

The results of the eight experiments performed are presented next. Figure 7.23 shows the ratio of the number of spores in probe and control as a function of average magnetic field at probe location. Vertical error bars are counting statistics, while the horizontal error bars represent the magnetic field intensity range during the experiment.

Experimental data shown in Figure 7.23 indicate the reduction in the number of spores in the probe sample with the decrease in magnetic field intensity in comparison with the control sample. The ratio comes to value 1 for a magnetic field intensity of 600 nT and higher. In other words, *B. subtilis* does not recognize the magnetic fields below 400 nT as an unfriendly extreme environmental condition; contrary to that, it seems that *B. subtilis* favors these conditions. The same is also seen in Figure 7.24, showing *B. subtilis* multiplication in 24 hours as a function of average magnetic field at probe location. Vertical error bars are counting statistics, while the horizontal error bars represent the magnetic field intensity range during the experiment. Open circles are multiplication values for experiments, while black circles correspond to controls.

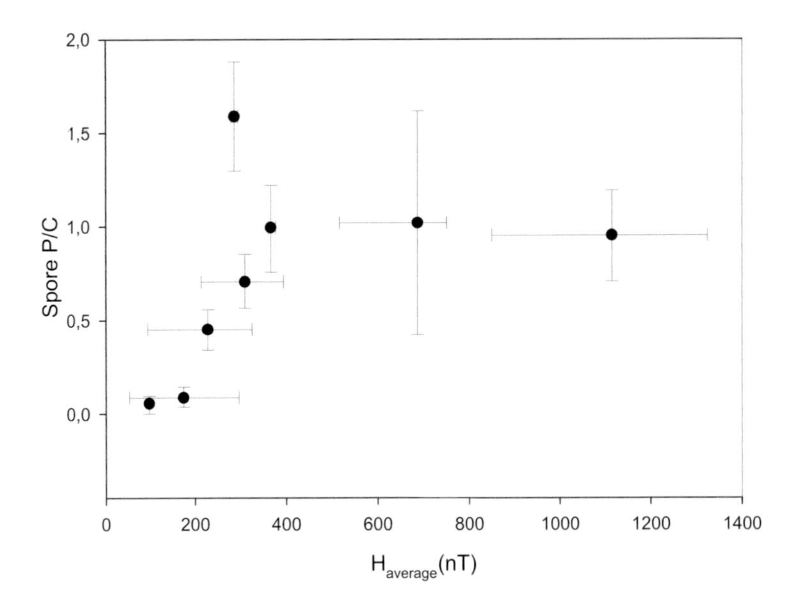

FIGURE 7.23 Ratio of the number of *Bacillus subtilis* spores in probe (P) and control (C) as a function of average magnetic field at probe location. Vertical error bars are counting statistics, while the horizontal error bars represent the magnetic field intensity range during the experiment.

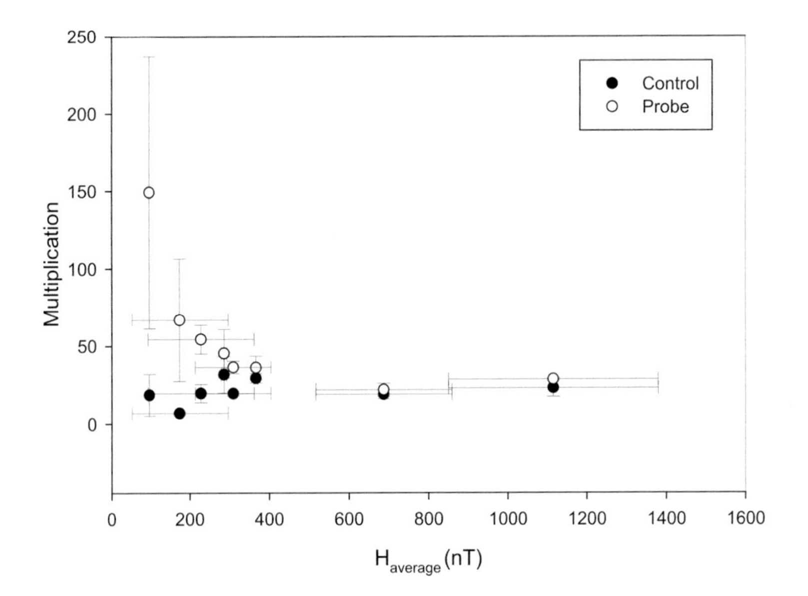

FIGURE 7.24 Vertical error bars are counting statistics, while the horizontal error bars represent the magnetic field intensity range during the experiment. Open circles represent experiment; black circles represent controls. The rate of bacterial multiplication was calculated as CFU/mL after the experiment divided by CFU/mL at the beginning of the experiment.

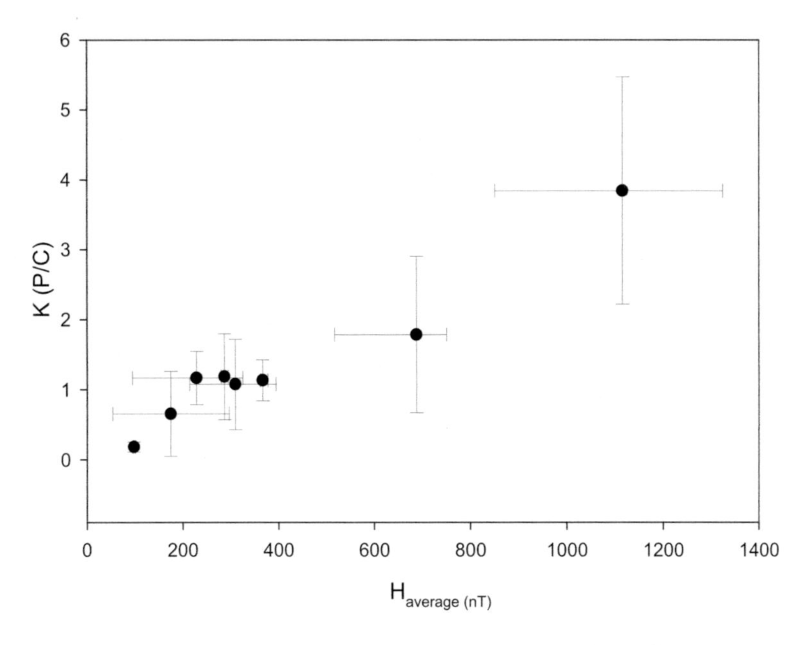

FIGURE 7.25 K concentration ratios (K in probe vs. K in control) as a function of magnetic field at probe location. Vertical error bars are counting statistics, while the horizontal error bars represent the magnetic field intensity range during the experiment.

Chemical element concentration measurements were attempted for some elements (Zn, Cu and Fe) by using EDXRF as an analytical method. Unfortunately, the attempt failed because of the low count rate when filtered probe material was exposed to X-rays. The K ratio values (P/C) were obtained and are shown in Figure 7.25 as a function of magnetic field intensity at probe location. Vertical error bars shown are counting statistics, while the horizontal error bars represent the magnetic field intensity range during the experiment. Experimental data show an increase in the K(P/C) ratio with the increase in the $H_{average}$ value.

7.3.7.2 Discussion and Speculations

We can expect that during the coming magnetic field reversal, the living species will be exposed to much smaller magnetic field intensity (in the nT–μT range) than the present one. The assumption is that during polarity reversals, the dipole component of the GMF is significantly reduced for periods of a few thousand years. Even this reduced magnetic field will show deviations around its average value of the same relative order of magnitude.

Also, the reported findings could have some consequences for planned man voyages into space, an environment with reduced magnetic field intensity. Namely, both the positive and negative aspects of the relationship between humans and bacteria should be considered. There are three different symbiotic relationships in which humans and bacteria coexist. The types of symbiosis are termed commensalism, mutualism and parasitism.

Each of the mentioned types will be affected by the observed effect of a weak magnetic field in addition to its possible effect on the human cell structure itself. Hong et al. (2009) reported the isolation and characterization of *B. subtilis* from the human gastrointestinal tract. Their study shows that *B. subtilis* has adapted to life within the digestive tract and should be considered gut commensal rather than solely soil microorganism.

Lefevre et al. (2017) have recently described *B. subtilis* CU1 as a probiotic strain with beneficial effects on immune health in elderly human subjects. Their work describes *Bacillus subtilis* CU1 as safe and well tolerated in the clinical subjects without undesirable physiological effects, liver and kidney function markers, complete blood counts, hemodynamic parameters and vital signs. Additionally, Ayala et al. (2017) found that *B. subtilis* significantly extended the lifespan of the model organism *Caenorhabditis elegans* (a worm). The prolongevity effect of *B. subtilis* was primarily due to a reduction in the insulin-like signaling, which is critical for the healthy longevity of human centenarians. These findings open the possibility to study if the observed effects of low magnetic field on *B. subtilis* could affect the extension of human life expectancy and dynamics of age-related diseases.

The experiments performed by Ramon et al. (1987) showed an increase in the growth of *B. subtilis* mutant strain FJ7 above controls, by exposing the bacterial culture to the magnetic field strength between 0.8 and 2.5 mT. Also, light microscopy and scanning electron microscopy demonstrated the morphology of controls to grow in microfiber of right-handed helix formation. In contrast, the field-exposed group showed little to no cohesion; the cells appeared to be homogeneously distributed throughout the sample. Their results suggest that the growth patterns of *B. subtilis* can be altered as a result of magnetic-field-induced effects.

Using the available EDXRF setup and the amount of material available for analysis, it was possible to determine only the concentrations of K in both control and probe samples. Potassium, K, is an essential element whose essentiality was established during the origin of life process based on its availability in the environment at that place and time (McClendon, 1976; Valković, 1990).

The essentiality of an element is defined by the dose–response curve, as shown in Figure 7.26.

The availability of element K was identical for the control and probes under all values of the magnetic field. Therefore, the only factor which could influence the P/C concentration ratios in experiments performed is the concentration factor dependence on magnetic field intensity, i.e., $F_E(H)$, which is defined as

$$F_K(H) = C_{K,probe} / C_{K,medium}$$

where $C_{K,\ probe}$ is the concentration of element K in probe microorganism, while $C_{K,\ medium}$ is its concentration in the growth medium. The same relation is valid for controls; therefore,

$$C_{K,\ probe} / C_{K,\ control} = F_K(H) \times C_{K,\ medium} / F_K(H_{Earth}) \times C_{K,\ medium} = F_K(H) \ / F_K(H_{Earth})$$

$$F_K(H) = \left(C_{K,\ probe} / C_{K,\ control}\right) \times F_K(H_{Earth})$$

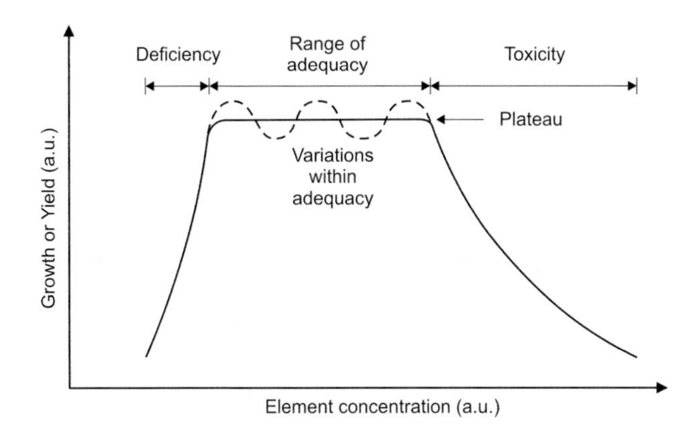

FIGURE 7.26 Dose–response curve for an essential element.

It is assumed that the value of $F_K(H_{Earth})$ has such value that concentrations of K, the investigated element, $C_{K,\ probe}$, is in the range of adequacy. The deviations from these values could bring the investigated species in a toxic or deficiency range.

Considering the rise in *B. subtilis* multiplication in 24 hours as a function of average magnetic field at probe location for $H<400$ nT, we could postulate that within this H-region, the values of $C_{K,\ probe}$ are in the range of adequacy. Chemical element K probably plays an important role in the process of spore formation; potassium content during growth and sporulation in *B. subtilis* has been first discussed by Eisenstadt (1972). The K concentration ratio, as shown in Figure 7.26, increases with the increase in the magnetic field intensity at probe location. This fact reflects the number of spores formed for these magnetic field intensity values, as shown in Figure 7.24.

Future experiments should be done with improved analytical techniques to enlarge the number of chemical elements measured. The variations in the Earth's magnetic field intensity will be compensated for to reduce the range of H-value in the experiments. For very low magnetic field intensity, small ΔH measurements should be done in search of the existence of possible resonance structure in $F_B(H)$ for chemical element boron, which could have significant consequences on the location of the origin of life environment.

References

Adamczyk, L., Agakishiev, G., Aggarwal, M. M., et al. 2013. Single spin asymmetry A_N in polarized proton-proton elastic scattering at $\sqrt{s}$=200 GeV. *Physics Letters B* 719:62–69.

Akimune, H., Fujimura, M. Fujiwara, M., et al. 2001. Evidence for a 3.8 MeV state in ^{9}B, *Physical Review C* 64:1.

Anderson, J. D., Wong, C., Pohl, B. A. and McClure, J. W. 1970. Fast neutron spectroscopy of the reaction ^{9}Be(p, n)^{9}B at 20 MeV. *Physical Review C* 2:319.

Arena, N., Cavallaro, S., Fazio, G. GIardina, G., ItalIano, A. and Mezzanares, F. 1988. Energy and width measurement of the ^{9}B first excited state observed by the ^{10}B(^{3}He, α)^{9}Be and ^{10}Be(^{3}He, α) ^{9}B(α) ^{5}Li reactions. *Europhysics Letters* 6:517.

ATSDR, The Agency for Toxic Substances and Disease Registry. 2009. Toxicological profiles for boron, pp. 59–65. https://www.atsdr.cdc.gov/toxprofiles/tp26-c7.pdf.

Audi, G., Wapstra, A. H. and Thibault, C. 2003. The AME 2003 atomic mass evaluation: (II). Tables, graphs and references. *Nuclear Physics A* 729:337–767.

Ayala, F. R., Bauman, C., Cogliati, S., Lenini, C., Bartolini, M. and Graul, R. 2017. Microbial flora, probiotics, *Bacillus subtilis* and the search for a long and healthy human longevity. *Microbial Cell* 4(4):133–136.

Baldwin, T. D., Catford, W. N., Mahboub, D. A., et al. 2012. First excited 12+ state in 9 B, *Physical Review C* 86:0343301.

Bartington Instruments Limited. 10 Thorney Leys Business Park, Witney, Oxon, OX28 4GG, England. (t: +44(0)1993 706565; e-mail: sales@bartington.com.

Batty, C. J., Gilmore, R. S. and Stafford, G. H. 1963. Measurement of the polarization in proton-proton scattering at 30 and 50 MeV. *Nuclear Physics* 45:481–491.

Bauer, R. W., Anderson, J. D. and Wong, C. 1964. A search for an excited state of ^{9}B near 1.7 MeV, *Nuclear Physics* 56:117.

Beck, R. and Wielebinski, R. 2013. *Magnetic Fields in Galaxies, Vol. 5, Chapter 13 in Planets, Stars and Stellar Systems.* Third updated version. Springer, Dordrecht.

Bellack, A. 2011. *Cell Architecture and Flagella of hyperthermophilic Archaea*, Dissertation. Fakultat fur Biologie und Vorklinische medizin der Universitat Regensburg, Germany.

Belyavskaya, N. A. 2004. Biological effects due to weak magnetic field on plants. *Advances in Space Research* 34:1566–1574.

Blackmond, D. G. 2010. *The Origin of Biological Homochirality. Cold Spring Harbour Perspectives in Biology* Deamer, D. and Szostak, J. W. (Eds.) Vol. 2, p. a002147. Cold Spring Harbor Laboratory Press, Cold Spring Harbor, NY, USA.

Bonilla, I., Garcia-González, M. and Mateo, P. 1990. Boron requirement in cyanobacteria: Its possible role in the early evolution of photosynthetic organisms. *Plant Physiology* 94(4):1554–1560.

Borlina, C. S., Weiss, B. P., Lima, E. A., et al. 2020. Reevaluating the evidence for a Hadean-Eoarchean dynamo. *Science Advances* 6:eaav9634.

British Geological Survey. 2018. Geomagnetism. Magnetic reversals. http://www.geomag.bgs.ac.uk/education/reversals.html.

Broggini, C., Canton, L., Fiorentini, G. and Villante, F. L. 2012. The cosmological ^{7}Li problem from a nuclear physics perspective. *Journal of Cosmology and Astroparticle Physics* 06:1–16. Doi:10.1088/1475-7516/2012/06/030.

Buchachenko, A. L. and Kuznetsov, D. A. 2008. Magnetic field affects enzymatic ATP synthesis. *Journal of the American Chemical Society* 130:12868–12869.

Buchachenko, A. L., Kuznetsov, D. A. and Breslavskaya, N. N. 2010. Ion-radical mechanism of enzymatic ATP synthesis: DFT calculations and experimental control. *The Journal of Physical Chemistry B* 114:2287–2292.

Buchachenko, A. L., Kouznetsov, D. A., Breslavskaya, N. N. and Orlova, M. A. 2008. Magnesium isotope effects in enzymatic phosphorylation. *The Journal of Physical Chemistry B* 112:2548–2556.

Buchachenko, A. L., Kouznetsov, D. A., Orlova, M. A. and Markarian, A. A. 2005. Magnetic isotope effect of magnesium in phosphoglycerate kinase phosphorylation. *Proceedings of the National Academy of Sciences of the United States of America* 102:10793–10796.

Burlein, M., Fortune, H. T., Kutt, P. H. and Gilman, R. 1988. Energies and widths of states in ^{9}B. *Physical Review C* 38:2078.

Byrd, R. C., Floyd, C. E., Guss, P. P., Murphy, K. and Walter, R. L. 1983. Cross-section measurement and Lane model analysis for the ^{9}Be(p, n) ^{9}B Reaction. *Nuclear Physics A* 399:94.

Cameron, A., Colgate, S. and Grossman, L. 1973. Cosmic Abundance of Boron. *Nature* 243:204–207.

Catford, W. N., Fifield, L. K., Reber, E. L., Kemper, K. W. and Brown, J. D. 1992. The missing 1/2+ mirror state in ^{9}B. *Nuclear Physics A* 550:517.

Chou, J. C., Fou, C. Lin, C., Song, P., Wen, M. and Liu, Y. 1978. Investigation of ^{9}B and ^{9}Be levels in a kinematically complete ^{9}Be(p, pn)^{8}Be experiment. *Journal of the Physical Society of Japan* 44:1–6.

Clarke, W. B., Koekebakker, M., Barr, R. D., Downing, R. G. and Fleming, R. F. Analysis of ultratrace lithium and boron by neutron activation and mass-spectrometric measurement of ^{3}He and ^{4}He. 1987. *International Journal of Radiation Applications and Instrumentation. Part A. Applied Radiation and Isotopes* 38:735–747.

Coe, R. S. and Glen, J. M. G. 2004. The complexity of reversals. http://geomaps.wr.usgs.gov/gump/people/jglen/pdfs/Coe&Glen_Chapman.

Coe, R. S., Hongre, L. and Glatzmaier, G. A. 2000. An examination of simulated geomagnetic reversals from a palaeomagnetic perspective. *Philosophical Transactions of the Royal Society A* 358:1141–1170.

Cox, A. 1969. Geomagnetic reversals. *Science* 163:237–245.

Crain, I. K. 1971. Possible direct causal relation between geomagnetic reversals and biological extinctions. *Bulletin of the Geological Society of America* 82:2603–2606.

Creanga, D. E., Poiata, A., Morariu, V. V. and Tupu, P. 2004. Zero-magnetic field effect in pathogen bacteria. *Journal of Magnetism and Magnetic Materials* 272–276:2442–2444.

De Santis, A., Tozzi, R. and Gaya-Pique, L. R. 2004. Information content and K-entropy of the present geomagnetic field. *Earth and Planetary Science Letters* 218:269–275.

Dhawi, F., Al-Khayri, J. M. and Hassan, E. 2009. Static magnetic field influence on elements composition in date palm. *Research Journal of Agriculture and Biological Sciences* 5(2):161–166.

Earl, A. M., Losick, R. and Kolter, R. 2008. Ecology and genomics of *Bacillus subtilis*. *Trends in Microbiology* 16:269–275.

Efros, V. D. and Bang, J. M. 1999. The first excited states of ^{9}Be and ^{9}B. *The European Physical Journal A* 4:33–39.

Eisenstadt, E. 1972. Potassium content during growth and sporulation in *Bacillus subtilis*. *Journal of Bacteriology* 112(1):264–267.

Engel, M. H. and Macko, S. A. 1997. Isotopic evidence for extraterrestrial non-racemic amino acids in the Murchison meteorite. *Nature* 389:265–268.

Farrow, E. F. and Hay, H. J. 1964. Search for a low lying 1/2+ level in ^{9}B. *Physics Letters* 11:50.

Fesenko, E. E., Mezhevikina, L. M., Osipenko, M. A., Gordon, R. Y. and Khutzian, S. S. 2010. Effect of the "zero" magnetic field on early embryogenesis in mice. *Electromagnetic Biology and Medicine* 29(1–2):1–8.

Fortune, H. T. and Sherr, R. 2013. Update on ^{9}B ($½^{+}$). *Nuclear Physics A* 898:78–83.

Fu, R. R., Weiss, B. P., Lima, E. A. et al. 2014. Solar nebula magnetic fields recorded in the Semarkona meteorite. *Science* 346(6213):1089–1092.

Goodman, E. M., Greenebaum, B. and Marron, M. T. 1995. Effects of electromagnetic fields on molecules and cells. *International Review of Cytology* 158:279–338.

Gul, K., Armitage, B. H. and Hooton, B. W. 1970. The $^{18}O(^{3}He, n)^{20}Ne$ and $^{7}Li(^{3}He, n)^{9}B$ *Physical Review A* 153:390.

Hays, J. D. 1971. Faunal extinctions and reversals of the Earth's magnetic field. *Bulletin of the Geological Society of America* 82:2433–2447.

Hong, H. A., Khaneja, R., Tam, N.M., Cazzato, A., Tan, S., Urdaci, M., Brisson, A., Gasbarrini, A., Barnes, I. and Cutting, S.M. 2009. *Bacillus subtilis* isolated from the human gastrointestinal tract. *Research in Microbiology* 160(2):134–143.

Hore, P. J. 2012. Are biochemical reactions affected by weak magnetic fields? *Proceedings of the National Academy of Sciences of the United States of America* 109(5): 1357–1358.

Islam, M. M. and Treacy, P. B. 1965. The first excited state of ^{9}B. *Nuclear Physics* 70: 236–240.

Iyengar, G. V., Clarke, W. B., Downing, R. G. and Fresenius, Z. 1990. Determination of boron and lithium in diverse biological matrices using neutron activation-mass spectrometry (NA-MS). *Analytical Chemistry* 338:562–566.

Johnson, F. M. 1972. Interstellar matter: Diffuse interstellar lines and porphyrins. *Annals of the New York Academy of Sciences* 187:186–206.

Kadija, K., Paic, G. and Antolkovic, B. 1987. ^{9}B excited states and analysis of the ^{9}Be(3He, t) spectra, *Physical Review C* 36:1269.

Karki, A. 2013. *A Study of ^{9}B Spectroscopy via the $^{9}Be(p, n)^{9}B$ Reaction using the Neutron Time-of-Flight Technique*. Thesis. Ohio University, Athens, OH.

Kmiecik, E., Tomaszewska, B., Wątor, K. and Bodzek, M. 2016. Selected problems with boron determination in water treatment processes. Part I: comparison of the reference methods for ICP-MS and ICP-OES determinations. *Environmental Science and Pollution Research* 23(12):11658–67.

Kobayashi, M., Soda, N., Miyo, T. and Ueda, Y. 2004. Effects of combined DC and AC magnetic fields on germination of hornwort seeds. *Bioelectromagnetics* 25:552–559.

Kratz, W. A. and Myers, J. 1955. Nutrition and growth of several blue-green algae. *American Journal of Botany* 42:282–287.

Kroepfl, J. J. and Browne, C. P. 1967. Energy level structure of ^{9}Be and ^{9}B. *Nuclear Physics A* 108:289.

Lamaze, G. P., Downing, R. G., Pilione, L., Badzian, A. and Badzian, T. 1993. Analysis of boron in CVD diamond surfaces using neutron depth profiling. *Applied Surface Science* 65/66: 587–592.

Lane, N, Allen, J. F. and Martin, W. 2010. How did LUCA make a living? Chemiosmosis in the origin of life. *BioEssays* 32:271–280.

Lane, N. and Martin, W. F. 2012. The origin of membrane bioenergetics. *Cell* 151:1406–1416.

Laurent-Pettersson, M., Delpech, B. and Theller, B. 1992. The mapping of natural boron in histological sections of mouse tissues by the use of neutron-capture radiography. *Journal of Histochemistry and Cytochemistry*. 24:939–950.

Lefevre, M., Racedo, S.M., Denayrolles, M., et al. 2017. Safety assessment of *Bacillus subtilis* CU1 for use as a prebiotic in humans. *Regulatory Toxicology and Pharmacology* 83:54–65.

Li, Y. and Héroux, P. 2014. Extra-low-frequency magnetic fields alter cancer cells through metabolic restriction. *Electromagnetic Biology and Medicine* 33(4):264–275.

Maffei, M. E. 2014. Magnetic field effects on plant growth, development, and evolution. *Frontiers in Plant Science* 5: 1–15 article 445.

Marion, J. B. and Levin, J. S. 1959. Investigation of the ^{9}Be(p, n) ^{9}B and ^{9}Be(p, $\alpha\gamma$) ^{6}Li reactions. *Physical Review* 115:144.

Martini, F. and Thellier, M. 1993. Boron distribution in parenchyma cells of clover leaves. *Plant Physiology and Biochemistry* 5:777–786.

Mateo, P., Bonilla, I., Fernandez-Valiente, E. and Sanchez-Maeso, E. 1986. Essentiality of boron for dinitrogen fixation in Anabaena sp. PCC 7119. *Plant Physiology* 81:430–433.

Matsumoto, T. and Aizawa, O. 1990. Prompt gamma-ray neutron activation analysis of boron-10 in biological materials. *International Journal of Radiation Applications and Instrumentation. Part A. Applied Radiation and Isotopes* A41:897–903.

Matsumoto, T., Aoki, M. and Aizawa, O. 1991. Phantom experiment and calculation for in vivo 10boron analysis by prompt gamma-ray spectroscopy *Physics in Medicine and Biology* 36:329–338.

May, P. 1997, Nov. Adenosine triphosphate – ATP. Chemistry in Britain. http://www.chm.bris.ac.uk/motm/atp/atp1.htm.

McClendon, J. H. 1976. Elemental abundance as a factor in the origin of mineral nutrient requirements. *Journal of Molecular Evolution* 8:175–195.

McDonough, W. F. 2001. The composition of the Earth. *International Geophysics* 76:3–23.

McGuire, B. A. and Carroll, P. B. 2016. Mirror asymmetry in life and in space. *Physics Today* 69(11): 86–87.

McGuire, B. A. Carroll, P. B., Loomis, R. A., Finneran, I. A., Jewell, P. R., Remijan, A. J. and Blake, G. A. 2016. Discovery of the interstellar chiral molecule propylene oxide (CH_3CHCH_2O). *Science* 352:1449–1452.

MEDA. 1999. *Compensation of Earth's Field with a Three-Axis Helmholtz Coil.* Macintyre Electronic Design Associates, Inc. (MEDA). Application Note AN-108. MEDA, Inc., Dulles, VA, USA.

Mitchell, P. 1961. Coupling of phosphorylation to electron and hydrogen transfer by a chemiosmotic type of mechanism. *Nature* 191:144–148.

Nagi, J. G. and Tiwari, R. K. 1983. Matching long term periodicity of geomagnetic reversals and galactic motions of the solar system. *Geophysical Research Letters* 10(8):713–716.

NASA. 2012. *Cosmic Dust Sample Investigator's Guidebook*. JSC-66466. NASA Johnson Space Center, Houston, TX.

Okeke, F. N. and Hamano, Y. 2000. Daily variations of geomagnetic H, D and Z-field at equatorial latitudes. *Earth Planets Space* 52:237–243.

Paris, M. W., Hale, G., Hayes, A. and Jungman G. 2013. R-matrix analysis of reactions in the ^{9}B compound system. arXiv:1304.3153, Nuclear Data Sheets 120:184–187.

Pillay, A. E. and Peisach, M. 1992. Some studies on nuclear methods for boron determination. *Nuclear Instruments and Methods in Physics Research Section B* 66:226–229.

Pizzarello, S. 2006. The chemistry of life's origin: A carbonaceous meteorite perspective. *Accounts of Chemical Research* 39:231–37.

Price, A. T. 1969. Daily variations of the geomagnetic field. *Space Science Reviews* 9:151–197.

Ramon, C., Martin, J. T. and Powell, M. R. 1987. Low-level, magnetic-field-induced growth modification of Bacillus subtilis. *Bio Electro Magnetics* 8(3):275–282.

Richter, I. 2010, Oct. 27. *Space Instrumentation: Measuring Magnetic Fields in Space.* Institute for Geophysics and Extraterrestrial Physics. TU, Braunschweig, MPS Lindau.

Richter, I. 2017. Modern magnetic field measurement devices. http://www.igep.tu-bs.de/lehre/skripten/mrode/mag_en_over.pdf.

Sah, R. N. and Brown, P. H. 1997. Boron determination – a review of analytical methods. *Microchemical Journal* 56:285–304.

Saji, Y. 1960. Energy spectrum and angular distributions of neutrons from the reaction $^{9}Be(p, n)^{9}B$ at 8 to 14 MeV of proton energies, *Journal of the Physical Society of Japan* 15:367.

Scholl, C., Fujita, Y., Adachi, T., et al. 2011. High resolution study of the $^{9}Be(^{3}He, t)^{9}B$ reaction up to the ^{9}B triton threshold. *Physical Review C* 84:0143081.

Semikhina, L. P. and Kiselev, V. F. 1988. Effect of weak magnetic fields on the properties of water and ice. *Russian Physics Journal* 31(5):351–354.

Shcherbakov, V. P., Solodovnikov, G. M. and Sycheva, N. K. 2002. Variations in the geomagnetic dipole during the past 400 million years (volcanic rocks). *Izvestiya, Physics of the Solid Earth* 38:113–119.

Sherr, R. and Bertsch, G. 1985. Coulomb energy systematics and the missing $J^{\pi} = \frac{1}{2}+$ state in ^{9}B. *Physical Review C* 32:1809–1816.

Slobodrian, R. J., Bichsel, H., Mckee, J. S. C. and Tiyol, W. F. 1967. High-resolution fast-neutron spectroscopy of the reaction $^{9}Be(p, n)^{9}B$ at 20 Me V. *Physical Review Letters* 19:595.

Sojo, V., Pomiankowski, A. and Lane, N. 2014. A bioenergetic basis for membrane divergence in archaea and bacteria. *PLOS Biology*, www.plosbiology.org, 12(8):e1001926.

Spencer, R. R., Floyd, G. C. and Young, T. E. 1960. Energy levels of ^{9}Be and ^{9}B, *Nuclear Physics A* 21:310.

Stock, D., Leslie, A. and Walker, J. 1999. Molecular architecture of the rotary motor in ATP synthase. *Science* 286:1700–1705.

Symons, G. D. and Treacy, P. B. 1962. Evidence for 1.7 MeV level in ^{9}B. *Physics Letters* 2:175.

Teranishi, E. and Furubayashi, B. 1964. Level width of the ground state of B9. *Physics Letters* 9(2):157–159.

Tilley, D. R., Kelley, J. H., Godwin, J. L., Millener, D. J., Purcell, J. E., Sheu, C. G. and Weller H. R. 2004. Energy levels of light nuclei A=8, 9, 10. *Nuclear Physics A* 745:155–362.

Tumanski, S. 2013. Modern magnetic field sensors – a review. Przeglad Elektrotechniczny R. 89 NR 10:1–12.

Valković, V. 1977. A possible mechanism for the influence of the geomagnetic field of the evolution of life. *Origins of Life* 8:7–11.

Valković, V. 1980. *Analysis of Biological Material for Trace Elements Using X-Ray Spectroscopy*. CRC Press, Boca Raton, FL, pp. 166–171.

Valković, V. 1990. *Origin of Trace Element Requirement by Living Matter. Symmetries in Science IV*, Gruber, B. and Yopp, J. H. (Eds.), Plenum Press, New York, pp. 213–242.

Valković, V., Sudac, D. and Obhodas, J. 2018. The role of 14 MeV neutrons in light element nucleosynthesis. *IEEE Transactions on Nuclear Science* 65(9): 2366–2371 and EPJ Web of Conferences 170: 01017.

Varga, A., Palmai, Z., Gugolya, Z., Gráczer, É., Vonderviszt, F., Závodszky, P., Balog, E. and Vas, M. 2012. Importance of aspartate residues in balancing the flexibility and fine-tuning the catalysis of human 3-phosphoglycerate kinase. *Biochemistry* 51 (51):10197–10207.

Additional reading

Abbott, J. J. 2015. Parametric design of tri-axial nested Helmholtz coils. *Review of Scientific Instruments* 86: 054701, 10 p.

Al-Khaza'leh, K. A. and Al-fawwaz, A. T. 2015. The effect of static magnetic field on *E. Coli, S. aureus*, and *B. Subtilis* viability. *Journal of Natural Sciences Research* 5(24):153–157.

Anderson, D. L., Cunningham, W. C. and Mackey, E. A. 1990. Boron analysis in biological material: microwave digestion procedure and determination by different methods. *Fresenius' Zeitschrift für Analytische Chemie* 338:554–558.

Barker, F. C. 1987. The first excited state of ^{9}B. *Australian Journal of Physics* 40:307–317.

Barnes, F. S. and Greenebaum, B. 2015. The effects of weak magnetic fields on radical pairs. *Bioelectromagnetics* 36(1):45–54.

Bertea, C. M., Narayana, R., Agliassa, C., Rodgers, C. T. and Maffei, M. E. 2015, Nov. 30. Geomagnetic field (Gmf) and plant evolution: Investigating the effects of Gmf reversal on arabidopsis thaliana development and gene expression. *Journal of Visualized Experiments (JOVE)* (105). Doi: 10.3791/53286.

Binhi, V. N. and Savin, A. V. 2003. Effects of weak magnetic fields on biological systems: Physical aspects. *Physics Uspekhi* 46(3):259–291. Translation from Russian: *Uspekhi Fizicheskih Nauk* 173(3):265–300.

Conley, C. C. 1970, Aug. 1970. *A Review of the Biological Effects of Very Low Magnetic Fields*, NASA Technical Note D-5902, Washington, DC.

Dhawi, F., Al-Khayri, J. M. and Hassan, E. 2009. Static magnetic field influence on elements composition in date palm (*Phoenix dactylifera* L.). *Research Journal of Agriculture and Biological Sciences* 5(2):161–166.

Fleischer, A., Titel, C. and Ehwald, R. 1998. The boron requirement and cell wall properties of growing and stationary suspension-cultured *Chenopodium album* L. Cells. *Plant Physiology* 117:1401–1410.

Hall, M., Turner, S., Harmon, S., et al. 2014. *Best Practice Guide for the Generation and Measurement of DC Magnetic Fields in the Magnetic Field Range of 1 nT to 1 mT.* National Physical Laboratory, Teddington, Middlesex. NPL Report Mat 65, p. 32.

Haneda, Y., Okada, M., Suganuma, Y. and Kitamura, T. 2020. A full sequence of the Matayama-Brunhes geomagnetic reversal in the Chiba composite section, Central Japan. *Progress in Earth and Planetary Science* 7:44 (p 22).

Kanokov, Z., Schmelzer, J. W. P. and Nasirov, A. K. 2009. On the influence of weak magnetic and electric fields on the fluctuations of ionic electric currents in blood circulations. arXiv:0904.1198 [physics.bio-ph].

Khater, Z. Z. K. and Ibraheim, M. H. 2015. Ecological studies on the effect of magnetic field on water. *International Journal of Current Research and Review* (IJCRAR) 3(8):262–279.

Kovacs, P. E., Valentine, R. L. and Alvarez, P. J. J. 1997. The effect of static magnetic fields on biological systems: Implications for enhanced biodegradation. *Critical Reviews in Environmental Science and Technology* 27(4):319–382.

Laskar, T., Alexander, K. D., Gill, R., et al. 2019. ALMA detection of a linearly polarized reverse shock in GRB 190114C. *The Astrophysical Journal Letters* 878:L26 (8 p).

Masood, S. 2017. Effect of weak magnetic field on bacterial growth. *Biophysical Reviews and Letters* 12:177–186.

Messiha, H. L., Wongnate, T., Chaiyen, P., Jones, A. R. and Scrutton, N. S. 2016. Magnetic field effects as a result of the radical pair mechanism are unlikely in redox enzymes. *Journal of the Royal Society Interface* 12:20141155. Doi: 10.1098/rsif.2014.1155.

Naito, M., Hirai, S., Mihara, M., Terayama, H., Hatayama, N., Hayashi, S., Matsushita, M. and Itoh, M. 2012. Effect of a magnetic field on drosophila under supercooled conditions. *PLOS ONE* 7(12):e51902.

Park, B.-J., Park, S.-J., Yoon, D.-N., Schouten, S., Sinninghe Damsté, J. S. and Rhee, S.-K. 2010. Cultivation of autotrophic ammonia-oxidizing archaea from marine sediments in coculture with sulfur-oxidizing bacteria. *Applied and Experimental Microbiology* 76(22):7575–7587.

Pazur, A. 2004, Nov. 30. Characterization of weak magnetic field effects in an aqueous glutamic acid solution by nonlinear dielectric spectroscopy and voltammetry. *BioMagnetic Research and Technology* 2:8, (11 p).

Peňuelas, J., Llusià, J., Martinez, B. and Fontcuberta, J. 2004. Diamagnetic susceptibility and root growth responses to magnetic fields in *Lens culinaris, Glicine soja*, and *Triticum Aestivum. Electromagnetic Biology and Medicine* 23(2):97–112.

Rogus, R., Harling, O. K., Olmez, I. and Wirdzek, S. 1992. Boron-10 prompt gamma analysis using diffracted neutron beam. In Allen, B. J. et al., (Eds.) *Progress in Neutron Capture Therapy for Cancer.* Plenum, New York, pp. 301–304.

Setlow, P. 1995. Mechanisms for the prevention of damage to DNA in spores of Bacillus species. *Annual Review of Microbiology* 49:29–54.

Sobotovich, E. V., Florinsky, I. V., Lysenko, O. B. and Grodzinsky, D. M. 2010. *Role of Isotopes in the Biosphere. Chapter in Man and Geosphere* Florinsky, I. V., (Ed.) Nova Science Publishers, Inc, pp. 33–68.

Sudac, D. 2018. Private communication.

Valet, J.-P., Meynadier, L. and Guyodo, Y. 2005. Geomagnetic dipole strength and reversal rate over the past two million years. *Nature* 435:802–805.

Volpe, P. 2003. Interactions of zero-frequency and oscillating magnetic fields with biostructures and biosystems. *Photochemistry and Photobiology Sciences* 2:637–648.

Wan, G.-J., Wang, W.-J., Xu, J.-J., Yang, Q.-F., Dai, M.-J., Zhang, F.-J., Sword, G. A., Pan, W.-D. and Chen, F.-J. 2015. Cryptochromes and hormone signal transduction under near-zero magnetic fields: New clues to magnetic field effects in rice planthopper. *PLOS One* 10(7):E0132966. Doi: 10.1371 /journal.pone.0132966.

Yamagishi, A., Kawaguchi, Y., Hashimoto, H., Yano, H., Imai, E., Kodaira, S., Uchihori, Y. and Nakagawa, K. 2017. Environmental data and survival data of *Deinococcus aetherius* from the exposure facility of the japan experimental module of the international space station obtained by the Tanpopo mission. *Astrobiology* 18(1), 6 p. Doi: 10.1089/ast.2017.1751.

Index

A

B

C

D

E

T

U

V

W

X

Z